Topics in Fluorescence Spectroscopy

Volume 11

Glucose Sensing

Topics in Fluorescence Spectroscopy

Edited by CHRIS D. GEDDES and JOSEPH R. LAKOWICZ

Volume 1: Techniques
Volume 2: Principles
Volume 3: Biochemical Applications
Volume 4: Probe Design and Chemical Sensing
Volume 5: Nonlinear and Two-Photon-Induced Fluorescence
Volume 6: Protein Fluorescence
Volume 7: DNA Technology
Volume 8: Radiative Decay Engineering
Volume 9: Advanced Concepts in Fluorescence Sensing
Part A: Small Molecule Sensing
Volume 10: Advanced Concepts in Fluorescence Sensing
Part B: Macromolecular Sensing
Volume 11: Glucose Sensing

Topics in Fluorescence Spectroscopy

Volume 11

Glucose Sensing

Edited by

CHRIS D. GEDDES

The Institute of Fluorescence
Medical Biotechnology Center
University of Maryland Biotechnology Institute
Baltimore, Maryland

and

JOSEPH R. LAKOWICZ

Center for Fluorescence Spectroscopy and
Department of Biochemistry and Molecular Biology
University of Maryland School of Medicine
Baltimore, Maryland

Springer

ISBN-10: 1-4899-9344-4
ISBN-13: 978-1-4899-9344-1

Printed on acid-free paper.

9 8 7 6 5 4 3 2 1

springeronline.com

PREFACE

As a common medical condition that produces excessive thirst, continuous urination and severe weight loss, Diabetes has interested medical researchers for over three millennia. Unfortunately it wasn't until the early 20th century that the prognosis for this condition became any better than it was 3000 years ago.

The term Diabetes was first used by Apollonius of Mephis around 230 BC, which in Greek means "To pass through" (Dia – through, betes – to go). Apollonius and his colleagues considered Diabetes as a disease of the Kidneys and subsequently recommended completely ineffective treatments, such as bloodletting or dehydration.

While the Ebers Papyrus, which was written around 1500 BC, excavated in 1862 AD from an ancient grave in Thebes, Egypt, described the first reference to what we now consider to be Diabetes Mellitus, it was physicians in India at around the same time that developed the first crude test for diabetes. They observed that the urine from people with diabetes attracted ants and flies. They subsequently named the condition "madhumeha" or "honey urine".

Our understanding, diet and our ability to monitor sugar levels for the management of diabetes has changed somewhat in the last couple of hundred years. Today a variety of sensing strategies exist, each with their own respective merits and many at different stages of commercialization. In this book, Volume 11 of the popular series "Topics in Fluorescence Spectroscopy", we have invited articles from leading authors in this important clinical sensing area, to provide a current opinion of glucose sensing and fluorescence-based monitoring technologies. The organization of the articles will hopefully also give readers a chance to visualize the extent of work currently being undertaken in this important area.

In closing, I would like to thank the authors for their excellent invaluable and timely contributions, which serve very well to demonstrate the applicability of fluorescence to this sensing area. In addition, many thanks to Dr Kadir Aslan for help in typesetting the volume.

Professor Chris D. Geddes
Baltimore, MD, USA
July 25th 2005

CONTRIBUTORS

Kadir Aslan. University of Maryland Biotechnology Institute, Institute of Fluorescence, Baltimore, MD.

Ramachandram Badugu. University of Maryland at Baltimore. Baltimore, MD.

Laurence I. Bosch. Department of Chemistry, University of Bath, Bath,UK.

Jason N. Camara. University of California - Santa Cruz, Santa Cruz, CA.

Frank E. Cappuccio. University of California - Santa Cruz, Santa Cruz, CA.

Marcella de Champdoré. Institute of Protein Biochemistry, Italian National Research Council, Naples, Italy.

Martin M.F. Choi. Department of Chemistry, Hong Kong Baptist University, Kowloon Tong, Hong Kong SAR.

David B. Cordes. University of California - Santa Cruz, Santa Cruz, CA.

Gerard L. Cote. Texas A&M University, Department of Biomedical Engineering, College Station, TX.

Sabato D'Auria. Institute of Protein Biochemistry, Italian National Research Council, Naples, Italy.

Axel Duerkop. Institute of Analytical Chemistry, Chemo- and Biosensors, University of Regens-burg, Regensburg, Germany

Jorge O. Escobedo. Department of Chemistry, Louisiana State University, Baton Rouge, LA.

Giovanni Ghirlanda. Institute of Protein Biochemistry, Italian National Research Council, Naples, Italy.

Soya Gamsey. University of California - Santa Cruz, Santa Cruz, CA.

Chris D. Geddes. University of Maryland Biotechnology Institute, Institute of Fluorescence, Baltimore, MD.

Takashi Hayashita. Department of Chemistry, Sophia University, Kioicho, Chiyoda-ku, Tokyo, JAPAN.

Michael D. Heagy, New Mexico Institute of Mining and Technology, Socorro, New Mexico.

Bennett L. Ibey. Texas A&M University, Department of Biomedical Engineering, College Station, TX.

Tony D. James. Department of Chemistry, University of Bath, Bath, UK.

M. M. Karim. Department of Chemistry, Kyungpook National University, Taegu, Korea.

Omar S. Khalil. Abbott Laboratories, Diagnostics Division, Abbott Park, IL.

Joseph R. Lakowicz. University of Maryland at Baltimore. Baltimore, MD.

S. H. Lee. Department of Chemistry, Kyungpook National University, Taegu, Korea.

Michael J. McShane. Biomedical Engineering, Institute for Micromanufacturing, Ruston, LA.

Antonietta Parracino. Institute of Protein Biochemistry, Italian National Research Council, Naples, Italy.

Michael V. Pishko. The Pennsylvania State University, Department of Chemical Engineering, University Park, PA.

Govind Rao. Center for Advanced Sensor Technology, Department of Chemical and Biochemical Engineering, University of Maryland Baltimore County, Baltimore, MD

Mosè Rossi. Institute of Protein Biochemistry, Italian National Research Council, Naples, Italy.

Oleksandr Rusin. Department of Chemistry, Louisiana State University, Baton Rouge, LA.

Viviana Scognamiglio. Institute of Protein Biochemistry, Italian National Research Council, Naples, Italy.

Zach Sharrett. University of California - Santa Cruz, Santa Cruz, CA.

Michael Schaeferling. Institute of Analytical Chemistry, Chemo- and Biosensors, University of Regensburg, Regensburg, Germany.

Jerome S. Schultz. University of California Riverside, CA.

Bakthan Singaram. University of California - Santa Cruz, Santa Cruz, CA.
Maria Staiano. Institute of Protein Biochemistry, Italian National Research Council, Naples, Italy.

Robert M. Strongin. Department of Chemistry, Louisiana State University, Baton Rouge, LA.

Jeff T. Suri. University of California - Santa Cruz, Santa Cruz, CA.

Iwao Suzuki. Graduate School of Pharmaceutical Sciences, Tohoku University, Aramaki, Aoba-ku, Sendai, JAPAN.

Leah Tolosa. Center for Advanced Sensor Technology, Department of Chemical and Biochemical Engineering, University of Maryland Baltimore County, Baltimore, MD

Praveen Thoniyot. University of California - Santa Cruz, Santa Cruz, CA.

Akiyo Yamauchi. Graduate School of Pharmaceutical Sciences, Tohoku University, Aramaki, Aoba-ku, Sendai, JAPAN.

Ritchie A. Wessling. University of California - Santa Cruz, Santa Cruz, CA.

Otto S. Wolfbeis. Institute of Analytical Chemistry, Chemo- and Biosensors, University of Regens-burg, Regensburg, Germany.

Xiao Jun Wu. Tian Kang-Yuan Biotechnology Company, Yingkou, Liaoning, China.

CONTENTS

1. *N*-PHENYLBORONIC ACID DERIVATIVES OF ARENECARBOXIMDES AS SACCHARIDE PROBES WITH VIRTUAL SPACER DESIGN 1

Micheal D. Heagy

1.1. INTRODUCTION 1
1.2. *N*-PHENYLBORONIC ACID ARENECARBOXIMIDES AS SACCHARIDE PROBES WITH VIRTUAL OR CO SPACER DESIGN 2
1.3. SUBSTITUENT EFFECTS ON MONOBORONIC ACID DERIVATIVES OF *N*-PHENYL-1, 8-NAPHTHALENEDICARBOXIMIDES 3
1.3.1. Synthesis 4
1.3.2. Photoelectrochemical Model 6
1.3.3. Saccharide Complexation Results 6
1.3.4. Conclusion 11
1.4 POSITIONAL ISOMERS OF NAPHTHALENE DICARBOXIMIDES 13
1.4.1. Ortho-substituted Monoboronic Acid Sensor 13
1.4.2. pH and Sugar Effects On The Fluorescent Decays 14
1.5. CONCLUSION 16
1.6. SUMMARY AND OUTLOOK 17
1.7. ACKNOWLEDGMENTS 17
1.8. REFERENCES 17

2. GENESIS OF FLUOROPHORE FORMATION IN MACROCYCLE SOLUTIONS AND THE DETECTION OF GLUCOSE AND RELATED SUGARS 21

Jorge O. Escobedo, Oleksandr Rusin, and Robert M. Strongin

2.1. INTRODUCTION 21
2.2. THE DISCOVERY OF NEW FUNCTIONAL FLUOROPHORES 21
2.3. THE MECHANISM OF SUGAR-INDUCED SIGNAL TRANSDUCTION 28
2.4. A HIGHLY CONVENIENT NEW AUTOMATED HPLC POST-COLUMN DETECTION METHOD FOR MONITORING GLUCOSE AND RELATED BIOMOLECULES 34
2.5. DETECTION OF GLUCOSE IN HUMAN BLOOD PLASMA AND PROGRESS TOWARDS CONCURRENT DETERMINATION OF GLUCOSE AND FRUCTOSE 38
2.6. CONCLUSION 42
2.7. ACKNOWLEDGMENTS 42
2.8. REFERENCES 42

3. TWO-COMPONENT OPTICAL SUGAR SENSING USING BORONIC ACID-SUBSTITUTED VIOLOGENS WITH ANIONIC FLUORESCENT DYES 47

David B. Cordes, Jeff T. Suri, Frank E. Cappuccio, Jason N. Camara, Soya Gamsey, Zach Sharrett, Praveen Thoniyot, Ritchie A. Wessling, and Bakthan Singaram

3.1. INTRODUCTION 47
3.2. BACKGROUND AND ILLUSTRATION OF TWO-COMPONENT GLUCOSE SENSING–PYRANINE (HPTS) AND BORONIC ACID-FUNCTIONALIZED VIOLOGEN (*o*-BBV^{2+}) 50
3.3. VARIATIONS IN VIOLOGEN QUENCHER – BIPYRIDINIUM AND PHENANTHROLINIUM QUENCHERS 56
3.4. VARIATIONS IN FLUORESCENT REPORTER-SULFONAMIDE DERIVATIVES OF HPTS 71
3.5. IMMOBILIZATION OF THE SENSING COMPONENTS – A GLUCOSE SENSITIVE THIN FILM HYDROGEL 80
3.6. SUMMARY AND FUTURE DIRECTIONS 83
3.7. REFERENCES 84

4. IMPLANTABLE CONCANAVLIN A BASED SENSORS FOR INTERSTITIAL FLUID GLUCOSE SENSING IN DIABETICS 89

Bennett L. Ibey, Michael V. Pishko, and Gerard L. Cote'

4.1. INTRODUCTION 89
4.1.1. Diabetes 89

4.1.2. Commercially Available Sensors ... 90
4.1.3. Other Approaches ... 90
4.2. CONCANAVALIN A AND DEXTRAN ... 92
4.2.1. Structure ... 92
4.2.2. Purification ... 93
4.2.3. Stability ... 94
4.2.4. Traditional uses of Con A ... 95
4.2.5. Dextran ... 96
4.3. FLUORESCENT BASED ASSAY ... 96
4.3.1. Fluorescence Resonance Energy Transfer ... 96
4.3.2. FRET Based Assay for Glucose Detection ... 98
4.4. SENSING MODALITIES ... 100
4.4.1. Hollow Fiber Based Sensors ... 100
4.4.2. Poly (ethylene glycol) Based Sensors ... 106
4.5. SUMMARY ... 111
4.6. REFERENCES ... 111

5. FLUORESCENCE BIOSENSORS FOR CONTINUOUSLY MONITORING THE BLOOD GLUCOSE LEVEL OF DIABETIC PATIENTS ... 117

Sabato D'Auria, Giovanni Ghirlanda, Antonietta Parracino, Marcella de Champdoré, Viviana Scognamiglio, Maria Staiano, and Mosè Rossi

5.1. INTRODUCTION ... 117
5.2. GLUCOSE OXIDASE FROM ASPERGILLUS NIGER ... 119
5.3. THERMOSTABLE GLUCOSE DEHYDROGENASE FROM THERMOPLASMA ACIDOPHILUM ... 121
5.4. A FLUORESCENCE COMPETITIVE ASSAY BY USING THE STABLE GLUCOKINASE ... 123
5.5. CONCLUSIONS ... 126
5.6. ACKNOWLEDGMENT ... 127
5.7. REFERENCES ... 127

6. MICROCAPSULES AS "SMART TATTOO" GLUCOSE SENSORS: ENGINEERING SYSTEMS WITH ENZYMES AND GLUCOSE-BINDING SENSING ELEMENTS ... 131

Michael J. McShane

6.1. THE "SMART TATTOO" CONCEPT ... 131
6.2. LBL NANOFILMS AND POLYELECTROLYTE MICROCAPSULES ... 134

6.3. GLUCOSE SENSORS FROM NANOENGINEERED CAPSULES 136
6.4. ENZYME-BASED MICROCAPSULE SENSORS 142
6.4.1. Design of Sensors through Mathematical Modeling 143
6.4.2. Experimental Validation of Enzyme-Loaded Microcapsule Sensors 154
6.5. GLUCOSE-BINDING PROTEINS IN MICROCAPSULES 156
6.6. CONCLUSIONS 160
6.7. ACKNOWLEDGMENT 161
6.8. REFERENCES 161

7. NON-INVASIVE MONITORING OF DIABETES 165

Omar S. Khalil

7.1. INTRODUCTION 165
7.2. SPECIFICITY OF NI GLUCOSE MEASUREMENTS 167
7.2.1. Tissue Spectroscopy 168
7.2.2. Infrared and Near Infrared Spectroscopy of Glucose 169
7.3. COMPARTMENTALIZATION OF GLUCOSE VALUES 172
7.4. CALIBRATION MODELS AND PATIENT-SPECIFIC CALIBRATION 172
7.4.1. Establishing a Standard Response Relationship Calibration Models 172
7.4.2. Prediction Parameters 173
7.5. THERMO-OPTICAL RESPONSE OF HUMAN SKIN 173
7.5.1. Thermo-Optical Response (THOR) Localized Reflectance Measurements 174
7.5.2. Effect of the Diabetic State on THOR of Human Skin 178
7.5.3. Qualitative Assessment of the Diabetic State from THOR of Human Skin Using Nonlinear Discriminant Functions for Data Analysis 180
7.5.4. Glucose Calibration Models from Thermo-Optical Response of Human Skin 185
7.5.5. Calibration Models at Fixed Source-Detector Distance 186
7.5.6. Glucose Prediction Models from Thermo-Optical Response of Human Skin 187
7.6. ENHANCING SPECIFICITY BY AFFINITY CAPTURE AGENTS 190
7.6.1. Fluorescence Measurements 191
7.6.2. Use Light Diffraction Elements 192
7.6.3. Specificity and Compartmentalization in Contact Lens Sensors 194
7.7. CONCLUSIONS 195
7.8. REFERENCES 196

8. OPTICAL ENZYME-BASED GLUCOSE BIOSENSORS 201

Xiao Jun Wu and Martin M.F. Choi

8.1. ABSTRACT 201
8.2. INTRODUCTION 201
8.3. OPTICAL TRANSDUCTION USING COMMON OPTICAL TRANSDUCERS 203
8.3.1. Absorbance Measurement 204
8.3.2. Chemiluminescence or Electrogenerated Chemiluminescence Measurement 205
8.3.3. Fluorescence Measurement 206
8.4. IMMOBILIZATION OF GLUCOSE OXIDASE 207
8.4.1. Immobilizing Enzyme on Surface 209
8.4.2. Enzyme Entrapped In or Buried Behind Thin Homogenous Film 210
8.4.3. Enzyme Entrapped Within Sol-Gel Matrix 212
8.4.4. Enzyme Immobilized By Nano-Technique and With Nanoparticles 213
8.4.5. Blend Composites of Immobilization Enzyme 214
8.5. CONSTRUCTION OF GLUCOSE BIOSENSOR 215
8.5.1. Bulky Bi-phase Bioreactor 215
8.5.2. Particle Packed Bioreactor 217
8.6. PERFORMANCE OF OPTICAL GLUCOSE BIOSENSOR 219
8.6.1. In Batch Operation 219
8.6.2. In Stop-Flow Mode 220
8.6.3. In Flow Mode 220
8.6.4. Flow Injection Analysis 222
8.6.5. Mixture of Organic Solvent and Aqueous Micelle Carriers 222
8.7. IMPLICATION OF THE DISSOLVED OXYGEN CONCENTRATION 223
8.8. ENHANCEMENT OF ENZYME STABILITY 224
8.9. ANALYTICAL FEATURE AND APPLICATION 226
8.10. CONCLUSION 228
8.11. ACKNOWLEDGMENT 229
8.12. REFERENCES 229

9. SACCHARIDE RECOGNITION BY BORONIC ACID FLUOROPHORE / CYCLODEXTRIN COMPLEXES IN WATER 237

Akiyo Yamauchi, Iwao Suzuki , and Takashi Hayashita

9.1. ADVANCES IN SYNTHETIC RECEPTORS FOR SACCHARIDES 237

9.2. SACCHARIDE RECOGNITION BY BORONIC ACID FLUOROPHORE / β-CYCLODEXTRIN COMPLEXES 242
9.2.1. Saccharine recognition by the C4-CPB/β-CyD Complex 243
9.2.2. Effect of Phenylboronic Acid Probe Spacer 251
9.3. FUTURE PERSPECTIVE OF SUPRAMOLECULAR CYCLODEXTRIN COMPLEX SENSORS 254
9.4. ACKNOWLEDGEMENT 256
9.5. REFERENCES 257

10. PLASMONIC GLUCOSE SENSING 259

Kadir Aslan, Joseph R. Lakowicz, and Chris D. Geddes

10.1. INTRODUCTION 259
10.2. OPTICAL PROPERTIES OF GOLD NANOPARTICLES 260
10.3. PREPARATION OF LIGAND FUNCTIONALIZED GOLD NANOPARTICLES FOR GLUCOSE SENSING AND OTHER APPLICATIONS 261
10.4. PLASMONIC GLUCOSE SENSING 264
10.4.1. Fabrication of Glucose-Sensing Aggregates 266
10.4.2. Glucose Sensing based on the Dissociation of Dextran-coated Gold Nanoparticle / Con A Aggregates 271
10.5. CONCLUSIONS AND FUTURE DIRECTIONS 280
10.6. ACKNOWLEDGMENTS 280
10.7. REFERENCES 280

11. OPTICALLY-BASED AFFINITY BIOSENSORS FOR GLUCOSE 283

Jerome S. Schultz

11.1. INTRODUCTION 283
11.2. DEVELOPMENT OF OPTICALLY BASED BIOSENSORS 285
11.3. COMPETITIVE OPTICAL AFFINITY GLUCOSE BIOSENSORS 289
11.4. DYNAMICS OF AFFINITY BIOSENSORS 298
11.5. RECENT DEVELOPMENTS IN AFFINITY BASED GLUCOSE BIOSENSOR SYSTEMS 299
11.6. INTEGRAL BIOSENSOR PROTEINS 304
11.7. ENHANCEMENTS OF OPTICAL GLUCOSE BIONSENSORS BASED ON AFFINITY PRINCIPLES 305
11.8. CONCLUSION 307
11.9. REFERENCES 307

12. RECENT CHEMILUMINESCENCE APPLICATIOINS FOR GLUCOSE SENSING 311

S. H. Lee and M. M. Karim

12.1. INTRODUCTION 311
12.2. CHEMILUMINESCENCE FOR THE DETERMINATION OF GLUCOSE 312
12.3. CHEMILUMINOGENIC BIOSENSOR 313
12.4. CAPILLARY ELECTROPHORESIS FOR GLUCOSE DETERMINATION 317
12.5. GLUCOSE SENSING: RECENT APPROACHES FOR CLINICAL USE IN DIABETES CARE 319
12.6. ACKNOWLEDGEMENT 321
12.7. REFERENCES 321

13. THE GLUCOSE BINDING PROTEIN AS GLUCOSE SENSOR 323

Leah Tolosa and Govind Rao

13.1. INTRODUCTION 323
13.2. PROTEIN ENGINEERING 324
13.3. POLARITY SENSITIVE PROBES 326
13.4. STRATEGIES FOR LOW-COST GLUCOSE SENSING 326
13.5. APPLICATION OF THE GBP-BASED BIOSENSOR IN CELL CULTURE 329
13.6. VALIDATION 330
13.7. REFERENCES 331

14. FLUORESCENT TICT SENSORS FOR SACCHARIDES 333

Laurence I. Bosch and Tony D. James

14.1. INTRODUCTION 333
14.2. RESULTS AND DISCUSSION 333
14.3. CONCLUSIONS 345
14.4. EXPERIMENTAL 345
14.4.1. General procedures 345
14.4.2. Preparation of the (2-boronobenzyl)-aniline (1) 346
14.4.3. Preparation of the (3-boronobenzyl)-aniline (2) 347
14.4.4. Preparation of the (4-boronobenzyl)-aniline (3) 347
14.4.5. Preparation of the (4-chloro-phenyl)-(2-boronobenzyl) amine (4) 348

14.4.6. Preparation of the (4-methoxy-phenyl)-(2-boronobenzyl) amine (5) 348
14.4.7. Preparation of the 2-(boronobenzyl)-naphthalen-1-yl-amine (6) 349
14.4.8. Preparation of the (2-boronobenzyl)-pyren-1-yl-amine (7) 349
14. 5. ACKNOWLEDGMENTS 350
14. 6. REFERENCES 350

15. GLUCOSE SENSING AND GLUCOSE DETERMINATION USING FLUORESCENT PROBES AND MOLECULAR RECEPTORS 351

Axel Duerkop, Michael Schaeferling, Otto S. Wolfbeis

15.1. ABSTRACT 351
15.2. INTRODUCTION 352
15.3. DETERMINATION OF GLUCOSE VIA GOx AND FLUORESCENT MOLECULAR PROBES 354
15.3.1. Determination of Glucose via Dissolved Glucose Oxidase and Changes in pH 355
15.3.2. Determination of Glucose via Dissolved Glucose Oxidase and Measurement of the Consumption of Oxygen 355
15.3.3. Determination of Glucose via Formation of Hydrogen Peroxide 356
15.4. CONTINOUS SENSING OF GLUCOSE USING FLUORESCENT SENSOR MEMBRANES 358
15.4.1. Sensing Glucose via pH Sensor Layers Using Immobilized GOx 359
15.4.2. Sensing Glucose via Oxygen Sensor Membranes 361
15.4.3. Sensing Glucose via Sensor Layers for Hydrogen Peroxide 363
15.5. SENSING GLUCOSE VIA METAL-LIGAND COMPLEXES WITH A BORONIC ACID AS THE RECOGNITION SITE 364
15.6. SENSING GLUCOSE VIA THIN FILMS OF ORGANIC CONDUCTING POLYMERS (OCPs) 365
15.7. SENSING GLUCOSE VIA OXYGEN-SENSITIVE NANOPARTICLES 367
15.8. SENSING GLUCOSE VIA THE INITRINSIC FLUORESCENCE OF GLUCOSE OXIDASE 367
15.9. IMAGING TECHNIQUES FOR GLUCOSE AND GLUCOSE OXIDASE 368
15.9.1. Time-Resolved Imaging of Glucose via Enzymatically Produced H_2O_2 369

15.9.2. Time-Resolved Imaging of Glucose Using Oxygen Sensors 370
15.9.3. Time-Resolved Imaging of Glucose Using Sensor Layers for Hydrogen Peroxide 371
15.9.4. Time-Resolved Imaging of the Activity of Glucose Oxidase 371
15.10. REFERENCES 373

16. BORONIC ACID-BASED FLUORESCENCE SENSORS FOR GLUCOSE MONITORING 377

Gurpreet Kaur, Na Lin, Hao Fang, and Binghe Wang

16.1. INTRODUCTION 377
16.2. FACTORS THAT AFFECT THE BINDING BETWEEN BORONIC ACIDS AND DIOLS 378
16.3. VARIOUS TYPES OF BORONIC ACID-BASED FLUORESCENT REPORTER COMPOUNDS AVAILABLE 380
16.3.1. Anthracene and Pyrene Based Fluorescent Reporter Compounds 381
16.3.2. Internal Charge Transfer Compounds 385
16.3.3. Polymer Based Sensors 386
16.3.4. Water Soluble Fluorescent Boronic Acid 387
16.3.5. Fluorescent Boronic Acids Suitable For Embedment in Contact Lens For Glucose Sensing 390
16.4. APPROACHES AVAILABLE FOR THE CONSTRUCTION OF THE APPROPRIATE 3-DIMENSIONAL SCAFFOLD FOR THE SELECTIVE RECOGNITION OF THE TARGET SUGAR 392
16.5. CONCLUSION 394
16.6. ACKNOWLEDGEMENT 394
16.7. REFERENCES 394

17. DEVELOPMENT OF SMART CONTACT LENSES FOR OPHTHALMIC GLUCOSE MONITORING 399

Ramachandram Badugu, Joseph R. Lakowicz, and Chris D. Geddes

17.1. INTRODUCTION 399
17.1.1. Design Concept 401
17.2. GLUCOSE SENSING USING BORONIC ACID PROBES IN SOLUTION 402
17.2.1. Boronic Acid Probes with Intramolecular Charge Transfer Mechanism 404

17.2.2. Boronic Acid Probes having Photoinduced Intramolecular Electron Transfer Mechanism, (PET) 406
17.3. LENS FEASIBITY STUDY 408
17.3.1. Lens Doping and Contact Lens Holder 408
17.3.2. Response of ICT Probes within the Contact Lens 409
17.3.3. Response of the PET Probes within the Contact Lens 412
17.3.4. Contact Lens Internal Properties and Effect on the Sensing Ability of the Probes 412
17.4. RATIONALE DESIGN OF NEW GLUCOSE SENSING PROBES 413
17.5. GLUCOSE SENSING PROBES BASED ON THE QUINOLINIUM MOIETY 414
17.5.1. Spectral Properties and Signaling Mechanism 415
17.5.2. Signaling Mechanism in Quinolinium Probes 416
17.5.2. Sugar Response of the Quinolinium Probes in Solution 417
17.5.3. Response of the New Glucose Signaling Probes in the Contact Lens 420
17.6. PROBE LEACHING, INTERFERENTS AND SHELF LIFE 422
17.7. FUTURE DEVELOPMENTS BASED ON THIS TECHNOLOGY 424
17.7.1. Continuous and Non-invasive Glucose Monitoring 424
17.7.2. Clinical Condition and Diagnosis from Tears 425
17.7.3. Drug Testing, Compliance and Screening 426
17.8. CONCLUDING REMARKS 426
17.9. ACKNOWLEDGEMENTS 427
17.10. REFERENCES 427

COLOR INSERT ....After 429

INDEX 431

N-PHENYLBORONIC ACID DERIVATIVES OF ARENECARBOXIMDES AS SACCHARIDE PROBES WITH VIRTUAL SPACER DESIGN

Michael D. Heagy*

1.1. INTRODUCTION

Because of their multiple stereocenters and ability to form either furanose or pyranose ring sizes, carbohydrates continue to present a challenge in both molecular recognition studies and chemosensory detection.[1-3] The importance of even simple monosaccharides in both biomedical and clinical research has promoted the development of improved detection methods. In addition to industrial applications such as monitoring sugar concentrations in beverages, quantitative analysis of saccharides is critically linked to certain disease therapies. For example, monitoring of glucose remains a key concern of diabetes management.[4] Currently, biosensors such as glucose oxidase have been applied in saccharide detection and offer a reliable method to detect glucose. However, some disadvantages such as the poor stability of the enzyme and consumption of substrate during the detection process represent some limitations to biosensing applications.[5] Although less selective than protein based biosensors, synthetic chemosensors offer advantages such as greater stability and reliability in a variety of conditions.

Phenylboronic acid has been widely utilized for the design of fluorescent chemosensors in the detection of saccharides over the past decade.[6-11] Its use in the chromatographic separation of sugars dates back even further.[12-13] The covalent interactions between phenylboronic acid and the hydroxyl groups of saccharides (1,2- or 1,3-diols) form a five- or six-membered ring.[14] Compared to other functional groups, this binding property allows boronic acid to be a

*Michael D. Heagy, New Mexico Institute of Mining and Technology, Socorro, New Mexico USA, 87801. Department of Chemistry, Jones Hall Rm 259; Tel: 505.835.5417, Email: mheagy@nmt.edu

competitive chelator for saccharides in water. For most saccharide probe designs, the receptor component is composed of a monoboronic acid or appropriately spaced bisboronic acid moiety. Numerous reports[15-19] have shown that bisboronic acid sensors can be configured to display higher selectivity for glucose over fructose with K_D values from 10^{-5} to 10^{-4} M, whereas monoboronic acid probes usually display weaker binding with K_D values from 10^{-3} to 10^{-2} M and an affinity for fructose that is approximately 100 times greater.[20-22] Although monoboronic acid sensors tend to be overlooked because of limited design geometries, recent reports have indicated that fluorophores may also have a role in determining saccharide selectivity.[23]

The typical fluorescent sensor design can be divided into three important parts. The first is the chelator group for the recognition step; the second is the fluorophore which gives the spectroscopic characteristics and the third; the mechanism producing the perturbation of the fluorophore upon chelation of the analyte.[24] Early designs for saccharide-sensors involved a common anthracene or similar stilbene-based fluorophore as the reporting unit with synthetic modification of the phenylboronic acid or methylene spacer that binds these two groups together. However, the use of such fluorophores frequently results in hydrophobic probes that necessitate the addition of organic cosolvents to increase their solubility in aqueous media.[25] To overcome this limitation, we have been developing monoboronic acid saccharide sensors which utilize more polar reporting groups based upon the naphthalic anhydride chromophore.[26] As a highly fluorescent and photostable molecule, 1,8-naphthalimide has attracted a number of investigations in recent years.[27-33] Derivatives of naphthalimide display useful photochemical properties and have been utilized as photoactivatable DNA-cleaving agents, fluorescent tags, and receptor antagonists.[34-36] Recently, *N*-phenylnaphthalimides have been found to exhibit dual fluorescence when appropriately substituted at both the *N*-phenyl ring and naphthalene π–system. These compounds display two clearly resolved emission bands in the visible region from the locally excited state and a strongly red-shifted band emitted by the internal charge transfer state.[37] These unique optical properties prompted us to explore *N*- phenyl derivatives of naphthalimide as fluorescent components for saccharide sensors.

1.2. *N*-PHENYLBORONIC ACID ARENECARBOXIMIDES AS SACCHARIDE PROBES WITH VIRTUAL OR C0 SPACER DESIGN

We center this review on a series of substituted 1,8-naphthalene dicarboximides as these systems represent a highly versatile set of monoboronic acid saccharide sensors. Their ease of synthesis and unusual photophysical properties make them interesting candidates as fluorescent molecular probes. The relatively understudied virtual spacer or C_0 design in these systems has been found to impart a significant quenching response in the presence of saccharides. Moreover, they operate in aqueous conditions at neutral pH; a prerequisite for most biological investigations. In the first section, we present the synthesis and optical features of several monoboronic acid probes and the effects that substituents located on the naphthalic ring have on their solubility, selectivity and fluorescent properties. In particular, we probe their fluorescence features through a number of substituted versions at either the 3 or 4- position of the naphthalene ring. Next, we investigate the interannular differences between the *N*-aryl component and the imide platform from the probable sugar binding paradigms. Finally, we examine the time-dependent fluorescence properties of the 2-and 3-phenylboronic group. These measurements provide a unique insight into the effects of positional isomers of the boronic acid.

1.3. SUBSTITUENT EFFECTS ON MONOBORONIC ACID DERIVATIVES OF *N*-PHENYL-1, 8-NAPHTHALENEDICARBOXIMIDES

A striking feature of the *N*-phenylnaphthalimides (compared to N-alkyl derivatives) is the presence of two emitting states, capable of yielding both short wavelength (SW) and long wavelength (LW) fluorescence.[38] Recent studies indicate that the geometry of the SW state is similar to that of the ground state whereas twisting of the phenyl group toward a coplanar conformation is thought to form the LW state.[39] The identification of the first and second excited singlet states is supported by Hückel MO calculations adapted from Berces *et. al* and carried out for the parent *N*-phenylnaphthalimide.[37] The calculated electron distributions for the relevant molecular orbitals are shown in Figure 1.1.

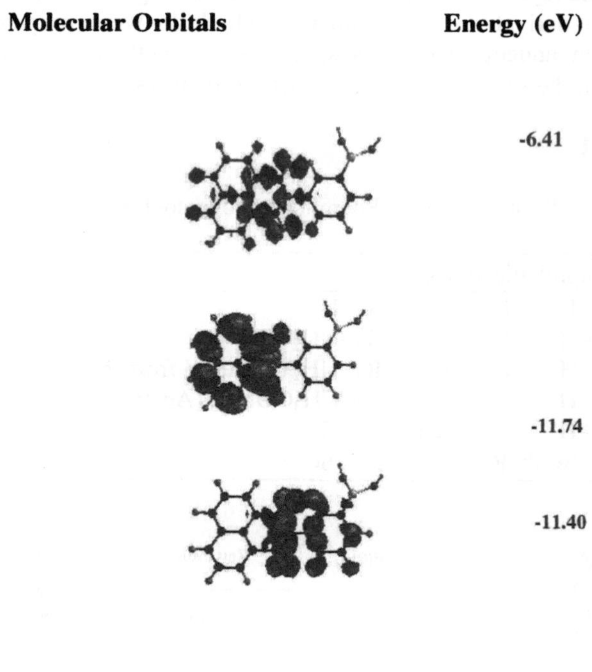

Figure 1.1. Relevant molecular orbitals of *N*-phenyl naphthalimide scaffold (adapted from reference 38).

Electron transfer from the HOMO to the LUMO occurs from the naphthalene moiety to the π^* orbitals of the carbonyl groups, in which the electrons of the aniline group do not participate. This excited state transition is expected to relax via radiative decay as SW emission. A comparison between the electron distributions in HOMO-1 and LUMO orbitals, however indicates that electron density shifts from the aniline moiety to the π^* orbitals of the carbonyl group. This S_2 state reverses the direction of the dipole moment in relation to the ground state thus giving rise to charge-transfer character (ICT) states responsible for LW fluorescence. In addition, substituent groups with electron releasing and electron withdrawing properties on the naphthalene moiety have been shown to significantly affect the photophysical behavior of these fluorophores.

1.3.1. Synthesis

To explore these effects, we synthesized a series of sensors **1-6**, using a variety of different fluorophores derived from 1,8-naphthalenedicarboxylic anhydride. Sensors **1**, **2**, **6** were prepared in single step reactions from commercially available naphthalic anhydrides. 3-aminophenylboronic acid was selected for the synthesis of all six sensors because of its low cost and commercial availability. The other sensors were obtained by reduction (**2** to **3**), acetylation (**3** to **4**) and introduction of methoxy substitution of *N*-(4'-bromo-1',8'-naphthaloyl)-3-aminophenyl boronic acid to give **5** (Scheme 1.1). Sulfo- and amino derivatives of naphthalic anhydride were prepared to increase solubility in aqueous solvent systems. Nitro, methoxy and acetamido were prepared for their potential charge transfer properties.

Scheme 1.1.

Table 1.1. Synthesis of 3-Phenylboronic acid-1,8-naphthalenedicarboimide derivatives

1,8 naphthalic anhydride			yield
1 R_1=H	R_2= H	R_3=H	50%
2 R_1=H	R_2= H	R_3=NO_2	67%
3 R_1=H	R_2= H	R_3=NH_2 (Reduced from **2**)	87%
4 R_1=H	R_2= H	R_3=CH_3CONH, (Acetylated from **3**)	91%
5 R_1=H	R_2= OCH_3	R_3=H,	40%
6 R_1=SO_3K	R_2= NH_2	R_3= SO_3K	91%

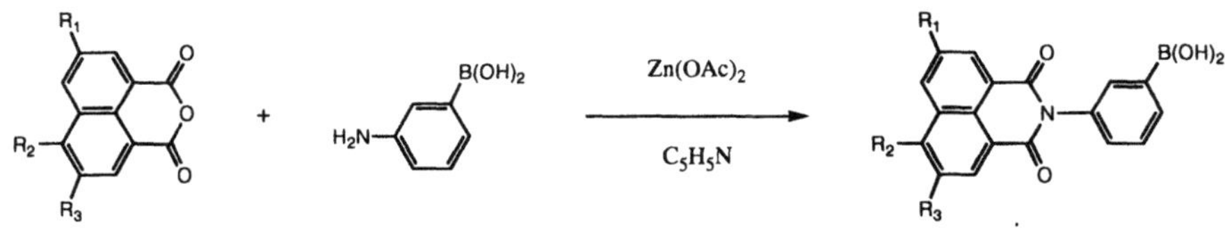

Table 1.2 summarizes the photophysical properties of the six compounds synthesized for this study. Sensors **2** and **3** display significant Stokes shifts and sensors **5** and **6** show relatively high quantum yields. In an effort to explain their optical features, we approximated the energy gap between HOMO to LUMO and HOMO-1 to LUMO for all six of the compounds synthesized in this study using the extended Hückel software from Chem 3D version 5.0.

Table 1.2. Photophysical properties of 3-Phenylboronic acid-1,8-naphthalimide and derivatives.

Entry	Sensor	λ_{ex}, nm	λ_{em}, nm	ϕ_F	ΔE HOMO-LUMO (eV)
1	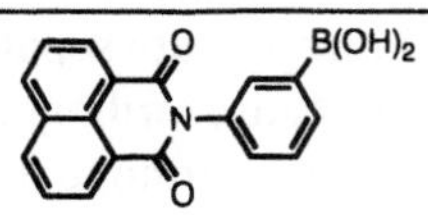	345	400	0.010	3.59
2	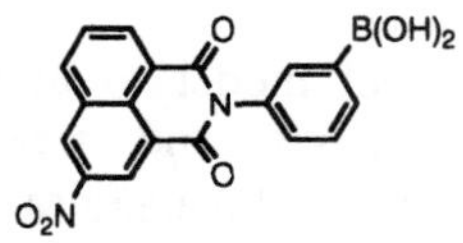	337	430/550	0.006	3.68
3	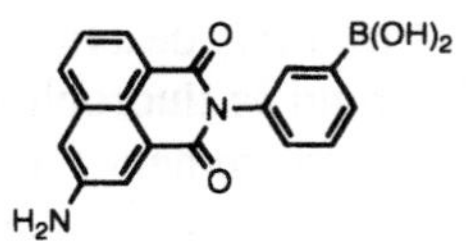	347	581	0.017	3.57
4	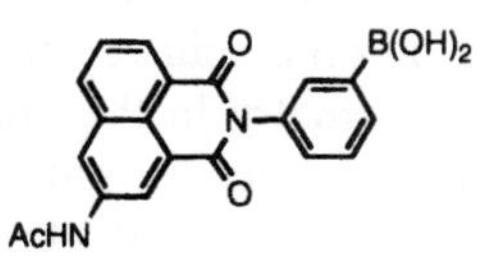	349	407	0.014	3.55
5	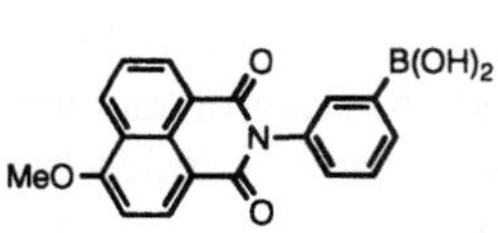	363	440	0.407	3.42
6	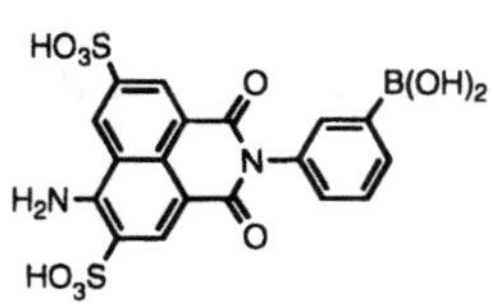	429	534	0.165	2.89

As shown in Table 1.2, probes **2** and **3** display significant Stokes shifts and sensors **5** and **6** show relatively high quantum yields. Probes **2** and **5** display the lowest and highest quantum yield, respectively. In addition, probe **2** displays two emission bands which are characteristic of *S*1 and *S*2 states. With the exception of entry **2**, the experimentally determined HOMO-LUMO energy gaps are in agreement with similarly substituted fluorophores.[40]

1.3.2. Photoelectrochemical Model

Photoinduced electron transfer has been widely used as a tool in the design of fluorescent sensors for saccharides. These fluorescent sensors are based on the boronate ester complex between carbohydrate and boronic acid receptor and typically display optical signals through changes in fluorescence intensity either through chelation enhanced-quenching (CHEQ), or chelation-enhanced fluorescence (CHEF). As with fluorescent chemosensors for ion detection, carbohydrate sensors usually consist of three parts; fluorophore, spacer and receptor. Fluorescent probes without spacers are far fewer in number and have been classified as orthogonal systems. Current examples include twisted biaryls where the π molecular orbitals of the fluorophore and receptor are separated due to steric hindrance between their σ frameworks.[24] Because orthogonality of their molecular orbitals is concomitant with geometric orthogonality, the signaling behavior of these systems has been interpreted as a PET process in a "fluor-spacer-receptor" assembly with a virtual C_0 spacer.

Here, our model deviates from the photophysical model to allow for conformational changes associated with sugar binding and treats the phenylboronic acid MO as a separate π system from the naphthalene MO.[41] X-ray crystal studies find that the plane of the *N*-phenyl group makes a dihedral angel of 69.4° with the plane of the naphthalimide.[42] From this description, we have a C_0 spacer design that places the naphthalimide group as fluorophore and phenylboronic acid moiety (receptor) in a non-planar arrangement. Because saccharide complexation alters the oxidation state of phenylboronic acid, our model proposes a photoelectrochemical mechanism similar to one generally accepted for ion-responsive probes. In this case, the boronic acid receptor has an sp^2-hybridized boron atom with a trigonal planar geometry in the unbound form. Upon complexation with saccharides, a boronate anion forms which possesses an sp^3-hybridized boron with tetrahedral geometry. This saccharide complex alters the orbital energy of the HOMO for the phenyl boronic π-system via occupation of the next highest molecular orbital as shown in Figure 1.2. Fluorescence quenching upon saccharide binding occurs since the electron transfer process (PET) from phenyl group to naphthalic imide becomes energetically favorable. This photoelectrochemical model is expected to agree with factors set forth in the Weller equation, where the free energy of electron transfer is given by $\Delta G_{ET} = -E_{S.fluor} - E_{red.fluor} + E_{ox.receptor} - E_{ion\ pair}$.[43,44] Thus any decrease in the singlet energy component ($E_{S.fluor}$) is expected to display a reduced PET quenching response.

1.3.3. Saccharide Complexation Results

From the model described above, we predict chelation enhanced quenching (CHEQ) to be the predominant signaling pathway. For chemosensors **1** and **2** this fluorescence quenching mechanism is supported by a pH dependent change in I_{max} and a significant shift of the pH- profile to lower pH- I_{max} region in the presence of saccharides. The pH-fluorescence profiles of **1** and **2** obtained from buffered solution are shown in Figures 1.3a and 1.4a, respectively.

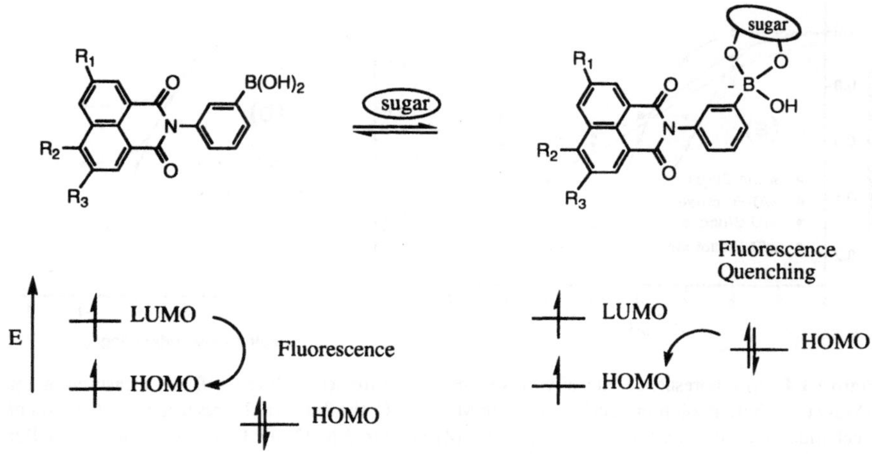

Figure 1.2. Photochemical model used in describing PET fluorescence quenching and interaction between saccharide complexation.

From figure 1.3a, probe **1** displays one emission band at 400 nm and a p*K* value calculated to be 7.7. In the presence of fructose, ester formation between probe **1** and fructose was observed as a function of pH to obtain a pK_a of hydroxyboronate nearly 2 pK_a units lower than unbound sensor. As shown in figure 1.4a, the higher degree of quenching in probe **2** compared to **1** is attributed to the greater E_{red} values which are characteristic of nitroaromatic compounds. The observed effect from this substituent leads to enhanced PET quenching relative to the parent probe **1**. Similar to probe **1**, the intensity changes recorded at 430 nm for **2** correlate with a mono-acid titration curve and pK_a of 8.0. The acidity of the boronic acid group increases in the presence of glucose, giving a pK_a value of 6.6

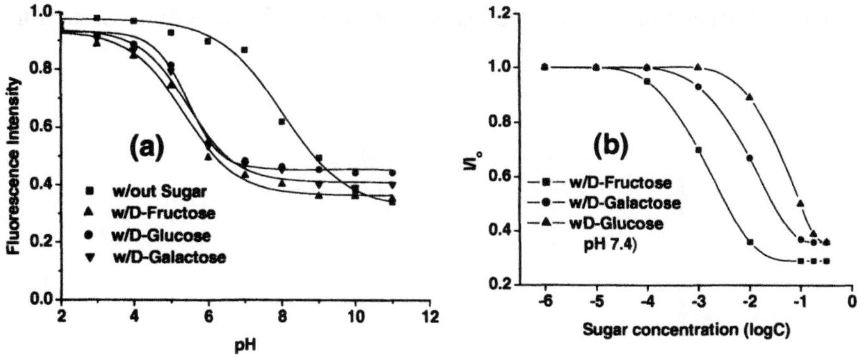

Figure 1.3. a) Fluorescence intensity versus pH profile for sensor **1** (3.0×10^{-5}M) measured in 1% DMSO (v/v) buffer solution, saccharides (0.05M) at 25°C. (b) Relative fluorescence as a function of saccharide concentration for sensor **1** measured in 1% DMSO (v/v) phosphate buffer (100mM), pH 7.4 at 25°C (λ_{ex} = 345nm, λ_{em} = 400nm)

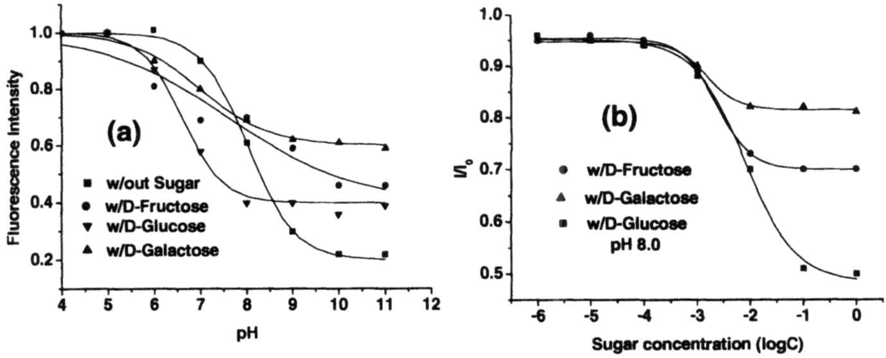

Figure 1.4. (a) Fluorescence intensity versus pH profile for sensor **2** (3.0×10^{-5}M) measured in 1% DMSO (v/v) buffer solution, saccharides (0.05M) at 25°C. (b) Relative fluorescence as a function of saccharide concentration for sensor **2** (3.0×10^{-5}M) measured in 1% DMSO (v/v) phosphate buffer (100mM), pH 7.4 8.0 at 25°C. (λ_{ex} = 337nm, λ_{em} = 430nm and 550nm)

Two unique features appear in both saccharide sensing systems relative to several other early saccharide probes at this time. Specifically, a minimum amount of organic cosolvent is required for these probes and secondly, both operate at relatively neutral pH. Such properties are attributed to the polarity of the naphthalimide platform as well as to the electron withdrawing nature of the imide functionality. In addition, probe **2** shows two emission bands (430/550 nm) found in figure 1.5 from a single excitation wavelength (337 nm). While this probe has a significantly lower quantum yield, the two well resolved bands indicate that the dual fluorescence signal is readily detected and may find use in practical applications such as ratiometric detection.

Next, we examined the selectivity of these sensors to common monosaccharides at neutral pH conditions. Figure 1.3b shows the relative fluorescence of **1** at 400 nm as a function of carbohydrate concentration. The decrease in fluorescence intensity (I in the presence of saccharide/ I_0 in the absence of saccharide) for this series is about 0.25. The selectivity of sensor **1**

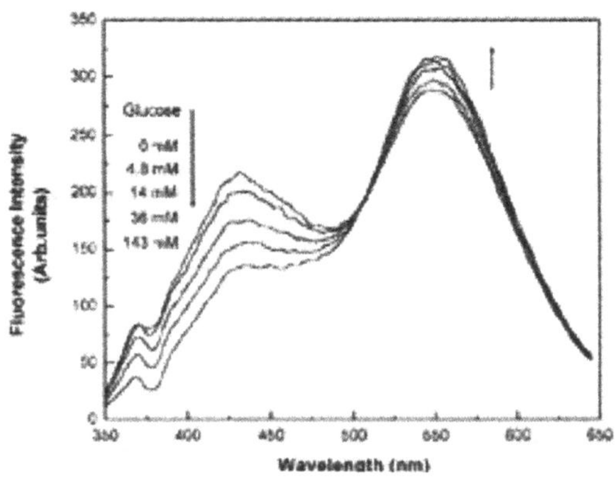

Figure 1.5. Fluorescence response for **2** (3.0×10^{-5}M) with glucose (0.05 M) measured in 1% DMSO (v/v) phosphate buffer (100mM), pH 8.0 at 25°C (λ_{ex} = 337nm).

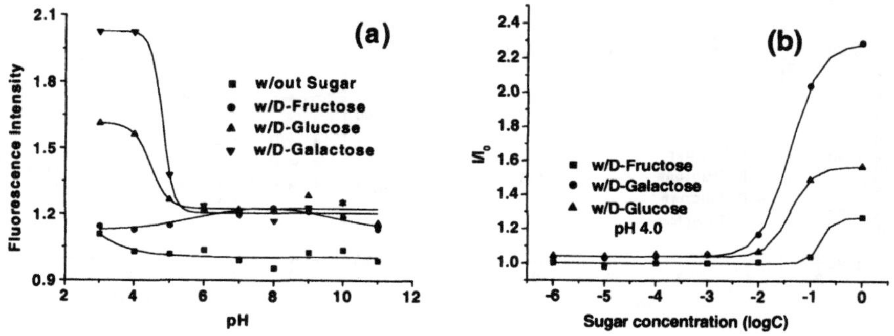

Figure 1.6: (a) Fluorescence intensity versus pH profile for **3** (3.0×10^{-5}M) measured in 1% DMSO (v/v) buffer solution, saccharides (0.05M) at 25°C. (b) Relative fluorescence as a function of saccharide concentration for sensor **3** (3.0×10^{-5}M) measured in 1% DMSO (v/v) phosphate buffer (100mM), pH 4.0 at 25°C. (λ_{ex} = 347nm, λ_{em} = 581nm)

compares with other monoboronic acid probes and shows the greatest association constant with D-fructose. On the basis of a 1:1 complex (obtained from Job's plot), the dissociation constant was found to be 1 mM for fructose while a higher K_d of 250 mM was calculated for glucose.

Despite a low quantum yield, sensor 2 shows two emission bands (430/550 nm) from a single excitation wavelength (337 nm). Figure 1.4b shows the fluorescence ratio (430/550 nm) as a function of carbohydrate concentration. Unexpectedly, sensor **2** displays the largest fluorescence decrease for glucose. In the presence of sugar, the fluorescence spectrum of **2** (Figure 1.5) shows a decrease in emission intensity occurs at 550 nm. Despite this unusual fluorescence response to glucose, the dissociation constants reflect the expected trend for boronic acid:saccharide complexes. For the conditional dissociation constant of **2** with glucose, Kd = 26 mM was measured at pH 8.0. A lower dissociation value for fructose was obtained at Kd = 2.1 mM. This higher pH was selected because it provided the maximum difference in chelation-enhanced quenching.

As the graph demonstrates in figure 1.6, sensor **3** displays marked fluorescence quenching response in the lower pH region between pH 3 to 5. Given this probe's dynamic response to galactose at low pH we titrated **3** with monosaccharides at pH 4. Here, the pH modulation of **3**'s ammonium group to amino is expected to alter its E_{red} values. Between pH of 3-5, the ammonium ion predominates in aqueous solution but as the ammonium group is converted to the free amine, less effective PET quenching occurs.

Probe **6** shows the least effective PET quenching of the four probes shown (figure 1.7). As evidenced from the HOMO-LUMO energy gaps in table 1.1, the 4-amino-3,6 -disulfo substituents raise the HOMO of the fluorophore which is expected to reduce PET from hydroxyboronate anion. A molecular modeling analysis which appears in the next section provides a possible rationale for the dynamic response to galactose relative to fructose and glucose.

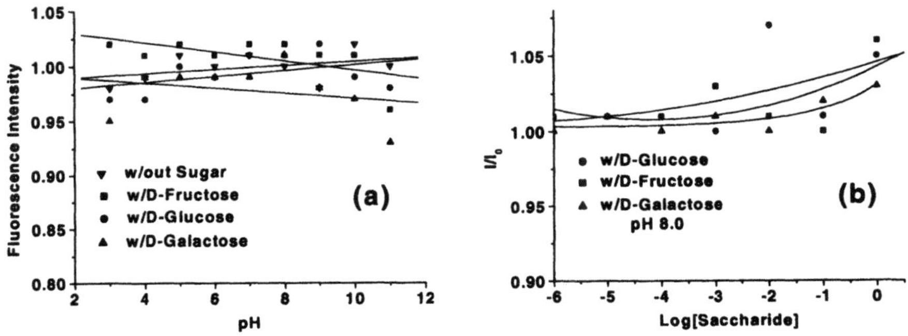

Figure 1.7: (a) Fluorescence intensity versus pH profile for **6** (3.0×10^{-5}M) measured in 1% DMSO (v/v) buffer solution, saccharides (0.05M) at 25°C. (b) Relative fluorescence as a function of saccharide concentration for sensor **6** (3.0×10^{-5}M) measured in 1% DMSO (v/v) phosphate buffer (100mM), pH 8.0 at 25°C. (λ_{ex} = 349nm, λ_{em} = 407nm)

Although probes **4-6** displayed unique optical properties such as **5** with good quantum yield (ϕ_F = 0.407), **6** with moderate quantum yield (ϕ_F = 0.165) and enhanced solubility in water, none of these sensors reported significant optical changes in the presence of saccharides. Results similar to those of **6** were also observed for sensors **4** and **5**. These results are consistent with a less negative ΔG_{ET} term from the Weller equation as electron donating substituent groups located on the naphthalene ring are expected to decrease the energy gap ($\Delta G_{S.fluor}$) between HOMO and LUMO. For the saccharide titrations with **3**, a fluorescence increase was observed only at high concentrations of sugar. While such low binding reflects the reduced affinity of trigonal boronic acid for sugars, a significantly enhanced fluorescence response for **3** with galactose is observed in the titration curve. To explain this unusual response, we compared MMX minimized geometries of these complexes using both tetrahedral and trigonal versions of boron (figure 1.8).[45] Using the predominant form of boron (tetrahedral) and free amino substituent in basic solution, Table 1.3 shows the calculated torsional angles along the biaryl bond for the known ester bonds of furanoid and pyranoid geometries.[46, 47]

As indicated in the second to last column for complexes of tetrahedral boron, minimized geometries display little variance (±2°) in interannular angle for all possible boronate esters of **3** (figure 1.8). Under acidic conditions where PET from a protonated amino substituent to the fluorophore is expected to suppressed, the geometries were minimized with trigonal boron as the organizational node (last column). A comparison between dihedral angles for these complexes reveals a reduced torsional angle for all complexes with both possible galactose complexes giving the closest interannular angle at 55°. Although this 5-6° angular difference between biaryl systems appears small, the fluorescence intensity of *N*-aryl carboximides has been shown to be particularly sensitive to dihedral angle. Because the probe **3**:galactose complex shows the largest fluorescence enhancement followed by glucose and fructose, these differences in torsional angles may account for the fluorescence titration data trends observed in Figure 1.7.

Table 1.3: MMX minimized N-imide/C-phenyl dihedral angles of **2** and **3**: saccharide esters

Saccharide Complex	Probable di- or triols in ester bonds	Dihedral angle (± 0.5°)		Dihedral angle (± 0.5°)	
		2(-NO$_2$) tetrahedral boron	**2(-NO$_2$)** trigonal boron	**3(-NH$_2$)** tetrahedral boron	**3(-NH$_3^+$)** trigonal boron
D-Fructose	β-2,3,6-furanoid	61°	-	60°	-
	β-2,3-furanoid	58°	59°	58°	57°
	β-3,6-furanoid	57°	61°	60°	57°
D-Glucose	1,2-furanoid	58°	59°	59°	57°
	1,2-pyranoid	57°	61°	60°	56°
D-Galactose	3,4-furanoid	59°	62°	61°	55°
	4,6-pyranoid	56°	59°	60°	55°

This fluorescence intensity dependence on torsional angle agrees with findings for all six probes (data not shown) where smaller dihedral angles were calculated for trigonal boron-containing probes relative to their tetrahedral boronate analogs. Finally, it should be mentioned that galactose did not give the smallest torsional angle when this conformational analysis was carried out with the other five probes.

An extension of this conformational analysis may account for the unusually large fluorescence change observed with probe **2** and glucose. Table 1.3 provides a comparative modeling study between nitro derivatives of **2** in the trigonal and tetrahedral state of phenylboronic acid. Although these interannular differences are less pronounced than those found in the case of sensor **3**, glucose shows a 4° increase in dihedral angle in its conversion from trigonal to tetrahedral boron. Thus, the increase in dihedral angle upon saccharide binding is expected to affect the fluorescence intensity. Again, the sensitivity of *N*-arylnaphthalimides to twist angle may be responsible for the dynamic response observed for glucose in pH 8.

1.3.4. Conclusion

In conclusion, a series of monoboronic acid fluorescent probes were synthesized based on the *N*-phenyl naphthalimide fluorophore, in most cases with one step. These carbohydrate molecular probes exhibit interesting photochemical and photophysical properties which provided useful insights into the relatively limited area of C_0 design chemosensor. While many saccharide probes include a spacer group, such as Wulff's use of a benzylic amine spacer to detect in neutral conditions, sensor **1** gives a large fluorescence response at pH 7.4 with no spacer.[48] To the best of our knowledge, sensor **1** displays the largest CHEQ response for a monotopic receptor in aqueous solution. The large fluorescence response lies within detection requirements for fructose but remains less sensitive to physiological level of glucose. Whereas several bisboronic acid sensors have been developed to provide a greater fluorescence

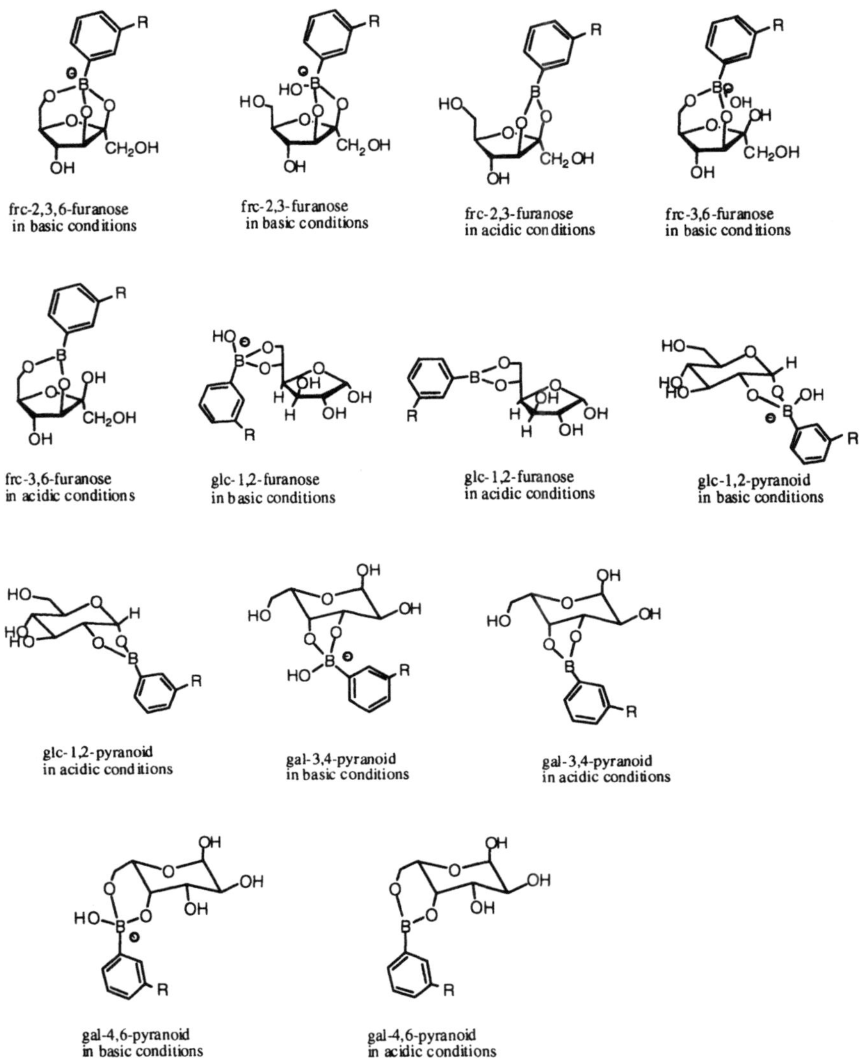

Figure 1.8. 1:1 Binding Motifs for Probable Boronate Ester Diol/Triol Complexes

response of glucose over fructose, sensor **2** represents the first monoboronic acid sensor to display higher sensitivity for glucose over fructose.[25,15-19] This anomalous chelation-enhanced quenching from **2** opposes the expected signal response involving monosaccharides. In this case, the fluorescence data show that the optical change which results from fructose complexation is weak relative to glucose. From the obtained K_d values, it appears that the binding affinity is not proportional to the observed optical change for this saccharide sensor. These findings suggest that other competing factors may be simultaneously operating, such as conformational dynamics between the phenylhydroxyboronate:saccharide complex and its influence on the excited state. Such geometrical changes are also attributed to sensor **3**'s unusual response which is specific to galactose. Probes **1-6** gave poor signal response upon titration with monosaccharides, however their substituent effects lend support to the chelation enhanced quenching model proposed for these systems. Based on this model a reduced PET effect is expected to be more pronounced in sensors **4** to **6**. In addition, the large Stoke shift and bright fluorescence that these compounds display may be useful in other applications such as saccharide labeling experiments.

1.4. POSITIONAL ISOMERS OF NAPHTHALENE DICARBOXIMIDES

1.4.1. Ortho-Substituted Monoboronic Acid Sensor

To observe the effects of positional isomers with these monoboronic acid sensors, **7** was prepared from 2-aminophenylboronic acid and 1,8-naphthalic anhydride. Sensor **7** possesses similar photophysical properties to isomer **5** with regard to its excitation and emission wavelengths. Any differences in its CHEQ response are therefore attributed to the placement of the boronic acid group.

Scheme 1.2.

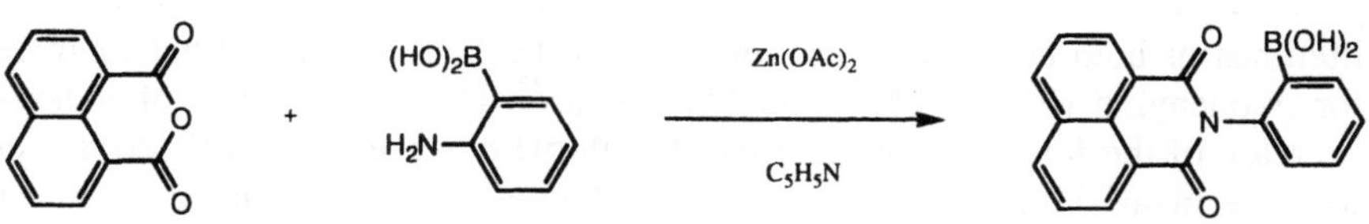

7

A pH vs fluorescence intensity profile (Figure 1.9a) demonstrated that PET signal transduction mechanism is operating as fluorescence intensity becomes suppressed upon ionization to the hydroxyboronate anion. This probe showed behavior similar to *meta*-phenylboronic acid based sensor giving the expected trend in saccharide selectivity with fructose>glucose>galactose (Figure 1.9b). K_D values for fructose and glucose were calculated to be 1.6 and 57 respectively.[49] A key difference between sensor **7** and **1**, however, is evident in the degree of CHEQ that these sensors display. *Ortho*-substituted probe **7** quenches fluorescence less effectively ($I/I_o = 0.4$) than *meta*-isomer **1** ($I/I_o = 0.2$).

The energy difference of their respective HOMO-LUMO gaps is expected to have little to no dependence on the position of the boronic acid goup. Based on our photoelectrochemical model for PET fluorescence quenching, these findings point to a less effective conversion of sensor **7** to its hydroxyboronate ester relative to **1**. Molecular modeling studies indicate that increased steric congestion is likely to be responsible for this weaker binding. These conclusions are further supported by time-dependent fluorescence spectroscopy presented in the next section.

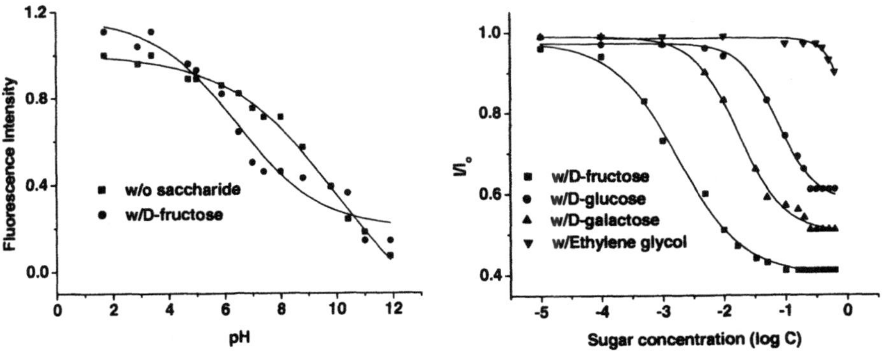

Figure 1.9. (a) Titration curve vs pH for **7** (6.0×10^{-3} M, λ_{exc} = 346 nm at 25 °C) (b) Titration curves against monosaccharides for **7**, measured in phosphate buffer (100 mM), pH 7.4 at 25 °C.

1.4.2. pH and Sugar Effects On The Fluorescent Decays

The pH effect on the fluorescent lifetime of 7 is shown in Figure 1.10 and the decay parameters are listed in Table 1.4. The fluorescent lifetime of **7** in aqueous solution is around 1 ns and does not show significant change with pH. Similar results were observed for the meta derivative (not shown). Fluorescence decays are multiexponential, characterized by a short component of a few picoseconds and a long range component of a few nanoseconds. Fluorescent lifetimes of both compounds are longer than the fluorescent lifetime reported for *N*-phenyl derivative of 1,8-naphthalimide.[37] The combination of a large decrease of the fluorescence intensity at high pH observed for both derivatives with the invariability of the fluorescence lifetime leads to a conclusion that a static quenching is at the origin of the important decrease of the fluorescence intensity.

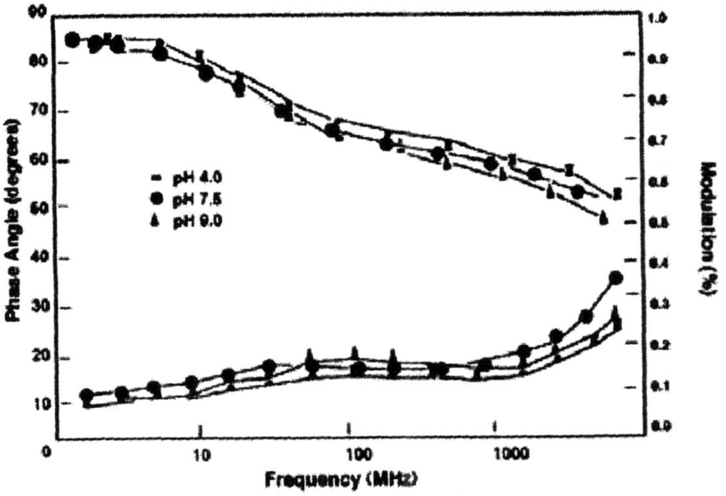

Figure 1.10. Effect of pH on the frequency decay profile of *ortho*-isomer **7** measured at room temperature (λ_{ex} = 310 nm)

Table 1.4. Fluorescence Intensity Decay Parameters for the Ortho and Meta Isomers

	τ_1 (ns)	τ_2 (ns)	α_1	α_2	f_1	f_2	τ_F (ns)
***o*-isomer**							
pH 4.0	0.72	3.70	0.01	0.004	0.17	0.26	1.11
pH 7.5	0.98	4.30	0.01	0.004	0.14	0.22	1.11
pH 9.0	0.75	4.8	0.02	0.003	0.23	0.23	1.30
+ fructose[a] (50 mM)	1.05	5.22	0.021	0.012	0.18	0.50	2.83
+ glucose[a] (200 mM)	1.18	5.92	0.019	0.005	0.21	0.28	1.91
***m*- isomer**							
pH 4.0	0.53	5.40	0.011	0.002	0.09	0.17	0.97
pH 7.5	1.25	6.83	0.005	0.007	0.008	0.18	1.36
pH 9.0	0.07	1.50	0.148	0.002	0.43	0.12	1.31
+ fructose[a] (50 mM)	1.63	6.58	0.005	0.003	0.10	0.21	1.61

[a] Phosphate buffer pH 7.5.

In contrast to the effect of pH, the presence of sugar induces a significant increase of the fluorescence lifetime of derivative **7** (Figure 1.11). The fluorescence lifetime changes are characterized by a decreased contribution of the short component and an increased contribution of the long component. Fluorescence decay results on the *ortho* derivative **7** in the presence of sugar do not correlate with a static quenching as observed for the pH effect. As shown in Table 1.4, sugar effects on the fluorescent lifetime of the *meta* isomer **1** are small. From this result, it seems that the position of the boronic acid group on the *N*-phenyl ring plays a role in the sugar response, but not in the pH response. As mentioned in the first section of this review, these findings suggest the presence of steric hindrance and/or participation of a charge transfer excited state. Because the photophysics of the *N*-phenyl derivatives of 1,8-naphthalimide are governed by a pseudo-Jahn-Teller effect, resulting in an important non-radiative internal conversion process, the presence of a sugar molecule bound to the boronic acid group at the *ortho* position is likely to lead to a rigidification of the molecule and thus to a decrease of the internal conversion process. On the other hand, the formation of the electron donor anionic form of the boronic acid group in the *ortho* position could lead also lead to a longer fluorescence lifetime,[37,38] but in this case a similar effect with pH is expected. In all cases, an increase of the fluorescent lifetime should be associated with an increase of the fluorescence intensity. Because we observed a decrease of the fluorescence intensity, it is probable that more than one process is involved in the photophysics of the probe in the presence of sugars.

1.5. CONCLUSION

The incorporation of the boronic acid group in the *meta* or *ortho* position does not lead to significant spectroscopic and photophysical changes in comparison with the parent *N*-phenyl analog. Both derivatives show a larger decrease of their fluorescence emission at high pH while no effect of the pH was observed on the fluorescence lifetime. This suggests the presence of a static quenching from the formation of the anionic form of the boronic acid group. Both compounds also show significant decrease of their fluorescence intensity in the presence of sugars, suggesting that they could be used as on-off probes in analytical devices for sugar signaling. A significant increase of the fluorescent lifetime was observed for the **7** in the presence of sugar while a relatively small effect was obtained for the *meta*–isomer. The decrease of the fluorescence appears to be correlated with a steric hindrance effect. Important changes in the phase angle and modulation in the presence of sugar show the potential of **7** for use as a sugar probe for fluorescence lifetime-based sensing.

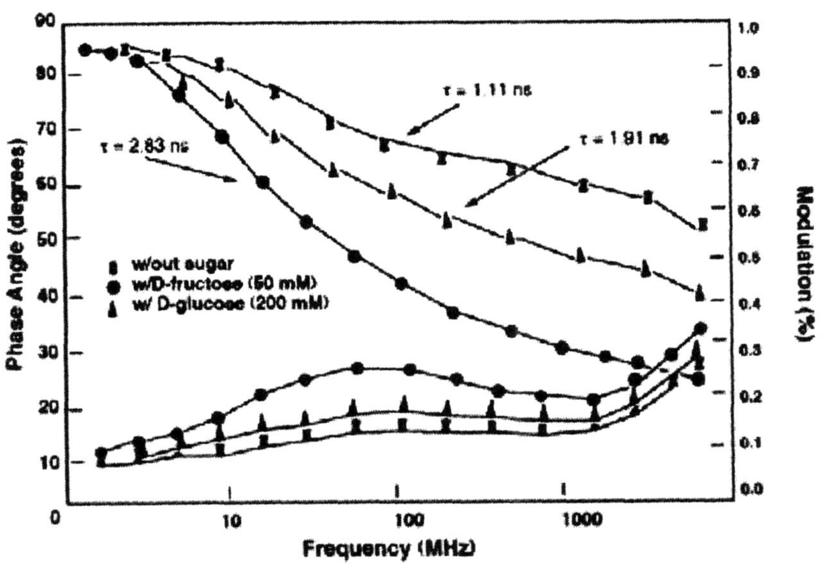

Figure 1.11. Effect of sugar on the frequency decay profile of *ortho*-isomer **7** measured in phosphate buffer (100 mM) pH 7.5 at room temperature (λ_{ex} = 310 nm)

1.6. SUMMARY AND OUTLOOK

Herein several approaches were described by which arenecarboximides have been employed as versatile fluorescent platforms for saccharide sensing. By investigating the effects of substituent groups as well as placement of the boronic acid a number of saccharide sensors were prepared with usually one step reactions. These approaches have resulted in the discovery of chemosensors with different saccharide selectivities. Though by no means an exhaustive sampling of this modular synthetic approach, the sensors that were generated exhibited interesting photochemical and photophysical properties and provided useful insights into the relatively limited area of C_0 design chemosensor.

Among several monoboronic probes we investigated, probe **1** displayed a large change in CHEQ signal response under physiological condition. Probe **2** exhibits dual emission with remarkable sensitivity for glucose relative to fructose and galactose through subtle changes in pH. At low pH sensor **3** signaled enhanced fluorescence (CHEF) in the presence of galactose. Probes **4**-**6** gave poor signal response upon titration with monosaccharides, yet their substituent effects lend support to the chelation enhanced quenching model proposed for these systems.

Our investigations into positional isomers of simple monoboronic acid probes gave additional evidence for the photoelectrochemical model involving CHEQ. A comparison between receptor platforms of the *ortho* and *meta*-phenylboronic groups (**7** and **1**, respectively) indicated significant differences in fluorescent lifetimes upon complexation with different sugars, particularly fructose. These findings suggest that the *ortho*-receptor group may have potential applications as a sugar probe for fluorescence lifetime-based sensing.

1.7. ACKNOWLEDGMENTS

The author wishes to thank the many students who have contributed to this area of research. Namely and in alphabetical order, the assistance of Devi Adhikari, Haishi Cao, Amanda Carnahan, Dalia Diaz, Susan Duong, Dustin English, Lisa Eskra, Justin Heynekamp, Fang Liang, Tom McGill and Qiang Li is gratefully acknowledged. It has been their enthusiasm and dedication which provides the driving force for our ongoing research in saccharide detection. MDH thanks Professors Chris Geddes and Joseph Lakowicz for their kind invitation to contribute to this annual review. Finally, we thank NIH for several years of support via its AREA grant program.

1.8. REFERENCES

1. A. P. Davis, R.A. Wareham, Carbohydrate recognition through noncaovalent interactions: A challenge for biomimetic and supramolecular chemistry, *Angew. Chem., Int. Ed. Engl* **35**, 2978-2996 (1996).
2. T. D. James, K. R. A. S. Sandanayake, S. Shinkai, Saccharide sensing with molecular receptors based on boronic acid, *Angew. Chem. Int. Ed. Engl.* **35**, 1910-1922 (1996).

3. H, Cao, M.D. Heagy, Fluorescent chemosensors for carbohydrates: A decade's worth of bright spies for saccharides in review. *J. Fluoresc* **14**, 569-584 (2004).
4. C. Meyerhoff, F.J. Mennel, F. Sternberg, E.F. Pfeiffer, Current status of the glucose sensor, *The Endocrinologist,* **6**, 51-58.
5. D.A. Gough, J.C. Armor, Development of the implantable glucose sensor – what are the prospects and why is it taking so long, *Diabetes,* **44**, 1005-1009, (1995).
6. N. DiCesare, J.R. Lakowicz, A new highly fluorescent probe for monosaccharides based on a donor-acceptor diphenyloxazole, *Chem. Commun.* 2022-2023 (2001).
7. S. Arimori, M.L. Bell, C.S. Oh, K.A. Frimat, T.D. James Modular fluorescent sensors for saccharides, *J. Chem . Soc. Perkin Trans. 1*, **6**, 803-808 (2002).
8. K.R.A.S Sandanayake, S. Shinkai, Novel molecular sensors for saccharides based on the interaction of boronic acids and amines- saccharide sensing in neutral water, *J. Chem. Soc. Chem. Commun.* **9**, 1083-1084 (1994).
9. B. Kukrer, E.U. Akkaya, Red to near IR fluorescent signaling of carbohydrates, *Tetrahedron Lett.* **40**, 9125-9128 (1999).
10. N. DiCesare, J.R. Lakowicz, Fluorescent probe for monosaccharides based on functionalized boron-dipyronemethane with a boronic acid group, *Tetrahedron Lett.* **42**, 9105-9108 (2001).
11. X. Gao, Y. Zhang, B. Wang, New boronic acid fluorescent reporter compounds. 2, A naphthalene-based on-off sensor functional at physiological pH, *Org. Lett.* **5** 4615-4618 (2003)
12. S. Soundararajin, M. Badawi, C.M. Kohlrust, J.H. Hageman, Boronic acids for affinity chromatography: Spectral methods for determinations of ionization and diol-binding constants, *Anal. Biochem*, **178**, 125-134 (1989).
13. R. Ramoncita, R. Maestas, J.R. Prieto, G.D. Kuehn, J.H. Hageman, Polyacrylamide-boronate beads saturated with biomolecules: A new general support for affinity chromatography of enzymes, *J. Chromatography,* **189**, 225-231, (1980).
14. J.C. Norrild, I. Sotofte, Design, synthesis and structure of new potential electrochemically active boronic acid-based glucose sensors, *J. Chem. Soc. Perkin Trans 2,* 303-311, (2002).
15. T.D. James, K.R.A.S. Sandanayake, S. Shinkai, A glucose selective molecular fluorescence sensor, *Angew. Chem. Int. Ed Engl.* **33**, 2207-2209 (1994)
16. T.D. James, K.R.A.S. Sandanayake, R.Iguchi, S. Shinkai, Novel saccharide-photoinduced electron transfer sensors based on the interaction of boronic acid and amine, *J. Am. Chem., Soc.* **117**, 8982-8987 (1995).
17. K.R.A.S. Sandanayake, T.D. James, S. Shinkai Two-dimensional photoinduced electron transfer (PET) fluorescence sensor, *Chem. Lett.* 503-504, (1995)
18. W. Yang, Y. He, D.G. Drueckhammer, Computer-guided design in molecular recognition: Design and synthesis of a glucopyranose receptor, *Angew. Chem. Int. Ed.* **40**, 1714-1717.
19. V.V. Karnati, X.M. Gao, S.H. Gao, W.Q. Wang, W. Ni, S. Sankar, B.H. Wang, A glucose-selective fluorescence sensor based on boronic acid-diol recognition, *Bioorg. Med. Chem. Lett.* **12**, 3373-3377 (2003).
20. N. DiCesare, J.R. Lakowicz, New color chemosensors for monosaccharides based on azo dyes, *Org. Lett,* **3**, 3891-3893 (2001)
21. N. DiCesare, J.R. Lakowicz, Chalcone-analogue fluorescent probes for saccharides signaling using the boronic acid group, *Tetrahedron Lett.* **43**, 2615-2618 (2002).
22. H. Cao, D. I. Diaz, N. DiCesare, J. R. Lakowicz, M. D. Heagy, Monoboronic acid sensor that displays anomalous fluorescence sensitivity to glucose, *Org. Lett.* **4**, 1503-1505 (2002).
23. S. Arimori, G.A. Consiglio, M.D. Phillips, T.D. James, Tuning saccharide selectivity in modular fluorescent sensors, *Tetrahedron Lett.* **44** 4789-4792.
24. R. A. Bissell, A. P. de Silva, H. Q. N. Gunaratne, P. L. M. Lynch, G. E. M. Maguire, K. R. A. S. Sandanayake, Molecular fluorescent signalling with fluor-spacer-receptor systems. Approaches to sensing and switching devices via supramolecular photophysics, *Chem. Soc. Rev.* 187-195 (1992).
25. H. Eggert, J. Frederiksen, C. Morin, J.-C. Norrild, A new glucose-selective fluorescent bis-boronic acid. first report of strong -furanose complexation in aqueous solution at physiological pH, *J. Org. Chem.* **64**, 3846-3852 (1999).
26. D. P. Adhikiri, M. D. Heagy, Fluorescent chemosensor for carbohydrates which shows large change in chelation-enhanced quenching, *Tetrahedron Lett.* **40**, 7893-7896 (1999).

27. R.W. Middleton, J. Parrick, E. D. Clarke, P. Wardman, Synthesis and fluorescence of *N*-substituted-1,8-naphthalimides, *J. Heterocyclic Chem.* **23**, 849-855 (1986).
28. J. Gawronski, K. Gawronska, P. Skowronek, A. Holmén, 1,8-Naphthalimides as stereochemical probes for chiral amines: A study of electronic transitions and exciton coupling, *J. Org. Chem.* **64**, 234-241 (1999).
29. K. Nakaya, K. Funabiki, H. Muramatsu, K. Shibata, M. Matsui, *N*-aryl-1,8-naphthalimides as highly sensitive fluorescent labeling reagents for carnitine, *Dyes and Pigments* 43, 235-239 (1999).
30. S. Chang, R. E. Utecht, D. E. Lewis, Synthesis and bromination of 4-alkylamino-*N*-alkyl-1,8-naphthalimides, *Dyes and Pigments* **43**, 83-94 (1999).
31. Gawronski, M. Kwit, K. Gawronska, Helicity Induction in a Bichromophore: A sensitive and practical chiroptical method for absolute configuration determination of aliphatic alcohols, *Org. Lett*, **4**, 4185-4188 (2002).
32. L. M. Daffy, A. P. D. de Silva, H. Q. N. Gunaratne, C. Huber, P. L. M. Lynch, T. Werner, O. S. Wolfbeis, Arenedicarboximide Building Blocks for Fluorescent Photoinduced Electron Transfer pH Sensors Applicable with Different Media and Communication Wavelengths, *Chem. Eur. J.* **4**, 1810-1815 (1998).
33. B. Ramachandram, A. Samanta, Modulation of metal-fluorophore to develop structurally simple fluorescent sensors for transition metal ions, *Chem. Commun.* 1037-1038 (1997)
34. I. Saito, M. Takayama, S. Kawanishi, Photoactivatable DNA-cleaving amino acids- highly sequence selective DNA photocleavage by novel L-lysine derivatives, *J. Am. Chem. Soc.* **117**, 5590-5591 (1995).
35. J.E. Rogers, S.J. Weiss, L.A. Kelly, Photoprocesses of naphthalene imide and diimide derivatives in aqueous solutions of DNA, *J. Am. Chem. Soc.* **122**, 427-436 (2000).
36. B. M. Aveline, S. Matsugo, R. W. Redmond, Photochemical Mechanisms Responsible for the Versatile Application of Naphthalimides and Naphthaldiimides in Biological Systems, *J. Am. Chem. Soc.* **119**, 11785-11795 (1997).
37. A. Demeter, T. Bérces, L. Biczók, V. Wintgens, P. Valat, J. Kossanyi, Comprehensive model of the photophysics of *N*-phenylnaphthalimides: The role of solvent and rotational relaxation, *J. Phys. Chem* **100**, 2001-2011 (1996).
38. V. Wintgens, P. Valat, J. Kossanyi, A. Demeter, L. Biczok, T. Berces, Spectroscopic properties of aromatic dicarboximides. Part 4. On the modification of the fluorescence and intersystem crossing processes of molecules by electron-donating methoxy groups at different positions. The case of 1,8-naphthalimides, *New J. Chem.*, **20**, 1149-1158 (1996).
39. P. Valat, V. Wintgens, J. Kossanyi, Temperature dependent behavior of the dual fluorescence of 2-(3-fluorophenyl)-2,3-dihydro-1H-benzo[f]isoindole-1,3-dione, *Helv. Chim. Acta,* **84**, 2813-2832 (2001)
40. I. Garbtchev, Tz. Philipova, P. Meallier, S. Guittoneau, Influence of Substituents on the Spectroscopic and Photochemical Properties of Naphthalimide Derivatives, *Dyes and Pigments*, **31**, 31-34 (1996).
41. A calculated energy barrier of rotation for sensor and sensor-saccharide complex was determined using PC model software (version 6.0). The sensor-saccharide complex displays a higher energy barrier (15.32 kcal/mole) than free sensor (14.41 kcal/mole). Although this difference appears small, the overall strain energy was much higher 140.3 kcal/mol). On this energy difference basis, the rotational energy barrier may prevent phenyl ring rotation.
42. Y. Dromzée, J. Kossanyi, V. Wintgens, P. Valat, L. Biczók, A. Demeter, T. Bérces, Crystal and molecular structure of *N*-phenyl substituted 1,2-, 2,3- and 1,8-naphthalimides, *Z. Kristallog.* **210**, 760-765 (1995).
43. Gibbs free energy term is divided by $-nF$ where E values are in units of volts.
44. A. Weller, Exciplex and radical pairs in photochemical eletron-transfer, *Pure and Appl. Chem.* **54** 1885-1888 (1982).
45. The default force field used in PCMODEL(version 8.0) is called MMX and is derived from MM2(QCPE-395, 1977) force field of N. L. Allinger, with the pi-VESCF routines taken from MMP1 (QCPE-318), also by N. L. Allinger.
46. J.H. Hageman, G.D. Kuehn, Boronic acid matrices for the affinity purification of glycoproteins and enzymes *Methods in Molecular Biology,* **11**, 45-71 (1992).
47. G.R. Kennedy, M.J. Row, The interaction of sugars with borate: an N.M.R. spectroscopic study *Carbohydrate Res.*, **28**, 13-19 (1973).

48. G. Wulff, Selective binding to polymers via covalent bonds: The construction of chiral cavities as specific receptor sites, *Pure Appl. Chem.* **54**, 2093-2102 (1982).
49. N. DiCesare, D. P. Adhikari, J. J. Heynekamp, M. D. Heagy, J. R. Lakowicz, Spectral properties of fluorophores combining the boronic acid group with electron donor or withdrawing groups. Implication in the development of fluorescent probes for saccharides, *J. Fluoresc.* **12**, 147-154 (2002).
50. H. Cao, T. McGill, M.D. Heagy, Substituent effects on monoboronic acid sensors for saccharides based on *N*-phenyl-1,8-naphthalenedicarboximides, *J. Org. Chem.* **69**, 2959-2966 (2004).

GENESIS OF FLUOROPHORE FORMATION IN MACROCYCLE SOLUTIONS AND THE DETECTION OF GLUCOSE AND RELATED SUGARS

Jorge O. Escobedo, Oleksandr Rusin, and Robert M. Strongin*

2.1. INTRODUCTION

The development of improved methods for glucose monitoring is a significant challenge. Diabetes and other glucose-related related disorders continue to afflict a significant proportion of the global population. Interdisciplinary research efforts are ongoing on the international scene.[1]

Our glucose sensing investigations began with the discovery of a series of unique functional chromophores and fluorophores. They are generated *in situ* from colorless materials. They bind and recognize glucose and other related specific sugars with relatively high selectivity. A promising assay which affords direct detection of glucose-selective fluorescence responses in human blood plasma has emanated from this work. A convenient automated HPLC postcolumn detection system has been developed which allows for the resolution and monitoring of glucose and other sugar-derived biomolecules in the visible spectral region. The detection of glucose in the presence of structurally related sugars such as fructose can be performed via ratiometric, dual wavelength monitoring. A mechanism based approach guides the development of appropriate experimental conditions for achieving desired selectivity.

2.2. THE DISCOVERY OF NEW FUNCTIONAL FLUOROPHORES

In 1871, Baeyer synthesized fluorescein, the prototypical xanthene dye.[2] Soon thereafter, he investigated the reaction of benzaldehyde and resorcinol in

* Department of Chemistry, Louisiana State University, Baton Rouge, LA 70803 USA.

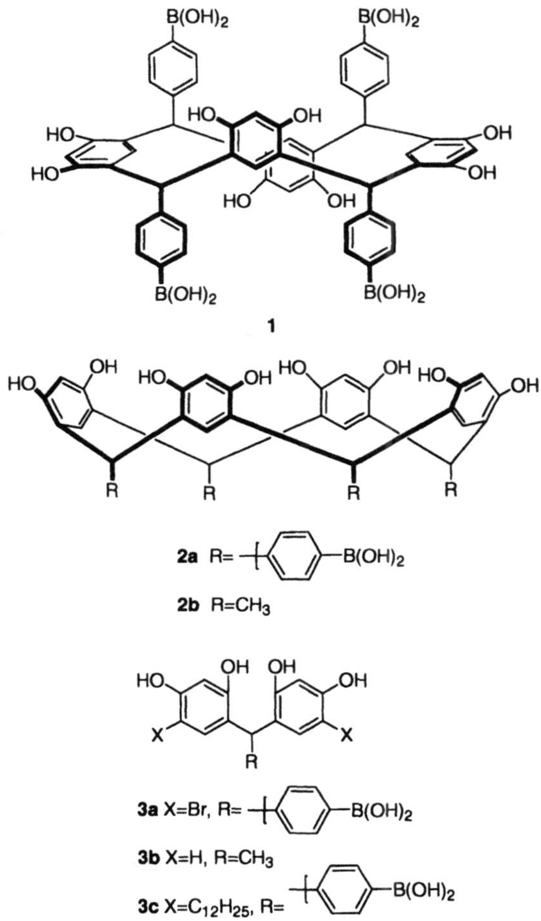

Figure 2.1. We previously reported that solutions containing resorcinarenes **1, 2a, 2b** and related condensation product **3a** exhibit significant color changes in the presence of sugars. Reprinted with permission from *J. Am. Chem. Soc.*, **124**(18), 5000-5009 (2002). Copyright 2002 American Chemical Society.

acidic media.[3] He obtained a reddish solution mixture which changed color to violet upon the addition of base. Baeyer's reaction resulted in the creation of the first resorcinarenes, cyclic tetramers of resorcinol.[4] However, resorcinarenes are colorless. His goal was to create new dyes. The trace amounts of dye materials he obtained were not investigated until our studies began in 1999.

The resorcinarenes have since greatly impacted the fields of molecular recognition, materials science and supramolecular chemistry.[4] They were among the first classes of compounds demonstrated to bind sugars in organic media, via hydrogen bonding and CH-π interactions with the resorcinol moieties comprising the macrocyclic framework.[5] Boronic acids are the basis of carbohydrate affinity chromatography. We reasoned that resorcinarenes containing boronic acids should embody excellent sugar binding agents. We thus synthesized **1** and **2a**.[6]

Compounds **1** and **2a** were synthesized in one step in combined 90 % yield. They were obtained via the HCl-catalyzed reaction of 4-formylphenyl boronic acid and resorcinol.[6] Facile separation of **1** and **2a** by fractional crystallization furnished white crystalline solids. X-ray crystallographic analysis of the half-methyl tetraboronate ester derivative of **1** showed that its solid state architecture embodied infinite, antiparallel two-dimensional network of macrocycles, each of which exhibited twelve intermolecular hydrogen bonds.[7]

Colorless DMSO solutions of recrystallized **1a** or **2a**, (5.2 mM), upon standing in solution for several hours or upon heating at 90 °C for 1 min, developed a pinkish-purple color. The color formation was monitored via the appearance of a λ_{max} at 535 nm and a less intense absorption at 500 nm, as well as new fluorescence maxima at λ_{ex} 525 nm and λ_{em} 570 nm.[8]

We observed eleven different solution colors in eleven different heated specific sugar solutions containing **1**.[8] The sugars studied included neutral carbohydrates, glucose phosphate isomers and carboxylic acid and amino sugars. The colorimetric responses were rapid, quantifiable and reproducible. Solutions containing **2b** exhibited relatively insignificant colorimetric responses in the presence of sugars compared to solutions containing the boronic acids (**1** and **2a**) or their acyclic boron-containing congener **3a**.[9]

Heating DMSO solutions of **1** in the dark and/or in degassed conditions inhibited solution color generation.[8] When the phenolic hydroxyls of **1** were acylated and the resultant derivative heated in a DMSO solution at reflux, the solution remained colorless.[9] Since light, O_2 and the resorcinol hydroxyls all serve to promote color formation, we hypothesized that oxidation of a resorcinol moiety to a quinone led to solution color formation.[8,9]

When resorcinol or benzeneboronic acid were heated in solution separately or as a mixture using the aforementioned conditions and corresponding stoichiometries, with and without added monosaccharides, no significant color formation occurred.[8] This demonstrated that the methine-bridged resorcinol/aldehyde condensation architectural motif was needed for chromophore formation and for an optical responses to sugars. Methine-bridged resorcinol condensation product substructures, of which **3a-c** are examples, have been previously described as reaction intermediates during xanthene dye formation (e.g., transformations such as **4** to **5**, n=m=0, Scheme 1).[10] These studies embodied the first investigations of the colorimetric properties of resorcinarenes solutions since Baeyer's initial report.[3]

Xanthenes include many of the oldest and most widely used dyes. Examples include fluorescein, rhodamine, **6a** and **6b** and ethyl eosin (**6c**) and a great variety of other related structures. Xanthenes typically exhibit two absorption maxima in the visible region.[11] The UV-Vis spectrum of **6b** (5.0 x 10-6 M) in 9:1 DMSO:H_2O is shown in Figure 2.3. It exhibits a λ_{max} at 530 nm

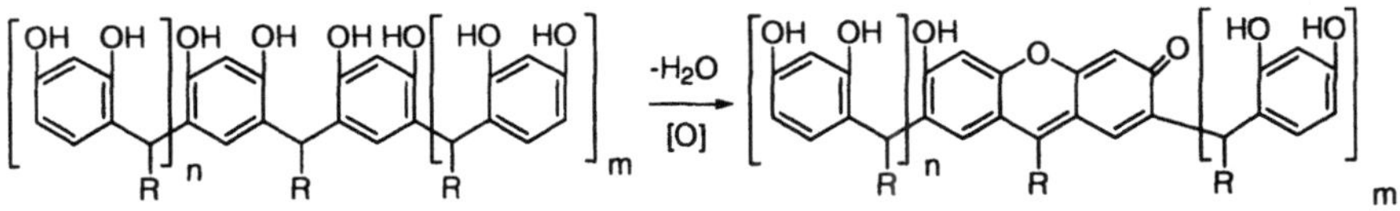

Scheme 2.1. Dehydration and oxidation of methine-bridged resorcinol oligomers leading to a xanthene. Reprinted with permission from *J. Am. Chem. Soc.*, **124**(18), 5000-5009 (2002). Copyright 2002 American Chemical Society.

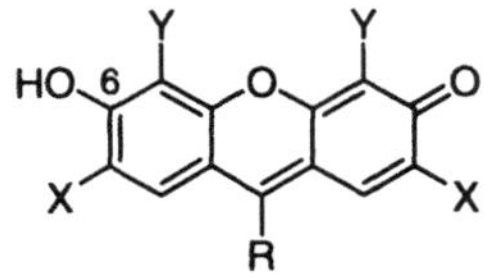

Figure 2.2. Reprinted with permission from *J. Am. Chem. Soc.*, **124**(18), 5000-5009 (2002). Copyright 2002 American Chemical Society.

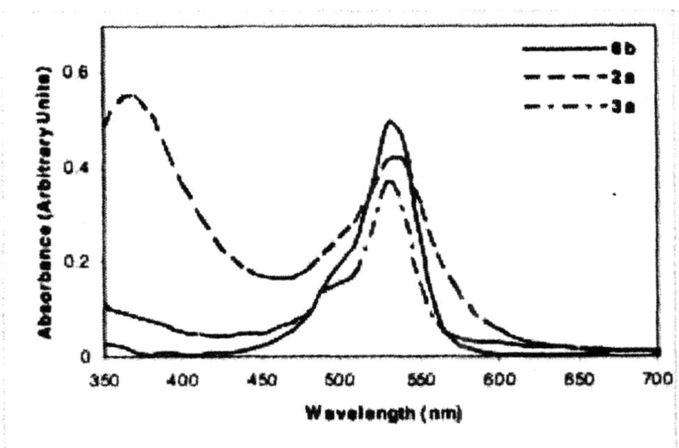

Figure 2.3. **2a** (1.0 mg) and **3a** (1.0 mg) each in 0.9 mL DMSO were heated to gentle reflux over two minutes and cooled to room temperature before 0.1 mL H_2O was added to each solution. The final concentrations of **2a** and **3a** in 9:1 DMSO:H_2O are 1.03×10^{-3} M and 1.96×10^{-3} M respectively. A solution of **6b** (5.0×10^{-6} M) was prepared at rt in 9:1 DMSO:H_2O. Reprinted with permission from *J. Am. Chem. Soc.*, **124**(18), 5000-5009 (2002). Copyright 2002 American Chemical Society.

and a less intense λ_{max} at 500 nm. Interestingly, the λ_{max} values and spectral features are strikingly similar to those observed for colored DMSO solutions of **1** as well as **2a**, **2b** and **3a** which we previously reported.[9]

Our simulations showed that formation of a planar xanthene within a tetrameric macrocycle would lead to an increase in strain energy of 34.2 kcal/mol. Prior studies of calixarenes also showed that xanthenes did not readily form in cyclic tetramers.[12] Ring opening to acyclic oligomers should precede xanthene formation in the resorcinarenes. The condensation reactions which form resorcinarenes are reversible under acidic conditions.[4]

The detailed mechanism of resorcinarene macrocycle genesis has been studied in detail by Weinelt and Schneider.[13] It was reported that **2b** and its stereoisomers interconverted via the intermediacy of acyclic oligomers; i.e., ring opening and ring reformation. Since resorcinarene ring opening had only been demonstrated in the presence of strong acid, acyclic oligomer formation in neutral mixed aqueous or neat DMSO solutions (our conditions) without added acid required further analysis.

The ^{1}H and ^{13}C NMR spectra of DMSO-d_6 solutions of **1** (5.2 mM), heated at 90 °C for 3 min exhibited no change compared to spectra of fresh, colorless solutions of **1** [8] Xanthene dyes, however, are highly absorbing compounds. Trace (ca. 0.5 % conversion; see, for example, the concentrations andabsorbances shown in Figure 2.2) amounts of xanthenes, below the NMR detection limit, would afford detectable solution colors, under our conditions. Relatively vigorous thermolytic conditions were necessary to achieve significant amounts of acyclic and colored products. Heating a DMSO (10 mL) solution of freshly prepared **2b** (100 mg, 18.4 mM) for 8 h at 120 °C followed by analysis via reversed phase HPLC led to the observation of numerous new products resulting in a highly complex HPLC trace. The new compounds represented a 74 % conversion of **2b** to products, based on relative peak areas.[14]

It is known that acid is produced from DMSO. The amount of acid formed is dependent on the presence of O_2 and peroxides.[15] Additionally, acid generation during DMSO decomposition has been inhibited by free radical scavengers.[15c] Under the thermolytic conditions described above, but in the presence of free radical scavengers (either BHT or phenothiazene, 10 mol %), less than 28 % conversion to new products from **2b** was observed.

Confirmatory evidence that acid forms under our neutral starting conditions was obtained via our isolation and characterization, via X-ray crystallography, of $(CH_3)_3S^+CH_3SO_3^-$.[16] The latter compound was obtained from a thermolysis reaction of **2b** in DMSO. Methanesulfonic, sulfenic and sulfinic acids (and several other products) are known to form via the free radical and acid promoted (propagated) decomposition of DMSO.[17]

Compound **7** (Figure 2.4), a rarely observed "diamond" resorcinarene stereoisomer,[18] isolated in 2.3 % yield from the thermolysis of **2b** in DMSO,

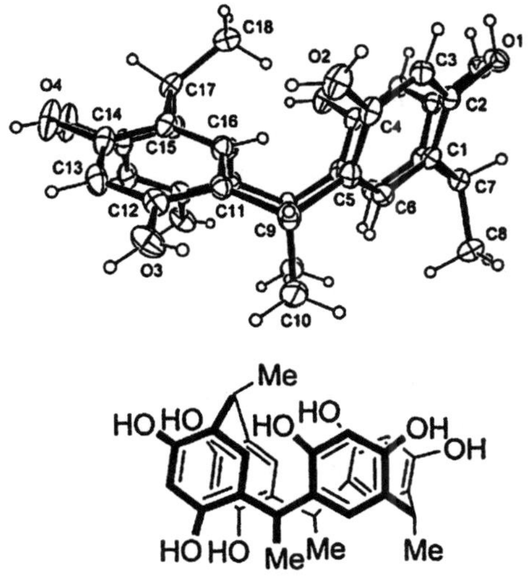

7

Figure 2.4. Compound **7** and ORTEP. Reprinted with permission from *J. Am. Chem. Soc.*, **124**(18), 5000-5009 (2002). Copyright 2002 American Chemical Society.

afforded evidence of resorcinarene bond breaking and reformation. If **7** were simply a conformer of **2b**, the methyl group (C18), would project outside, rather than above, the plane of the macrocycle's cavity.[14]

Upon thermolysis of **2b** we found that acyclic products could be semi-purified and isolated. A key product, **3b** was identified in a broad HPLC fraction eluting from 18-20 min. The ^{1}H NMR spectrum and MALDI MS afforded evidence for **3b**.[13] The ^{1}H NMR spectrum of the HPLC isolate overlayed with that of **3b** (independently synthesized in our laboratory) confirmed the assignment (Figure 2.4). The production of compounds **3b** and **7** under our conditions linked our investigations and the prior acid-catalyzed macrocycle condesation kinetic studies of Schneider.[13]

Larger acyclic oligomers were also produced under our conditions. At least five sets of doublets appeared between 0.72 and 1.53 ppm in the ^{1}H NMR spectrum of each of two flash column fractions, corresponding to various methyl group resonances.

In addition, the MALDI MS of other fractions showed peaks for higher homologs of **3b** (entries 1 and 2, Table 2.1). MALDI MS evidence also suggested the formation of xanthene materials not previously reported in previous fragmentation and equilibration studies of **2b**(entries 3-6, Table 2.1).[19, 20]

Heating a solution of **3b**, the parent acyclic unoxidized homolog attained via ring-opening of **2b,** afforded the corresponding xanthene. The ^{1}H NMR spectrum of the product mixture revealed resorcinol as the predominant (90 %)

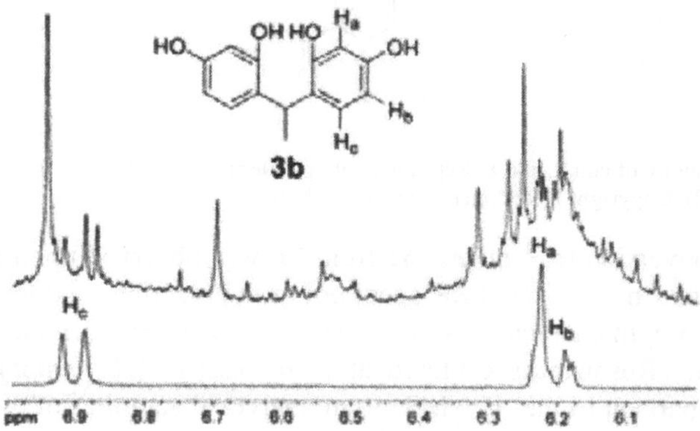

Figure 2.5. Top: expansion of a ^{1}H NMR spectrum of semipurified thermolysis reaction products of **2b** showing the formation of **3b**. Bottom: expansion of a ^{1}H NMR spectrum of pure **3b**. Reprinted with permission from *J. Am. Chem. Soc.*, **124**(18), 5000-5009 (2002). Copyright 2002 American Chemical Society.

product as well as minor conversion to 2,4-dihydroxyacetophenone **8** (ratio of integrals of resorcinol triplet 6.94 ppm to **8** doublet at 7.76 ppm is 153:1, CH_3OD) and very small traces of xanthene **6a** (d, 7.65 ppm).

Table 2.1. MALDI MS evidence for the formation of acyclic oxidized and unoxidized products from the thermolysis of **2b**. Reprinted with permission from *J. Am. Chem. Soc.*, **124**(18), 5000-5009 (2002). Copyright 2002 American Chemical Society.

Entry	Structure	TLC R_F	(*m/z*) calcd	(*m/z*) obsd
1	**4**, R=Me, m=1, n=0	0.29	382.41	381.89
2	**4**, R=Me, m=3, n=2	0.44	926.36	926.28
3	**6a**	0.44	226.23	225.61
4	**5**, R=Me, m+n=4	0.26	906.01	906.33
5	**5**, R=Me, m+n=3	0.84	770.79	770.82
6	**5**, R=Me, n=1, m=0	0.79	362.51	361.38

The production of resorcinol and **8** is consistent with the reversible condensation mechanism of resorcinarenes in acidic media.[13] This result also complemented our related finding that 4-formylphenylboronic acid is produced from **3c**.[21] Moreover, in the presence of acid, addition of H_2O to the methine carbon of **4** (R=Ar, n=0, m=0) followed by elimination is known as a key intermediate step in the synthesis of xanthenes.[22]

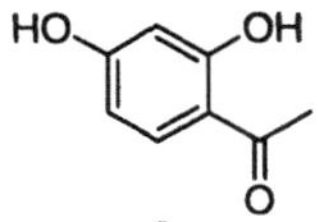

8

Figure 2.6. Structure of compound **8**. Reprinted with permission from *J. Am. Chem. Soc.*, **124**(18), 5000-5009 (2002). Copyright 2002 American Chemical Society.

Better conversion to xanthene **6a** from **3b** was observed upon limiting the heating time to 2 h. The ^{1}H NMR spectrum of the reaction mixture exhibited a doublet at 7.65 ppm characteristic of **6a** with improved S/N compared to the 28 h experiment. Resonances centered at 5.26, 6.49 and 6.60 ppm were also present. Comparison to the ^{1}H NMR of an analytical sample[22] of **6a** confirmed its assignment. Heating a solution of **3b**, H_2SO_4 and $K_2S_2O_8$ in MeOH at reflux for 2h afforded the most efficient conversion (4 % yield) of **3b** to **6a** we have attained to date.

The color observed in Baeyer's resorcinol and benzaldehyde condensation reaction[3] was thus caused by the presence of trace amounts of highly absorbing xanthenes. He had indeed synthesized fluorescein[2] analogs, albeit as minor products. In our studies, we have found that functional xanthenes form in solutions containing resorcinarene macrocycles (Scheme 2.1). An O_2-induced radical decomposition of DMSO leads to the *in situ* formation of strong acid. The acid forms catalyzes a reverse condensation reaction. This leads to acyclic oligomers. The oligomers readily undergo oxidation reactions assisted by the presence of acid and peroxide. Acyclic xanthene formation from macrocyclic resorcinarenes is summarized in Scheme 2.1.

2.3. THE MECHANISM OF SUGAR-INDUCED SIGNAL TRANSDUCTION

Resorcinol and congeners have played a role in tests for sugars dating to the 19th century. In 1887 Seliwanoff reported a resorcinol assay which led to related methods still in use today.[23] These methods can suffer from selectivity issues, relatively harsh conditions and tedium.[24] In the 1990's, significant progress was made towards milder detection of monosaccharides. This was due in large part to the pioneering efforts of Shinkai and coworkers. Their studies were based primarily on aniline-functionalized azo dyes containing appended arylboronic acids.[25]

The study of glycobiology is of great current interest.[26] Oligosaccharides reside on glycoproteins and cell surfaces. Most of the recent progress towards the detection of sugars has, however, involved monosaccharide analysis.[25,27] The natural occurence of the bewildering variety of oligosaccharides amplifies the problems associated with monosaccharide analysis.

The colorimetric detection of mono- and oligosaccharides eluting from chromatographic columns is a major challenge in sugar analysis.[28] The classical methods used in automated HPLC post-column systems do not efficiently allow for the detection of oligosaccharides containing more than three residues. For example, it has been reported that maltohexaose's response is just 18 % of that observed for the same weight of glucose.[28] Importantly, complete hydrolysis to monosaccharides or covalent attachment of the oligosaccharides to a chromophore or fluorophore, which can lead to diminished separation, is required.[28,29]

Dyes containing boronic acid moieties are known to exhibit fluorescence and, less commonly, colorimetric changes in the presence of sugars.[25] When saccharides condense with boronic acids, they form cyclic boronates. When part of the boronate ring structure, the boron's Lewis acidity is heightened.[30] As a result, after binding with saccharides, sp^2-hybridized neutral boron becomes an sp^3-hybridized anion due to the addition of H_2O or HO^- as a fourth ligand (Figure 2.8). Several authors have reported that signal transduction promoted by sugar binding to chromophores or fluorophores containing boronic acid moieties are due to the changes in boron hybridization from the neutral sp^2 species to an sp^3-hybridized charged complex.[24,30b]

Our NMR studies showed that in the presence of **2a** (40 mM), D-fructose-2-^{13}C (1 equiv) in 9:1 DMSO-d_6:D_2O several new ^{13}C-2 resonances corresponding to cyclic sugar boronic esters (Figure 2.7) appear.[14] The chemical shifts obtained (Table 2.2) are in striking agreement with those

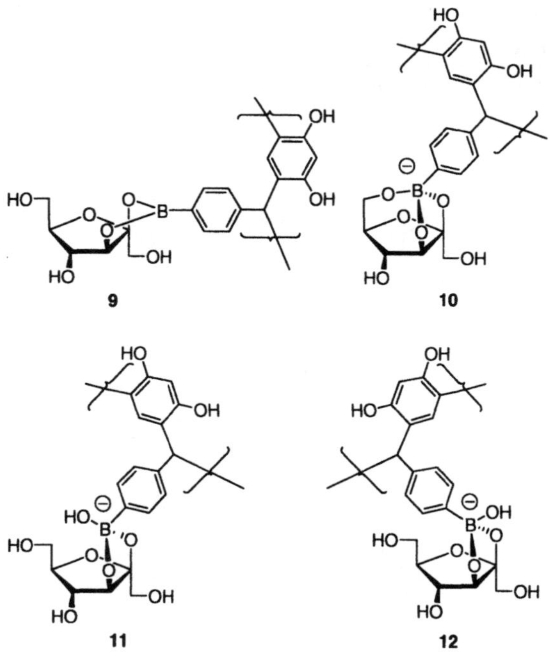

Figure 2.7. Structures of the four **2a**:β-D-fructofuranose complexes (**9-12**) observed in 9:1 DMSO-d_6:D_2O. Boronates **11** and **12** are "exo" and "endo" isomers exhibiting ^{13}C-2 chemical shifts of equal intensity at 114.4 and 114.3 ppm, respectively. Reprinted with permission from *J. Am. Chem. Soc.*, **124**(18), 5000-5009 (2002). Copyright 2002 American Chemical Society.

observed by Norrild in his studies of model *p*-tolylboronic acid- sugar complexes.[31] Importantly, we observed resonances for four anionic complexes, three of which are related to known β-D-fructofuranose esters and shown in Figure 2.7. An upfield shift of the ^{11}B NMR resonance connotes conversion of sp^2-hybridized neutral boron to an sp^3-hybridized boronate anion.[32] ^{11}B NMR investigations confirmed the formation of sp^3-hybridized boronate anions upon fructose complexation, consistent with our ^{13}C NMR results.[14]

Table 2.2. Reprinted with permission from *J. Am. Chem. Soc.*, **124**(18), 5000-5009 (2002). Copyright 2002 American Chemical Society.

	^{13}C-2 chemical shift (ppm)	
Fructose mutarotational isomer bound to boron	Literature value[c] (*p*-tolylboronic ester)	Experimental value[d] (**2a** boronic ester)
β-D-fructofuranose	113.4[a]	113.8 (**10**)
	114.7[a]	114.4 (**11**)
	114.6[a]	114.3 (**12**)
	115.2[a]	115.5 (**9**)
β-D-fructopyranose	105.0[b]	105.0
	107.6[b]	107.7
α-D-fructofuranose	99.8[b]	99.9
α-D-fructopyranose	92.6[b]	93.2

[a]In D_2O at pD = 11-12. [b]In DMSO-d_6. [c]Reference 31. [d]Experimental values are obtained in 9:1 DMSO-d_6:D_2O.

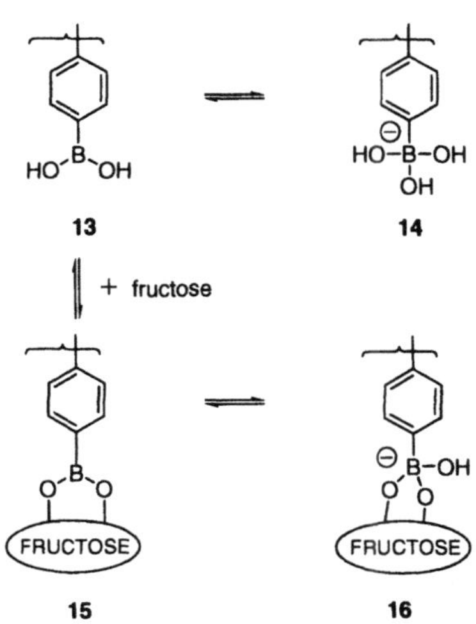

Figure 2.8. Equilibria for solutions of **2a** and D-fructose. Reprinted with permission from *J. Am. Chem. Soc.*, **124**(18), 5000-5009 (2002). Copyright 2002 American Chemical Society.

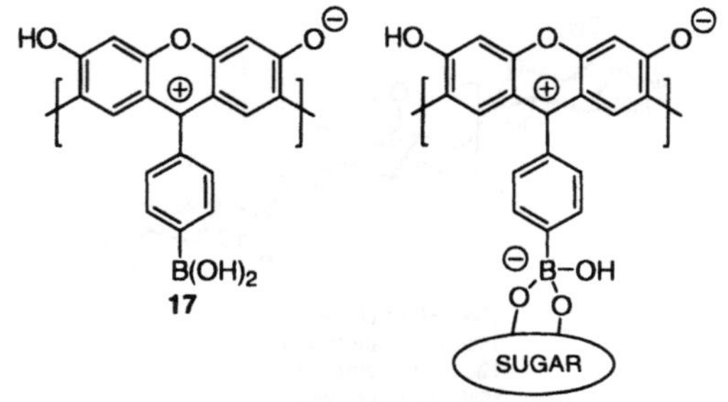

17 18

Figure 2.9. Reprinted with permission from *J. Am. Chem. Soc.*, **124**(18), 5000-5009 (2002). Copyright 2002 American Chemical Society.

Anionic boronate formation lowers the pK_a of **5**. One might also visualize this by analyzing the resonance forms of xanthenes **17** and **18**. Structure **18** has a more stable cation than **17**, rendering the C-6 hydroxyl of **18** relatively less ionizable (Figure 2.9). Anionic boron serves as an excellent σ-bond donor.

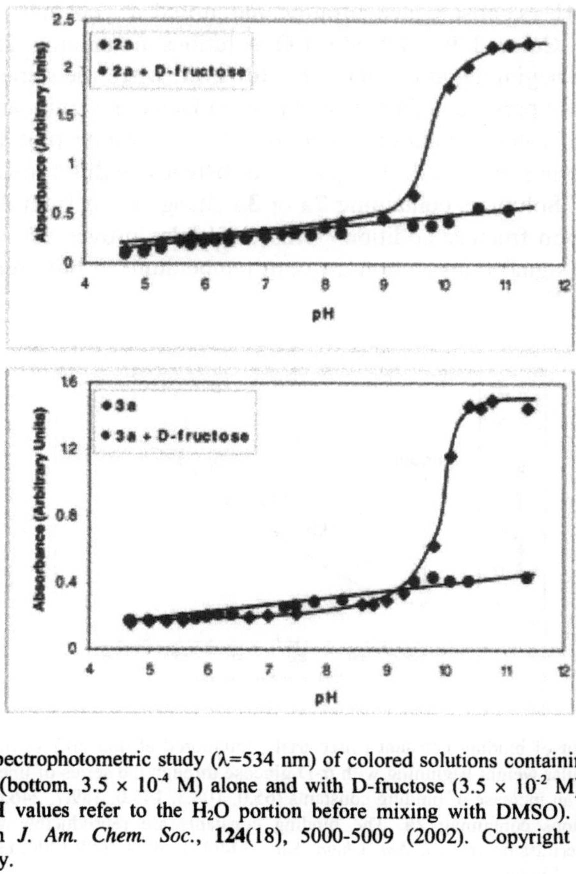

Figure 2.10. Spectrophotometric study (λ=534 nm) of colored solutions containing **2a** (top, 3.5 × 10^{-4} M) and **3a** (bottom, 3.5 × 10^{-4} M) alone and with D-fructose (3.5 × 10^{-2} M) in buffered 9:1 DMSO:H_2O (pH values refer to the H_2O portion before mixing with DMSO). Reprinted with permission from *J. Am. Chem. Soc.*, **124**(18), 5000-5009 (2002). Copyright 2002 American Chemical Society.

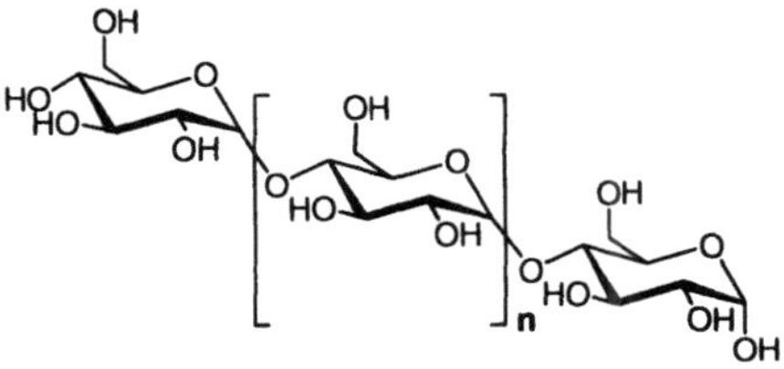

19a n=0 maltose
19b n=1 maltotriose
19c n=2 maltotetraose
19d n=3 maltopentaose
19e n=4 maltohexaose
19f n=5 maltoheptaose

Figure 2.11. Structures of maltodextrines **19a-f**. Reprinted with permission from *J. Am. Chem. Soc.*, **124**(18), 5000-5009 (2002). Copyright 2002 American Chemical Society.

We have reported increased absorbance at 532 nm as a function of raising the solution pH of colored aqueous DMSO solutions of resorcinol condensation products.[22] Diminished absorbance in pH-controlled media in the presence of saccharides should thus result if sugar complexation lowers the dye pK_a. The results of a pH vs. absorbance titration monitored at 534 nm are shown in Figure 2.10. Colored 9:1 DMSO:H_2O solutions containing **2a** exhibit an intensity onset beginning at ca. pH = 9. However, when the same procedure is performed in the presence of fructose, the absorbance is significantly lessened. The titration of colored solutions of **3a** results in a similar plot as that for **2a** solutions showing that both receptors consistently exhibit pKa depression (Figure 2.10). Solutions containing **2a** or **3a** change color from dark brown to pale orange upon fructose addition. This result also proves that we can attain sugar-induced signal transduction at room temperature in buffered media with our techniques.[33]

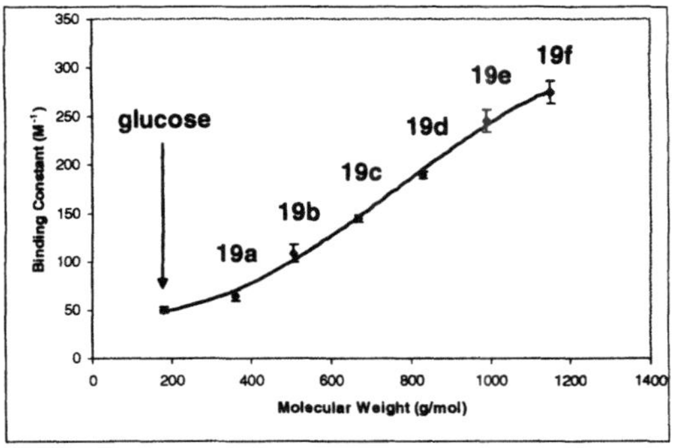

Figure 2.12. Plot of binding constant (three trials, monitored at 460 nm) vs molecular weight. Increasing molecular weight beginning with α-D-glucose through the series of linear maltodextrins (**19a-f**) results in an increase of binding constants (65.0, 108.3, 145.0, 189.7, 246.0 and 275.0 M^{-1}) in colored solutions containing **2a**. The binding constants are reproducible within ± 10 %. Reprinted with permission from *J. Am. Chem. Soc.*, **124**(18), 5000-5009 (2002). Copyright 2002 American Chemical Society.

Prior studies (using fluorescence detection) showed that binding constants for the interaction of oligosaccharides with other fluorophore-appended arylboronic acids increased with larger oligosaccharide size.[32] We thus applied our methodology to the problem of simple detection of linear maltodextrins **19a-f** (Figure 2.11). The maltodextrins are neutral oligosaccharides composed of glucose residues with α-1,4-glycoside linkages. Maltodextrins **19a-f** (0-7.5 ×

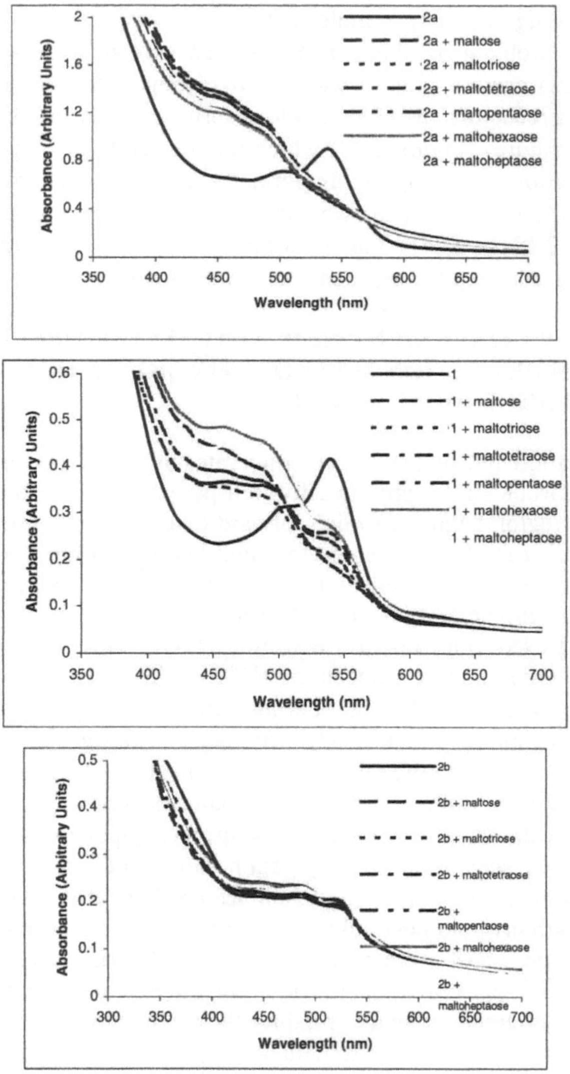

Figure 2.13. UV-Vis spectra of solutions of **2a**, **1** and **2b** heated alone or in the presence of **19a-f**. The boronic acid-containing receptors **1** and **2a** exhibit larger absorbance responses to oligosaccharides. Reprinted with permission from *J. Am. Chem. Soc.*, **124**(18), 5000-5009 (2002). Copyright 2002 American Chemical Society.

10^{-2} M) were dissolved in 0.5 mL carbonate buffer (2.5×10^{-2} M, pH = 11) and mixed with 0.5 mL DMSO colored preheated solutions of 2a (5.0×10^{-4} M) respectively to furnish a 1:1 DMSO:H_2O solvent system. The solution colors changed from orange to yellow upon sugar addition.

We thus can detect neutral oligosaccharides in buffered media at room temperature by visual inspection. Apparent equilibrium constants (K_{exp}(app)) were determined. Enhanced binding, rather than diminished (*vide supra*) affinity correlating with larger molecular weight was observed (Figure 2.12). The important problem of diminished absorbance correlating with increasing glucose oligomer chain length[28,29] was thus overcome in the case of our boronic acid-containing receptors. More dramatic absorbance changes were observed at elevated temperatures (Figure 2.13) which are amenable to HPLC postcolumn detection conditions (*vide infra*). Importantly, compound **2b**, which does not contain boronic acid moieties, showed insignificant responses to the presence of sugars (Figure 2.13, bottom). This is further strong evidence of the key role of boronic acid-based receptors in glucose oligomer monitoring.

2.4. A HIGHLY CONVENIENT NEW AUTOMATED HPLC POST-COLUMN DETECTION METHOD FOR MONITORING GLUCOSE AND RELATED BIOMOLECULES

The separation and analysis of carbohydrates by HPLC is common practice. The fact that carbohydrates are virtually transparent in the visible and most of the ultraviolet region hampers straightforward detection with common UV-Vis and fluorescence detectors.[24] Direct detection below 210 nm severely restricts solvent choice. Refractivity detection has low sensitivity and is readily affected by changes in temperature. It can be used only under isocratic conditions. Detection by pulsed amperometric detection (PAD) usually necessitates high pH. Mass spectrometric detection (HPLC-MS), requires sophisticated equipment.[24] Evaporative light scattering detection (ELSD) offers a limited choice of buffers and may require advanced detector design for molecules with a lower molecular weight range that can evaporate along with the mobile phase. The system may also require relatively high maintenance.[34]

Pre-column derivatization methods using chromopohores or fluorophores can hamper sugar separation as conjugation can significantly affect retention times. Post-column derivatization techniques are thus attractive.[24,34-35] Older post-column carbohydrate post-column reactions required reagent delivery and detection systems which were incompatible with certain solvents needed for separations, exhibit peak broadening and needed to be corrosion-resistant due to the common requirement of acid to produce the detectable species.[34,35]

Milder post-column reactions with fluorogenic compounds are known.[37] Many of the reagents used are selective only for specific groups of compounds. The reactions are irreversible. The use of a synthetic reagent which produces signal transduction via non-covalent or reversible covalent binding, could allow for greater selectivity and the recovery of scarce or expensive analytes.

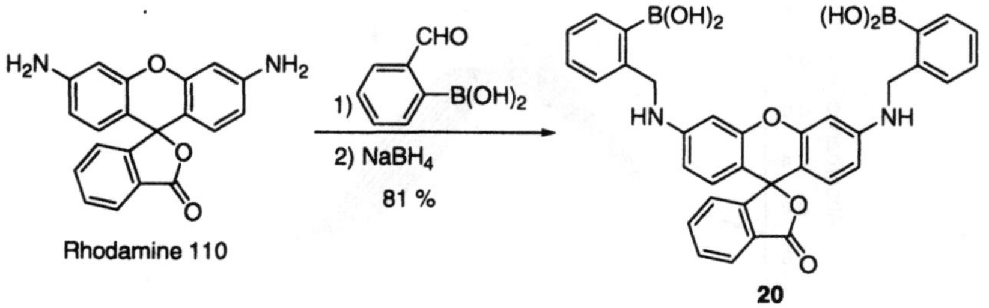

Scheme 2.2. Synthesis of **20**. Reprinted with permission from *Org. Lett.* **5**(26), 5007-5010 (2003). Copyright 2003 American Chemical Society.

We designed receptor **20** based on the high molar absortivity in the visible region of rhodamine.[37] Additionally, it has the potential for favorable boron-nitrogen interactions to promote sugar-boronate formation at neutral pH.[38] Compound **20** was readily synthesized as shown in Scheme 2. It is selective for fructose over glucose in solution, in keeping with the behavior of most boronic acid derivatives.[38] A blue shift of the λ_{max} from 550 nm to 530 nm upon addition of fructose or glucose was observed (Figure 2.14). The apparent equilibrium constant for the interaction of **20** and D-fructose is 3,806 M^{-1}; for D-glucose the value is 375 M^{-1}. The result is consistent with a B-N interaction. The aniline nitrogen lone pair interaction with boron increases upon sugar binding. Its interaction with the conjugated xanthene moiety is concomitantly diminished, affording the observed blue shift.

Fructose is an interferent of concern in glucose monitoring as most boronic acid compounds have higher affinities to fructose than to glucose. High D-fructose intake is associated with hypertriglyceridaemia, atherosclerosis[39] and insulin resistance.[40] Non-enzymatic glycosidation product formation from fructose is of concern.[41] The determination of fructose levels in human plasma is hampered because of interference from excess (ca. 100-fold) glucose.[42] Thus, levels of fructose reported suffer from Interlaboratory imprecisionr.[43] GC or HPLC determinations typically require tedious sample preparation.[43]

Mixtures of D-glucose and D-fructose can be readily monitored via automated post-column HPLC detection using **20** as the postcolumn reagent.[44] A significant peak for D-fructose was clearly observed even in the presence of a 100-fold excess of D-glucose (Figure 2.15). Importantly, at physiologically relevant proportions, both sugars were easily resolved and detected in the visible spectral region (Figure 2.13).

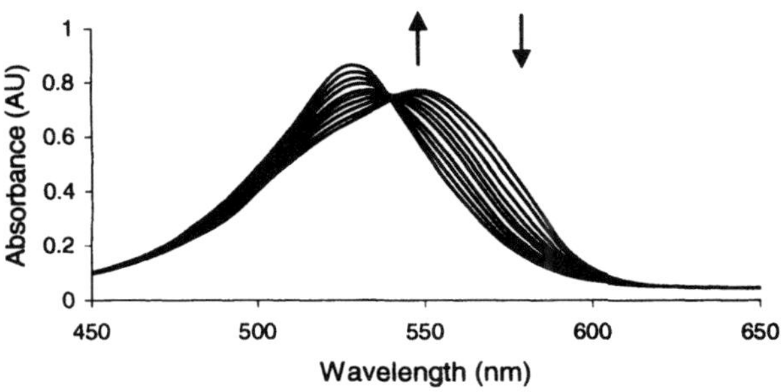

Figure 2.14. UV-Vis spectra of solutions of **20** (1.64×10^{-5} M, 0.16 M pH 9.5 carbonate buffer in 1:2 MeOH:H_2O) upon titration with D-fructose. The final concentration of fructose was 8.33×10^{-4} M. Reprinted with permission from *Org. Lett.* **5**(26), 5007-5010 (2003). Copyright 2003 American Chemical Society.

The glucose oligomers (maltodextrins) can also be readily detected. In contrast, and as noted above, many currently used oligosaccharide HPLC detection methods require prior complete hydrolysis to monosaccharides or pre-column derivatizations.[28,44]

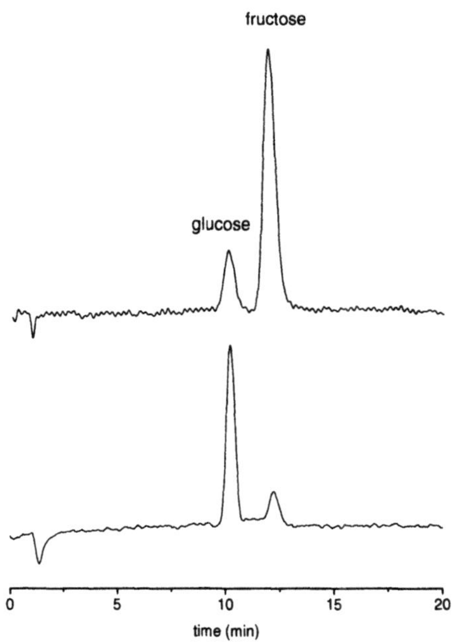

Figure 2.15. Top: chromatogram of a 1:1 mixture of D-fructose and D-glucose (20.0 μg, λ=560 nm). Bottom: chromatogram of a mixture of D-fructose (4.5 μg) in the presence of a 100-fold excess of D-glucose. Reprinted with permission from *Org. Lett.* **5**(26), 5007-5010 (2003). Copyright 2003 American Chemical Society.

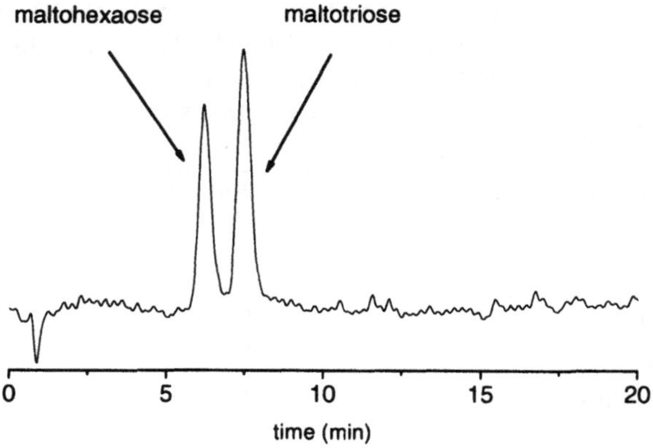

Figure 2.16. Chromatogram of a 1:1 mixture of maltohexaose and maltotriose (80 μg, λ=560 nm). Reprinted with permission from *Org. Lett.* **5**(26), 5007-5010 (2003). Copyright 2003 American Chemical Society.

The boronic acid-functionalized xanthenes allowed us to generate strong colorimetric responses for glucose oligosaccharides (*vide supra*).[12] Under similar conditions (reactor temperature set to 94 °C) to those used for the glucose and fructose, mixtures of maltotriose and maltohexaose (80 μg) were monitored (Figure 2.16) at 560 nm using rhodamine derivative **20**. Figure 2.17 and Table 2.2 further attest to the general use of this approach in monitoring and resolving glucose from selected potential related interferents and biomolecules of interest, such as sialic acid and GM1.[45-50]

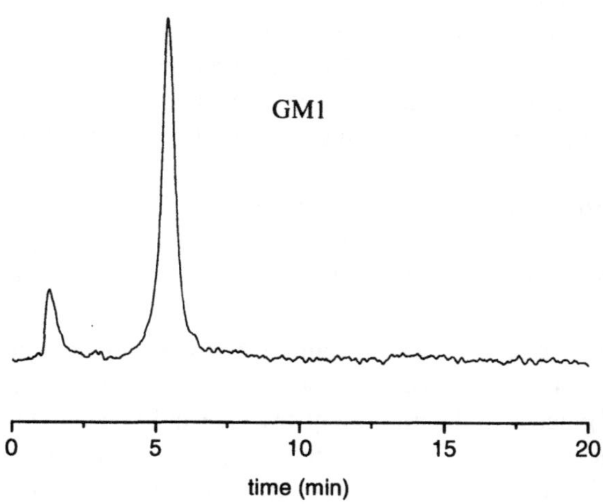

Figure 2.17. Chromatogram of GM1 (220 μg, λ=520 nm). Reprinted with permission from *Org. Lett.* **5**(26), 5007-5010 (2003). Copyright 2003 American Chemical Society.

Table 2.3. Retention times (min) and limits of detection (LOD, µg) for various carbohydrates via spectrophotometry. The Table shows the potential for achieving significant LODs using fluorescence detection (in progress). Reprinted with permission from *Org. Lett.* **5**(26), 5007-5010 (2003). Copyright 2003 American Chemical Society..

carbohydrate	retention time	LOD
D-fructose	10.1	2.3
D-glucose	12.0	7.1
maltohexaose	6.0	30.1
maltotriose	7.3	35.8
N-acetylneuraminic acid	7.2	45.4
GM1	5.5	26.8

2.5. DETECTION OF GLUCOSE IN HUMAN BLOOD PLASMA AND PROGRESS TOWARDS CONCURRENT DETERMINATION OF GLUCOSE AND FRUCTOSE

In keeping with our studies involving monitoring glucose with good selectivity in the presence of other sugars, we recently focused our attention on the direct detection of glucose and fructose in mixtures and in human blood plasma. We have found that glucose can be selectively detected in human blood plasma by fluorescence. In contrast, glucose and fructose may be detected concurrently via a spectrophotometric method using ratiometric, dual wavelength monitoring.

Colored solutions of **1** at room temperature change color from pink-purple to orange-yellow upon addition of 1 equiv fructose. Other saccharides such as glucose, sucrose, maltose, lactose, xylose, and glucose, even added in significant excess, exhibited no color change under these mild conditions.[52]

The color change observed corresponds to a ratiometric absorbance decrease at 536 nm and increase at 464 nm (Figure 2.18). The absorbance changes exhibited a linear dependence on fructose concentration with good correlation (R = 0.9079 and 0.9419 at each of the two wavelengths, respectively).

Addition of 1 equiv fructose at room temperature to a colored solution of **1** resulted in an 8.6 % increase in absorbance at 464 nm. Addition of 100 equivalents of glucose to the mixture of fructose and **1** resulted in no significant spectral change at 464 nm. When a second equivalent of fructose was added to the mixture of **1**, fructose and glucose, an increase in absorbance was clearly observed. Thus, at 464 nm, only fructose is detected, even in the presence of 100 equiv glucose.

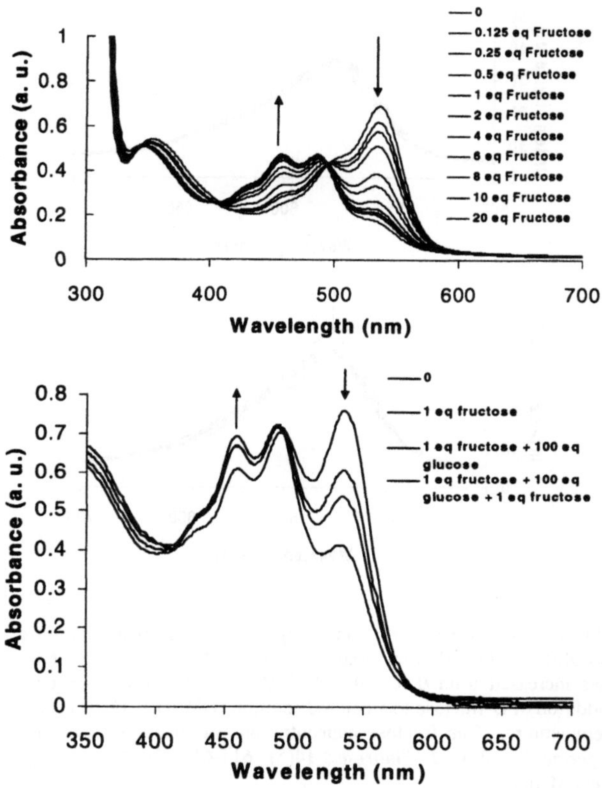

Figure 2.18. Upper: Addition of fructose to a preheated (3.0 min at reflux) solution of **1** (5.2×10^{-3} M) in DMSO at room temperature affords concentration-dependent absorbance changes at 464 nm and 536 nm. Lower: UV-Vis spectra of a 9:1 DMSO:H_2O solution containing **1** (5.2×10^{-3} M) pre-heated (1.5 min at reflux) (i) alone, (ii) upon addition of 1 equiv fructose at room temperature which produces an absorbance increase at 464 nm and a corresponding decrease at 536 nm, (iii) upon addition of 100 equiv glucose which produces no absorbance change at 464 nm but a decrease at 536 nm and (iv) upon addition of a second equivalent of fructose which affords a further absorbance increase at 464 nm and decrease at 536 nm. Reprinted with permission from *J. Fluoresc.*, **14**(5), 611-615 (2004). Copyright 2004 Springer Science + Business Media, Inc.

In contrast, the absorbance of the solution of **1** at 536 nm was diminished 20 % upon treatment with 1 equiv fructose. The addition of 100 equivalents of glucose to this latter mixture decreased the absorbance by 12 %. Addition of another equivalent of fructose lowered the absorbance again, by 24 % (Figure 2.18). At 536 nm, both glucose and fructose are detectable.

These effects can be useful for the indirect determination of glucose in samples containing fructose and glucose. Such a procedure can be performed via analysis of the ratio of the absorbance at 536 nm (both glucose and fructose

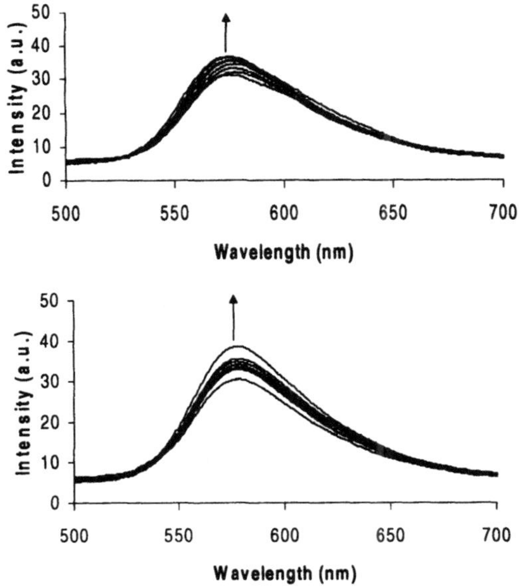

Figure 2.19. Upper: Fluorescence emission changes produced upon addition of D-glucose to a colored solution (DMSO:H_2O 9:1) containing 1 (5.0×10^{-3} M) at room temperature. The glucose concentration was increased from 0 to 7.4×10^{-4} M. Lower: Fluorescence emission changes produced upon addition of D-fructose to a colored solution (DMSO:H_2O 9:1) containing **1** (5.0×10^{-3} M) at room temperature. The fructose concentration was increased from 0 to 1.8×10^{-3} M. Reprinted with permission from *J. Fluoresc.*, **14**(5), 611-615 (2004). Copyright 2004 Springer Science + Business Media, Inc.

promote signal changes) to the absorbance at 464 nm (signaling promoted by fructose only), for instance, after the fructose concentration is determined at 464 nm.

The high selectivity of **1** for fructose is consistent with the reported binding constants for saccharides.[21] Our UV-Vis studies require high concentrations of saccharides. They may be applied towards the analysis of sugars in samples such as syrups, caramels and honey. For other samples (e.g. naturally-occurring biological liquids, many pharmaceuticals, etc.), where the concentration of sugars is significantly lower, solutions of **1** can be successfully used for direct monitoring of glucose in the range of physiological levels via fluorescence spectroscopy, as described below.

The fluorescence emission spectra of the colored DMSO solutions of **1** and added glucose or fructose are shown in Figure 2.19. As the sugar concentration was increased, an emission increase was promoted by both saccharides. In contrast to the UV-Vis studies, a significant signal change was generated by glucose, using similar concentrations of fructose.

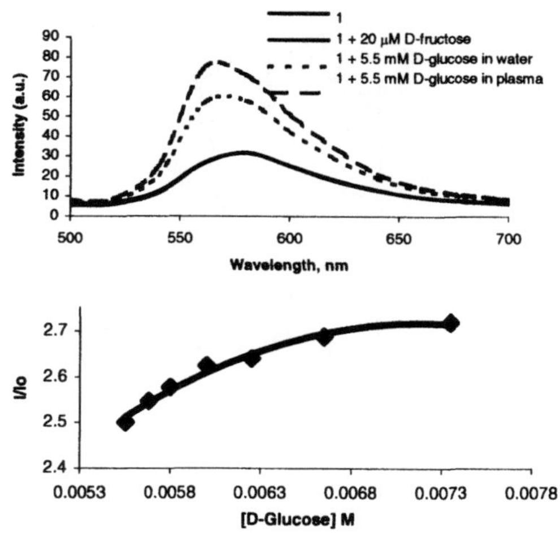

Figure 2.20. Upper: Fluorescence emission spectra produced by (i) a preheated (3 min at reflux) colored solution (DMSO:H_2O 9:1) containing 1 (5.0 x 10^{-3} M) at room temperature, (ii) the same conditions but in the presence of 20 µM D-fructose, added at room temperature, which affords no observable change in emission, (iii) the same conditions as (i) but with added D-glucose (5.3 µM) which promotes an emission increase and (iv) same conditions as (iii) but in deproteinized human blood plasma instead of H_2O, which exhibits an emission increase in response to added glucose. Lower: Concentration-dependent emission changes produced via room temperature additions of D-glucose to a 9:1 DMSO:plasma solution containing **1**. Reprinted with permission from *J. Fluoresc.*, **14**(5), 611-615 (2004). Copyright 2004 Springer Science + Business Media, Inc.

The level of D-fructose in human blood plasma is ca. 50 µM. This is significantly lower than that of D-glucose (the ratio of glucose to fructose in blood plasma is *ca.* 100:1). Fluorescence emission spectra of colored solutions of **1** in DMSO, upon addition of fructose (the fructose concentration range studied is 20 µM to 100 µM), exhibited no detectable fructose-promoted emission. Healthy levels of D-glucose are *ca.* 5 mM. The emission spectra of glucose (5.5 x 10^{-3} M) and **1** (1.0 x 10^{-3} M) in DMSO:H_2O, 9:1, as well as in a 9:1 DMSO:plasma solution, are shown in Figure 2.20. Emission increases due to the presence of glucose were observed in both cases as a function of glucose levels (Figure 2.20).

The fluorescence studies showed that boronic acid-functionalized xanthenes may detect glucose in blood plasma without interference from fructose, within the range of their physiologically relevant concentrations. The enhanced fluorescence emission of **1** promoted by glucose, in comparison to its weaker UV-Vis response relative to fructose, may be attributed to chelation by neighboring boronic acids of **5**. It is well known that glucose can be bound simultaneously by two neighboring boronic acids. In the case of xanthene fluorophores, this results in scaffold rigidification. Such rigidification effects have previously demonstrated in enhancing fluorescence emission.[38,51] Figure 2.21 shows an energy-minimized model of a xanthene fluorophore binding D-glucose as the bis-boronate ester.

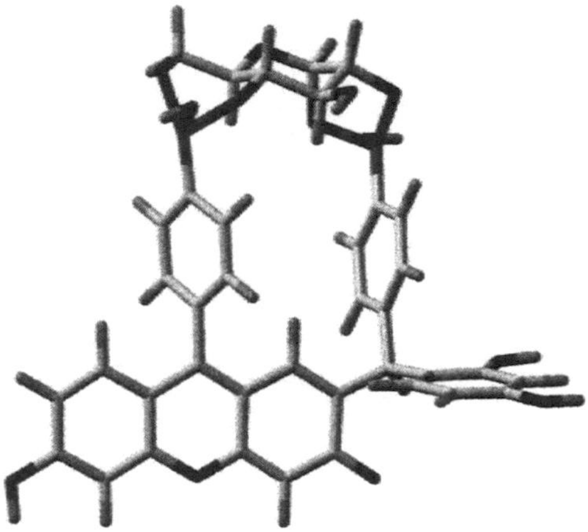

Figure 2.21. Energy-minimized structure of the bis-boronate complex formed between **5** and glucose.

2.6. CONCLUSION

Baeyer synthesized xanthene dyes during his resorcinol-benzaldehyde condensation, albeit at low levels compared to the major product macrocycles. We found that these resorcinarenes macrocycles form acyclic oligomers which convert to trace amounts of highly absorbing chromophoric and fluorescent materials. When functionalized with boronic acids, xanthenes can be used as colorimertic and fluorescent reagents that are highly versatile. They are useful in applications such as convenient automated post-column detection systems for monitoring glucose and related sugars in the visible spectral region. They allow selective glucose detection in human blood plasma by fluorescence emission, and ratiometric monitoring of sugar mixtures by spectrophotometry. The further study of the mechanism of signal transduction and the development of convenient glucose detection protocols in challenging biological matrices is ongoing in our laboratory.

2.7. ACKNOWLEDGMENTS

We are very grateful to the National Institutes of Health for supporting this research via grant 8R01EB002044-03.

2.8. REFERENCES

1. *J. Fluoresc.* **14**(5), 611-615 (2004) entire issue.
2. A. von Baeyer, *Chem. Ber.* **5**, 255 (1871).

3. (a) A. von Baeyer, Ueber die verbindungen der aldehyde mit den phenolen, *Ber. Dtsch. Chem. Ges.* 5, 25 (1872). (b) A. von Baeyer, Ueber die verbindungen der aldehyde mit den phenolen, *Ber. Dtsch. Chem. Ges.* **5**, 280 (1872).
4. (a) H.-J. Schneider and U. Schneider, Supramolecular chemistry, part 53. The host-guest chemistry of resorcinarenes, *J. Inclusion Phenom. Macrocyclic Chem.* **19**(1-4), 67-83 (1994). (b) Container Molecules and Their Guests, Cram, D. J.; Cram, J. M., The Royal Society of Chemistry: Cambridge, U.K. 1994. (c) J. C. Sherman, Carceplexes and hemicarceplexes - molecular encapsulation - from hours to forever, *Tetrahedron* **51**(12), 3395-3422 (1995). (d) P. Timmerman, W. Verboom and D. N. Reinhoudt, Resorcinarenes, *Tetrahedron* **52**(8), 2663-2704 (1996). (e) A Jasat and J. C. Sherman, Carceplexes and hemicarceplexes, *Chem. Rev.* **99**(4), 931-967 (1999). (f) D. M. Rudkevich and J. Rebek, Deepening cavitands, *Eur. J. Org. Chem.* **9**, 1991-2005 (1999).
5. Y. Aoyama, Y. Tanaka and S. Sugihara, Molecular recognition. 5. Molecular recognition of sugars via hydrogen-bonding interaction with a synthetic polyhydroxy macrocycle, *J. Am. Chem. Soc.* **111**(14), 5397-5404 (1989).
6. P. T. Lewis, C. J. Davis, M. Saraiva, D. Treleaven, T. McCarley and R. M. Strongin, Tetraarylboronic acid resorcinarene stereoisomers. Versatile new substrates for divergent polyfunctionalization and molecular recognition, *J. Org. Chem.* **62**(18), 6110-6111 (1997).
7. C. J. Davis, P. T. Lewis, D. R. Billodeaux, F. R. Fronczek, J. O. Escobedo and R. M. Strongin, Solid-state supramolecular structures of resorcinol-arylboronic acid compounds, *Org. Lett.*, **3**(16), 2443-2445 (2001).
8. C. J. Davis, P. T. Lewis, M. E. McCarroll, M. W. Read, R. Cueto and R. M. Strongin, Simple and rapid visual sensing of saccharides, *Org. Lett.* **1**(2), 331-334 (1999).
9. P. T. Lewis, C. J. Davis, L. A. Cabell, M. He, M. W. Read, M. E. McCarroll and R. M. Strongin, Visual sensing of saccharides promoted by resorcinol condensation products, *Org. Lett.* **2**(5), 589-592 (2000).
10. R. N. Sen and N. N. Sinha, Condensations of aldehydes with resorcinol and some other aromatic hydroxy compounds, *J. Am. Chem. Soc.* **45**(12), 2984-2996 (1923).
11. S. N. Gupta, S. M. Linden, A. Wrzyszczynski, D. C. Neckers, Light-induced spectral changes in Rose Bengal endcapped polystyrene, *Macromolecules* **21**(1), 51-55 (1988).
12. K. Agraria and S. E. Biali, Intramolecular Ar-O-Ar bond formation in calixarenes, *J. Org. Chem.* **66**(16), 5482-5489 (2001).
13. F. Weinelt and H.-J. Schneider, Mechanisms of macrocycle genesis. The condensation of resorcinol with aldehydes, *J. Org. Chem.* **56**(19), 5527-5535 (1991).
14. M. He, R. J. Johnson, J. O. Escobedo, P. A. Beck, K. K. Kim, N. N. St. Luce, C. J. Davis, P. T. Lewis, F. R. Fronczek, B. J. Melancon, A. A. Mrse, W. D. Treleaven and R. M. Strongin, Chromophore formation in resorcinarene solutions and the visual detection of mono- and oligosaccharides, *J. Am. Chem. Soc.* **124**(18), 5000-5009 (2002).
15. (a) B. T. Gillis and P. E. Beck, Formation of tetrahydrofuran derivatives from 1,4-diols in dimethyl sulfoxide, *J. Org. Chem.* **28**(5), 1388-1390 (1963). (b) T. M. Santususso and D. Swern, Acid catalysis as a basis for a mechanistic rationale of some dimethyl sulfoxide reactions, *Tetrahedron Lett.* **15**(48), 4255-4258 (1974). (c) J. Emert, M. Goldenberg, G. L. Chiu, and A. Valeri, Synthesis of dibenzyl ethers via the dehydration of benzylic alcohols in dimethyl sulfoxide, *J. Org. Chem.*, **42**(11), 2012-2013 (1977).
16. F. R. Fronczek, R. J Johnson and R. M. Strongin, Trimethylsulfonium methanesulfonate, *Acta Cryst.* **E5**, 0447-0448 (2001).
17. (a) C.-T. Chen and S.-J. Yan, Acid catalysis as a basis for a mechanistic rationale of some dimethyl sulfoxide reactions, *Tetrahedron Lett.* **10**(44), 3855-3856 (1969). (b) D. L. Head and C. G. McCarty, *Tetrahedron Lett.* **14**(16), 1405-1408 (1973).
18. For the structures of the different possible resorcinarene stereoisomers see reference 4d.
19. In the previous work (reference 11), acyclic oligomeric products (**3b** and two stereoisomeric trimeric compounds, three resorcinol rings, **4**, R=Me, m=1, n=0, Scheme 1) were isolated and characterized. Higher order acyclic oligomers (e.g., pentamers and hexamers) were also observed as major reaction products. Methyl ^{1}H NMR resonances, appearing as several doublets between 0.7 and 2.0 ppm (CH_3OD) that corresponded to neither **3b**, **4** (R=Me, m=1, n=0), or resorcinarene macrocycles, thus were assigned to acyclics with five or more resorcinol moieties.

20. Flash column chromatography and TLC analysis of the thermolysis products of **2b** were complicated by the multiple product formation and fraction streaking.
21. F. R. Fronczek, N. N. St. Luce and R. M. Strongin, Space-group revision for 4-formylphenylboronic acid, *Acta Cryst.* **C57**, 1423-1425 (2001).
22. R. N. Sen and N. N. Sarkar, The condensation of primary alcohols with resorcinol and other hydroxy aromatic compounds, *J. Am. Chem. Soc.*, **47**(4), 1079-1091 (1925).
23. For example: (a) T. Seliwanoff, Notiz über eine frunchtzucker-reaction, *Chem. Ber.* **20**, 181 (1887). (b) M. Bial, Über die diagnose der pentoseurie mit dem von mir angegebenegen reagents, *Dtsch. Med. Wehschr.* 253 (1902). (c) R. G. Kulka, Colorimetric estimation of ketopentoses and ketohexoses, *Biochem J.* **63**, 542 (1956).
24. M. F. Chaplin in: *Monosaccharides/Carbohydrate Analysis. A Practical Approach,* edited by M. F. Chaplin and J. F. Kennedy (Oxford University Press, Oxford, 1994), pp 1-40.
25. Review: (a) T. D. James, K. R. A. S Samankumara Sandanayake and S. Shinkai, Saccharide sensing with molecular receptors based on boronic acid, *Angew. Chem. Int. Ed. Engl.* **35**(17), 1911-1922 (1996). More recent examples of boronic acid-based dyes and optical sensing of neutral sugars: (b) K. Koumoto, M. Takeuchi and S. Shinkai, Design of a visualized sugar sensing system utilizing a boronic acid azopyridine interaction, *Supramol. Chem.* **9**(3), 203 (1998). (c) K. Koumoto and S. Shinkai, Colorimetric sugar sensing method useful in "neutral" aqueous media, *Chem. Lett.* **8**, 856-857 (2000). (d) C. J. Ward, P. Patel, P. R. Ashton and T. D. James, A molecular colour sensor for monosaccharides, *Chem. Commun.* **3**, 229-230 (2000). (d) N. DiCesare and J. R. Lakowicz, New color chemosensors for monosaccharides based on Azo dyes, *Org. Lett.* **3**(24), 3891-3893 (2001).
26. Recent examples: a) C. R. Bertozzi and L. L. Kiessling, Chemical glycobiology, *Science* **291**(5512), 2357-2364 (2001). (b) J. Axford, The impact of glycobiology on medicine, *Trends Immunol.* **22**(5), 237-239 (2001), and references cited therein.
27. Solution color changes in response to monosaccharides, maltose, maltotriose and maltoteraose using a resorcinarene-methyl red dye competition process: O. Rusin and V. Kral, 1,1'-Binaphthyl substituted calixresorcinols-methyl red complexes: receptors for optical saccharide sensing, *Tetrahedron Lett.* 42(25), 4235-4238 (2001).
28. J. F. Kennedy and G. Pagliuca in: *Oligosaccharides/Carbohydrate Analysis. A Practical Approach,* edited by M. F. Chaplin and J. F. Kennedy (Oxford University Press, Oxford, 1994), pp 46-62.
29. Review: J. M. LoGuidice and M. Lhermitte (1996). HPLC of oligosaccharides in glycobiology *Biomed. Chromatogr.* **10**(6), 290-296.
30. (a) J. P. Lorand and J. D. Edwards, Polyol complexes and structure of the benzeneboronate ion, *J. Org. Chem.* **24**, 769-774 (1959). (b) J. Yoon and A. W. Czarnik, Fluorescent chemosensors of carbohydrates. A means of chemically communicating the binding of polyols in water based on chelation-enhanced quenching, *J. Am. Chem. Soc.* **114**(14), 5874-5875 (1992).
31. J. C. Norrild and H. Eggert, Boronic acids as fructose sensors. Structure determination of the complexes involved using (1)J(CC) coupling constants, *J. Chem. Soc., Perkin Trans.* **2**(12), 2583-2588 (1996).
32. Y. Nagai, K. Kobayashi, H. Toi and Y. Aoyama, Stabilization of sugar-boronic esters of indolylboronic acid in water via sugar indole interaction - a notable selectivity in oligosaccharides, *Bull Chem. Soc. Japan* **66**(10), 2965-2971 (1993).
33. A known method for the determination of resorcinarene pK_a values in a mixed aqueous solvent system (H.-J. Schneider, D. Güttes, U. Schneider, Host guest chemistry. 15. Host guest complexes with water-soluble macrocyclic polyphenolates including induced fit and simple elements of a proton pump, *J. Am. Chem. Soc.* **110**(19), 6449-6454 (1988).) more appropriate for potentiometric titrations was applied to colored solutions of **2a** and **3** with and without added fructose. Unfortunately, solubility problems precluded us from obtaining pK_a data via this method
34. Review: B. Herbreteau, Review and state of sugar analysis by high-performance liquid-chromatography, *Analusis* **20**(7), 355-374 (1992).
35. Reviews: (a) S. Honda, Postcolumn derivatization for chromatographic analysis of carbohydrates, *J. Chromatogr. A* **720**(1-2), 183-199 (1996). (b) S. Honda, High-performance liquid chromatography of mono- and oligosaccharides, *Anal. Biochem.* **140**(1), 1-47 (1984).

36. For example: K. Mopper, R. Dawson, G. Liebezeit, and H. P. Hansen, Borate complex ion exchange chromatography with fluorimetric detection for determination of saccharide ethylenediamine, *Anal. Chem.* **52**(13), 2018-2022 (1980).
37. S. Miljanić, Z. Cimerman, L. Frkanec, M. Žinić, Lipophilic derivative of rhodamine 19: characterization and spectroscopic properties, *Anal. Chim. Acta* **468**(1), 13-25 (2002).
38. T. D. James and S. Shinkai, Artificial receptors as chemosensors for carbohydrates, *Top. Curr. Chem.* **218**, 159-200 (2002). (b) W. Wang, X. Gao and B. Wang, Boronic acid-based sensors, *Curr. Org. Chem.* **6**(14), 1285-1317 (2002).
39. I. MacDonald, Diet and human atherosclerosis carbohydrates, *Adv. Exp. Med. Biol.* **60**, 57-64 (1975).
40. A. W. Thorburn, L. H. Storlein, A. B. Jenkins, S. Khouri and E. W. Kreagen, Fructose-induced invivo insulin resistance and elevated plasma triglyceride levels in rats, *Am. J. Clin. Nutr.* **49**(6), 1155-1163 (1989).
41. (a) H. F. Bunn and P. J. Higgins, Reaction of monosaccharides with proteins - possible evolutionary significance, *Science* **213**, 222 (1981). (b) A. C. Burden, Fructose and misleading glycosylation data, *Lancet* **2**(8409), 986-986 (1984). (c) G. Suarez, R. Rajaram, A. L. Oronsky and M. A. Gawinowicz, Nonenzymatic glycation of bovine serum albumin by fructose (fructation). Comparison with the Maillard reaction initiated by glucose, *J. Biol. Chem.* **264**(7), 3674-3679 (1989).
42. B. R. Pettit, G. S. King and K. Blau, The analysis of hexitols in biological fluid by selected ion monitoring, *Biomed. Mass Spectrom.* **7**(7), 309-313 (1980).
43. E. Pitkänen and T. Kanninen, Determination of mannose and fructose in human plasma using deuterium labeling and gas chromatography/mass spectrometry, *Biol. Mass Spec.* **23**(9), 590-595 (1994).
44. Reviews: (a) H. A. Bardelmeijer, J. C. M. Waterval, H. Lingeman, R. van't Hof, A. Bult and W. J. M. Underberg, Pre-, on- and post-column derivatization in capillary electrophoresis, *Electrophoresis*, **18**(12-13), 2214-2227 (1997). (b) J. M. LoGuidice and M. Lhermitte, HPLC of oligosaccharides in glycobiology, *Biomed. Chromatogr.* **10**(6), 290-296 (1996).
45. Reviews: (a) E. M. J. Schutter, J. J. Visser, G. J. Vankamp, S. Mensdorffpouilly, W. Vandijk, J. Hilgers and P. Kenemans, The utility of lipid-associated sialic-acid (lasa or lsa) as a serum marker for malignancy - a review of the literature, *Tumor Biol.* **13**(3), 121-132 (1992). (b) J. Roth, Cellular sialoglycoconjugates - a histochemical perspective, *Histochemical J.* **25**, 687-710 (1993). (c) A.Kobata, Glycobiology - an expanding research area in carbohydrate-chemistry, *Acc. Chem. Res.* **26**(6), 319-324 (1993). (d) G. Reuter and H.-J. Gabius, *Biol. Chem. Hoppe-Seyler* **377**(6), 325-342 (1996).
46. (a) R. Schauer, S. Kelm, G. Rerter, P. Roggentin and L. Shaw, in: *Biology of the Sialic Acids*, edited by A. Rosenberg (Plenum, New York, 1995), pp. 7. (b) Y. Nagai and M. Iwamori, in: *Biology of the Sialic Acids*, edited by A. Rosenberg (Plenum, New York, 1995), pp.197.
47. L. Warren, The thiobarbituric acid assay of sialic acids, *J. Biol. Chem.* **234**(8), 1971-1975 (1959).
48. L. Svennerholm, Quantitive estimation of sialic acids: II. A colorimetric resorcinol-hydrochloric acid method, *Biochim. Biophys. Acta*, **24**, 604-611 (1957).
49. T. Hikita, K. Tadano-Aritomi, N. Tanaka, H. Toyoda, A. Suzuki, T. Toida, T. Imanari, T. Abe, Y. Yanagawa and I. Ishizuka, Determination of *N*-acetyl- and *N*-glycolylneuraminic acids in gangliosides by combination of neuraminidase hydrolysis and fluorometric high-performance liquid chromatography using a GM3 derivative as an internal Standard, *Anal. Biochem.* **281**(2), 193-201 (2000).
50. R. L. Mattoo and S. Roseman, Quantitative determination of sialic acid in the monosialoganglioside, GM1, by the thiobarbituric acid method, *Anal. Biochem.* **246**(1), 30-33 (1997).
51. (a) M. Takeuchi, S. Yoda, T. Imada and S. Shinkai, Chiral sugar recognition by a diboronic-acid-appended binaphtyl derivative through rigidification effect, *Tetrahedron* **53**(25), 8335-8348 (1997). (b) M. Takeuchi, T. Mizuno, H. Shinmori, M. Nakashima and S. Shinkai, Fluorescence and CD spectroscopic sugar sensing by a cyanine-appended diboronic acid probe, *Tetrahedron*, **52**(4), 1195-1204 (1996).
52. O. Rusin, O. Alpturk, M. He, J. O. Escobedo, S. Jiang, F. Dawan, K. Lian, M. E. McCarroll, I. M. Warner and R. M. Strongin, Macrocycle-Derived functional xanthenes and progress towards concurrent detection of glucose and fructose. *J. Fluoresc.* **14**(5), 611-615 (2004).

TWO-COMPONENT OPTICAL SUGAR SENSING USING BORONIC ACID-SUBSTITUTED VIOLOGENS WITH ANIONIC FLUORESCENT DYES

Modulated quenching with viologens as a method for monosaccharide detection

David B. Cordes, Jeff T. Suri, Frank E. Cappuccio, Jason N. Camara, Soya Gamsey, Zach Sharrett, Praveen Thoniyot, Ritchie A. Wessling, and Bakthan Singaram*

3.1. INTRODUCTION

The design and synthesis of chemosensors for biologically important molecules has developed into a major research area. Progress has been driven by advances in the analytical capabilities of biologists and chemists and from medical professionals whose practices lay increasing emphasis on accurately monitoring a patient's biochemical balance. In particular, medical providers are interested in the accurate and real-time measurement of blood glucose levels for diabetic patients. The shortcomings of existing glucose detection methods available to both patient and clinician have inspired the development of many new glucose sensing strategies over the past fifteen years. Many of these are fluorescence-based systems that rely on the use of glucose oxidase. A large number of non-enzymatic glucose sensing systems, however, have also been developed.[1] Many of these non-enzymatic systems rely on a signal derived from the modulated emission of a fluorophore. Such modulation can be made to occur upon glucose binding to a boronic acid receptor.

The impetus for boronic acid-derived glucose detection systems derives from the pioneering work of Yoon and Czarnik. In the early 1990's, these

* Prof. Bakthan Singaram, University of California - Santa Cruz, Santa Cruz, CA, 95060, singaram@chemistry.ucsc.edu

workers reported the first scientific study in which a boronic acid receptor was combined with fluorescence reporter to act as a glucose sensor.[2] In this initial sensor configuration, a boronic acid was directly attached to an anthracene fluorophore. Saccharide binding causes a change in the geometry and electronics of the boron resulting in a diminished fluorescence emission. Additional glucose sensing systems, such as those prepared by Suenaga and coworkers,[3,4] also relied on this configuration by directly attaching boronic acid receptors to common fluorophores such as stillbene, pyrene, and naphthalene. Other workers have had success with variations on this basic theme by attaching a boronic acid receptor to a fluorescent dye via a linker arm. In these systems, glucose binding can cause a fluorescence modulating process, such as photoinduced electron transfer (PET), to be enhanced or interrupted. The classic example of such a system is that developed by Shinkai and coworkers, who attached a boronic acid receptor to an anthracene fluorophore by way of an amine linker.[5, 6] In the absence of saccharide, the amine is able to quench the anthracene fluorescence through what is believed to be a PET mechanism. When sugar binds to the boron, however, the Lewis acidity of the boron is increased, causing it to coordinate to the amine nitrogen, thereby deactivating the PET mechanism. Thus, the fluorescence intensity increases as a function of glucose concentration. Numerous clever variations have been made on this sensing motif.

From a system design point-of-view, the practice of coupling the modulation of fluorescence quenching with glucose binding to a boronic acid has been exploited for some time, but these systems relied exclusively on the use of a single sensing moiety containing both receptor and fluorophore. Our approach to monosaccharide sensing, however, has relied on a two-component design in which the fluorophore is quenched by a physically separate boronic acid-substituted viologen receptor. Monosaccharide binding to the boronic acid of the viologen diminishes the efficiency with which the fluorescence emission is quenched. Thus, the intensity of fluorescence emission can be correlated with monosaccharide concentration. This two-component approach removes the synthetic difficulties associated with combining the fluorophore and receptor in a single sensor molecule and allows for considerable versatility in choosing each component. For example, the structure of the receptor can be modified to improve quenching ability or to provide selectivity for one monosaccharide over another without having to modify the structure of the dye. At the same time, fluorophore units can be interchanged without any synthetic modification of the receptor, making it possible to use numerous, commercially available fluorescent dyes. This is a considerable advantage of the two-component approach, since synthetic transformation can cause unwanted changes in the photophysical properties of the dye. A schematic representation of a typical one-component system is shown in Figure 3.1 A and our two-component system is given in figure 3.1 B.

Our lab has pursued this two-component approach to glucose sensor design and explored the use of a variety of dye and quencher molecules in aqueous solution. We have generally designed our glucose sensors for *in vivo* applications. Accordingly, we have carried out most of our experiments at physiological pH of 7.4 and in open air. As we continue to refine our system for medical applications, we are gradually adding physiological constraints to our experimental designs. We have also successfully immobilized our two sensing components in a thin film hydrogel to prepare a continuous glucose sensing system. Further, we have studied the mechanism responsible for the signal transduction.

The following review follows, more or less chronologically, the progress of our research in the glucose sensing field over the past five years. The review is organized into five parts (Sections 3.2-6):

3.2. Background and Illustration of Two-Component Glucose Sensing – Pyranine (HPTS) and a Boronic Acid-Substituted Viologen (o-BBV^{2+})

3.3. Variations in Viologen Quencher - Bipyridinium and Phenanthrolinium Quenchers

3.4. Variations in Fluorescent Reporter – Sulfonamide Derivatives of HPTS

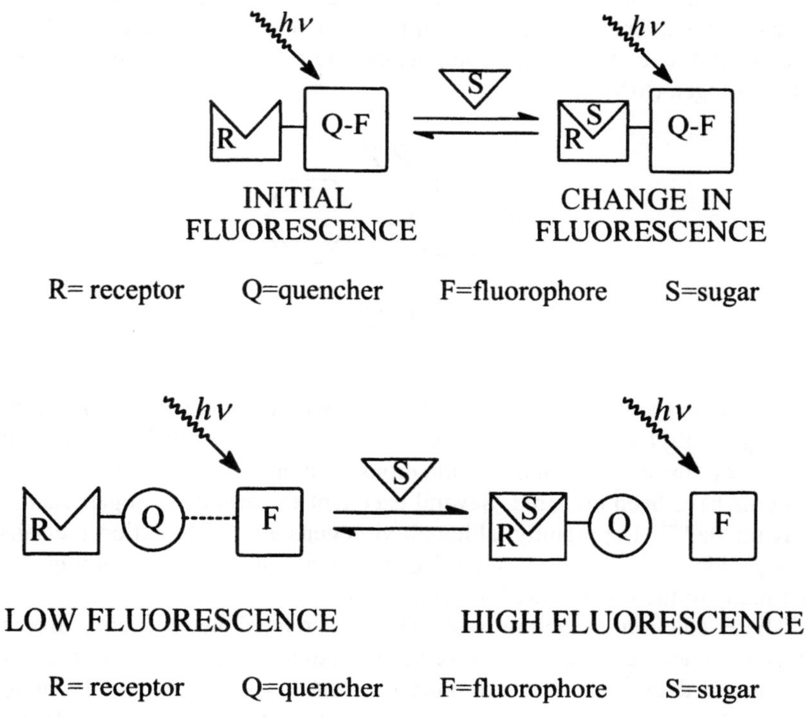

Figure 3.1 **A**. One-component saccharide sensing system. **B.** Two-component saccharide sensing system.

3.5. Immobilization of the Sensing Components - A Glucose Sensitive Thin Film Hydrogel
3.6. Summary and Future Directions

Interested readers are encouraged to consult the original papers cited in this review for further details of the work. Several excellent, recent reviews of boronic acid-based fluorescent sensing systems are already available.[7, 8] Therefore, citation of other work in this field is only made where it connects with our studies.

3.2. BACKGROUND AND ILLUSTRATION OF TWO-COMPONENT GLUCOSE SENSING – PYRANINE (HPTS) AND BORONIC ACID-FUNCTIONALIZED VIOLOGEN (*o*-BBV^{2+})

In our two-component saccharide sensing system, boronic acid-functionalized viologen quenchers are used as the saccharide receptor unit.[9-12] In this system, the viologen is first used to quench the fluorescence of the dye, then the fluorescence is recovered as glucose binds to the boronic acids and diminishes the quenching efficiency of the viologen. Viologens themselves are quaternized 4,4'-bypyridinium compounds, the most well-known of which is methylviologen (MV^{2+}).

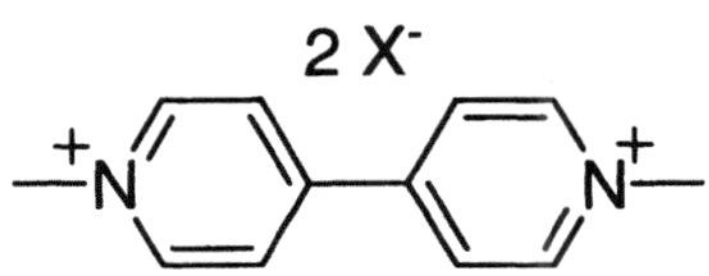

MV^{2+}

Viologens are characterized by their activity as good electron acceptors.[13] Although methyl viologen (MV^{2+}) is the most widely studied and most often utilized of these compounds, numerous viologens with different redox potentials have been prepared. Several excellent reviews of viologen chemistry are available.[14-17] Importantly, although viologens are by definition the salts of 4,4'-bipyridinium,[18] several other bipyridinium and phenanthrolinium salts share many of the properties of the viologens.

These electron accepting viologens and related heterocyclic, aromatic, nitrogen-containing compounds have been found to quench the luminescence of both simple dyes[19-21] and macromolecular systems. [22-25] The ability of these compounds to quench fluorescence stems at least partly from the fact that they contain two positive charges. The positive charges of the viologen facilitate two possible processes: electron transfer and Coulombic attraction. The relatively modest fluorescence quenching of an anionic dye molecule by nonionic

4,4'-dipyridyl follows simple Stern-Volmer kinetics and appears to occur exclusively through dynamic, or collisional quenching. Quenching with a cationic viologen, however, is dramatically enhanced as a result of electrostatic attraction. This results in static quenching through complex formation between dye and quencher.[26] Wang and co-workers have shown this to be the case in the fluorescence quenching of a water-soluble macromolecule, poly[5-methoxy-2-(4-sulfobutoxy)-1,4-phenylenevinylene], with 4,4'-dipyridyl and 4,4'-dipyridinium salts.[27]

Research on glucose sensors based on boronic acid functionalized viologens was initiated in our laboratory in 1999. Our goal in this research was to design a fluorescence sensor that could operate under physiological conditions. In a breakthrough experiment carried out that year, the quenching efficiency of a boronic acid-functionalized viologen was shown to be reduced by binding to a diol. We determined that this change in quenching efficiency could be used in combination with a fluorescent dye to signal glucose.

Recognizing the limitations of sensors in which the dye, quencher and receptor were combined in a single molecule, we chose to pursue the two-component approach described earlier: one component being the fluorophore, the other a quencher to which a boronic acid receptor was attached. Among the various quenchers screened in the early stages of this project, viologens stood out as superior quenchers with the ability to quench the fluorescence of many dyes.

Since the intent was to make an implantable sensor, much of the early work was done with polymeric components. However, the complexity of the system led us to conduct a parallel investigation of water soluble analogs which could more easily be studied in solution. The commercially available dye pyranine, 8-Hydroxy-1,3,6-pyrenetrisulfonic acid trisodium salt, was selected as a reference dye for this work. Pyranine, hereafter HPTS, is a well-studied dye that has found use in numerous sensor applications,[28-32] particularly in the studies of Wolfbeis and co-workers.[33-36] HPTS is a photostable, anionic and highly water soluble green-fluorescing dye with a quantum yield of nearly one. These characteristics made it an ideal dye for our studies.

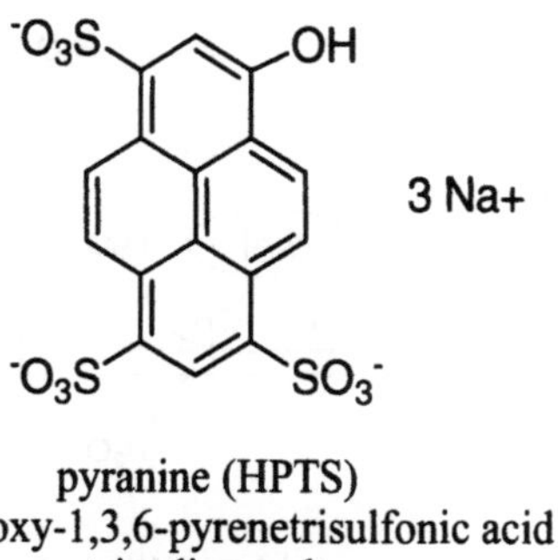

pyranine (HPTS)
8-Hydroxy-1,3,6-pyrenetrisulfonic acid
trisodium salt

While this research was in progress, Baptista and co-workers reported a study of complex formation between HPTS and methyl viologen (MV^{2+}) in which process the fluorescence of HPTS was quenched.[19] They attributed the complexation to electrostatic attraction between cationic quencher and anionic dye and offered convincing absorbance data to support this explanation. Clarke and co-workers have observed similar complex formation between viologens and anionic porphyrin dyes.[37] This work suggested a mechanistic basis for our fluorescence-based glucose sensing system in which boronic acid receptors were attached to the viologen quencher.

The key step in this sensing process is, of course, the binding of sugars by the boronic acids of the viologen quencher. Arylboronic acids are well known for their ability to reversibly bind monosaccharides.[38] Significant equilibria between arylboronic acid and generic diols are shown in scheme 3.1.

We had hypothesized that the degree to which the fluorescence of a dye was quenched by the viologen could be modulated by glucose binding to boronic acids appended to the viologen. We began screening various quenchers and found that benzyl viologen (BV^{2+}) quenched HPTS effectively (Figure 2).

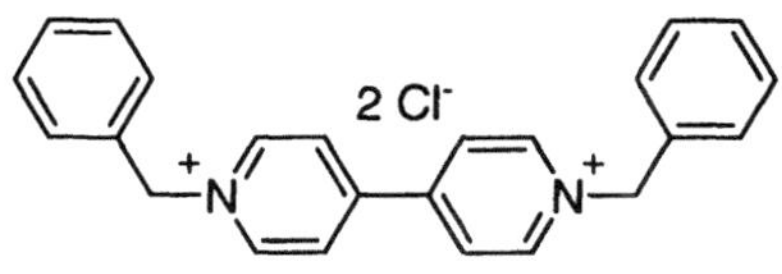

BV^{2+}

This result encouraged us to proceed with the synthesis of boronic acid substituted viologens beginning with 4,4'-*N,N'*-bis(benzyl-2-boronic acid)-bipyridinium dibromide (*o*-BBV^{2+}) and study their ability to quench HPTS. Synthetically, the preparation of *o*-BBV^{2+} is very similar to that of benzyl viologen (BV^{2+}) (Scheme 2).[39]

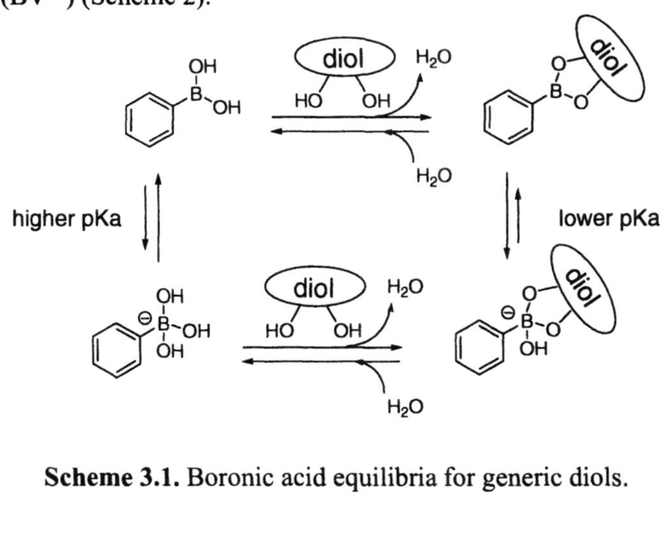

Scheme 3.1. Boronic acid equilibria for generic diols.

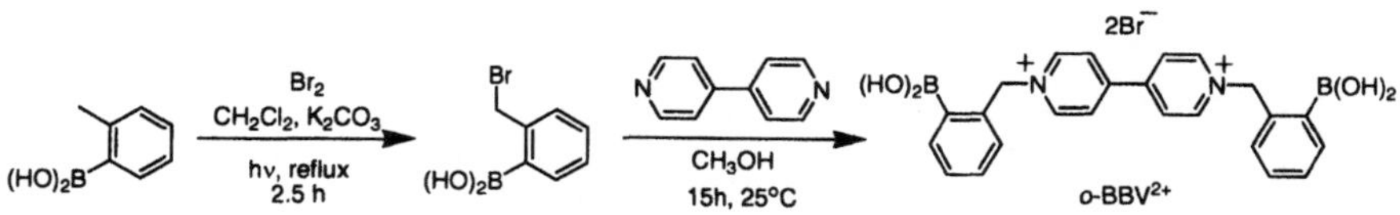

Scheme 3.2. Synthesis of o-BBV^{2+}. For synthetic details see reference #10.

Because neither the mono- nor the bis-boronic acid substituted viologen was expected to demonstrate selectivity for glucose we chose to study the bis-boronic acid due to its expected ease of preparation and enhanced water solubility. Its ability to quench pyranine fluorescence was then investigated and found to be comparable to that of benzyl viologen (BV^{2+}).

In their study of the complexation between HPTS and MV^{2+}, Baptista's group also noted that the Stern-Volmer analysis of this quenching often provided superlinear plots that were best explained by the operation of two independent quenching mechanisms. Generally, we have found that there appear to be two quenching mechanisms at work in nearly all of the viologen-HPTS interactions we have studied. The primary contribution to fluorescence quenching is from a static quenching process in which a non-fluorescent ground state complex is formed between dye and quencher. A second process known as dynamic quenching occurs due to collisions between dye and quencher in which the fluorophore is deactivated through interaction with the colliding quencher molecule.[26] In our study we use Eq. 1 which has been developed based on a "sphere of action" quenching model[19, 40] and which provides quenching constants for both static (K_{sv}) and dynamic (V) processes where F_o is the initial fluorescence and F is the fluorescence after addition of quencher.

$$F_o/F = (1 + K_{sv}[Q])e^{V[Q]} \qquad (1)$$

Comparison of quenching efficiency for BV^{2+} and o-BBV^{2+} indicated that the addition of the boronic acids to o-BBV^{2+} had little effect on the quenching interaction. This similarity was not always observed in comparison of other viologens with their substituted analogues (See section 3.3.). The Stern-Volmer plots for MV2+, BV2+ and o-BBV2+ are given in Figure 3.2.

Having established the quenching ability of o-BBV^{2+}, monosaccharide sensing was examined utilizing the o-BBV^{2+}/HPTS system. Our first systematic optical sugar detection studies were done utilizing solutions of HPTS (1×10^{-5} M) in pH 7.4 buffer that had been partially quenched with o-BBV^{2+}.

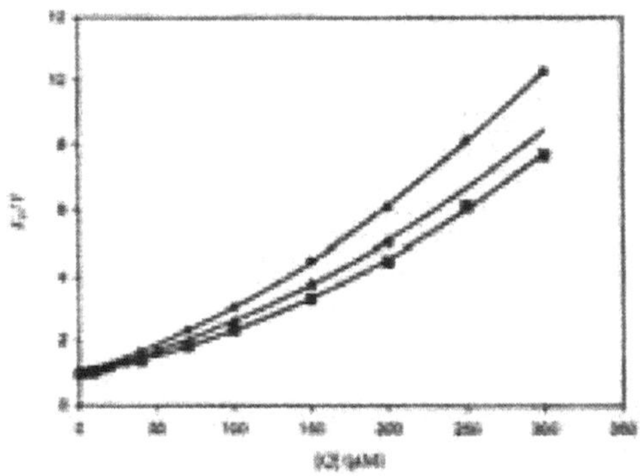

Figure 3.2. Stern-Volmer plots of HPTS (1 x 10^{-5} M) fluorescence quenched by •, BV^{2+}; ▲, MV^{2+}; and ■, *o*-BBV^{2+} at pH 7.4.

The molar ratio of quencher to dye was 30:1. Glucose was then added in small aliquots to achieve a range of concentrations from 0 to 1800 mg/dL (100 mM) (Figure 3.3).

As seen in Fig. 3.3, addition of glucose to an aqueous solution of *o*-BBV^{2+} and HPTS led to a significant increase in fluorescence intensity, whereas BV^{2+}/HPTS showed no response. Thus, addition of 360 mg/dL glucose resulted in approximately a twofold signal increase.[9] Fructose and galactose gave a fivefold and three-and-one-half fold increase in signal, respectively. The observed selectivity for fructose is in line with that observed for other monoboronic acid derivatives reported in the literature. Generally, the selectivity observed in monoboronic acid receptors tends to mirror the 33:3:1 ratio of association constants for fructose, galactose, and glucose observed by Springsteen and Wang for simple phenylboronic acid.[41]

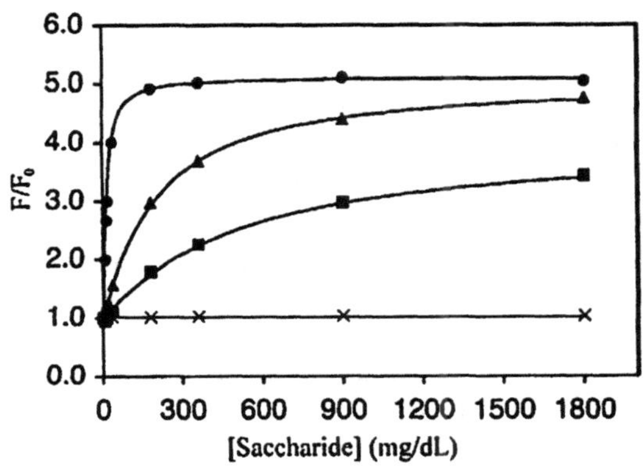

Figure 3.3. Relative fluorescence emission of HPTS (1×10^{-5} M, λ_{ex} = 461.8 nm, λ_{em} = 511 nm); *o*-BBV^{2+} (3×10^{-4} M) system as a function of saccharide concentration in 0.1 ionic strength aqueous phosphate buffer pH 7.4 (•, fructose; ▲, galactose; ■ glucose; ×, glucose (*o*-BBV^{2+} replaced with BV^{2+} (3×10^{-4} M) to demonstrate lack of sensitivity without boronic acid functionality)). Note: 180 mg/dL = 10 mM saccharide.

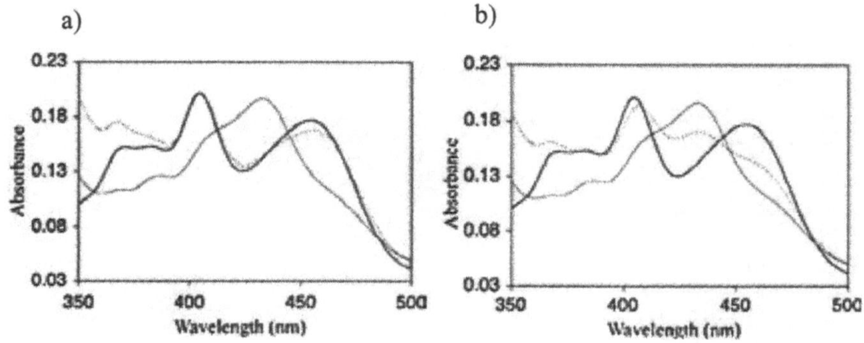

Figure 3.4. a) UV–vis absorption spectra of HPTS (1×10^{-5} M) — ; HPTS (1×10^{-5} M) with *o*-BBV^{2+} (3×10^{-4} M) —; HPTS (1×10^{-5} M) with *o*-BBV^{2+} (3×10^{-4} M) and fructose (1800 mg/dL) --.
b) UV–vis absorption spectra of HPTS (1×10^{-5} M) —; HPTS (1×10^{-5} M) with *o*-BBV^{2+} (3×10^{-4} M) —; HPTS (1×10^{-5} M) with *o*-BBV^{2+} (3×10^{-4} M) and glucose (1800 mg/dL) --.

Our studies also found that the combination of viologens with HPTS caused a non-fluorescent complex to form. Baptista saw evidence for this complexation in the UV-visible absorbance spectra and we saw similar changes in the absorbance spectra as pyranine was quenched with *o*-BBV^{2+} (Figure 3.4).

At pH 7.4, both of the distinctive absorbance bands at 404 and 454 nm are diminished as quencher is added. Simultaneously, a new absorbance band, that of the putative complex, appears at ~438 nm. Significantly, as sugar was added and the fluorescence emission recovered, the sugar binding event appeared to effect a dissociation of the complex and a return to the uncomplexed state as evidenced by changes in the absorbance spectra. In the case of fructose (Figure 3.4 a), the return to the initial uncomplexed pyranine spectrum appears nearly quantitative. With glucose (figure 3.4 b), however, where the binding is expected to be considerably weaker than with fructose, the original spectral pattern is only partly recovered. This indicates a smaller equilibrium population of the ester form of the boron in glucose solution relative to a fructose solution of the same concentration.

When conducting quenching or sugar sensing studies in our system we have observed that changes in the fluorescence emission are generally paralleled by changes in the UV-visible absorbance spectra of the fluorescent reporter. The correlation we have found between the absorbance and fluorescence data is consistent with tight complex formation between quencher and dye. The presence of isosbestic points in the absorbance spectra of HPTS at 412 and 450 nm during titration with *o*-BBV^{2+} are strongly indicative of a single complex being formed between the dye and the quencher with what we speculate is 1:1 stoichiometry (Figure 3.5).

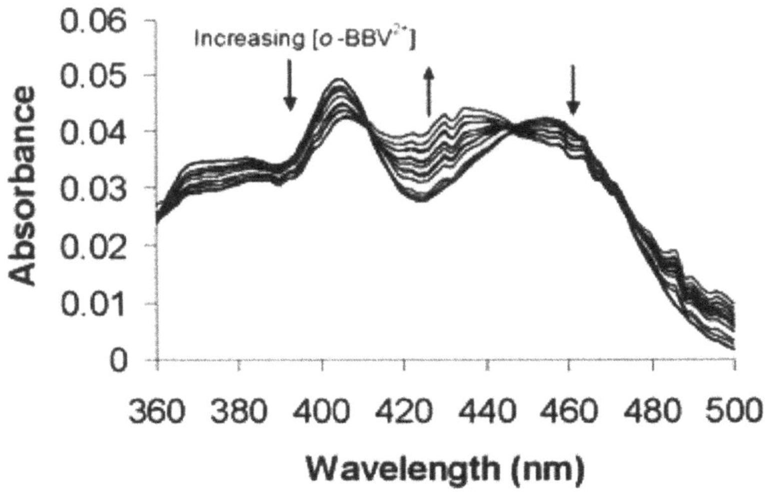

Figure 3.5. Changes in UV-vis absorbance spectra of HPTS (4 x 10^{-6} M) as titrated by *o*-BBV^{2+} at pH 7.4

This dissociation between dye and quencher that occurs when monosaccharide is added appears to be the result of electronic and/or steric changes that take place in the viologen receptor upon monosaccharide binding. We hypothesize that these changes cause a diminished electrostatic attraction between quencher and dye resulting in a fluorescence increase. The putative general sensing mechanism is represented in scheme 3.3 with the viologen quencher *o*-BBV^{2+}.

To summarize, we developed a glucose sensing system in which signal transduction derives from two separate reversible complexation reactions. The first is complexation of a fluorescent dye with a viologen which quenches the fluorescence. The second is the binding of the sugar with the boronic acid receptor of the viologen which causes the initial dye:quencher complex to dissociate and fluorescence emission to recover. It is essentially a system based on competitive binding. Additional studies were undertaken to explore the mechanistic details of our sensing system and to optimize signal strength and glucose sensitivity in the physiological range. The next series of studies were carried out on quencher variations.

3.3. VARIATIONS IN VIOLOGEN QUENCHER – BIPYRIDINIUM AND PHENANTHROLINIUM QUENCHERS

After our initial work with *o*-BBV^{2+}, the use of the *meta*- and *para*-boronic acid substituted analogues was investigated. Since our initial report in 2002, Lakowicz and coworkers made use of the *para*-substituted derivative in a

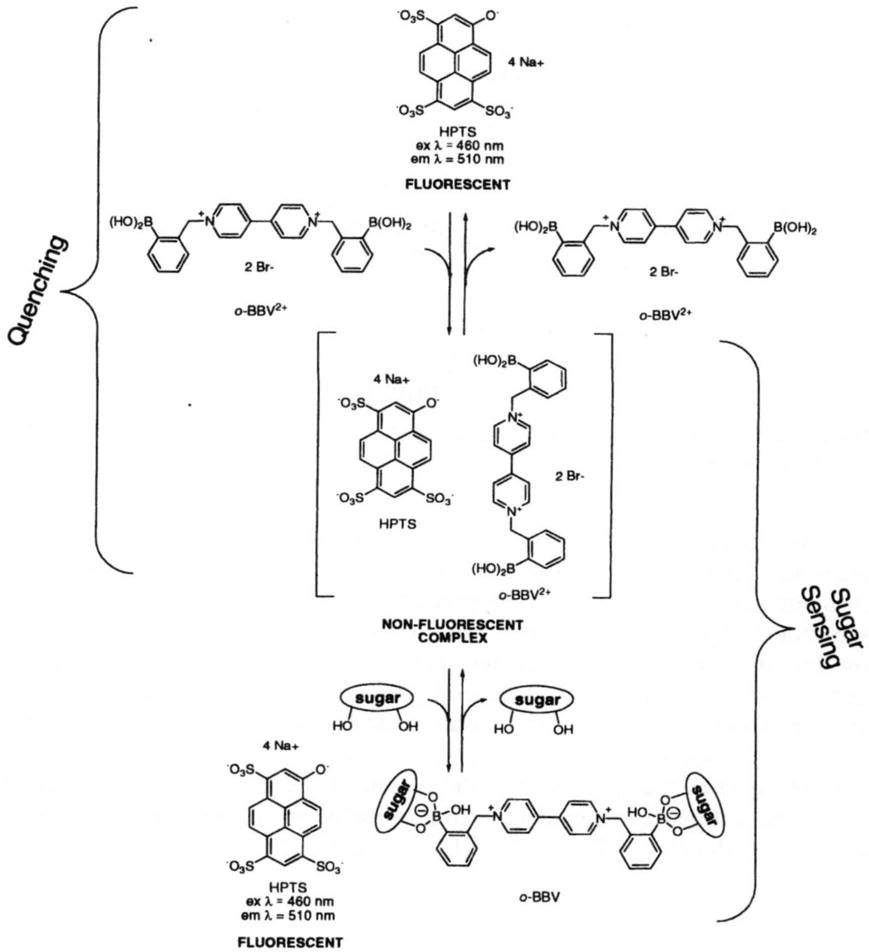

Scheme 3.3. General sensing scheme illustrated with HPTS and *o*-BBV^{2+}.

two-component glucose sensing system to provide amplified quenching of the fluorescence of an anionic poly(phenylene ethynylene) dye.[25] The relative fluorescence increase obtained on addition of glucose to this quenched fluorophore was quite large. For our own studies, the syntheses of these compounds were achieved in the same manner as the *ortho*-substituted compound as shown in Scheme 2. We were interested in the effects that these changes would have on both quenching and sugar sensing. First, the ability *m*-BBV^{2+} and *p*-BBV^{2+} to quench HPTS fluorescence was evaluated (Figure 3.6).

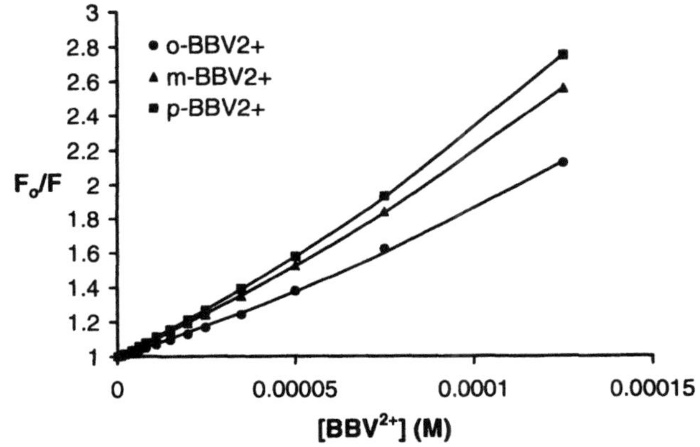

Figure 3.6. Stern-Volmer plots of HPTS (4×10^{-6} M) fluorescence quenched by *o*-BBV^{2+}, *m*-BBV^{2+}, *p*-BBV^{2+} in pH 7.4 buffer.

As indicated in figure 3.6, we found only minor differences in quenching efficiency for these three bipyridinium quenchers with the *para*-substiuted viologen providing the best results followed by *meta*- and then *ortho*-substituted BBVs. Interestingly, we found similar changes in the UV-vis spectra when each of these quenchers was used, indicating that formation of a non-fluorescent complex was a general phenomenon. In each case, however, the perturbation of the spectra and absorbance maxima of the quencher:dye complex was slightly different, suggesting that the substitution pattern on the viologen quencher directly affects the nature of the complex in a significant way. Such differences are evident in comparing the absorbance spectra of HPTS as it is titrated with *o*-BBV^{2+} and *m*-BBV^{2+} (Figure 3.7).

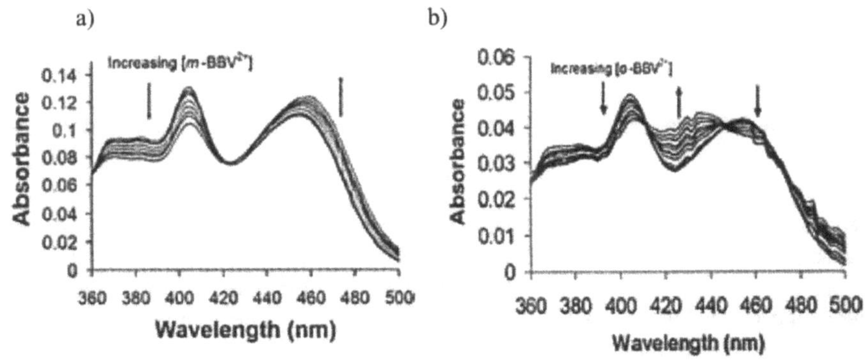

Figure 3.7. a) UV-visible absorbance spectra of HPTS (1×10^{-5} M) during titration with *m*-BBV^{2+}; b) UV-visible absorbance spectra of HPTS (4×10^{-6} M) during titration with *o*-BBV^{2+}. Both titrations were performed in pH 7.4 phosphate buffer.

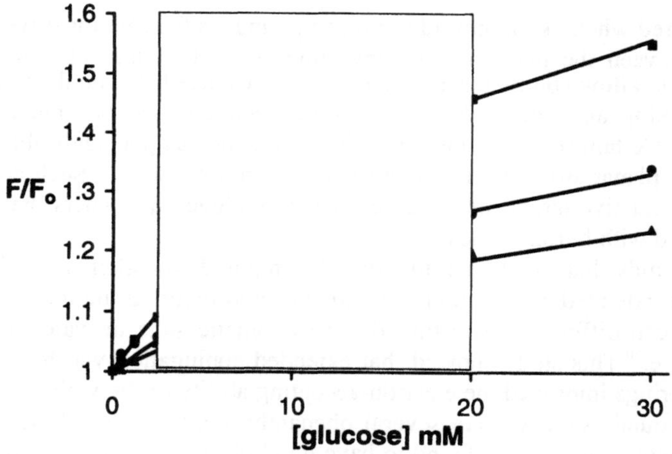

Figure 3.8. Glucose response of *o*, *m*, and *p*-BBV^{2+} in combination with HPTS (4×10^{-6} M) at the same quencher:dye of ratio of 30:1 in pH 7.4 aqueous solution. Physiological glucose range is boxed.

When these three compounds were tested for their response to glucose we found a fair degree of variation in the results, with *o*-BBV^{2+} giving the greatest signal. Surprisingly, this was the compound that had provided the weakest quenching among the three. The *meta*- and *ortho*-substituted viologens gave a smaller, but still significant response (Figure 3.8).

Several important observations emerged from these quenching and sugar response experiments. First, in comparing structurally and electronically similar viologen-receptors, we realized that the best quencher in a series is not necessarily expected to have the best response to changes in saccharide concentration. In this study of *ortho*-, *meta*-, and *para*-substituted compounds there was actually an inverse correlation between quenching efficiency and glucose sensing ability. Secondly, because the quenching and sugar sensing interactions are interdependent by virtue of their dependence on quencher concentration, the ratio of quencher-to-dye is a critical parameter. The relationship between a viologen-receptor's ability to both quench fluorescence and bind monosaccharides through optimization of the ratio between quencher and dye is among several areas of active research in our labs.[42, 43] Monosaccharide binding studies were carried out using fructose and galactose as well, and no difference was observed with respect to the selectivity among the three variously substituted viologens. Instead they all demonstrated essentially the same selectivity as observed for simple phenylboronic acid. We had hoped that variation in the substitution pattern might provide the correct geometry for bidentate binding of the glucose. Such tailor-made spacing between boronic acid receptors to provide saccharide selectivity has been achieved by other groups in several one component systems.[5, 6, 44-49] This approach is widely perceived as a promising way in which to achieve selectivity for a particular monosaccharide. In this regard, one of the difficulties encountered when using bipyridinium compounds is the rotational freedom that exists between the two rings of bipyridinium itself. This rotation would be expected to allow continuous alteration in the distance between the two boronic acids making any cooperative binding between the two boronic acids very difficult. Certain related compounds, however, such as phenanthrolinium salts, possess a planar structure in which rotation is not allowed. Such compounds became attractive targets as the core for new viologen quenchers that could be substituted with boronic acids.

In a study that caught our attention, Huenig and coworkers investigated the effects of extended conjugation in viologens, and reported the redox potentials of seventeen different *N*-substituted heteroaromatic salts in water, DMF, and acetonitrile.[50] This study showed that extended conjugation via the addition of aromatic rings improved the electron-accepting ability of the viologens. Among the compounds studied were several phenanthrolinium salts. These three ring aromatic compounds were found to have good electron accepting abilities. One particular compound, *N,N'*-dimethyl-4,7-phenanthrolinium tetrafluoroborate gave a redox potential of -490 mV (vs AgCl) in water for a single electron transfer. This is significantly more positive than the value of -640 mV recorded for MV^{2+}.

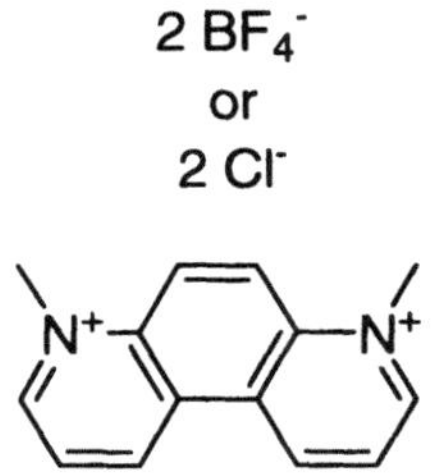

N,N'-dimethyl-4,7-phenanthrolinium

While it appeared that 4,7-phenanthrolinium salts are better electron acceptors than bipyridium salts, we found only one report of a 4,7-phenanthrolinium salt being used as a fluorescence quencher. Ihara and co-workers used a porphyrin dye in combination with the chloride salt of *N,N'*-dimethyl-4,7-phenanthrolinium to determine base mismatching in a DNA sequence.[51] In this study, the phenanthrolinium salt was thought to quench the fluorescence of the porphyrin via a photoinduced electron transfer mechanism. The ability of 4,7-phenanthrolinium salts to accept electrons and quench fluorescence prompted us to study the quenching ability of these compounds and compare them to our previously reported bipyridinium-based quenchers.

We anticipated that as better electron acceptors, the phenanthrolinium-derived viologens might show greater quenching. We were also interested in modifying the viologens by the attachment of a boronic acid group and determining how their quenching ability was modulated in the presence of monosaccharides. Thus, we prepared *N,N'*-bisbenzyl-4,7-phenanthrolinium dibromide (PV^{2+}), *N*-(benzyl-2-boronic acid)-4,7-phenanthrolinium bromide

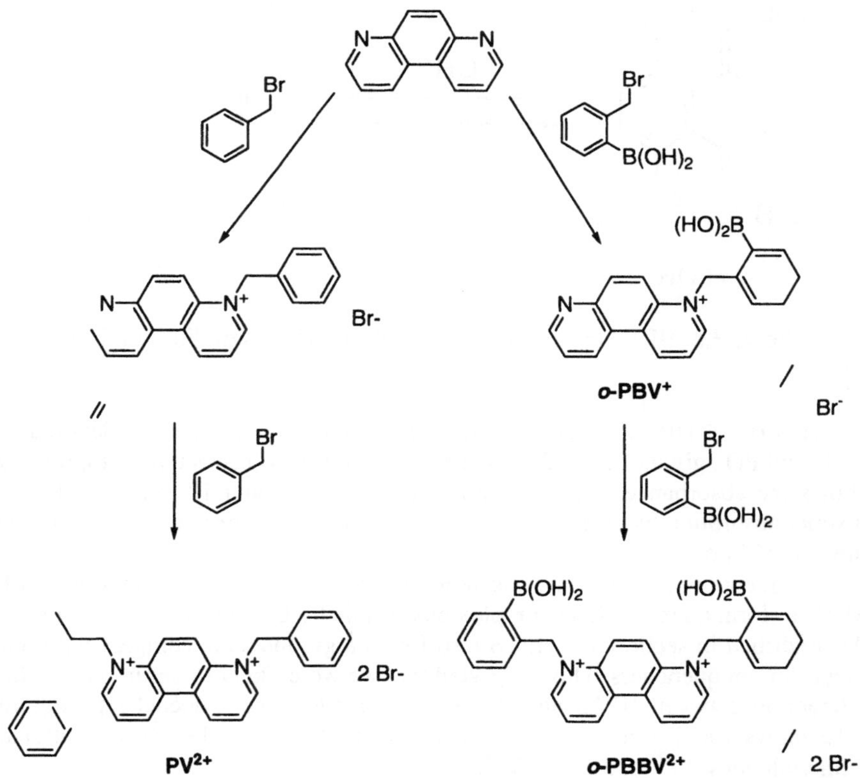

Scheme 3.4. Synthesis of phenanthrolinium quenchers. Synthetic details are provided in supporting information of reference #10.

(o-PBV$^+$), and *N,N'*-bis(benzyl-2-boronic acid)-47 -phenanthrolinium dibromide (o-PBBV^{2+}) (Scheme 3.4) and compared their ability to quench the fluorescence of pyranine and act as sugar sensors with that of the previously studied bipyridinium quenchers.

The synthesis of the model compound PV^{2+} and the monosubstituted boronic acid phenanthrolinium o-PBV$^+$ were achieved in good yield without significant difficulty. The bis-substituted compound, o-PBBV^{2+}, however could only be obtained in pure form after reverse phase chromatography on a reaction mixture that contained both mono- and bis-substituted products.

In our study of phenanthrolinium quenchers we also looked at the effect of solution pH. We studied the behavior of the quenchers at pH 3, 7.4, and 10 in combination with HPTS. HPTS is itself a pH sensitive dye and its spectral profile changes quite dramatically in the range around physiological pH. The ground-state p*K*a of the phenolic group in pyranine is 7.2.[34, 36, 52] At pH 3, pyranine is entirely in its protonated POH form, while at pH 10 it exists in its unprotonated form PO$^-$ (Scheme 3.5).

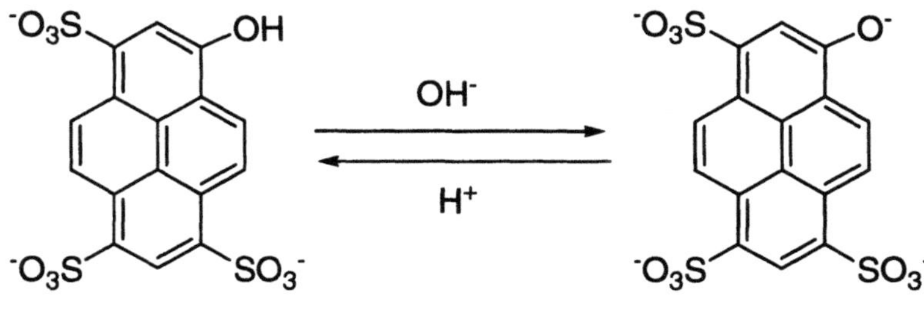

Scheme 3.5. HPTS in its acid (POH) and base (PO^-) equilibrium forms.

At pH 7.4 HPTS exists as a mixture of two different species, denoted as POH and PO^-, giving rise to the 404 and 454 nm bands, respectively. Figure 3.9 shows the absorbance of pure pyranine at pH 3, 7.4, and 10. At pH 7.4, two distinct absorption bands are apparent in the blue region, one at 404 nm and the other at 454 nm.

As mentioned earlier, it was reported that HPTS reacts reversibly with MV^{2+} to form a ground state complex observable by UV-visible spectroscopy.[19] As described in section 3.2, HPTS also forms a ground state complex with our bipyridinium quenchers. This suggested that we were likely to observe a similar interaction between HPTS and the new phenanthrolinium quenchers. Figure 3.10 shows the changes in HPTS absorbance at pH 7.4 as the dye was titrated with viologens PV^{2+} and *o*-$PBBV^{2+}$.

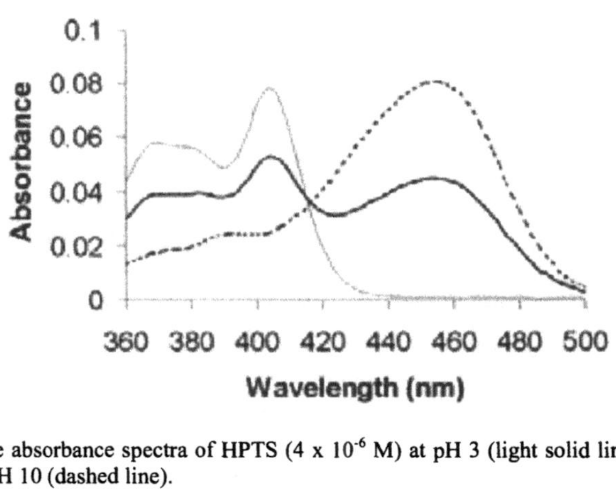

Figure 3.9. UV-visible absorbance spectra of HPTS (4 x 10^{-6} M) at pH 3 (light solid line), pH 7.4 (bold solid line), and pH 10 (dashed line).

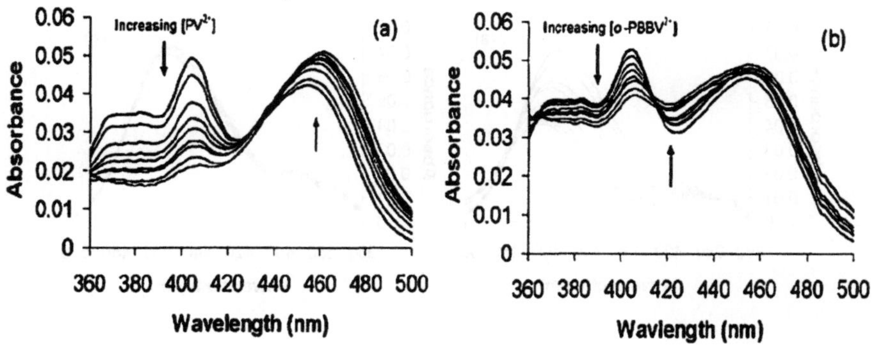

Figure 3.10. UV-visible absorbance spectra of HPTS (4 x 10^{-6} M) with increasing amounts of a) PV^{2+}, b) *o*-$PBBV^{2+}$ in a pH 7.4 phosphate buffer.

As was anticipated, significant changes occurred in the POH and PO^- bands of the absorption spectra as HPTS was titrated with the phenanthrolinium quenchers. For PV^{2+} and *o*-$PBBV^{2+}$ a decrease in the 404 nm band is apparent. For PV^{2+} an isosbestic point is observed at 434 nm, for *o*-$PBBV^{2+}$ one is observed at 416 nm.

To gain a better understanding of the perturbed spectra, experiments were carried out at pH 3 and 10 as well as pH 7.4. At pH 3, addition of each of the phenanthrolinium quenchers caused changes in the UV spectra of HPTS that were consistent with complex formation between dye and quencher. At pH 10, however, only the non-functionalized phenanthrolinium PV^{2+} appeared to form a complex with HPTS. This is consistent with a model of charge induced sensing as depicted in scheme 3. That is, at pH 3 all the phenanthrolinium quenchers are expected to have a net charge of 2+ and thus maintain an electrostatic attraction for the anionic dye. When the pH is raised to 10, however, the boronic acids convert to their anionic "ate" configuration. Thus at pH 10, boronic acid substituted quenchers such as the bipyridinium *o*-BBV^{2+} and phenanthrolinium *o*-$PBBV^{2+}$ would exist in their neutral, zwitterionic form. In this state they do not appear to possess a strong attraction for anionic HPTS as evidenced by the unchanged absorbance spectra during titration (Figure 3.11).

On the other hand, PV^{2+}, which is unsubstituted, should still have a net charge of 2+ at pH 10. Thus, even at high pH, it will maintain an electrostatic attraction for HPTS and, indeed, complex formation is strongly suggested by the UV-visible spectra. Association constants were determined for complex formation between quenchers and HPTS at each of the three pH levels studied. Association constants were calculated using Benesi-Hildebrand[53] plots for each derivative at the different pHs and assuming 1:1 binding stoichiometry (Table 3.1).

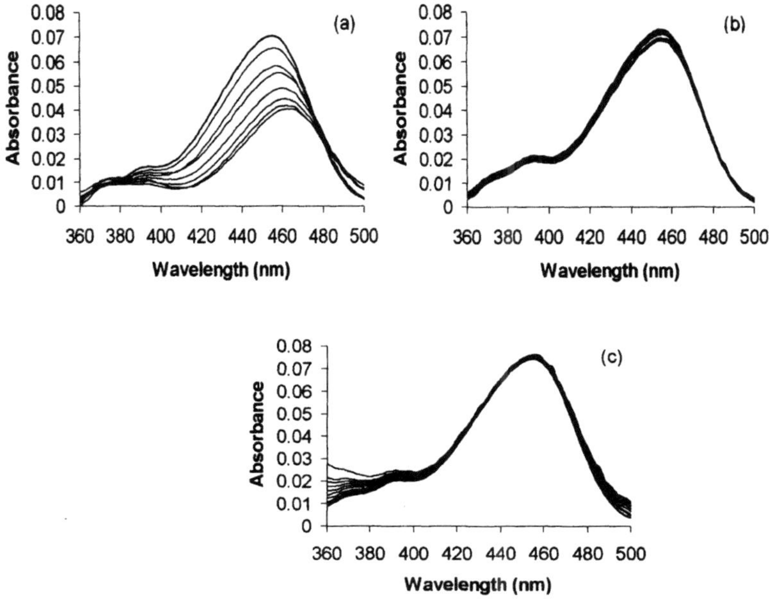

Figure 3.11. UV-visible absorbance spectra of HPTS (4 x 10^{-6} M) with increasing amounts of a) PV^{2+}, b) *o*-$PBBV^{2+}$, and c) *o*-BBV^{2+} in a pH 10 carbonate buffer.

Table 3.1. Association Constants (K_S) Determined for HPTS (4 x 10^{-6} M) complexed with PV^{2+}, *o*-$PBBV^{2+}$, and *o*-BBV^{2+} at different pH levels.

	pH 3	pH 7.4	pH 10
	$K_s(10^4\ M^{-1})$	$K_s(10^4\ M^{-1})$	$K_s(10^4\ M^{-1})$
***o*-BBV^{2+}**	4.9 ± 0.3	1.6 ± 0.2	0
***o*-$PBBV^{2+}$**	7.7 ± 0.5	3.1 ± 0.5	0
PV^{2+}	7.7 ± 0.6	6.0 ± 0.7	4.6 ± 0.8

We investigated the ability of PV^{2+} and *o*-$PBBV^{2+}$ to quench the fluorescence of HPTS under steady-state conditions. At pH 3 and 7.4, significant fluorescence quenching occurred for both compounds. At pH 10 a significant amount of quenching is apparent with PV^{2+}; however, this is not the case for the boronic acid-substituted derivatives, which show a large decrease in the quenching. Stern-Volmer plots of quenching at pH 7.4 and pH 10 are shown for comparison in Figure 3.12. The static and dynamic constants are summarized in Table 3.2.

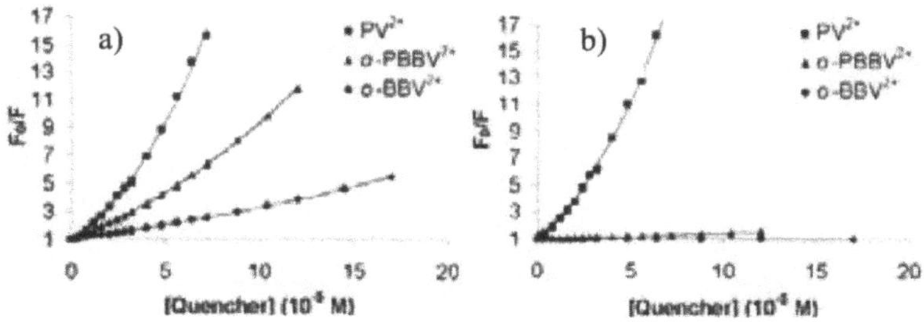

Figure 3.12. Stern-Volmer plots of HPTS (4 x 10^{-6} M) with increasing amounts of PV^{2+}, o-$PBBV^{2+}$, and o-BBV^{2+} at a) pH 7.4, and b) pH 10, λ_{em} = 510 nm.

Table 3.2. Static (K_s) and Dynamic (V) Quenching constants of the Viologens PV^{2+}, o-$PBBV^{2+}$, and o-BBV^{2+} with HPTS (4 x 10^{-6} M) in buffered 0.1 Ionic Strength Solutions at Different pH Levels. For complete results, including those at pH 3, see reference #10.

	pH 7.4		pH 10	
	K_s (10^4 M^{-1})	V (10^4 M^{-1})	K_s (10^4 M^{-1})	V (10^4 M^{-1})
***o*-BBV^{2+}**	1.6 ± 0	2.1 ± 0	0	0.076 ± 0.001
***o*-PBBV^{2+}**	4.9 ± 0	0.5 ± 0	0	0.4 ± 0.1
PV^{2+}	8.6 ± 0.5	1.1 ± 0	11 ± 0	1.0 ± 0.1

As described in the previous discussion, there is a loss of electrostatic attraction of these compounds for HPTS at high pH when boron develops a negative charge. This situation results in no static quenching and corresponds to what is observed in the UV-visible data, that is, the lack of complex formation. Because no shift in λ_{max} was observed in the excitation spectra of pyranine upon the addition of the quencher (data not shown), the fluorescence is solely due to the population of uncomplexed fluorophores.[26]

As predicted according to our mechanistic hypothesis, we found that the quenching efficiencies of the phenanthrolinium viologens were diminished by addition of even a small amount of glucose. Glucose was added to a dye solution to provide a concentration of 5 mM glucose, then the fluorescence was quenched with the phenanthrolinium compound. The quenching observed in the presence of glucose was then compared to that in the absence of glucose (Figure 3.13).

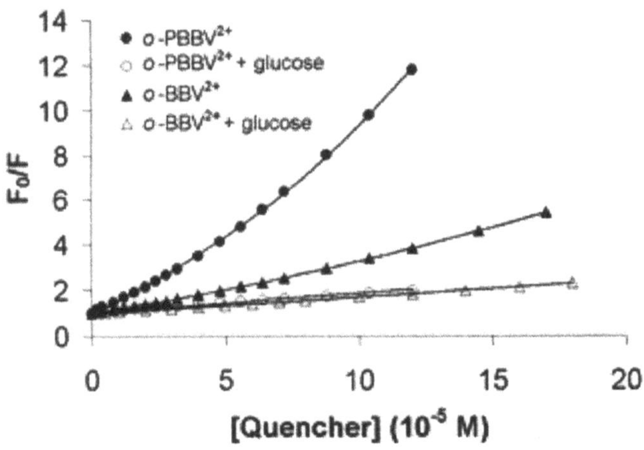

Figure 3.13. Stern-Volmer plot of HPTS (4 x 10^{-6} M) with *o*-PBBV^{2+} and *o*-BBV^{2+} with and without 5 mM glucose in a pH 7.4 phosphate buffer. λ_{ex} = 470 nm, λ_{em} = 510 nm.

Identical behavior had been observed with *o*-BBV^{2+}, but the difference in quenching efficiency between *o*-PBBV^{2+} with and without glucose present was considerably larger than in the case of *o*-BBV^{2+}. This led us to expect a large glucose response when using the phenanthrolinium compounds to measure changes in glucose solution.

Based on the observed glucose sensitivity of our 4,7-phenanthroline-derived viologen quencher, *o*-PBBV^{2+}, we had expected that it would provide a strong signal in response to changes in glucose concentration across the physiological range. To determine the effect of glucose on the quenching ability of viologens, titrations of HPTS were carried out with *o*-PBBV^{2+} and *o*-BBV^{2+} at pH 7.4 in the presence of glucose. In order to obtain maximum glucose sensitivity, however, we first had to determine the optimal dye-to-quencher ratio.

Early in our studies we had found that a potential benefit of the two-component system was the ability to vary the ratio of quencher-to-dye in order to optimize the magnitude of the sensing response. Initially, we arrived at an optimal quencher-to-dye ratio by comparing fluorescence emission of a quenched solution of HPTS in the absence of glucose with that observed in the presence of 5 mM glucose.[10] The difference between these two was then taken as a percentage of normal unquenched fluorescence emission of HPTS to give a "percent recovery".

Using this method, it was determined that maximum fluorescence recovery occurred for *o*-PBBV^{2+} at a quencher-to-dye ratio of 18:1. The Job-type plot used to make this determination plots quencher-to-dye ratio against fluorescence recovery of the original HPTS emission in the presence of 5 mM glucose (figure 3.14).

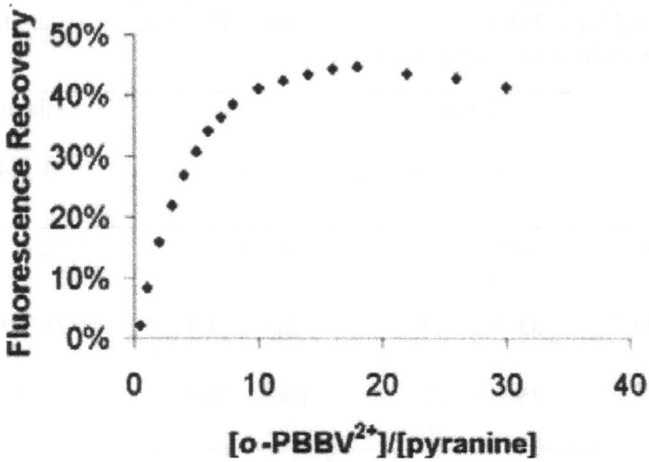

Figure 3.14. Percent fluorescence recovery of HPTS (4 x 10^{-6} M) in the presence of glucose (5 mM) upon the addition of different concentrations of *o*-$PBBV^{2+}$ in pH 7.4 buffer . λ_{ex} = 470 nm, λ_{em} = 510 nm.

Using the ratios determined from these studies, we carried out the monosaccharide binding studies. As suggested by the quencher-to-dye ratio study, the strongest signal was observed for *o*-$PBBV^{2+}$ which gave a large fluorescence recovery for glucose across the physiological range compared to the monocationic *o*-PBV^{+}. Binding curves from these studies are given in figure 3.15 and apparent saccharide binding constants from this data are given in table 3.3.

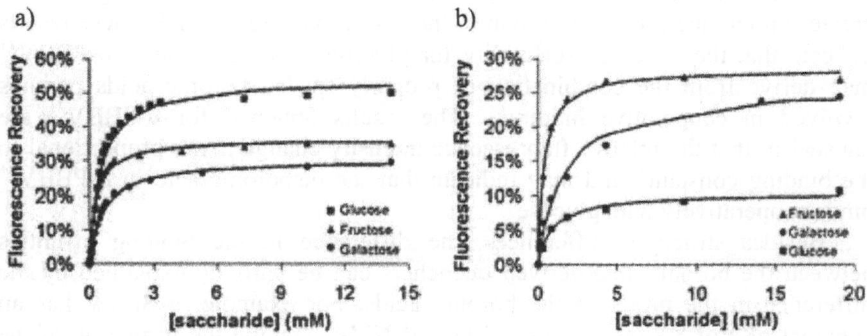

Figure 3.15. Percent fluorescence recovery of HPTS (4 x 10^{-6} M) in the presence of quencher and upon the addition of different monosaccharides: a) HPTS/*o*-$PBBV^{2+}$ and b) HPTS/*o*-PBV^{+} in pH 7.4 buffer. λ_{ex} = 470 nm, λ_{em} = 510 nm.

Table 3.3. Apparent monosaccharide binding/association constants (K_A) determined for *o*-PBBV^{2+}, and *o*-PBV^{+} from fluorescence data. Data from *o*-BBV^{2+} is included for comparison.

	Fructose	Glucose	Galactose
	K_A (M^{-1})	K_A (M^{-1})	K_A(M^{-1})
***o*-BBV^{2+}**	2600 ± 100	43 ± 3	167 ± 4
***o*-PBBV^{2+}**	3300 ± 200	1800 ± 200	1600 ± 100
***o*-PBV^{2+}**	2500 ± 200	1000 ± 200	600 ± 90

With regards to binding selectivity determined from the apparent binding constants, while *o*-BBV^{2+} demonstrated selectivity in the commonly observed binding order of fructose > galactose > glucose,[38, 41] the binding order for *o*-PBBV^{2+} is fructose > glucose > galactose. The binding constant of *o*-PBBV^{2+} for fructose is approximately only twice that of glucose. The structural differences between *o*-BBV^{2+} and *o*-PBBV^{2+} may play a part in their relative abilities to bind glucose. For example, as discussed earlier, the spacing between the bisboronic acids in *o*-PBBV^{2+} may lead to cooperative binding, where two boronic acids bind to one glucose molecule. To probe the binding mode of *o*-PBBV^{2+}, we carried out studies with the monoboronic acid derivative *o*-PBV^{+}. The binding constants of *o*-PBV^{+} were approximately half those of *o*-PBBV^{2+}, consistent with only possessing half the number of binding sites. The selectivity of *o*-PBV^{+} was found to be in the same order as *o*-PBBV^{2+}, though with a greater preference for fructose than in the bis-boronic compound. These results indicate that the observed selectivity for glucose over galactose in *o*-PBBV^{2+} may derive from the combination of properly spaced boronic acids perhaps involved in cooperative binding. The results obtained for *o*-PBBV^{2+} are unusual in that the relative fluorescence intensity change is not proportional to the binding constant[53] and may indicate that the bisboronic acid in *o*-PBBV^{2+} binds cooperatively with glucose.

Besides structural differences, the difference in the binding affinities between the boronic acid-derived quenchers can be partially explained by the difference in the p*K*a's of the boronic acids. For example, *o*-BBV^{2+} has an apparent p*K*a of 8.8, as determined by ^{11}B NMR, making the formation of the boronate complex with glucose unfavorable at pH 7.4. By comparison, the apparent p*K*a of *o*-PBBV^{2+} is 7.6, giving rise to a more favorable binding at pH 7.4 and, as a result, a larger binding constant. The effect of boronic acid pKa on monosaccharide binding is an area of active research for many workers in the field and our own studies on this important parameter are continuing.

The study of phenanthrolinium quenchers was one of the first to demonstrate the generality and versatility of our system. Encouraged by this study and our freedom to substitute one viologen for another, we focused on preparing a quencher for our sensing system that could be copolymerized or otherwise immobilized in a sensing polymer. We are ultimately interested in preparing a sensing system that can be immobilized in some biocompatible polymer matrix that is permeable to glucose. Thus, the synthesis of polymerizable boronic acid-substituted phenanthrolinium monomers was explored, however, synthesis of these compounds initially proved challenging. In contrast, the polymerizable analogs of the boronic acid-substituted bipyridinium compounds were found to be more readily obtained for use in our sensing systems. Before immobilizing any components in a polymer matrix, we ordinarily carry out solution studies first for comparison with other compounds. These solution studies are reported below, while the use of these compounds in polymeric form is described in section 3.5.

Polymeric viologens composed of 4,4′-dipyridinium units are widely reported in the literature. The viologen moieties have been incorporated into polymers as part of the main chain,[54-57] or as pendent groups,[58-64] and have also been incorporated as crosslinkers.[65] Previously it had been shown in our laboratory that boronic acid functionalized viologens with polymerizable groups could be prepared in a two-step process.[39] Thus, *m*-SBV^{2+}, the first polymerizable viologen monomer that was studied in detail, was prepared by treatment of 4,4′-dipyridyl with 4-vinylbenzyl chloride and subsequent reaction with 3-bromomethylphenylboronic acid.

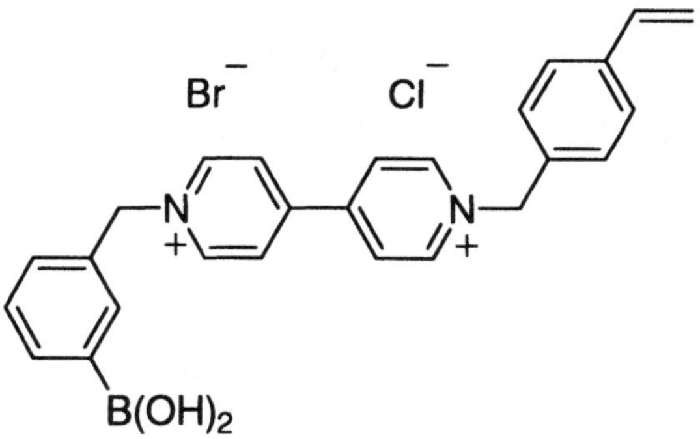

***m*-SBV^{2+}**

This compound contains all the properties it needs in order to function as a receptor-quencher. In addition it is polymerizable. However, it only contains one boronic acid, and our preliminary studies indicated that it had lower glucose sensitivity compared to the bisboronic acid substituted analog *m*-BBV^{2+}.[39] This difference in glucose sensing between mono- and bis-boronic acid substituted compounds was also obvious in the case of *o*-PBV^{+} and *o*-PBBV^{2+}, in which case the mono-substituted compound had a fluorescence recovery that was only

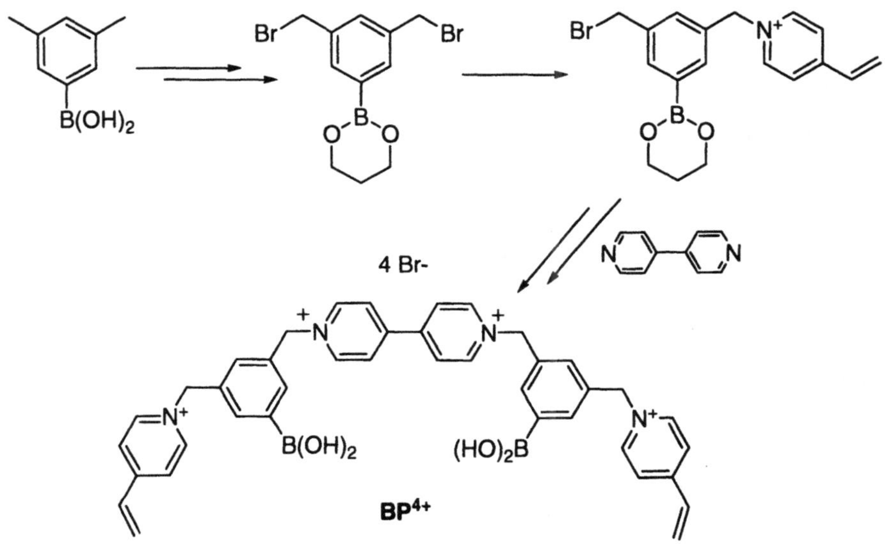

Scheme 3.6: Synthesis of BP^{4+}. Synthetic details are provided in reference #11.

~1/10 that of the bis-substituted compound upon addition of glucose. The weak response of *m*-SBV^{2+} inclined us towards pursuing a bis-boronic acid substituted bipyridinium quencher that could be immobilized in a hydrogel polymer. To this end, the bipyridinium called BP^{4+} was prepared (Scheme 3.6).

Stern-Volmer quenching studies and glucose response studies were carried out with BP4+ and HPTS in pH 7.4 solution and the results are given in Figure 3.16.

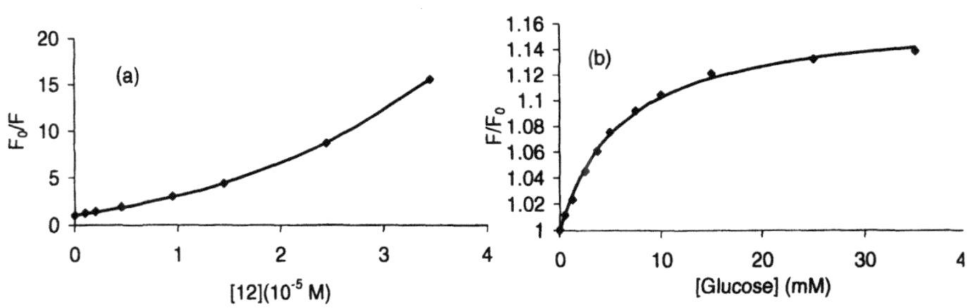

Figure 3.16: (a) Stern-Volmer Plot of HPTS (4×10^{-6} M) showing the quenching efficiency of BP^{4+} λ_{ex} = 470 nm and λ_{em}= 510 nm; (b) Relative fluorescence intensity of HPTS (4×10^{-6} M) with BP^{4+} (4×10^{-5} M) and increasing [glucose].

The Stern Volmer quenching constants for BP^{4+} were determined to be K_s = 130,000 M^{-1} and V = 31,000 M^{-1}, and the apparent glucose binding constant was K = 157 M^{-1} at a quencher-to-dye ratio of 10:1.[66] The quenching by this polymerizable bisboronic acid quencher, BP^{4+}, was quite impressive when compared to all the other quenchers studied thus far. The static quenching constant of 130,000 M^{-1} is nearly twice that of the phenanthrolinium *o*-$PBBV^{2+}$ and nearly ten times that of the bipyridinium *o*-BBV^{2+} (K_s = 16,000 M^{-1}). We believe that the simple addition of the two extra positive charges significantly enhances the electrostatic attraction of the quencher for the anionic dye. This provides for more efficient quenching since it renders a greater percentage of the fluorophores non-fluorescent through complexation. The results of the initial glucose sensing experiment with BP^{4+}, which was performed at the less than optimal quencher-to-dye ratio of 10:1, showed moderate glucose sensitivity with a signal stronger than that of *m*-SBV^{2+}, but considerably weaker than observed in *o*-$PBBV^{2+}$. Importantly, however, BP^{4+} provided us with a very satisfactory quencher monomer for immobilization in a sensing hydrogel polymer.

3.4. VARIATIONS IN FLUORESCENT REPORTER – SULFONAMIDE DERIVATIVES OF HPTS

While studies were being conducted on quenchers, parallel experiments were performed to optimize the fluorescent dye component. In our early studies all of the dyes used were derivatives of our model compound, HPTS. Since we were interested in immobilizing our anionic fluorescent component, many of our very earliest studies were directed towards methods for attaching polymerizable groups to the HPTS core without causing disadvantageous changes in the photophysical properties of the dye.

We soon found that simple addition of a polymerizable group, such as methacryl, to HPTS through functionalization of the phenol group would not suit our needs. The preparation of test compound HPTSAc by treatment of HPTS with acetic anhydride and a catalytic amount of NaOAc for 48 h gave the product in high yield.[11]

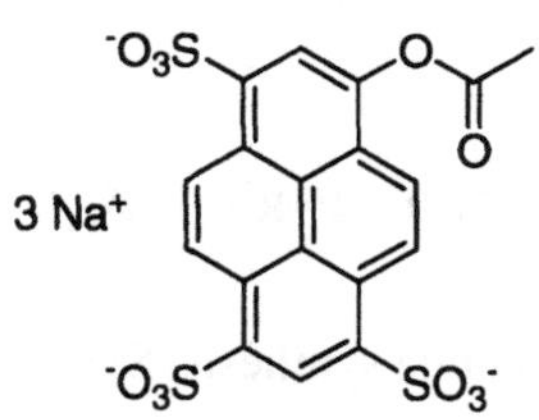

HPTSAc

When we tested the photophysical properties of this acetate, however, we found that substitution at the phenol had completely removed the absorbance band at 454 nm associated with the phenolate, or PO^- form. The behavior of HPTSAc was not entirely unexpected as the methoxy substituted analogue of HPTS, methoxypyrene trisulfonic acid (MPTS), also has its absorbance maximum at fairly short wavelength of 404 nm. This required us to use a shorter wavelength of 416 nm rather than the usual 460 nm for fluorescence excitation of HPTSAc. As with HPTS itself, HPTSAc emission still occurs at 510 nm.

Since we were interested in using a longer wavelength for emission we decided to instead pursue the installation of polymerizable groups through substitution at the sulfonic acid position. In an approach modeled on that of Wolfbeis,[33] we planned to first convert the acids to their corresponding sulfonyl chlorides and then to attach the polymerizable groups through conversion of the sulfonyl chlorides to sulfonamides. This strategy was used to prepare a number of model compounds and polymerizable analogs several of which are indicated in Scheme 3.7.

Though HPTS is commonly used in aqueous solution, a variety of techniques have been employed to immobilize it in one form or another. For example, it has been bonded to a water insoluble polyamine by direct reaction of the latter with a sulfonyl chloride intermediate.[36] HPTS also forms water insoluble complexes with cationic polyelectrolytes.[30] Finally, it can be rendered

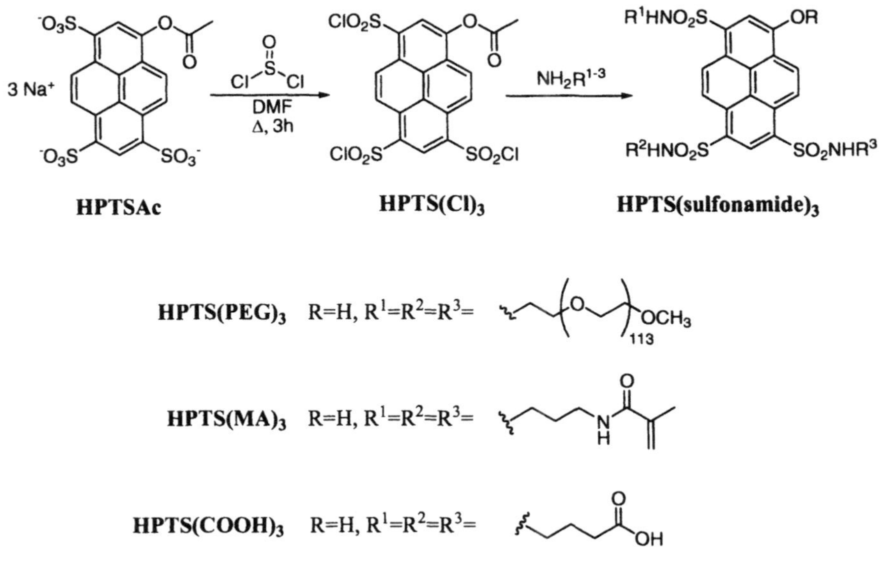

Scheme 3.7. HPTS trisulfonamide synthesis. Synthetic details are provided in reference #12.

water insoluble by amidation with a hydrophobic amine[33] or by forming a water insoluble salt with a hydrophobic quaternary ammonium counterion.[31]

Because we needed to incorporate both dye and quencher in specific ratios as well as insure complete conversion to a hydrophilic trisulfonamide that would be resistant to leaching, none of these procedures was deemed suitable for our purposes. We chose instead to make a soluble polyethylene glycol derivative, HPTS(PEG)$_3$, that could be immobilized in a semi-interpenetrating polymer network (semi-IPN). When HPTS(PEG)$_3$ was used in conjunction with *m*-BBV^{2+} in solution studies it showed a weak quenching interaction when compared to the interaction between *m*-BBV^{2+} and HPTS. This is reflected in the more than ten-fold difference in the Stern-Volmer static quenching constants from *m*-BBV^{2+} quenching HPTS(PEG)$_3$ (K_s=750 M^{-1}) versus the same viologen quenching simple HPTS (K_s=8700 M^{-1}).

HPTS(PEG)$_3$ was also tested for quenching and sugar sensing in combination with the phenanthrolinium compound *m*-PBBV^{2+} and its polymerizable monoboronic acid analogue *m*-PSBV^{2+}.[66]

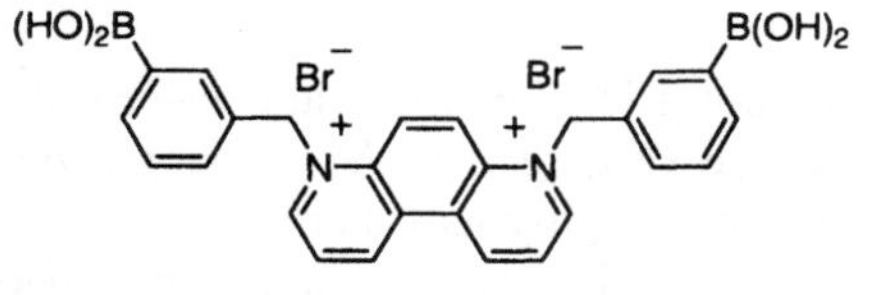

***m*-PBBV^{2+}**

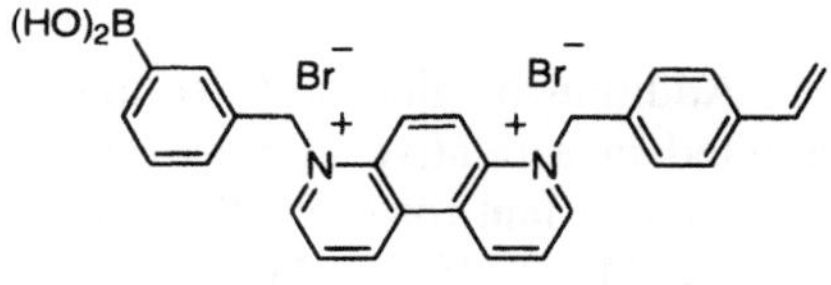

***m*-PSBV^{2+}**

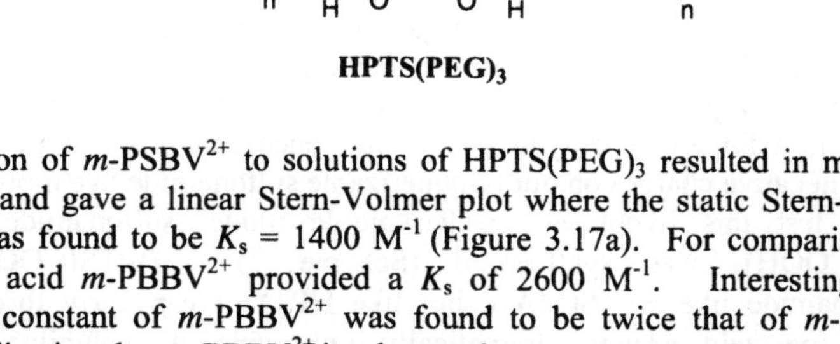

HPTS(PEG)$_3$

Addition of *m*-PSBV^{2+} to solutions of HPTS(PEG)$_3$ resulted in moderate quenching and gave a linear Stern-Volmer plot where the static Stern-Volmer constant was found to be K_s = 1400 M^{-1} (Figure 3.17a). For comparison, the bisboronic acid *m*-PBBV^{2+} provided a K_s of 2600 M^{-1}. Interestingly, the quenching constant of *m*-PBBV^{2+} was found to be twice that of *m*-PSBV^{2+} perhaps indicating that *m*-PBBV^{2+} is a better electron acceptor. Furthermore, no ground state complex was observed when HPTS(PEG)$_3$ was mixed with the different quenchers as determined by UV-visible measurements, indicating that the quenching occurs solely in the excited state (i.e., there is no static quenching). The failure of this non-ionic dye to form a complex is consistent with the charge induced sensing mechanism described in Scheme 3, though the polymeric nature of the dye may also be a factor.

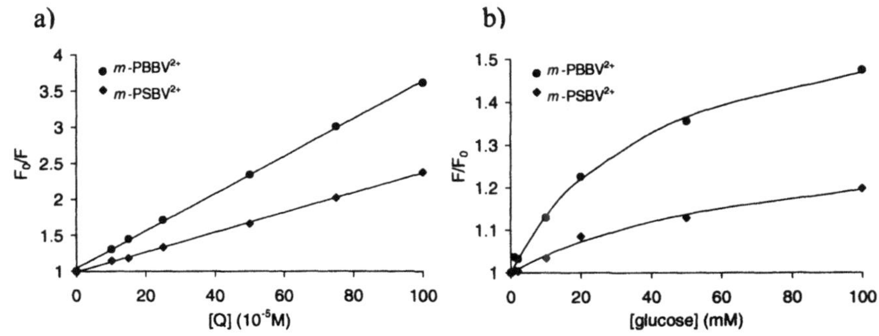

Figure 3.17. a) Stern-Volmer plot of HPTS(PEG)$_3$ with m-PBBV^{2+} and m-PSBV^{2+} in aqueous solution at pH 7.4.
b) Relative fluorescence change of HPTS(PEG)$_3$ (1×10^{-5} M) with m-PBBV^{2+} or m-PSBV^{2+} (100 mM) upon addition of glucose. $\lambda_{ex} = 470$ nm and $\lambda_{em} = 534$ nm.

Addition of glucose to solutions containing HPTS(PEG)$_3$ and m-PSBV^{2+} resulted in a relative increase in fluorescence (Figure 3.17 b). The glucose binding constant for m-PSBV^{2+} was found to be roughly half that of m-PBBV^{2+} (14 ± 3 M^{-1} v. 25 ± 3 M^{-1}, respectively). From this data it appears that two boronic acids are better than one in the receptor-quencher. Further work on the use of HPTS(PEG)$_3$ was abandoned when we observed that in some cases it appeared to be leaching out of IPN hydrogels in which we had hoped it had been immobilized.

We then shifted our attention to immobilization of the sensing components through covalent bonds in a hydrogel matrix. This required the preparation of appropriate monomers, then combining these monomers with polyethyleneglycol-dimethacrylate (PEGDMA) and hydroxyethylmethacrylate (HEMA) to form a hydrogel through radical polymerization. The HPTS derivative HPTS(MA)$_3$ was selected as the polymerizable dye. Unfortunately, many of our efforts to work with this compound in solution proved difficult as it possessed very poor water solubility. Although we later had limited success using it in a functional polymer, this experience convinced us of the need to include negative charges on our polymerizable sulfonamide dye monomers.

To test this hypothesis, water-soluble model sulfonamide derivative HPTS(COOH)$_3$ was synthesized (Scheme 7). HPTS(COOH)$_3$ is a trisulfonamide like HPTS(MA)$_3$, but like HPTS it can carry three negative charges on acid groups, dramatically improving its solubility in water. Quenching and glucose response studies with this compound were run in solution with m-BBV^{2+}. The results in Figure 3.18 a) show that with the anionic HPTS(COOH)$_3$ the quenching is more effective compared to HPTS(PEG)$_3$ and is in fact comparable to HPTS itself. The Stern Volmer quenching constants for this negatively charged sulfonamide in combination with m-BBV^{2+} were determined to be K_s = 12,500 M^{-1} and V = 660 M$^{-1.}$ In

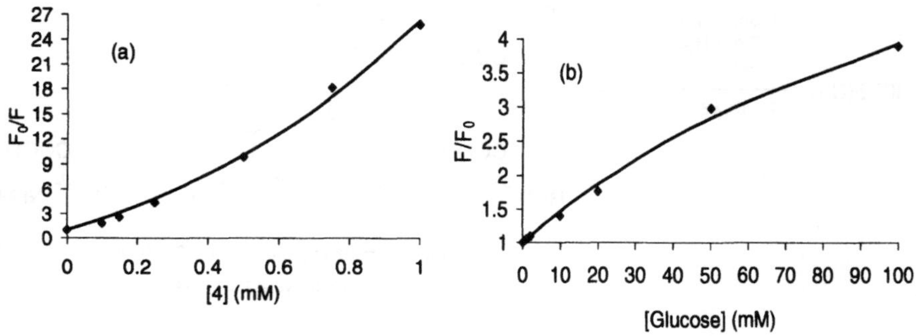

Figure 3.18. (a) Stern-Volmer Plot of HPTS(COOH)$_3$ (1×10^{-5} M) showing the quenching efficiency of *m*-BBV^{2+}, (b) Relative fluorescence intensity of HPTS(COOH)$_3$ (1×10^{-5} M) with *m*-BBV^{2+} (1 mM) and increasing [glucose]; λ_{ex} = 470 nm and λ_{em} = 515 nm.

addition, there is a broad range of glucose response, with a fairly linear response over the biologically significant range (Figure 3.18 b). The apparent glucose binding constant was $K = 7\ M^{-1}$.

These three test compounds suggested to us two absolutely essential characteristics for any dye that we wish to use in an immobilized system: These were:

a) The dye must bear anionic groups. These provide both the desired water solubility for non-PEG dyes and also provide for greater electrostatic attraction and quenching interaction with the cationic viologen.

b) The dye should have a polymerizable group such as vinyl, acryl or methacryl that can be used to immobilize it in a hydrogel matrix through free radical polymerization.

An HPTS derivative was envisioned that contained polymerizable groups and negative charges. It was desirable to utilize the sulfonamide chemistry that was already established and somehow incorporate negative charges into the molecule. The synthesis of a suitable dye molecule was successfully carried out in our laboratory and is illustrated in Scheme 3.7.

Formation of the sulfonamide was expected to lower the pKa of the dye's phenolic group. This is considered advantageous as it increases the fluorescence intensity by increasing the percentage of dye in its deprotonated form. This lowered pKa is also advantageous as it reduces the sensitivity to slight pH changes around the level of 7.4. Importantly, when they are directly attached to a fluorophore, the change from sulfonic acid to sulfonamide is also commonly accompanied by a red shift in both the absorbance and fluorescence emission maxima of the affected fluorophore. In the sulfonamide substituted compounds we prepared, we typically observed shifts of 10-30 nm for fully sulfonamidated derivatives.

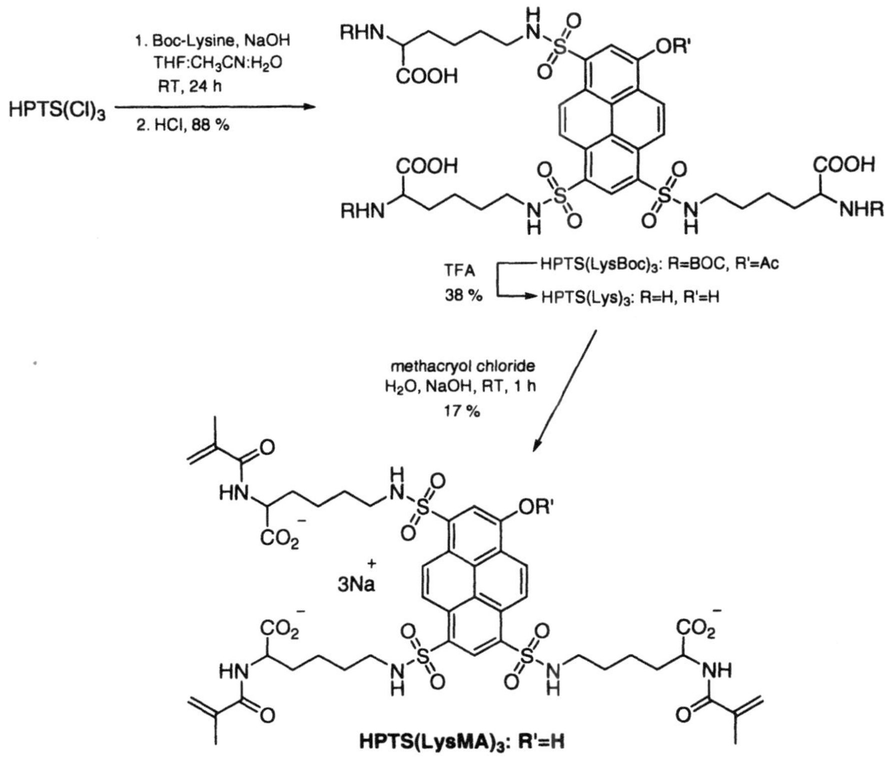

Scheme 3.7. Synthesis of HPTS(LysMA)$_3$. Synthetic details are given in reference #11.

For comparison, the absorption and emission spectra of the different HPTS derivatives are given in Figure 3.19. Relative to HPTS, the absorption and emission bands are red shifted for the trifunctionalized HPTS derivatives. HPTS displays absorption bands at 404 nm and 454 nm, whereas HPTS(PEG)$_3$ and HPTS(LysMA)$_3$ display only one band at 484 nm and 494 nm respectively. The band at 404 nm is due to the POH form of HPTS (Scheme 5) and at pH 7.4 is comparable in intensity to the PO^- band at 454 nm. HPTS(PEG)$_3$ and HPTS(LysMA)$_3$ display no POH band and apparently exist exclusively in the PO^- form at pH 7.4.

We chose to run similar solution studies with this new polymerizable dye and the polymerizable quencher BP^{4+} prior to running immobilization experiments. In contrast to HPTS(PEG)$_3$, HPTS(LysMA)$_3$ has negative charges.

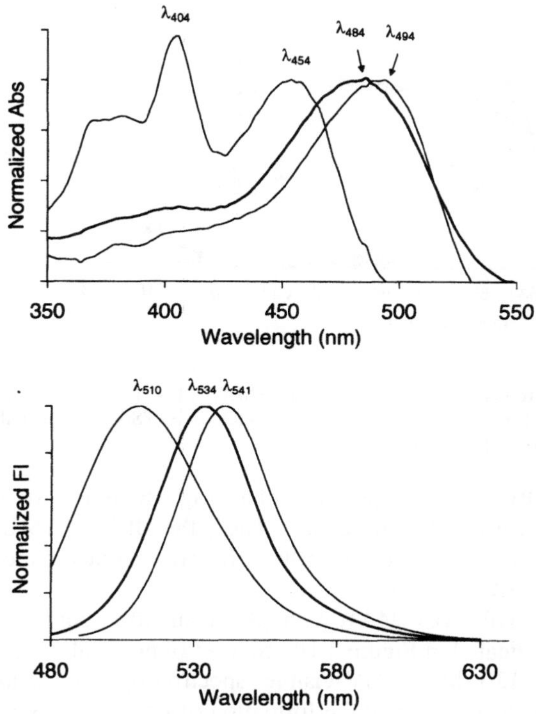

Figure 3.19. Absorbance (top) and emission (bottom) spectra of the different HPTS derivatives. HPTS has absorption maxima at 404 and 454 nm with emission at 510 nm. HPTS(PEG)3 and HPTS(LysMA)$_3$ have absorption maxima, respectively, at 484 and 494 nm, with emission at 534 and 541 nm.

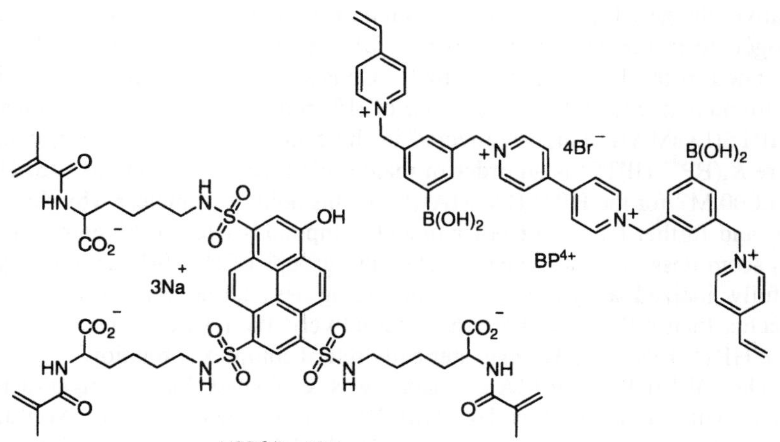

HPTS(LysMA)$_3$

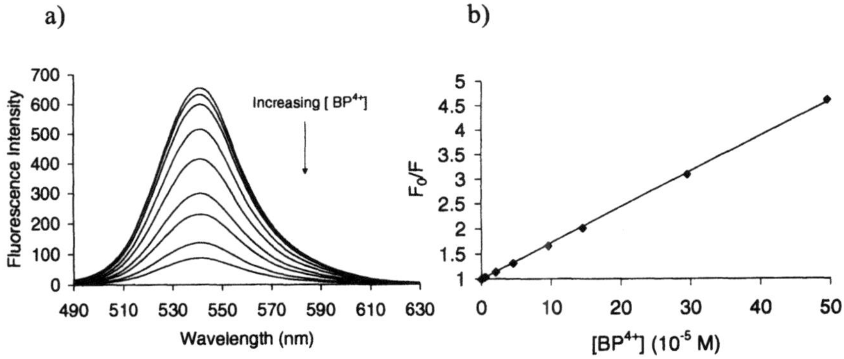

Figure 3.20. Fluorescence spectra of $HPTS(LysMA)_3$ (4×10^{-6} M) with increasing [BP^{4+}] in aqueous solution at pH 7.4 (left); Stern-Volmer plot of $HPTS(LysMA)_3$ with increasing [BP^{4+}] (right). λ_{ex} = 470 nm and λ_{em} = 540 nm.

In addition, BP4+ has four positive charges, which is two more than any other viologen we had tested so far. Thus, the $BP^{4+}/HPTS(LysMA)_3$ system was expected to show significantly different characteristics than the *m*-$PSBV^{2+}/HPTS(PEG)_3$ system.

Titration of $HPTS(LysMA)_3$ with BP^{4+} resulted in significant fluorescence quenching as indicated in Figure 3.10. Stern-Volmer analysis gives K_s = 12,000 M^{-1} and V = 130 M^{-1}. UV-visible spectroscopic measurements of this interaction also clearly showed the formation of a non-fluorescent quencher:dye complex. Obviously, the increased number of charges in both the dye and quencher molecule, relative to $HPTS(PEG)_3$ and *m*-$PSBV^{2+}$, dramatically changes the quenching.

Comparison of the BP^{4+}/HPTS system with $BP^{4+}/HPTS(LysMA)_3$ system reveals a marked difference in the dyes. Although both HPTS derivatives have negative charges, the electrostatic attraction between BP^{4+} and HPTS is much stronger than the attraction between BP^{4+} and HPTS-$(LysMA)_3$; this was evidenced in the UV-visible spectra.[66] Consistent with this observation, BP^{4+} was found to quenche the fluorescence of HPTS much more efficiently than that of $HPTS(LysMA)_3$. This is reflected in the Stern-Volmer quenching constant where $K_s(BP^{4+}/HPTS)$ is an order of magnitude greater (130,000 M^{-1} for HPTS v. 14,000 M^{-1} for the $HPTS(LysMA)_3$). At this point it is unclear why this is the case, and further studies of other model compounds need to be carried out. It does seem possible that the carboxylic acid groups of $HPTS(LysMA)_3$ may not be fully ionized at pH 7.4, thus providing less attraction for the cationic quencher than fully ionized HPTS. Alternatively, the lysine-methacryl subunits of the $HPTS(LysMA)_3$ dye may sterically retard complex formation.

The $BP^{4+}/HPTS(LysMA)_3$ system was evaluated for its sugar sensing ability. Titrating $HPTS(LysMA)_3$ with BP^{4+}, in the presence of 5 mM glucose, decreased the quenching efficiency of BP^{4+} giving $K_s = 7.2 \times 10^3$ M^{-1}. To illustrate the difference in quenching, the Stern-Volmer plots in the absence and presence of glucose are given in Figure 3.21.

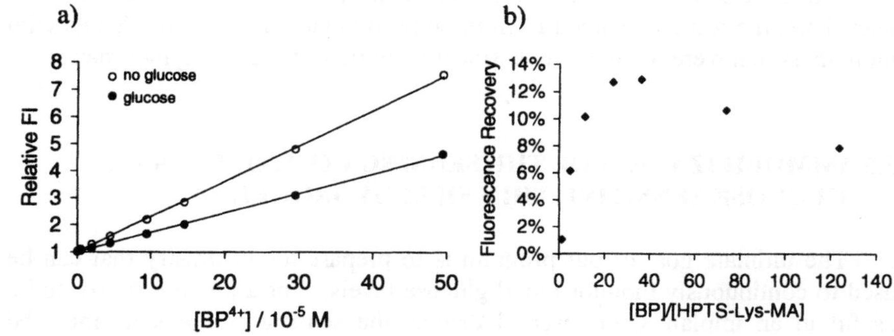

Figure 3.21. a) Stern-Volmer plots of the BP^{4+}/HPTS(LysMA)$_3$ system with and without 5 mM glucose in aqueous solution at pH 7.4; b) Quencher sensitivity plot showing the quencher:dye ratio that gives the optimum signal change for 5 mM glucose.

Figure 3.21 also illustrates the optimum ratio of BP^{4+}/HPTS(LysMA)$_3$ that gives the greatest fluorescence recovery upon addition of 5 mM glucose. As indicated, the curve reaches its maximum of 13 % recovery at a quencher/dye ratio of 30:1. This corresponds to a [BP^{4+}] of 1.2×10^{-4} M and an [HPTS(LysMA)$_3$] of 4.6×10^{-6} M.

Using this optimum quencher-to-dye ratio, the relative affinity of BP^{4+} towards different saccharides was then determined. Titration of solutions containing HPTS(LysMA)$_3$ and BP^{4+} were carried out with fructose, glucose, and galactose and binding constants for each sugar were calculated from the binding isotherms (Figure 3.22). The order of affinity was fructose (2100 ± 100 M^{-1}) > glucose (280 ± 20 M^{-1}) > galactose (210 ± 10 M^{-1}). Phenyl boronic acids normally show a 3:1 affinity for galactose over glucose; however, in this study BP^{4+} showed comparable binding to both glucose and galactose.

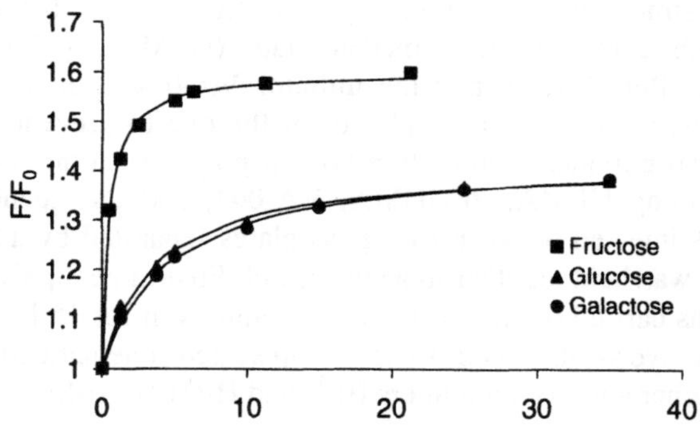

Figure 3.22. Relative fluorescence change of HPTS(LysMA)$_3$ (4×10^{-6} M) with BP^{4+} (1.2×10^{-4} M) upon addition of saccharide. λ_{ex} = 470 nm and λ_{em} = 540 nm.

While we continue to study variations in design and application of dyes and quenchers, the results obtained from these preliminary studies provided us with monomers that were ready for assessment in an immobilized polymer matrix.

3.5. IMMOBILIZATION OF THE SENSING COMPONENTS – A GLUCOSE SENSITIVE THIN FILM HYDROGEL

The ultimate goal of our program is to prepare the chemistry that can be used to continuously monitor blood glucose levels. For a glucose sensor to be useful in an implanted or inserted device, the sensing components must be immobilized to allow real-time monitoring. To that end, we had prepared monomers for immobilization in a thin film glucose permeable hydrogel. This approach met our needs with respect to material biocompatibility, ease of synthesis and evaluation. For *in vivo* use, a glucose monitoring device must operate at physiological temperature, ionic strength, and pH. Mimicking these physiological conditions, we tested several quencher-dye combinations in an immobilized monosaccharide sensing system before settling on boronic-acid-functionalized quencher BP^{4+} and anionic dye HPTS(LysMA)$_3$ as our best candidates. We found that this combination functioned well to provide continuous glucose sensing under physiological conditions.

The covalent attachment of the monomers BP^{4+} and HPTS(LysMA)$_3$ to a hydrogel polymer was effected through free radical polymerization. For sensing applications, the polymer used for component immobilization must be highly permeable to water and allow for the free movement of glucose in and out of the polymer matrix. Polymers that possess these types of properties are termed hydrogels. Hydrogels are polymers that swell substantially, but do not dissolve in water. They are generally cross-linked networks of hydrophilic homopolymers or copolymers. Depending on their chemical and physical structure, hydrogels have been classified as neutral, ionic, or interpenetrating networks (IPNs). Techniques for their preparation are well established[67-71] We relied on the simple method of copolymerization of our dye and quencher monomers with 2-hydroxyethyl methacrylate (HEMA), polyethyleneglycol dimethacrylate (PEG-DMA), and the initiator VA-044. In all of our early polymerizations, we protected the phenol of the dye to prevent its interfering with free radical polymerization. In a typical polymerization, a 50% aqueous solution containing HEMA, PEGDMA, VA-044, and the quencher and dye monomers was injected between two glass plates separated by a teflon spacer. This assembly was held together in a steel mold. Free-radical polymerization of the mixture was carried out at 40 °C in a vacuum oven for 15 h to produce the desired sensing hydrogel. This process is represented schematically in scheme 8 using the quencher and dye monomers BP^{4+} and HPTS(LysMA)$_3$.

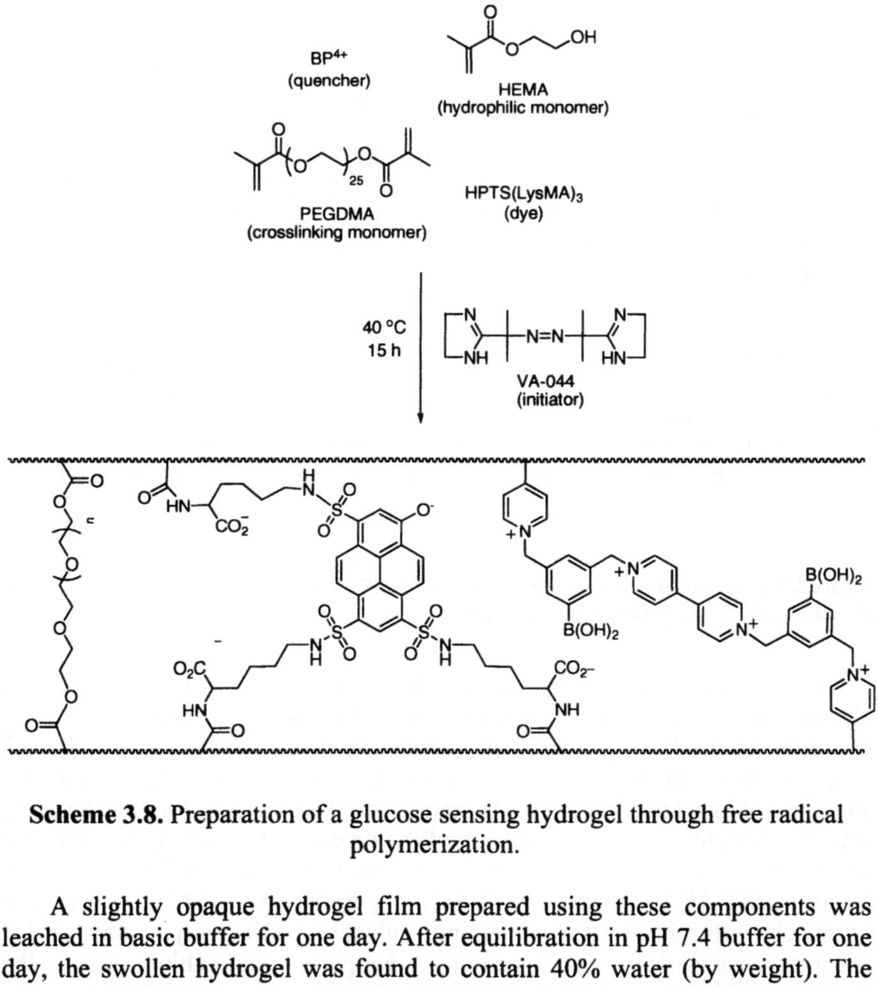

Scheme 3.8. Preparation of a glucose sensing hydrogel through free radical polymerization.

A slightly opaque hydrogel film prepared using these components was leached in basic buffer for one day. After equilibration in pH 7.4 buffer for one day, the swollen hydrogel was found to contain 40% water (by weight). The deprotected film was subsequently tested for its ability to continuously detect both increases and decreases in monosaccharide concentration. The hydrogel was mounted into a flow cell and phosphate buffer of ionic strength 0.1 M was circulated through the cell. The hydrogel film was excited at 470 nm by front-face illumination while the emission at 540 nm was monitored over time. The temperature was kept constant at 37 °C. After a stable baseline had been obtained, the buffer solution was replaced with saccharide solution and the change in fluorescence intensity was measured. As indicated in Figure 3.23, the infusion of glucose into the hydrogel prepared using BP^{4+} and $HPTS(LysMA)_3$ resulted in stepwise changes in fluorescence intensity that were dependent on sugar concentration. Importantly, the sensor detected glucose in the physiological range of 2.5–20 mM (45-360 mg/dL) and changes in fluorescence were completely reversible.[11]

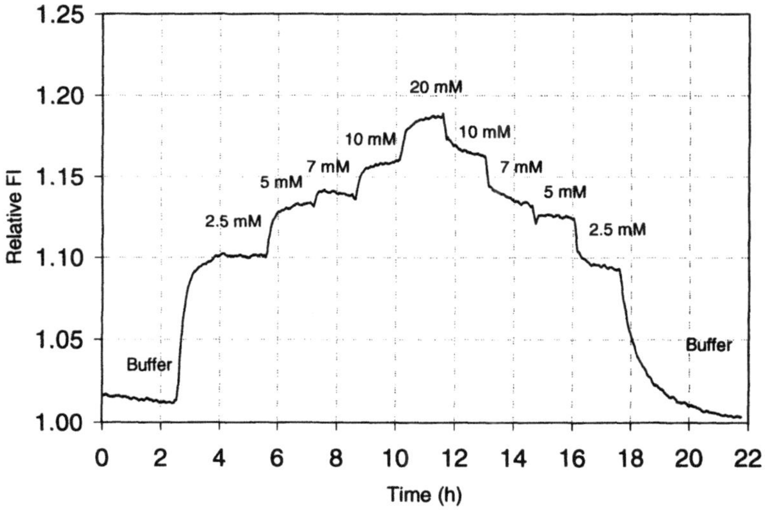

Figure 3.23. Continuous glucose sensing profile of a 0.002" hydrogel containing BP^{4+} and HPTS $(LysMA)_3$ at 37 °C. λ_{ex} = 470 nm and λ_{em} = 540 nm.

Using this hydrogel, similar profiles were obtained when measuring changes in concentration of fructose and galactose. The apparent binding constants of the hydrogel for each sugar were determined. Interestingly, the order of selectivity for these three saccharides changed once the components were immobilized in a polymer. The immobilized system was found to be more selective for glucose and less selective for fructose and galactose than the free components; the relative affinities in the hydrogel were determined to be: fructose (666 M^{-1}) > glucose (333 M^{-1}) > galactose (111 M^{-1}).[11] As we are ultimately interested in obtaining glucose selectivity, this 6:1:3 fructose:galactose:glucose selectivity compares favorably with the approximately 10:1:1 selectivity observed using the same monomers in aqueous solution.[66] Such differences in binding between solution and immobilized systems are another area of continuing research in our labs.

Depending on their method of preparation and degree of crosslinking, hydrogels can mimic the behavior of polymer chains in solution. For while long-range diffusion can not occur, a high degree of segmental mobility is still possible and at the local level functional groups attached to the polymer chains are free to move about and interact. This movement appears to occur in the hydrogels we have prepared by the method described here. Since the quenching mechanism appears to be predominately static, the dye and quencher units in the polymer chain must have the freedom to associate and dissociate within the polymer matrix depending on the local saccharide concentration. Through a reversible electrostatic interaction within its polymer matrix, the chemosensor depicted in Scheme 3.8 is able to detect glucose in the physiologically important range of 2.5-20 mM (45-360 mg/dL) while operating under physiological conditions (37 °C, 0.1 M ionic strength, pH 7.4). Importantly, the fluorescence signal obtained is completely reversible, which allows real-time monitoring of glucose levels. Further studies are underway to determine the scope of this approach in continuous glucose monitoring.

3.6. SUMMARY AND FUTURE DIRECTIONS

In summary, we have demonstrated a general monosaccharide sensing system using a range of fluorescent HPTS derivatives in combination with boronic acid-substituted viologens. The system functions at pH 7.4 and is sensitive to changes over the physiologically important concentration range of 2.5-20 mM (45-360 mg/dL). Mechanistically, in the absence of saccharide, a non-fluorescent ground-state complex forms through ionic association between the dye and the viologen boronic acid, resulting in a greatly reduced fluorescence. The electrostatic attraction between the dye and the quencher is lost, however, upon the binding of the viologen to monosaccharide causing the dye-quencher complex to dissociate. This produces a fluorescence recovery that is directly dependent on the saccharide concentration.

In order to improve performance parameters such as sensitivity and selectivity and to optimize monomers for use in an immobilized sensing system, we carried out solution studies of various dye-quencher combinations. Our studies suggest that maintaining negative charges on the dye component is critical both for good aqueous solubility and to provide for electrostatic attraction to the viologen quencher. We also found utility in the use of sulfonamide derivatives as our fluorescent dyes. These compounds provided longer wavelengths for excitation and emission and this chemistry offered us a good degree of synthetic leeway in preparing HPTS(LysMA)$_3$, a dye with both negative charges and polymerizable groups,.

In our studies of various quenchers we found that the ability to strongly quench the fluorescence of anionic dyes was a general property of all the viologen derivatives studied. Further, this characteristic appears to be general for other related compounds such as the phenanthrolinium salts. With regard to monosaccharide selectivity, we were unable to achieve glucose selectivity in the bipyridinium compounds simply by altering the substitution pattern of the boronic acids. The phenanthrolinium salts such as *o*-PBBV^{2+}, however, did provide a selectivity for glucose that was considerably enhanced over that observed for simple phenylboronic acids. Additionally, we found that more charges on the viologen quenchers improved quenching ability considerably and produced a more tightly bound non-fluorescent ground-state complex with the dyes. In particular, the quadruply charged polymerizable viologen BP^{4+} demonstrated very strong quenching and moderate glucose sensitivity.

Using various dye and quencher monomers, we immobilized our monosaccharide sensing system in thin film hydrogels and assessed them for continuous glucose monitoring. The saccharide-sensitive hydrogels we were able to prepare could detect glucose in the physiologically important range of 2.5-20 mM (45-360 mg/dL) and operate under physiological conditions (37 °C, 0.1 M ionic strength, pH 7.4). Importantly, the fluorescence signal obtained was demonstrated to be completely reversible, thus allowing for real-time monitoring of glucose levels.

We are currently pursuing research in a number of areas related to sensor biocompatibility, signal sensitivity, and monosaccharide selectivity. The generality of this two-component sensing system and the ease with which different quenchers and fluorescent dyes can be introduced allows for a wide range of system configurations to be considered. We are actively investigating the use of new dye-quencher combinations for use in glucose sensing hydrogels and are exploring the application of this two-component sensing system towards other analytes.

3.7. REFERENCES

1. For a recent review see: T. D. James, and S. Shinkai, in: *Host-Guest Chemistry,* vol. 218, edited by S. Penadés. (Springer-Verlag, Heidelberg, 2002), pp. 159-200.
2. J. Yoon, and A. W. Czarnik, Fluorescent Chemosensors of carbohydrates - a means of chemically communicating the binding of polyols in water based on chelation-enhanced quenching, *J. Am. Chem. Soc.* **114**(14), 5874-5875 (1992).
3. H. Suenaga, H. Yamamoto, and S. Shinkai, Screening of boronic acids for strong inhibition of the hydrolytic activity of alpha-chymotrypsin and for sugar sensing associated with a large fluorescence change, *Pure Appl. Chem.* **68**(11), 2179-2186 (1996).
4. H. Suenaga, M. Mikami, K. Sandanayake, and S. Shinkai, Screening of fluorescent boronic acids for sugar sensing which show a large fluorescence change, *Tetrahedron Lett.* **36**(27), 4825-4828 (1995).
5. T. D. James, K. Sandanayake, and S. Shinkai, A glucose-selective molecular fluorescence sensor, *Angew. Chem., Int. Ed. Engl.* **33**(21), 2207-2209 (1994).
6. T. D. James, K. Sandanayake, R. Iguchi, and S. Shinkai, Novel saccharide-photoinduced electron-transfer sensors based on the interaction of boronic acid and amine, *J. Am Chem. Soc.* **117**(35), 8982-8987 (1995).
7. H. S. Cao, and M. D. Heagy, Fluorescent chemosensors for carbohydrates: A decade's worth of bright spies for saccharides in review, *J. Fluoresc.* **14**(5), 569-584 (2004).
8. H. Fang, G. Kaur, and B. H. Wang, Progress in boronic acid-based fluorescent glucose sensors, *J. Fluoresc.* **14**(5), 481-489 (2004).
9. J. N. Camara, J. T. Suri, F. E. Cappuccio, R. A. Wessling, and B. Singaram, Boronic acid substituted viologen based optical sugar sensors: modulated quenching with viologen as a method for monosaccharide detection, *Tetrahedron Lett.* **43**(7), 1139-1141 (2002).
10. J. T. Suri, D. B. Cordes, F. E. Cappuccio, R. A. Wessling, and B. Singaram, Monosaccharide detection with 4,7-phenanthrolinium salts: Charge-induced fluorescence sensing, *Langmuir* **19**(12), 5145-5152 (2003).
11. J. T. Suri, D. B. Cordes, F. E. Cappuccio, R. A. Wessling, and B. Singaram, Continuous glucose sensing with a fluorescent thin-film hydrogel, *Angew. Chem., Int. Ed.* **42**(47), 5857-5859 (2003).
12. F. E. Cappuccio, J. T. Suri, D. B. Cordes, R. A. Wessling, and B. Singaram, Evaluation of pyranine derivatives in boronic acid based saccharide sensing: Significance of charge interaction between dye and quencher in solution and hydrogel, *J. Fluoresc.* **14**(5), 521-533 (2004).
13. P. M. S. Monk, *The Viologens: physicochemical properties, synthesis, and applications of the salts of 4,4'-bipyridine* (John Wiley & Sons, New York, 1998).
14. L. A. Summers, *The Bipyridinium Herbicides* (Academic Press, London, 1980).
15. L. A. Summers, *Adv. Heterocyc. Chem.* **35**, 281-374 (1984).

16. A. J. Bard, A. Ledwith, and H. J. Shine, *Adv. Phys. Org. Chem.* **13**, 155-278 (1976).
17. C. L. Bird, and A. T. Kuhn, *Chem. Soc. Rev.* **10**(1), 49-82 (1981).
18. International Union of Pure and Applied Chemistry, Compendium of Chemical Terminology. http://www.iupac.org/publications/compendium; (accessed January 2005).
19. E. B. de Borba, C. L. C. Amaral, M. J. Politi, R. Villalobos, and M. S. Baptista, Photophysical and photochemical properties of pyranine/methyl viologen complexes in solution and in supramolecular aggregates: A switchable complex, *Langmuir* **16**(14), 5900-5907 (2000).
20. K. Nakashima, and N. Kido, Fluorescence quenching of 1-pyrenemethanol by methylviologen in polystyrene latex dispersions, *Photochem. Photobiol.* **64**(2), 296-302 (1996).
21. Z. G. Zhao, T. Shen, and H. J. Xu, Photoinduced interaction between eosin and viologen, *J. Photochem. Photobiol., A* **52**(1), 47-53 (1990).
22. B. S. Gaylord, S. J. Wang, A. J. Heeger, and G. C. Bazan, Water-soluble conjugated oligomers: Effect of chain length and aggregation on photoluminescence-quenching efficiencies, *J. Am Chem. Soc.* **123**(26), 6417-6418 (2001).
23. D. L. Wang, X. Gong, P. S. Heeger, F. Rininsland, G. C. Bazan, and A. J. Heeger, Biosensors from conjugated polyelectrolyte complexes, *Proc. Natl. Acad. Sci. U.S.A.* **99**(1), 49-53 (2002).
24. L. H. Chen, D. W. McBranch, H. L. Wang, R. Helgeson, F. Wudl, and D. G. Whitten, Highly sensitive biological and chemical sensors based on reversible fluorescence quenching in a conjugated polymer, *Proc. Natl. Acad. Sci. U.S.A.* **96**(22), 12287-12292 (1999).
25. N. DiCesare, M. R. Pinto, K. S. Schanze, and J. R. Lakowicz, Saccharide detection based on the amplified fluorescence quenching of a water-soluble poly(phenylene ethynylene) by a boronic acid functionalized benzyl viologen derivative, *Langmuir* **18**(21), 7785-7787 (2002).
26. J. R. Lakowicz, *Principles of Fluorescence Spectroscopy* (Kluwer Academic/Plenum Publishers, New York, 1999).
27. D. L. Wang, J. Wang, D. Moses, G. C. Bazan, and A. J. Heeger, Photoluminescence quenching of conjugated macromolecules by bipyridinium derivatives in aqueous media: Charge dependence, *Langmuir* **17**(4), 1262-1266 (2001).
28. K. Kano, and J. H. Fendler, Pyranine as a sensitive pH probe for liposome interiors and surfaces - pH gradients across phospholipid vesicles, *Biochim. Biophys. Acta* **509**(2), 289-299 (1978).
29. H. R. Kermis, Y. Kostov, P. Harms, and G. Rao, Dual excitation ratiometric fluorescent pH sensor for noninvasive bioprocess monitoring: Development and application, *Biotechnol. Prog.* **18**(5), 1047-1053 (2002).
30. M. F. Choi, Spectroscopic behaviour of 8-hydroxy-1,3,6-pyrenetrisulphonate immobilized in ethyl cellulose, *J. Photochem. Photobiol., A* **104**(1-3), 207-212 (1997).
31. Y. Yilmaz, Fluorescence study on the phase transition of hydrogen-bonding gels, *Physical Review E* **66**(5) (2002).
32. C. Bankert, S. Hahn, and H. Hui Optical fiber pH microsensor and method of manufacture. US Patent # 5,378,432, January 3, 1995.
33. G. J. Mohr, T. Werner, and O. S. Wolfbeis, Application of a novel lipophilized fluorescent dye in an optical nitrate sensor, *J. Fluoresc.* **5**(2), 135-138 (1995).
34. O. S. Wolfbeis, E. Furlinger, H. Kroneis, and H. Marsoner, Fluorimetric Analysis .1. A study on fluorescent indicators for measuring near neutral (physiological) pH-values, *Fresenius Zeitschrift Fur Analytische Chemie* **314**(2), 119-124 (1983).
35. H. Offenbacher, O. S. Wolfbeis, and E. Furlinger, Fluorescence optical sensors for continuous determination of near-neutral pH values, *Sensors and Actuators* **9**(1), 73-84 (1986).
36. S. G. Schulman, S. X. Chen, F. L. Bai, M. J. P. Leiner, L. Weis, and O. S. Wolfbeis, Dependence of the fluorescence of immobilized 1-hydroxypyrene-3,6,8-trisulfonate on solution pH - extension of the range of applicability of a pH fluorosensor, *Anal. Chim. Acta* **304**(2), 165-170 (1995).
37. S. E. Clarke, C. C. Wamser, and H. E. Bell, Aqueous complexation equilibria of meso-tetrakis(4-carboxyphenyl)porphyrin with viologens: Evidence for 1 : 1 and 1 : 2 complexes and induced porphyrin dimerization, *J. Phys. Chem. A* **106**(13), 3235-3242 (2002).
38. J. P. Lorand, and J. O. Edwards, Polyol complexes and structure of the benzeneboronate ion, *J. Org. Chem.* **24**(769-774 (1959).
39. J. N. Camara (2003). The application of boronic acid functionalized viologens in the quantitative determination of polyhydroxy compound concentrations under physiological conditions, University of California at Santa Cruz, Santa Cruz, CA.
40. I. M. Frank, and S. I. Vavilov, *Z. Phys. Chem. (Munich)* **69**, 100 (1931).

41. G. Springsteen, and B. H. Wang, A detailed examination of boronic acid-diol complexation, *Tetrahedron* **58**(26), 5291-5300 (2002).
42. D. B. Cordes, S. Gamsey, Z. Sharrett, A. Miller, P. Thoniyot, R. A. Wessling, and B. Singaram, Fluorescence sensing of monosaccharides with boronic acid substituted bipyridinium quenchers and pyranine: the effects of quencher charge, *Submitted for Publication* (2005).
43. D. B. Cordes, A. Miller, S. Gamsey, Z. Sharrett, P. Thoniyot, R. A. Wessling, and B. Singaram, Optical glucose detection across the visible spectrum using anionic fluorescent dyes and a viologen quencher in a two-component saccharide sensing system, *Submitted for Publication* (2004).
44. S. Arimori, M. L. Bell, C. S. Oh, K. A. Frimat, and T. D. James, Modular fluorescence sensors for saccharides, *Chem. Commun.* **18**, 1836-1837 (2001).
45. S. Arimori, S. Ushiroda, L. M. Peter, A. T. A. Jenkins, and T. D. James, A modular electrochemical sensor for saccharides, *Chem. Commun.* **20**, 2368-2369 (2002).
46. H. Eggert, J. Frederiksen, C. Morin, and J. C. Norrild, A new glucose-selective fluorescent bisboronic acid. First report of strong alpha-furanose complexation in aqueous solution at physiological pH, *J. Org. Chem.* **64**(11), 3846-3852 (1999).
47. V. V. Karnati, X. M. Gao, S. H. Gao, W. Q. Yang, W. J. Ni, S. Sankar, and B. H. Wang, A glucose-selective fluorescence sensor based on boronic acid-diol recognition, *Bioorg. Med. Chem. Lett.* **12**(23), 3373-3377 (2002).
48. T. D. James, K. Sandanayake, and S. Shinkai, Chiral Discrimination of Monosaccharides Using a Fluorescent Molecular Sensor, *Nature* **374**(6520), 345-347 (1995).
49. W. Yang, H. He, and D. G. Drueckhammer, Computer-guided design in molecular recognition: Design and synthesis of a glucopyranose receptor, *Angew. Chem., Int. Ed.* **40**(9), 1714-1718 (2001).
50. S. Huenig, J. Gross, E. F. Lier, and H. Quast, *Justus Liebigs Ann. Chem.* **2**, 339-358 (1973).
51. T. Ihara, J. Takata, and M. Takagi, Novel method of detecting mismatch on DNA by using photo-induced electron transfer through DNA, *Anal. Chim. Acta* **365**(1-3), 49-54 (1998).
52. J. J. Politi, and A. M. C. Fernandez, *J. Photochem. Photobiol., A* **104**(1-3), 165-172 (1997).
53. K. A. Connors, *Binding Constants: The Measurement of Complex Stability* (John Wiley & Sons, Inc., New York, 1987).
54. M. S. Simon, and P. T. Moore, Novel Polyviologens - Photochromic redox polymers with film-forming properties, *J. Polym. Sci., Part A: Polym. Chem.* **13**(1), 1-16 (1975).
55. J. B. Schlenoff, D. Laurent, H. Ly, and J. Stepp, Redox-active polyelectrolyte multilayers, *Adv. Mater. (Weinheim, Ger.)* **10**, 347-349 (1998).
56. K. Ito, and H. Ohno, Polyether/salt hybrid .5. Phase and bulk electrochemical response of viologens having poly(ethylene oxide) chain, *Polymer*. **38**(4), 921-926 (1997).
57. T. Endo, A. Kameyama, Y. Mambu, Y. Kashi, and M. Okawara, Synthesis of polyethers containing viologen moiety and their application to electron-transfer catalyst, *J. Polym. Sci., Part A: Polym. Chem.* **28**(9), 2509-2516 (1990).
58. K. Ageishi, T. Endo, and M. Okawara, Reduction of aldehydes and ketones by sodium dithionite using viologens as electron-transfer catalyst, *J. Polym. Sci., Part A: Polym. Chem.* **21**(1), 175-181 (1983).
59. I. Druta, E. Avram, and V. Cozan, Polymers with pendent functional groups IV. The reaction of chloromethylated polystyrene with N-phenacyl-4,4'-bipyridinium bromides, *Eur. Polym. J.* **36**(1), 221-224 (1999).
60. M. Okawara, T. Hirose, and N. Kamiya, Preparation of polymers containing viologen structure and their redox behavior, *J. Polym. Sci., Part A: Polym. Chem.* **17**(3), 927-930 (1979).
61. Y. Nambu, K. Yamamoto, and T. Endo, Selective formation of polymer-bound monomeric and dimeric viologen cation radicals by choice of pendant group linkage, *J. Chem. Soc. Chem. Commun.* **7**, 574-576 (1986).
62. H. Kamogawa, and S. Amemiya, Organic-solid photochromism by photoreduction mechanism - polar aprotic viologen copolymers, *J. Polym. Sci., Part A: Polym. Chem.* **23**(9), 2413-2423 (1985).
63. A. Kameyama, Y. Nambu, and T. Endo, Electron relay in polymer systems - effective electron-transfer from dihydronicotinamide to viologen moieties in their copolymers, *J. Polym. Sci., Part A: Polym. Chem.* **30**(6), 1199-1202 (1992).

64. F. T. Liu, X. D. Yu, and S. B. Li, Structural characterization of ethyl cellulose and its use for preparing polymers with viologen moieties, *J. Polym. Sci., Part A: Polym. Chem.* **31**(13), 3245-3249 (1993).
65. Y. Saotome, T. Endo, and M. Okawara, Syntheses of Some Novel Polymers Containing the Viologen Structure and Their Application to Electron-Transfer Reactions in Heterophases, *Macromolecules* **16**(6), 881-884 (1983).
66. J. T. Suri (2003). Monosaccharide detection with boronic acid substituted viologens and fluorescent dyes: towards the *in vivo* sensing of glucose, University of California at Santa Cruz, Santa Cruz.
67. A. M. Usmani, and N. Akmal, *Diagnostic Biosensor Polymers* (American Chemical Society, Washington, D.C., 1993).
68. J. P. Montheard, M. Chatzopoulos, and D. Chappard, 2-Hydroxyethyl methacrylate (HEMA): chemical properties and applications in biomedical fields, *J. Macromol. Sci., Rev. Macromol. Chem. Phys.* **C32**(1), 1-34 (1992).
69. A. Mathur, S. K. Moorjani, and A. B. Scranton, Methods for Synthesis of Hydrogel Networks: A Review, *J. Macromol. Sci., Rev. Macromol. Chem. Phys.* **C36**(2), 405-430 (1996).
70. P. G. Edelman, and J. Wang, *Biosensors and Chemical Sensors: Optimizing Performance through Polymeric Materials* (American Chemical Society, Washington, D.C., 1992).
71. N. Akmal, and A. M. Usmani, *Polymers in Sensors, Theory and Practice* (American Chemical Society, Washington, D.C., 1998).

IMPLANTABLE CONCANAVLIN A BASED SENSORS FOR INTERSTITIAL FLUID GLUCOSE SENSING IN DIABETICS

Bennett L. Ibey, Michael V. Pishko, and Gerard L. Cote'*

4.1. INTRODUCTION

4.1.1. Diabetes

Currently, 6.2% of the United States population has been diagnosed with diabetes[1]. Of the estimated 17 million diabetics, 90% of them are insulin resistant[2]. Diabetes is commonly split into two categories depending on the cause of the disease. Type 1 diabetes, traditionally termed juvenile, is marked by a failure of the B-cells in the endocrine portion of the pancreas to properly synthesize insulin resulting in improper control of blood glucose levels. Type 2 diabetes, traditionally classified as late onset, develops when organs and cells in the body become immune to native insulin resulting in dramatic oscillations in blood sugar[2-4]. Type 1 diabetics regulate their blood sugar by either consuming a glucose rich food, if blood sugar is low (< 70 mg/dL) or intravenously injecting insulin if blood sugar is too high (> 150 mg/dL)[1]. Type 2 diabetics, being resistant to insulin therapy, must regulate their blood sugar through dietary control and through the aid of commercially available drug treatments[1]. All diabetics are recommended to monitor their blood sugar often to avoid immediate threats attributed to high or low blood sugar such as dizziness, coma, or even death[5]. New research has shown that continuous monitoring of blood sugar with tight blood glucose control (as close to normal as possible, 70-150 mg/dL) can substantially reduce the risk of developing afflictions associated with prolonged misregulation of blood sugar[5]. These afflictions include kidney,

* Bennett L. Ibey and Gerard L. Coté, Department of Biomedical Engineering, Texas A&M University, College Station, Texas 77843. Michael V. Pishko, Department of Chemical Engineering, The Pennsylvania State University, University Park, Pennsylvania, 16802.

liver, and heart disease, amputations due to poor circulation and blindness due to retinopathy[5].

4.1.2. Commercially Available Sensors

The commercial blood glucose sensors for diabetics are based on the enzymes glucose oxidase or PQQ-dependent glucose dehydrogenase in combination with electrochemical mediators[6,7]. In the presence of glucose, these enzymes oxidize glucose, transferring the electrons to the mediators. The reduced mediators are then oxidized at an electrode surface, producing a current or charge proportional to the amount of glucose within the blood volume[6,7]. These sensors typically have an error of prediction of less than 10 mg/dL blood glucose concentration[7]. They however, have the disadvantage of requiring blood from the patient through either a forearm or finger-prick resulting in pain to the patient and also a potential risk of infection[8-11]. The pain associated with and patient involvement required to use these sensors along with the cost of disposables restricts the use of these sensors for near continuous or frequent blood glucose monitoring[9,12]. Newer sensors based on reverse iontophoresis (electrical stimulation of the epidermis resulting in leakage of interstitial fluid) have been shown to be less accurate than traditional direct sensing approaches, but have been approved for trend monitoring of blood glucose by the FDA[13-14]. Many studies have shown that the interstitial fluid (ISF), while not containing the same amount of glucose as the blood, correlates well with a time lag of approximately 5 minutes. This lag is due to the physiological separation between blood and ISF and the associated mass transfer of glucose between the two compartments[13-14]. This technology has been used on dogs, cats and horses and shown to perform well and alleviate the stress caused from using traditional blood glucose devices[14]. It has also been used widely on humans, but reported to cause skin irritation due to repeated electrical stimulation resulting in inability to wear the monitor over long periods of time[15-16]. This sensor requires roughly 20 minutes to acquire a single reading; as such the measurement is an average over this acquisition time. It also requires calibration, is only capable of being used for up to 12 hours, and results in significant skin irritation with repeated use[15-16].

4.1.3. Other Approaches

4.1.3.1. Non-invasive Modalities

Sensing approaches are being investigated that are based on absorption of infrared radiation and rotation of polarized light by glucose because they possess the potential to be completely non-invasive, which could result in painless continuous glucose monitoring. Glucose, like all molecules, absorbs light within the mid and near infrared wavelength range in specific regions due to chemical composition[17-24]. This trait can be very beneficial because unlike previously described sensors this absorption allows for direct measurement of glucose based on inherent properties of the substance. Implementation of this technology has proven difficult due to heavy water absorption within the mid-infrared regime and confounding absorptions from other molecules within the near infrared regime requiring complex statistical analysis[25]. Polarized light has been used for over a century to measure the sugar concentration in cane syrup by taking advantage of the chiral nature of glucose[26-28]. Chiral substances such as glucose rotate plane polarized light as a function of their concentration. Using this phenomenon, researchers have attempted to couple polarized light across the aqueous humor of the eye, known to contain glucose concentration proportional to the blood with a time lag[29]. This technology has been shown to work *in vitro* for biological concentrations of glucose and also in a rabbit model, but remains very sensitive to motion artifact[28].

4.1.3.2. Minimally Invasive Modalities

Use of a minimally invasive technology for determination of blood glucose has been the pursuit of many research groups. McShane *et al.* are investigating the implantation of a polymeric sensor based upon monolayer self assembly that uses glucose oxidase similarly to the commercial blood based sensor. This sensor contains glucose oxidase immobilized onto a polymer coated nanoparticle which, when implanted into the dermal layer of the skin, will react with native glucose and generate hydrogen peroxide. This reaction consumes oxygen to cause an increase in fluorescent emission of an embedded ruthenium organometallic complex known to be quenched in the presence of oxygen. This technology has been shown to work *in vitro*, but requires not only the presence of oxygen, but also consumes glucose which may result in ambiguous readings due to reduction in local glucose concentration around the sensor[30]. A second technology that shows promise in sensing glucose is a hydrogel contact lens which has been altered to contain a special molecule called boronic acid. Boronic acid is a compound which binds glucose reversibly and through use of fluorescent probes or photonic crystals, this binding can be monitored and indirectly correlated to glucose concentration[31-35]. This sensor has targeted the measurement of glucose in tears but they are an order of magnitude smaller than that of blood, and have yet to be proven to correlate to blood glucose, especially when tearing is induced. This sensor has also been shown to work *in vitro*, but again it is unclear whether this sensor will work *in vivo* where eye hydration, environment, and other confounders can affect results[36-40].

Multiple research groups are investigating the use of fluorescent monitoring of glucose lectin binding, specifically Concanavalin A (Con A)[41-43]. Fluorescent monitoring of a binding reaction is typically based on either a quenching reaction or a fluorescence resonance energy transfer (FRET) reaction resulting in both a change in intensity and lifetime of the emission. Intensity monitoring has the advantage of being very straight forward requiring limited technology, but is vulnerable to system artifacts due to the inherent sensitivity of fluorescent based detection systems[42-43]. Lifetime monitoring is advantageous in that it is independent of intensity variations and therefore less likely to be confounded by systemic artifacts. However, this technology requires either a sophisticated light source or complex electrical components to measure lifetimes on the order of nanoseconds in length[44-46].

One potential technology for blood glucose monitoring based on fluorescent intensity detection is an implantable chemical assay based on a competitive binding reaction between the protein Con A, dextran, and glucose. The envisioned sensor would be implanted superficially into the skin tissue and an optical probe would be designed to noninvasively monitor the implanted-sensor response to change in interstitial glucose concentration known to be comparable to that of the blood[47,48]. This chapter will focus on the description and explanation of Con A based glucose sensors for use in implantable blood glucose monitoring devices for diabetics. A description of Con A, the glucose binding assay, previous sensing modalities, and current technologies will be discussed.

4.2. CONCANAVALIN A AND DEXTRAN

Concanavalin A is a protein that was originally isolated from the Jack bean by Sumner and Howell in 1935[49]. Extensive research has been performed that uses Con A specifically for its ability to agglutinate various biologically relevant complexes such as erythrocytes, glycoproteins, and starches. For biologists, physiologists, and chemists, this functionality has enabled Con A to be used for various applications from purifying compounds containing carbohydrate molecules to fluorescently tagging cellular components[49]. This section will describe the structure, purification, stability, and practical uses of Con A as it applies to biological and chemical systems.

4.2.1. Structure

Con A is a unique protein because it does not contain any nucleic acids, carbohydrate, or lipid components, but is essentially made up only of 237 amino acids resulting in a molecular weight of 25,500 Da[49-50]. Each monomer protein measures 42 X 39 Å across and 42 Å high and is predominately composed of two large β sheets. A simplified version of a Con A tetramer is shown below in Figure 4.1. Two metal binding sites are present on each monomer (separated by 4.6 Å) along with a single sugar binding site (reported to be 10 Å from the Mn^{2+} binding site). The first metal site (M1) can bind transition metals, most commonly Mn^{2+}, and the second site (M2) binds Ca^{2+} preferentially. Studies have shown that the presence of Mn^{2+} at the M1 binding site is required for binding of Ca^{2+} to the M2 binding site[49-50].

It has been hypothesized that metal binding alters the geometric structure of the sugar binding site supporting evidence which concluded that metals are required for sugar binding by Con A.The sugar binding site is approximately 3.5 to 6 Å wide and 7 Å high. It has been shown that this site binds well with 5 and

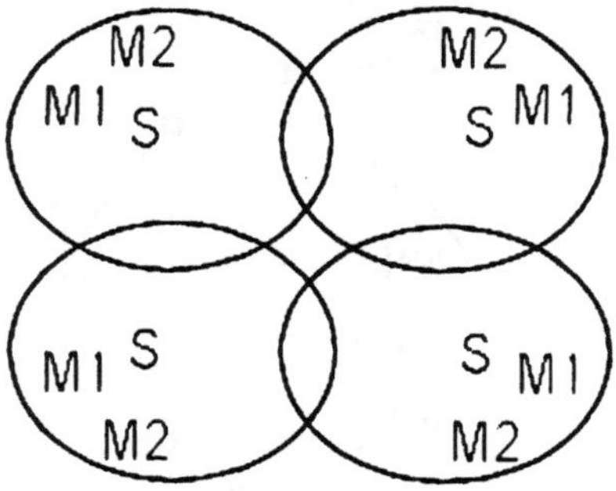

Figure 4.1. Simplified schematic of a Con A tetramer showing the relative placement and order of the three primary binding structures. Each of the monomer subunits binds metals and sugar independent of the other three[49].

Table 4.1. Association constants of sugar to Con A [51]

Sugar	K (L/mol)
Methyl α-D-mannopyranoside[a]	20,600
Methyl α-D-glucopyranoside[a]	4,940
D- fructose[a]	1,370
D-glucose[b]	400
70, 000 MW dextran[b]	15,000

[a] pH=6.2 at 2°C
[b] pH=7.2 at 27°C

6 carbon sugars which contain the required equilateral hydroxyl groups at carbons C3 and C4 and an unmodified hydroxymethyl group at C6. The four monomer binding sites have also been proven to be independent of each other meaning that sugar binding to one site will not affect the sugar binding characteristics of adjacent monomers. In Table 4.1, the relative binding affinity for common saccharide molecules is presented. The structure of glucose is shown below with the appropriate labels to demonstrate the required chemical structure for binding (Figure 4.2)[49-50].

4.2.2. Purification

Con A was first purified by Sumner and Howell in 1936 by crystallization[49]. This process was abandoned in 1965 when the use of cross linked dextrans as a means of purification through affinity binding was discovered by Agrawal and Goldstein[49]. By using the inherent saccharide binding characteristics of Con A towards sugar-like molecules, a means for separation of the protein from the jack bean was devised. This procedure used cross–linked dextrans (sephadex) beads to extract the Con A from the macerated bean components by passing the solution through a column, allowing Con A to bind to the sephadex material. such as mannose or glucose. The simple sugar would then be removed through

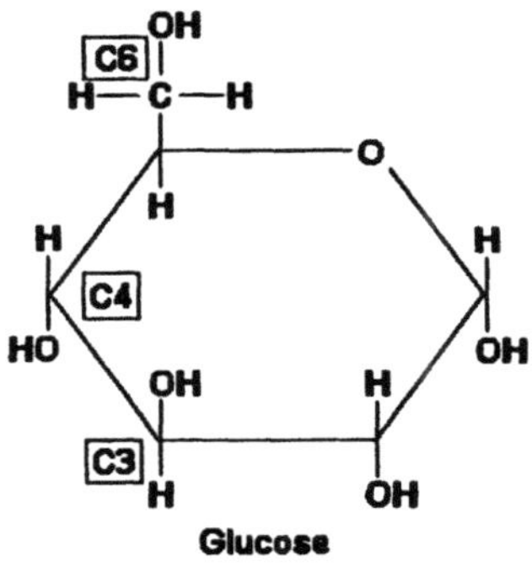

Figure 4.2. Structural diagram of glucose illustrated the structure necessary for binding to Con A protein. The C3, C4, and C6 have been shown to require this specific configuration to bind to the Con A sugar binding site.

dialysis or gel filtration[51]. The concentration of the eluted protein could be determined by measuring ultraviolet light absorption at 280 nm[51].

4.2.3. Stability

In its native form Con A is a tetramer consisting of two dimers which have partially overlapped and created a series of weak hydrogen bonds. The stability of the tetramer form heavily depends upon temperature and pH due to weakness of these bonds. In temperatures between 20-40° C and pH between 6-8 it has been shown that the protein prefers a tetramer state[49,52]. At low temperatures (0-4° C) and lower pH's (4-6) the protein is predominantly in dimer formation due to the breaking of the weak hydrogen bounds between the overlapping dimers. At temperatures above 50° C the protein denatures resulting in a complete loss of binding capability. The ionic strength of the environment surrounding the protein has also been shown to have a stabilizing affect on the protein at strengths greater than 0.3M[49,52].

Con A in native solution (pH 7, 27°C) is dominantly in a tetramer state and is susceptible to aggregation leading to eventual precipitation. To stabilize the Con A protein in solution, modifications of the e-amino group by succinylation have been shown to inhibit the formation of tetramers thus greatly reducing aggregation[53,54]. It has also been shown that this process does not inhibit the sugar binding characteristics of the protein, specifically its ability to bind sugars and to sephadex. However, in a study by Meadows *et al.* it was shown that succinylation did not reduce the rate or prevalence of inactivation of the sugar binding capabilities of Con A. Aliquots of purified and succinylated Con A were stored for 7 days at 4, 25, and 37 degrees C and the percent deactivated, still active, and precipitated protein was measured. A reduction in binding activity of 60% was seen for both pure and succinylated Con A. The most pronounced difference between the two forms was a substantially limited amount of aggregation in the succinylated Con A[54].

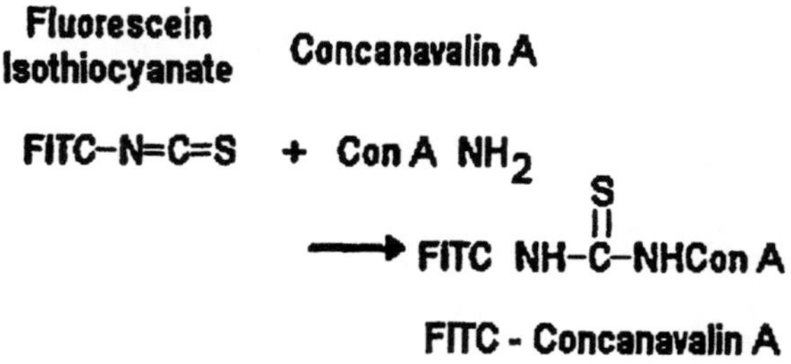

Figure 4.3. Fluorescein isothiocyanate binding reaction with Con A amino group forming FITC-ConA. The isothiocyanate compound reacts with the amino group of the target molecule (Con A) resulting in the formation of a thiourea group[56].

4.2.4. Traditional uses of Con A

Con A has been widely used in biological and chemical research. In microscopy, Con A is used as a target molecule for attaching dyes and markers to specific regions of cells and sub-cellular components[55]. Con A has been used as a linker molecule to bind markers to glycoproteins within the cell membrane. Fluorescent tags, specifically fluorescein isothiocyanate (FITC) are also routinely attached to Con A allowing for fluorescent detection of cell membranes, micro-projections in lymphocytes, and other surface structures. Fluorescent isothiocyanate molecules are attached to amino groups contained within amino acids through the formation of a thiourea compound as depicted below in Figure 4.3[49,55,56]. This process is performed by mixing proteins and isothiocyanate complexes at a basic pH (9). Con A has also been used as an indirect measurement of cell structures and surface chemistry through titrations in conjunction with anti-Con A antibodies resulting in numerous immunological and pharmacological applications[49,56].

Immobilization of Con A onto agarose has been investigated for use in separation of molecules containing sugar residues. Agarose beads (Sepharose) have been used to immobilize Con A through reaction using cyanogens bromide which allows Con A molecules to bind to the agarose particles[49]. Another method for immobilization is performed by treating a polyacrylamide surface with nitrous acid enabling amino groups within the Con A protein to bind to the surface. This process was reported to have better stability and less non-specific saccharide binding as compared to the sepharose based beads. Both of these methods create compounds which can be loaded into chromatography columns or gels and used to isolate saccharide molecules such as glycoproteins and starches giving separation scientists a powerful tool[49].

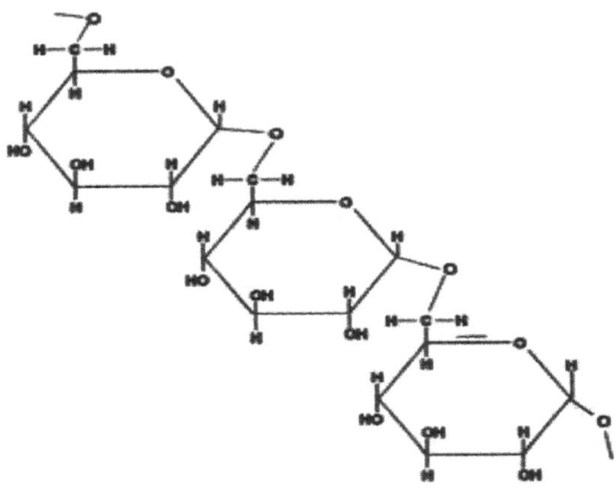

Figure 4.4. Molecular structure of dextran showing polysaccharide chain. This figure illustrates three saccharide connected as part of a large dextran chain. Dextran molecules range in size from 15 to 1000 monosaccharides in length[57].

4.2.5. Dextran

Dextran is a hydrophilic polysaccharide chain made up of α-D-1,6-glucose molecules and a large portion of D-glucopyransoyl residues. The molecular weight can vary from 3 to 2000 kDa and the degree of branching can range from 5% to 50%.[56] Dextrans are commonly used in biological applications such as tracing neuronal projections to track cell lineage following mitosis due to its inherently low toxicity and resistance to cleavage. For these applications, dextran is labeled with a fluorophore by reacting amino dextran molecules with isothiocyanate based fluorescent molecules through a process analogous to the previously described protein labeling. The chemical structure of dextran is shown below in Figure 4.4[57].

4.3. FLUORESCENT BASED ASSAY

4.3.1. Fluorescence Resonance Energy Transfer

Fluorescence is the process in which an excited molecule releases energy in the form of photon energy. Typically, the molecule is raised to an excited state via the absorption of a high energy photon[58-59]. The process is illustrated below in a Jablonski diagram (Figure 4.5). In this diagram, an incident photon is absorbed by the molecule via stimulated absorption and raised to an excited energy state (S_o). Through intramolecular vibrational relaxation, the molecule relaxes down to a lower energy excited state (S_1). Energy is then released through spontaneous emission in the form of a lower energy photon (longer wavelength) returning the molecule to its unexcited state[58-59].

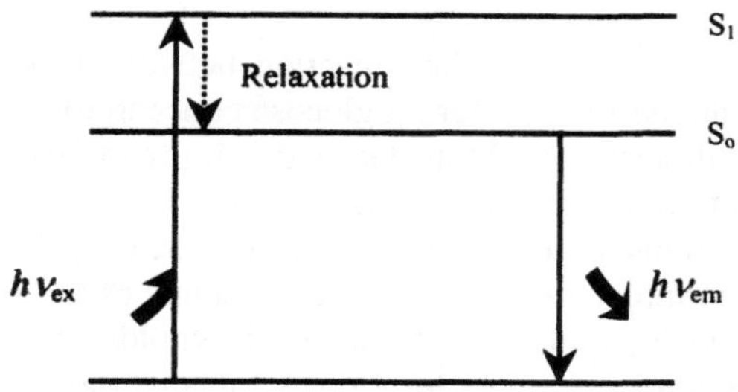

Figure 4.5. Jablonski diagram illustrating spontaneous emission from a fluorescent molecule. Light is absorbed via stimulated absorption raising the molecule to an excited state (S1). The molecule then relaxation via vibrational relaxation to a lower excited state (S0) and fully relaxes through a spontaneous emission of a photon.

Two fluorescent dyes contained within a small volume can act independently by absorbing photon energy and emitting fluorescent photons. However, if the two fluorophores have overlapping absorption and emission spectra and are chemically restrained within 100Å of each other, a special non-radiative transfer of absorbed photon energy can occur. This transfer of energy is called Fluorescence Resonance Energy Transfer (FRET) and is depicted below in a Jablonski diagram (Figure 4.6). In this reaction an incident photon is absorbed by the "donor" fluorophore, defined as the fluorophore with a higher energy emission spectrum, resulting in excitement of that molecule to a high energy state. Following relaxation, energy is transferred non-radiatively to an "acceptor" molecule and released as photon energy within the emission range of the acceptor. A FRET reaction can occur between two identical (homogenous) or two different (heterogeneous) fluorescent molecules[59-61].

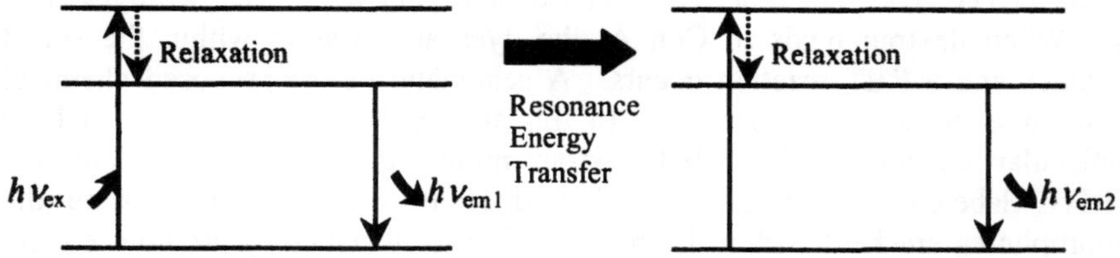

Figure 4.6. Above is a simplified Jablonski which illustrates the energy pathways taken during a FRET reaction. A stimulated absorption occurs followed by vibrational relaxation and nonradiative transfer of energy to a neighboring fluorophore. The neighboring fluorophore then emits a photon within its emission spectra to fully relax.

4.3.2. FRET Based Assay For Glucose Detection.

The competitive binding reaction between Con A and FITC labeled dextran as a method for monitoring glucose concentrations within the interstitial fluid was first described by Mansouri and Schultz in 1979[51]. A formula for the proposed reaction is shown in Equation 1. It can be seen that these two competing reactions are reversible and depend heavily on the concentrations of each saccharide molecule. The binding constants of the following reaction can be calculated using Equations 2 and 3. Reynolds *et al.* used this theory to generate association constants for Con A towards dextran (15,000 L/mol) and glucose (400 L/mol) using Con A sepharose[51]. It can be seen that the interaction between the Con A and dextran is much stronger than the interaction between glucose and Con A. This large difference has been hypothesized to be due to multivalent binding of the Con A to the large dextran molecules[51].

$$\begin{aligned} &ConA + gluc \leftrightarrow ConA - gluc \\ &ConA + FITCdextran \leftrightarrow ConA - FITCdextran \end{aligned} \tag{1}$$

$$\frac{[ConA][gluc]}{[ConA - gluc]} = K_G \tag{2}$$

$$\frac{[ConA][dextran]}{[ConA - dextran]} = K_D \tag{3}$$

The most common assay for glucose sensing using Con A is based on a FRET reaction between fluorescein isothiocyanate (FITC) labeled dextran and tetramethyl-rhodamine isothiocyanate (TRITC) labeled Con A. The absorbance and emission spectra for FITC and TRITC are shown in Figure 4.7[62]. It can be seen that the emission of the FITC dye heavily overlaps the absorption spectrum of the TRITC dye. The reported Förster distance between these two dyes is 54 Å. When dextran binds to Con A, the dyes are brought within the stated distance and a RET reaction occurs. A schematic of the proposed chemical reaction is shown in Figure 4.8. In this figure, a representation of a high molecular weight FITC labeled dextran molecule is shown with multiple TRITC labeled Con A tetramers bound to it. In the bound state the fluorophores are held within the required Forster distance to promote energy transfer. Following introduction of glucose into the system, the large dextran molecule is displaced resulting in a increase in the separation distance. Furthermore, there is a loss of energy transfer between the two dye molecules. Figure 4.9 shows sample spectra illustrating the FRET phenomena by plotting the FITC dextran with the addition of TRITC Con A, and the assay with

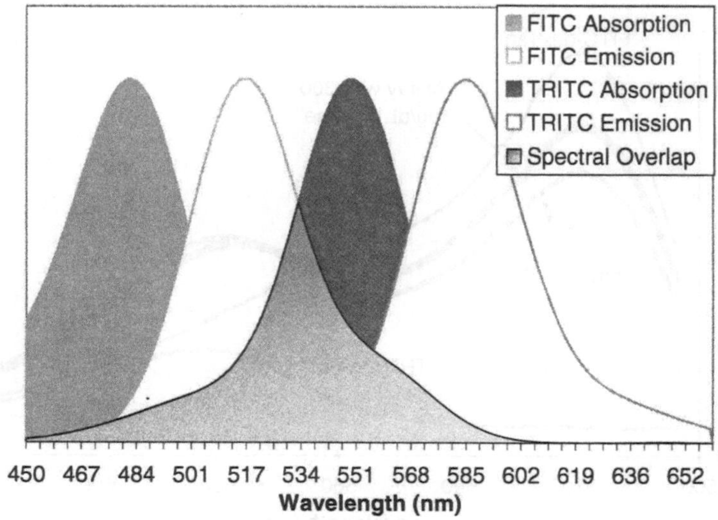

Figure 4.7. Absorption and Emission Spectra of FITC and TRITC. It can been seen from this figure that the emission spectra of FITC dye heavily absorbs the absorption spectra of TRITC meaning that this pair has potential to exhibit a FRET reaction[62].

addition of glucose to the solution. As would be expected from a FRET reaction, the fluorescent emission from the FITC dextran is reduced while the TRITC Con A emission increases upon creation of the assay. Upon introduction of glucose this transfer is disrupted and the FITC emission is increased, while the emission of the TRITC dye is reduced.[42-43]

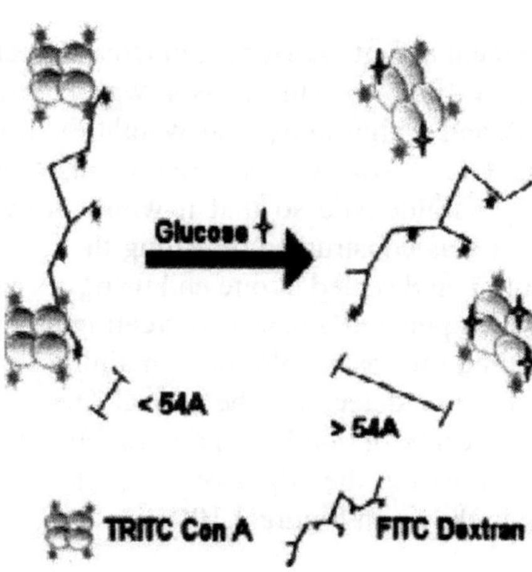

Figure 4.8. FRET reaction between FITC labeled dextran and TRITC Labeled Con A. In the absence of glucose the Con A molecules are bound to the long dextran chain and FRET occurs due close proximity between the dyes. In the presence of glucose the dextran is displaced resulting in a loss of the FRET reaction[42,43].

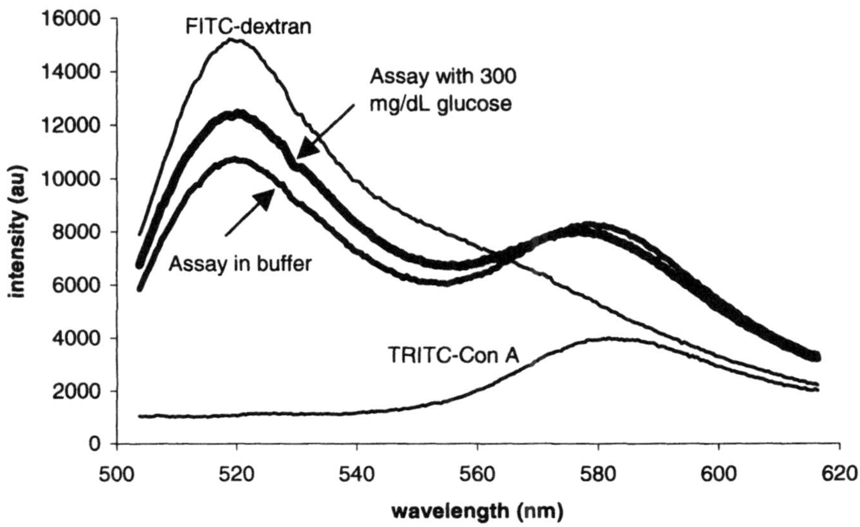

Figure 4.9. Plot of the emission spectrum of FITC and TRITC dyes illustrating the formation of Con A-dextran complex through a FRET reaction and disruption of the FRET reaction through addition of glucose[43].

4.4. SENSING MODALITIES

4.4.1. Hollow Fiber Based Sensors

4.4.1.1 Previous Work

The first fluorescent affinity sensor for glucose detection was devised by Mansouri and Schultz in 1982[51]. This sensor was based upon the affinity reaction between Con A and a high molecular weight dextran (70kD). Using a hollow fiber membrane, Con A was immobilized by adsorbing the protein onto the hollow fiber with glutaraldehyde so that it would irreversibly bind to the luminal side. The sensor was constructed by filling the hollow fiber membrane with FITC dextran solution and sealed at one end using a sealant. A raw optical fiber was inserted into the open end and sealed creating an optical path for light to travel from the fiber into the assay solution containing fluorescent molecules and back through the fiber to a detector. The optical fiber was designed to have a low numerical aperture resulting in a light path traveling through the center of the solution and not illuminating the edges of the hollow fiber. A diagram of the sensor construction is shown in Figure 4.10[51].

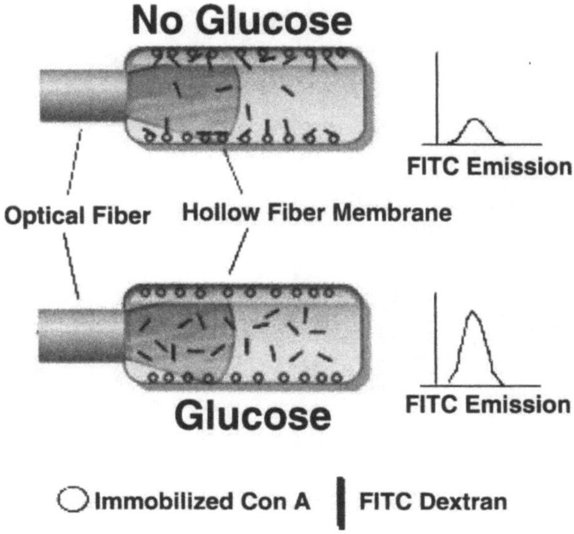

Figure 4.10. Diagram of the single fluorophore hollow fiber sensor devised by Mansouri and Schultz[51].

The sensor was placed into a buffer solution for testing and showed minimal fluorescence emission from FITC-dextran due to binding to the immobilized Con A and removal from the excitation path. Upon placement of the fiber sensor into a glucose enriched buffer, a rise in fluorescent emission was seen due to displacement of the FITC dextran by glucose. This increase in fluorescent emission was directly proportional to the amount of glucose which had diffused into the hollow fiber. This glucose concentration was shown to equilibrate with the surrounding media's concentration within 5-10 minutes. The assay was theoretically determined to respond within seconds whereas the diffusion time for glucose into the membrane was calculated to be on the order of minutes proving to be the limiting factor of the sensor's response. The largest overall response to glucose seen in this sensor was approximately 60% over a range of 0-10mM of glucose. This sensor modality proved the feasibility of the Con A/dextran system as a sensing element for biological concentrations of analytes such as glucose. The limitation of this sensor was the lack of an internal reference making the sensor unusable in long term studies due to drift which required recalibration using known standards[51].

Due to the lack of an internal reference and in hope of avoiding the immobilization procedure of the previous modality, which proved to be complicated and hard to repeat, a new assay was created by Meadows and Schultz that used a FRET based approach[42] (Figure 4.11). By labeling Con A with TRITC and placing it into a hollow fiber with FITC labeled dextran, a quenching of the FITC dye was realized. This quenching was due to a development of a FRET reaction between the FITC and TRITC dye resulting in

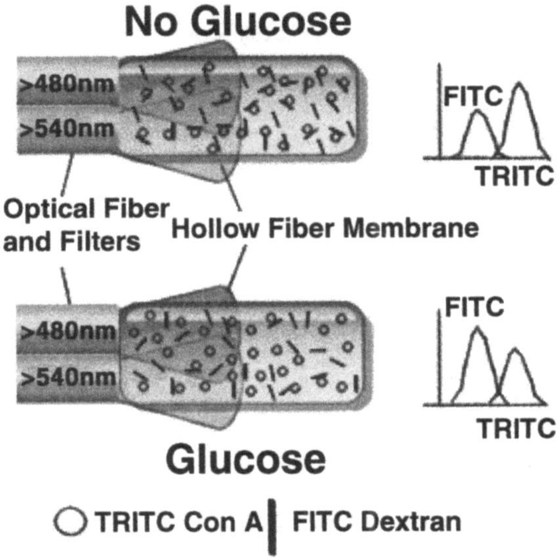

Figure 4.11. Diagram of the FRET based hollow fiber sensor devised by Meadows *et al*[54].

a loss of emission of the FITC dye due to the energy transfer. The TRITC dye was reported to show very little increase in intensity and was therefore used as an internal reference. The sensing system was modified to include two excitation sources which were swapped on and off alternately to preferentially excite the FITC and TRITC dye separately. The TRITC dye acted as an internal reference recording any loss of protein due to denaturation and mechanical disturbance in the excitation and detection pathways. The overall response of the sensor was improved to 60% recovery of 50% quenching of FITC over a range of 0-900 mg/dL glucose with a similar response time as the first sensor primarily due to the diffusion constant of the hollow fiber membrane[42].

This sensor modality was further modified by including two fibers, one for each dye into the end of the hollow fiber membrane resulting in an increased sensitivity and elimination of the need to alternate the excitation pathways[54]. This modification allowed for a continuous reference of the TRITC dye so that variations could be removed accurately. The new continuous ratiometric system lowered the detection limit of FITC emission to 0.05 μg/ml, which improved upon the 4 mg/dl reported in the single fluorophore experiments. In this work, the effectiveness of Con A as a quencher and the long term stability of the protein were also investigated. It was found that commercially purchased labeled protein was very poor in quenching the FITC fluorescence due to poor labeling ratios and a large amount of inactive protein. By labeling the protein prior to implementing the sensor a much larger quenching of the FITC protein was seen and also long term stability of the protein was obtainable by addition of a succinyl group to the protein rendering it in a permanent dimer state at room temperature. This dimerization of the protein was shown to not limit the saccharide binding of the protein, but did limit the aggregation due to the protein's inability to form a tetramer or multimer. However, this step did not eliminate inactivity of the protein. This inactivity was reported to affect about 40% of dissolved protein which severely limited the response of the sensor probe[54].

Ballerstadt *et al.* attempted to modify this sensor in two distinct ways[62]. The first way was to completely remove the need to either immobilize the Con A onto the hollow fiber membrane or fluorescently label it through a chemical process, since the later might alter the binding characteristics of the protein and/or damage it. The devised system used the binding of the Con A tetramer to tether the two fluorescently (FITC, TRITC) labeled polymer spheres, modified with sugar ligands, together and in turn create a FRET reaction between the two dyes. The second modification was to use oblique excitation by passing light (495nm) directly through the hollow fiber membrane rather than coupling the raw fiber into the end of the membrane (Figure 4.12). This advancement is very important because it removes the need for an indwelling catheter type of sensor in which the raw fiber had to penetrate into the skin. This newly devised sensor, however, was only able to achieve a 20% change in fluorescent quenching and only a 15% change over the entire glucose range. This sensor also proved to have stability issues due to the deactivation of the protein binding characteristics resulting in loss of activity to glucose over time[62].

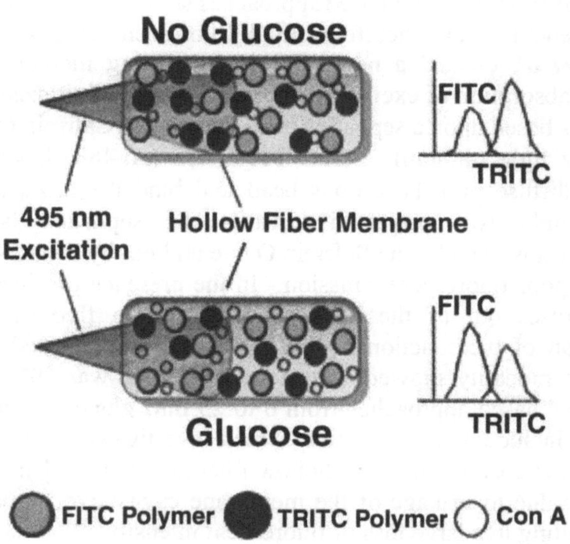

Figure 4.12. Diagram of FRET based system using unmodified unlabeled Con A and two fluorescently tagged polymer spheres grafted with a specific sugar ligand as a novel assay system for glucose sensing[62]

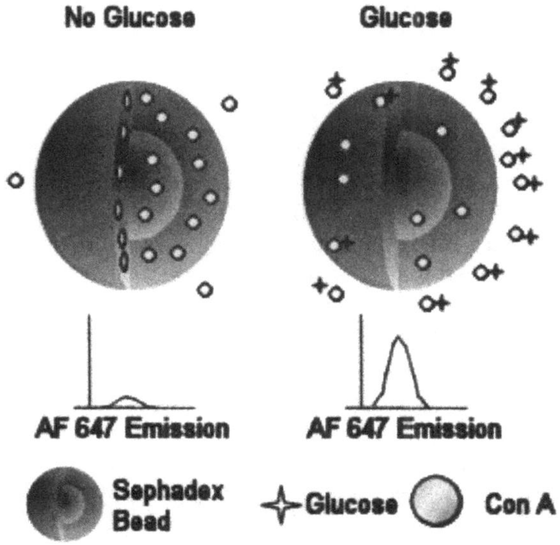

Figure 4.13. Schematic showing the premise of the novel sensor devised by Ballerstadt which relied on preferential absorption of excitation light by an absorbing molecule to limit fluorescent emission of bound Con A[64].

Since previous FRET based approaches showed potential, but were limited by poor quenching and therefore small fluorescent change due to glucose, Ballerstadt *et al.* created a new fluorescent sensing modality that relies on nonspecific absorption of excitation light resulting in minimized fluorescence[63]. The sensor is based upon a sephadex bead doped with Safrinin O, which heavily absorbs blue light (488nm). Alexa Fluor 488 (AF488) labeled Con A was allowed to diffuse into the porous bead and bind the cross linked dextrans. Excitation light was projected towards the sephadex beads and was preferentially absorbed by the Safrinin O dye and not by the AF488 fluorophore resulting in poor fluorescent emission. In the presence of glucose, the labeled Con A diffuses out of the beads and is free to fluoresce. A pictorial representation of the functionality of the sensor is presented in Figure 4.13. This sensing modality showed a dynamic range that was 50% better than that seen in FRET based approaches from 0 to 25 mM glucose. This research was furthered to include an *in vitro* study which investigated the long term stability of Con A while encapsulated in a hollow fiber membrane. This research proved inconclusive due to leakage of the membrane cavity over the course of many months resulting in a large loss of fluorescent intensity[63,64].

4.4.1.2. Current Modalities

Ballerstadt *et al.* continued to pursue their technology using a hollow fiber membrane, but returned to a FRET based sensing approach by using a macroporous Sepharose bead to immobilize the QSY21 labeled Con A protein because they believed an immobilized protein would add stability to the system[65]. Alexa Fluor 647 (AF647) labeled dextran binds to the Con A bead in the absence of glucose, generating a FRET reaction between the two dyes, and displaces with the addition of glucose (Figure 4.14). Also included in this sensing technology is a separate hollow fiber sensor to serve as a reference using polysulfone fibers labeled with LD800. This sensing technology shows only about a 20% change due to glucose binding. This current modality for glucose sensing shows promise since it contains dyes which emit light into the red and near infrared regime. This emission allows it to be more viable *in vivo* since the longer wavelengths provide deeper penetration through skin with little autofluorescence from the tissue. However, the drawbacks of this system include the need for an external reference dye platform within the sensor probe namely a second hollow fiber containing nonreactive molecules. This may prove difficult to correlate to the primary sensor due to differing microenvironments at the implantation site, most notably optical pathlength. Also, in its current state, this sensor may prove difficult to implant due to its multiple element construction and unconfirmed longevity and biocompatibility of implanted hollow fiber membranes[65].

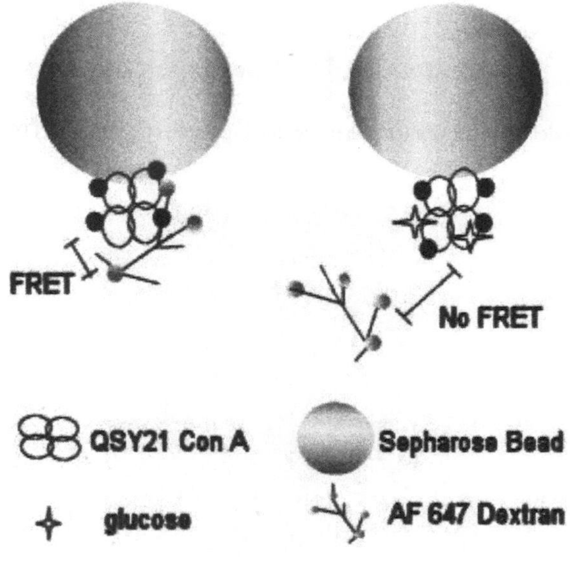

Figure 4.14. Schematic of the sepharose immobilized Con A assay developed by Ballerstadt *et al* [65]

4.4.2. Poly(ethylene glycol) Based Sensors

4.4.2.1 Previous Modalities

Russell *et al.* developed an implantable glucose sensor by encapsulating TRITC Con A and FITC dextran into a poly(ethylene glycol) (PEG) hydrogel[43]. Poly(ethylene glycol) hydrogels have been shown to be very biocompatible and are currently used for drug delivery devices and for coating orthopedic implants to inhibit thrombosis[66-75]. PEG has also been shown to work very well for *in situ* glucose biosensors acting as a barrier between the body fluids and electrode[76]. The envisioned sensor is a collection of PEG hydrogel spheres which contain the functional glucose assay (Figure 4.15)[43]. Poly(ethylene) glycol diacrylate, when polymerized using a high intensity UV light, forms a hydrogel which will swell when placed into a buffered solution. This

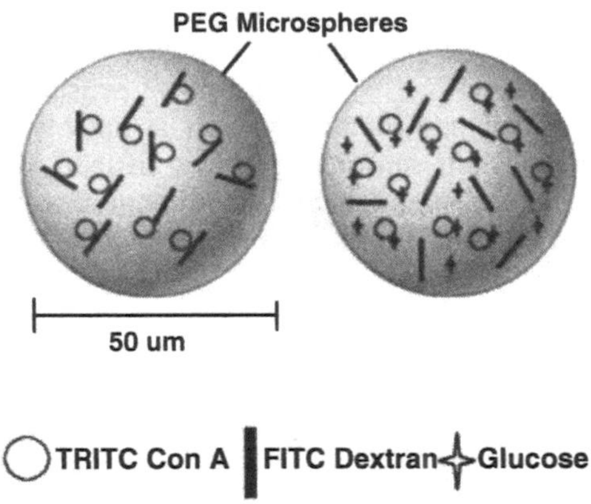

Figure 4.15. A model of the PEG microsphere containing the FITCdextran TRITC Con A glucose assay[43]

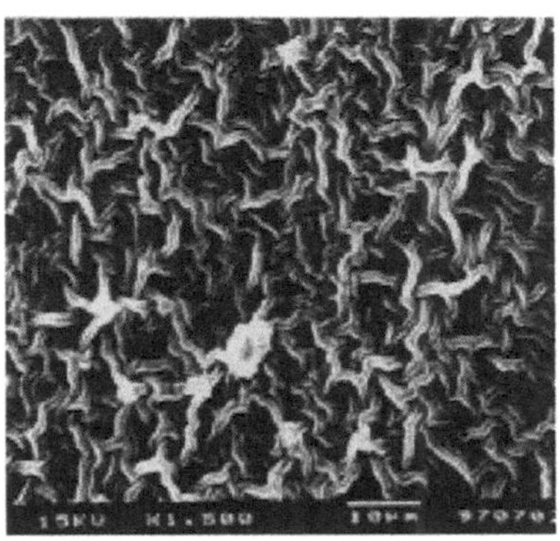

Figure 4.16. A scanning electron microscope image of a PEG hydrogel[43]

swelling increases the mesh size within the hydrogel allowing for diffusion of small molecules such as glucose while inhibiting the loss of large molecules such as Con A and dextran[77-78,43]. To construct the PEG spheres, the Con A/dextran assay was mixed with α-acryloyl, ω-N-hydroxysuccinimidyl ester of PEG-propionic acid, a compound which acts to bind the Con A to the polymer backbone to inhibit leaching. Following the chemical binding, the material is polymerized via cross-linking using a long wavelength UV source ensnaring the dextran molecules and protein within the hydrogel. Figure 4.16 is a scanning electron microscope image of a PEG hydrogel displaying the porous structure important for this application[43].

The ideal sensor will be placed just underneath the epidermal layer of the skin allowing free exchange of analytes within the interstitial fluid[79,43]. Figure 4.17 shows the ideal placement of the implanted sensor. Using a custom fiber optic probe, excitation light can be delivered and fluorescent emission can be detected through the skin. In 1999, Russell *et al.* showed that the traditional FRET assay originally devised by Schultz could be encapsulated into a PEG microsphere and still retain functionality. The assay was made with TRITC labeled Con A and 2000 kDa MW FITC labeled dextran. Preliminary results showed a 35% change across the 0-1000 mg/dL range of glucose concentration and a 20% change across 0-2000 mg/dL range for mannosylated FITC dextran. The response time of the sphere from 0-200mg/dL glucose was shown to be about 15 minutes comparable to to the hollow fiber membrane technology. It was also discovered that the optimum mass ratio between FITC dextran and TRITC Con A was 100:1 resulting in a 60% difference in fluorescence due to a 800 mg/dL insult[43].

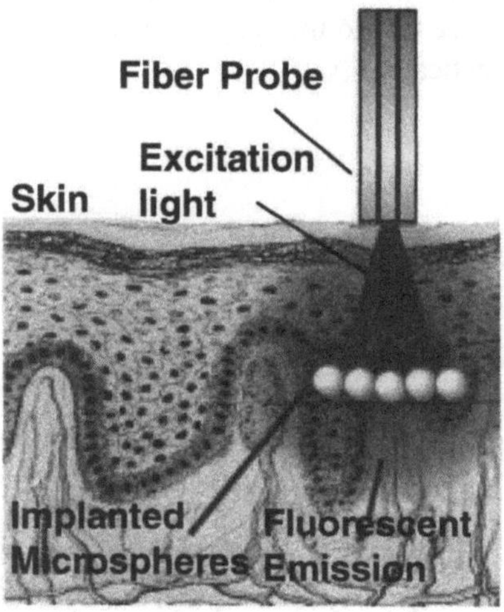

Figure 4.17. A schematic illustrating the ideal sensor placement in the dermal tissue and the proposed detection system using a fiber optic probe.

4.4.2.2 Fluorescent light through the skin

Further research focused on the evaluation of the ability of fluorescent light to penetrate through the dermal tissue. In an attempt to model the interaction of light with the dermal tissue and fluorescent implant, Monte Carlo photon tracing simulations were run to isolate both the anticipated signal and the effect of the spheroid geometry of the devised sensor implant[80]. McShane *et al.* devised a modification of the original MLMC code derived by Wang *et al.* that included the ability to generate a fluorescent reaction based upon the quantum efficiency of dye at a single wavelength[80-81]. Using this code, they were able to generate prediction about tissue depths and signal-to-noise which were confirmed through experimental results using dyed microspheres in porcine tissue. This code was rewritten to take into account the effect of geometry of the spheres since it would affect the collection efficiency of the fiber probe through increased scatter. It was concluded that two main factors influence the ability to retrieve fluorescent light from the dermal tissue layer. The first factor was based upon the thickness of the homogenous sensor slab used in the simulation. Due to the complicated nature of modeling spheres, a simple slab was used. It was shown that increasing the thickness of the slab and therefore the amount of the dye, had a profound affect on signal to noise ratio of the fluorescent light to background autofluorescence of the dermal tissue components. Within this study it was also determined that fluorescent emission from within the skin had a very broad distribution upon arriving at the surface as compared to standard diffuse reflectance, presumably due to the isotropic nature of spontaneous emission. The second main factor was distance of the collection optics from the excitation optics. It was shown that as the detection optics are further separated from the excitation locale, the signal to noise ratio defined as the fluorescence due to the sensor divided by the autofluorescence of the tissue dropped dramatically.

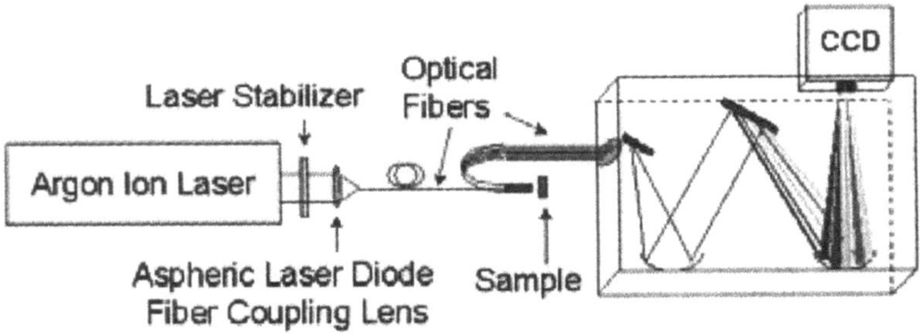

Figure 4.18. In vitro and in vivo optical system composed of a argon ion laser source, laser stabilizer, custom fiber optic probe, and a CCD based spectrometer[84].

An optical system was devised by McShane *et al.* to investigate the ability of the sensor's fluorescent emission to pass through multiple dermal tissue layers[82]. The optical setup consisted of a blue argon laser, a custom fiber optic probe, and a spectrometer (Figure 4.18)[82]. Polystyrene spheres containing FITC and TRITC dyes were purchased and implanted into the dorsal tissue of rats. This study confirmed the model prediction proving that at shallow implantation depths, fluorescent emission could be adequately captured and quantified. O'Neal furthered this study by investigating the effect of implantation depth on signal to noise by generating phantom PEG slabs which contained both FITC and TRITC dextran at a thickness of 1mm using the polymerization protocol outlined by Russell *et al.*[83]. These slabs were covered with thicknesses of excised porcine tissue ranging from 0-800 μm. These studies showed that, despite a significant lose of signal, emission peaks were attainable up to 700μm without baseline correction.

4.4.2.3. Current Progress

As outlined above, the traditional FITC/TRITC based assay could be successfully contained within a PEG hydrogel microsphere and that sensor functionality could be retained despite noticeable attenuation in dynamic range. Through modeling of light penetration in dermal tissue and validation of that model through *in vivo* and *in vitro* skin studies, the feasibility of an implantable sensor was recognized. By implanting the microspheres into a rat model, the biocompatibility of the sensor as well as its *in vivo* functionality was explored[84]. It was shown that at an implantation depth of 500μm fluorescent spectra using the FITC/TRITC (blue/green/orange) wavelengths could be obtained through baseline subtraction and that a glucose response in the spheres was visible through a bolus injection of glucose into the tail vein of a rat (Figure 4.19)[84].

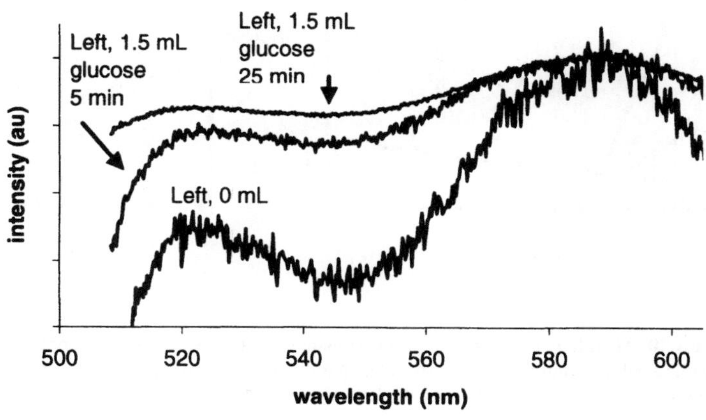

Figure 4.19. Preliminary results obtained from spheres implanted into the abdominal dermal tissue of a hairless rat. Baseline readings were taken and verified using a commercial glucose meter. A bolus injection of glucose was administered into the tail vein and spectra were recorded using the fiber optic detection system[85].

In this work, however, it was evident that the autofluorescence of the tissue created a large baseline signal that required subtraction of an adjacently acquired tissue spectra to visualize the peak of interest. Histological sectioning of the normal tissue and tissue containing spheres were taken at three weeks and reviewed by a pathologist. The results showed little to no adverse affect on the surrounding tissue and no noticeable immune response to the implanted spheres. Microscopic pictures of the histology preparation are shown in Figure 4.20.

It was found that the excitation wavelength of the traditional sensor was not optimum for penetration of dermal tissue resulting in low sensitivity in the presence of high tissue autofluorescence. Ibey *et al.* compared the traditional FITC dextran TRITC Con A spheres with AF594 dextran AF647 Con A spheres through *in vitro* porcine skin studies similar to that previously described by O'Neal[85]. It was hypothesized that the longer wavelength dyes would improve the signal-to-noise due to reduced absorption and scattering of longer wavelengths in biological tissues and reduced autofluorescence of the tissues. Using excised tissue it was shown that the longer wavelength dyes indeed had less susceptibility to implantation depth and a much higher signal to noise given the same excitation power. It was also shown that spheres containing new dyes were functional and compared well to that of the previous sensor reported by Russell *et al.*[85,43].

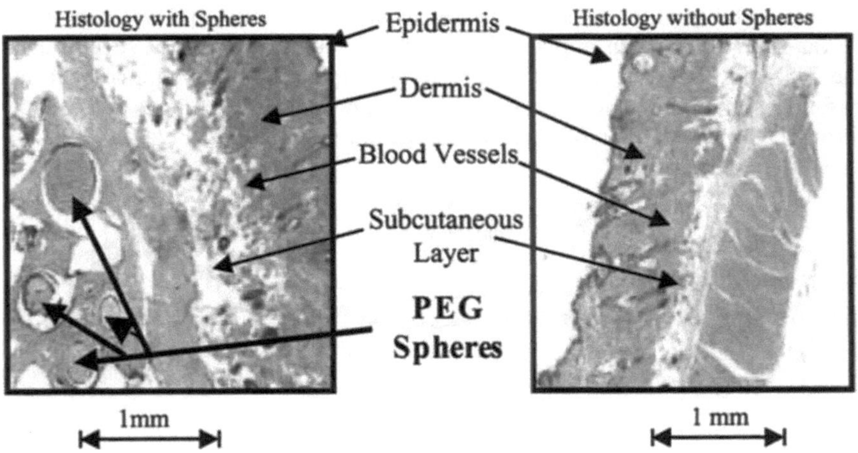

Figure 4.20. Histological sectioning of the implant site of the spheres and a control site allowed for immunological screening of the sensor site. No evidence of an immune response was seen at the implantation site despite overshooting the target implant depth[85]. (see the color insert after p. 429.)

4.5. SUMMARY

The focus of the chapter was an overview of the protein Con A as a sugar binding protein for use in implantable fluorescent glucose monitoring systems. An overall description of the protein, the assay, and the fluorescent properties used in indirectly measuring the binding and unbinding of the protein was presented. Two sensing approaches using this protein were highlighted and described in depth, namely hollow fiber based sensing and polymer encapsulation based sensing. The hollow fiber sensor is limited in its application due to the uncertainty of the implantable lifetime of the membrane. In its current form, the sensor may need a material coating to protect it from the body's immune system which would slow the diffusion of glucose and the time response of the sensor. The PEG sensor, while showing good results for biocompatibility, lacks a quick response which is limited by the current microsphere size and mesh size dictated by the molecular weight of the PEG. The long term activity of this type of sensor is also unknown as well as the degree of leeching and stability of the ensnared Con A. The PEG and the hollow fiber sensors both suffer from a very small overall spectral response to glucose leading to low signal-to-noise ratios. Current research groups have replaced assay components and are attempting to yield larger fluorescent emission changes and improve long term assay stability to overcome this obstacle. Therefore, despite some obstacles facing this approach, glucose sensing through the implantation of an affinity based assay still remains promising due to its potential for continuous and pain free monitoring of glucose.

4.6. REFERENCES

1. Centers for Disease Control and Prevention, National diabetes fact sheet: general information and national estimates on diabetes in the United States, 2000, Atlanta, GA: U.S. (Department of Health and Human Services, Centers for Disease Control and Prevention, 2002).
2. F. H. Martini, *Fundamentals of Anatomy & Physiology* (Prentice Hall, New Jersey, 1995).
3. L.C. Junqueira, J. Carneiro, and R.O. Kelly, *Basic Histology: 9th Edition* (Appleton & Lange, Connecticut, 1998).
4. L.S. Costanzo, *Physiology* (W. B. Saunders Company, Pennsylvania, 1998).
5. Diabetes Research Group, The effects of intensive treatment of diabetes on the development and progression of long-term complications in insulin-dependent diabetes mellitus, *New England Journal of Medicine* **329**, 977-986 (1993).
6. H.K. Naito, Y.S. Kwak, C. Cottingham, Accuracy of the One Touch II whole blood glucose analyzer when used by analysts with diverse technical backgrounds, *J Fam Pract.* **37**(2) 153-157 (1993).
7. S. Skeie, G. Thue, K. Nerhus and S. Sandberg, Instruments for Self-Monitoring of Blood Glucose: Comparisons of Testing Quality Achieved by Patients and a Technician, *Clinical Chemistry* **48**, 994-1003 (2002).
8. J. Pickup, L. McCartney, O. Rolinski, and D. Birch, In vivo glucose sensing for diabetes management: progress towards non-invasive monitoring, *Biomedical Journal* **13,** 1-4 (1999).
9. G. L. Cote', Noninvasive and minimally invasive optical monitoring technologies, *American Society of Nutritional Science*, 1596-1604 (2001).
10. T. Koschinksy and L. Heinemann, Sensors for glucose monitoring: technical and clinical aspects, *Diabetes Metab Res Review* **17**, 113-123 (2001).

11. D. C. Klonoff, Noninvasive blood glucose monitoring, *Diabetes Care*, *20* (1997).
12. J. C. de Graff, G.J. Hemmes, T. Bruin, D.T. Ubbink, R. P. J. Michels, J. H. M. Jacobs, and G. T. B. Sanders, Influence of repetitive finger puncturing on skin perfusion and capillary blood analysis in patients with diabetes mellitus, *Clinical Chemistry* **45**(12), 2200-2206, (1999).
13. T. J.. Bohannon and N. J. Potts, Measurement of glucose in diabetic subjects using noninvasive transdermal extraction, *National Medicine*, **1**(11), 1132-1133 (1995).
14. G. Rao, R.H. Guy, P. Glikfeld, W.R. LaCourse, L. Leung, J. Tamada, M.O. Potts, and N. Azimi, Reverse Iontophoresis: Noninvasive glucose monitoring in vivo in humans, *Pharmaceutical Research*, **12**(12), 1869-1873 (1995).
15. P. Connolly, C. Cotton, and F. Morin, Opportunities at the skin interface for continous patient monitoring: A reverse iontophoresis model tested on lactate and glucose, *IEEE transactions on nanobioscience* **1**(1), 37-41 (2002).
16. T. Nunnold, S. R. Colberg, M.T. Herriott, and C.T. Somma, Use of the noninvasive GlucoWatch Biographer during exercise of varying intensity, *Diabetes Technology & Therapeutics*, **6**(4), 454-462 (2004).
17. S. F. Malin, T. L. Ruchti, T. B. Blank, S. N. Thennadil, and S. L. Monfre, Noninvasive prediction of glucose by near-infrared diffuse reflectance spectroscopy, *Clinical Chemistry*, **45**(9), 1651-1658 (1999).
18. J.J. Burmeister, and M.A. Arnold, Evalution of measurement sites for noninvasive blood glucose sensing with near-infrared transmission spectroscopy, *Clinical Chemistry*, **45**(9), 1621-1627 (1999).
19. S. Yeh, C. F. Hanna, and O. S. Khalil, Monitoring blood glucose changes in cutaneous tissue by temperature-modulated localized reflectance measurements, *Clinical Chemistry*, **49**(6), 924-934 (2003).
20. R. Vonach , J. Buschmann, et al., Applicaton of Mid-infrared transmission spectroscopy to the direct determination of glucose in whole blood, *Applied Spectroscopy*, **52**(6), 820-822 (1998).
21. K. H. Hazen, M. A. Arnold, and G. W. Small, Measurement of glucose and other analytes in undiluted human serum with near-infrared transmission spectroscopy, *Analytica Chimica Acta* **371**, 255-267 (1998).
22. G. W. Small, M. A. Arnold, and L. A. Marquardt, Strategies for coupling digital filtering with partial least-squares regression - application to the determination of glucose in plasma by Fourier-transform near-infrared spectroscopy, *Analytical Chemistry* **65**, 3279, (1993).
23. K.H. Hazen, M.A. Arnold, and G.W. Small, Measurement of glucose and other analytes in undiluted human serum with near-infrared transmission spectroscopy, *Analytica Chimica Acta* **371**, 225-267, (1998).
24. K. H. Hazen, M. A. Arnold, and G. W. Small, Measurement of glucose in water with first-overtone near-infrared spectra, *Applied Spectroscopy*, **52**, 1597, (1998).
25. M. A. Arnold, Non-invasive glucose monitoring, *Current Opinion in Biotechnology*, **7**, 46-49 (1996).
26. B. D. Cameron and G.L. Cote', Noninvasive glucose sensing utilizing a digital closed-loop polarimetric approach, *IEEE Transactions on Biomedical Engineering* **44**(12), 1221-1227 (1997).
27. B. D. Cameron, J. S. Baba, and G. L. Cote', Optical polarimetry applied to the development of a noninvasive in vivo glucose monitor, *SPIE-BiOS 2000 Conference Proceedings*, V3923.
28. J. S. Baba, B. D. Cameron, S. Theru , and G. L. Cote', Effect of temperature, pH, and corneal birefringence on polarimetric glucose monitoring in the eye, *Journal of Biomedical Optics*, **7**(3), 321-328 (2002).
29. R. R. Ansari, S. Bockle, and L. Rovati, New optical scheme for a polarimetric-based glucose sensor, *Journal of Biomedical Optics*, **9**(1), 103-115 (2004).
30. M. J. McShane, Potential for glucose monitoring with nanoengineered fluorescent biosensors, Diabetes *Technology & Therapeutics* **4**(4), 533-538 (2002).
31. N. DiCesare and J. R. Lakowicz, A new highly fluorescent probe for monosaccharides based on a donor-acceptor diphenyloxazole, *Chemcomm Communication*, 2022-2023 (2001).
32. J.C. Norrild and I. Sotofte, Design, synthesis and structure of new potential electrochemically active boronic acid-based sensors, *Perkin Transactions* **2**, 303-311 (2002).
33. Z. Murtaza, L. Tolosa, P. Harms, and J. R. Lakowicz, On the possibility of glucose sensing using boronic acid and a luminescent ruthenium metal-ligand complex, *Journal of Fluorescence* **12**(2), 187-192 (2002).

34. J. T. Suri, D. B. Cordes, F. E. Cappuccio, R. A. Wessling, and B. Singaram, Continous glucose sensing with a fluorescent thin-film hydrogel, *Angewandte Chemie* **42**, 5857-5859 (2003).
35. V. L. Alexeev, A. C. Sharma, A. V. Goponenko, S. Das, I. K. Lednev, C. S. Wilcox, D. N. Finegold, and S. A. Asher, High ionic strength glucose-sensing photonic crystal, *Analytical Chemistry* **75**, 2316-2323, (2003).
36. R. Badugu, J. R. Lakowicz, and C. D. Geddes, A glucose sensing contact lens: a non-invasive technique for continuous physiological glucose monitoring, *Journal of Fluorescence* **13**(5), 371-374 (2003).
37. R. Badugu, J. R. Lakowicz, and C. D. Geddes, Noninvasive continuous monitoring of physiological glucose using a monosaccharide-sensing contact lens, *Analytical Chemistry* **76**, 610-618 (2004).
38. M. D. Philips and T. D. James, Boronic acid based modular fluorescent sensors for glucose, *Journal of Fluorescence* **14**(5), 549 559 (2004).
39. H. Fang, G. Kaur, and B. Wang, Progress in boronic acid-based fluorescent glucose sensors, *Journal of Fluorescence* **14**(5), 481-489, (2004).
40. R. Badugu, J. R. Lakowicz, and C. D. Geddes, Opthalmic glucose sensing: a novel monosaccharide sensing disposable and colorless contact lens, *Analyst* **129**, 516-521 (2004).
41. J. R. Lakowicz and B. Maliwal, Optical sensing of glucose using phase-modulation fluorimetry, *Analytica Chimica Acta* **271**, 155-164 (1993).
42. D. L. Meadows and J. S. Schultz, Design, Manufacture and Characterization of an Optical-Fiber Glucose Affinity Sensor-Based on an Homogenous Fluorescence Energy-Transfer Assay System, *Analytica Chimica Acta* **280**(1), 21-30 (1993).
43. R .J. Russell, M. V. Pishko, C. C. Gefrides, M. J. McShane, and G. L. Coté , A fluorescence-based glucose biosensor using concanavalin A and dextran encapsulated in a poly(ethylene glycol) hydrogel, *Analytical Chemistry* **71**, 3126-3132 (1999).
44. L. Tolosa, H. Malak, G. Roab, and J. R. Lakowicz, Optical assay for glucose based on the luminescence decay time of the long wavelength dye Cy5, *Sensors and Actuators B-Chemical.* **45** (2), 93-99 (1997).
45. L. Tolosa, H. Szmacinski, G. Roab, and J. R. Lakowicz, Lifetime-based sensing of glucose using energy transfer with a long lifetime donor, *Analytical Biochemistry* **250**, 102-108, (1997).
46. O. J. Rolinski, D. J. S. Birch, L. J. McCartney, and J. C. Pickup, A time-resolved near-infrared fluorescent assay for glucose: opportunities for trans-dermal sensing, *Journal of Photochemistry and Photobiology B: Biology* **54**, 26-34 (2000).
47. E. Kulcu, J. A. Tamada, G. Reach, R. O. Potts, M. L. J. Lesho, Physiological Differences between interstitial glucose and blood glucose measure in human subjects, *Diabetes Care* **26**(8), 2405-2409 (2003).
48. B. Aussedat, M. Dupire-Angel, R. Gifford, J. C. Klein, G. S. Wilson, and G. Reach, Interstitial glucose concentration and glycemia: implications for continuous subcutaneous glucose monitoring, *Am J Physiol Endocrinol Metab* **278,** 716–728 (2000).
49. H. Bittiger and H.P. Schnebli, *Concanavalin A as a tool* (John Wiles & Sons, New York, 1976).
50. T. K. Chowdhury, and A. K. Weiss, *Advances in experimental medicine and biology: Concanavalin A*, (Plenum Press, New York, 1974).
51. J. S. Schultz, S. Mansouri, and I. J. Goldstein, Affinity sensor: a new technique for developing implantable sensors for glucose and other metabolites, *Diabetes Care* **5**(3), 245-253 (1982).
52. D. L. Meadows and J. S. Schultz, Fiber-optic biosensors based on fluorescence energy transfer, *Talanta* **35**(2), 145-150 (1988).
53. D. L. Meadows, A fiber optic biosensor for glucose monitoring based on fluorescence energy transfer, (Ph.D. Dissertation, The University of Michigan, 1988).
54. C. Huet, M. Lonchampt, M. Huet, and A. Bernadac, Temperature effects on the concanavalin A molecule and on concanavalin A binding, *Biochimica et Biophysica Acta.* **365**, 28-39 (1974).
55. A. Clark and M. Denborough, The interaction of Concanavalin A with blood-group-substance glycoprotiens from human secretions, *Biochemistry* **121**, 811 (1971).
56. Chapter 2 — Thiol-Reactive Probes, *Handbook of Fluorescent Probes and Research Products (8th Edition)* (Molecular Probes, Oregon, 2002).
Also available online: http://www.probes.com/handbook/index.html.

57. Chapter 14 —Fluorescent Tracers of Cell Morphology and Fluid Flow, *Handbook of Fluorescent Probes and Research Products (8th Edition)* (Molecular Probes, Oregon, 2002). Also available online: http://www.probes.com/handbook/index.html.
58. P. Atkins, *Physical Chemistry: 5th Edition* (W.H. Freeman and Company, New York, 1994).
59. J. R. Lakowicz, Chapter 13: Energy Transfer, *Principles of Fluorescence Spectroscopy, 2nd Edition* (Kluwer Academic / Plenum Publishers, New York, 1999).
60. D.L. Andrews and A. A. Demidov, *Resonance Energy Transfer* (John Wiley & Sons, New York, 1999).
61. P. Wu and L. Brand, Resonance energy transfer: methods and applications, *Analytical Biochemistry* **218**, 1-13 (1994).
62. Chapter 1 — Fluorophores and Their Amine-Reactive Derivatives, *Handbook of Fluorescent Probes and Research Products (8th Edition)* (Oregon, Molecular Probes, 2002). Also available online: http://www.probes.com/handbook/index.html.
63. R. Ballerstadt and J. S. Schultz, Competitive-binding assay method based on fluorescence quenching of ligands held in close proximity by a multivalent receptor, *Analytica Chimica Acta* **345**, 203-212 (1997).
64. R. Ballerstadt, and J. S. Schultz, A fluorescence affinity hollow fiber sensor for continuous transdermal glucose monitoring, *Analytical Chemistry* **72** (17), 4185-4192 (2000).
65. R. Ballerstadt, A. Polak, A. Beuhler, J. Frye, In vitro long-term performance study of a near infrared fluorescence affinity sensor for glucose monitoring, *Biosensors and Bioelectronics* **19**, 905-914 (2004).
66. R. Ballerstadt, A. Gowda, and R. McNichols, Fluorescence Resonance Energy Transfer-Based Near-Infrared Fluorescence Sensor for Glucose Monitoring, *Diabetes Technology & Therapeutics* **6**(2), (2004).
67. S. Chowdhury and J. Hubbell, Adhesion prevention with ancrod released via a tissue-adherent hydrogel, *Journal of Surgical Research* **61**, 58-64 (1996).
68. J. Westand and J. Hubbell, Photopolymerized hydrogel materials for drug delivery applications, *Reactive Polymers* **25**, 139-147 (1995).
69. C. P. Pathak, A. S. Sawhney, and J. A. Hubbell, Rapid photopolymerization of immunoprotective gels in contact with cells and tissue, *Journal of the American Chemistry Society* **114**, 8311-8312 (1992).
70. G. M. Cruise, O. D. Hegre, F. V. Lamberti, S. R. Hager, R. Hill, D. S. Sharp, and J. A. Hubbell, In vitro and in vivo performance of porcine islets encapsulated in interfacially photopolymerized poly(ethylene glycol) diacrylate membrane, *Cell Transplantation* **8**, 293-306 (1999).
71. J. West and J. Hubbell, Separation of the arterial wall from blood contact using hydrogel barriers reduces intimal thickening after balloon injury in the rat: the roles of medial and luminal factors in arterial healing, *Proceedings of the National Academy of Science USA.* **93**(23), 13188-13193 (1996).
72. A. S. Sawhney, C. P. Pathak, J. J. van Rensburg, R. C. Dunn, and J. A. Hubbell, Optimization of photopolymerized bioerodible hydrogel properties for adhesion prevention, *Journal of Biomedical Materials Research.* **28**, 831-838 (1994).
73. J. Hill-West, S. M. Chowdhury, M. J. Slepian, J. A. Hubbell, Inhibition of thrombosis and intimal thickening by in situ photopolymerization of thin hydrogel barriers, *Proceedings of the National Academy of Science USA.* **91**, 5967-5971 (1994).
74. K. L. Prime, and G. M. Whitesides, Self-Assembled Organic Monolayers: Model Systems for Studying Adsorption of Proteins at Surfaces, *Science* **252**, 1164-1167 (1991).
75. P. Drumheller and J. Hubbell, Densely crosslinked polymer networks of poly(ethylene glycol) in trimethylolpropane triacrylate for cell resistant surfaces, *Journal of Biomedical Materials Research.* **29**, 207-215 (1995).
76. A.P. Christopher, R.E. Connor, and A. Heller, Biocompatible, glucose-permeable hydrogel for in situ coating of implantable biosensors, *Biomaterials* **18**, 1665-1670 (1997).
77. M. B. Mellott, K. Searcy, and M. V. Pishko, Release of protein from highly cross-linked hydrogels of poly(ethylene glycol) diacrylate fabricated by UV polymerization, *Biomaterials.* **22**(9), 929-941 (2001).
78. R. J. Russell, A. C. Axel, K. L. Shields, and M. V. Pishko, Mass transfer in rapidly photopolymerized poly(ethylene glycol) hydrogels used for chemical sensing, *Polymer* **42**, 4893-4901 (2001).

79. M. J. McShane, R.J. Russell, M.V. Pishko, and G.L. Coté, Glucose monitoring using implanted fluorescent microspheres, *IEEE Engineering in Medicine and Biology Magazine* **19** (6), 36-45 (2000).
80. M. J. McShane, S. Rastegar, M.V. Pishko, and G.L. Cote, Monte carlo modeling for implantable fluorescent analyte sensors, *IEEE Transactions on Biomedical Engineering* **47**(5), 624-632 (2000).
81. L. H. Wang, S. L. Jacques, and L.-Q. Zheng, MCML - Monte Carlo modeling of photon transport in multi-layered tissues, *Computer Methods and Programs in Biomedicine* **47**, 131-146 (1995).
82. M. J. McShane, D. P. O'Neal, R. J. Russell, M. V. Pishko, and G. L. Coté, Progress toward implantable fluorescence-based sensors for monitoring glucose levels in interstitial fluid, *SPIE Proceedings* **3923**, 78-87 (2000).
83. D. P. O'Neal, M. J. McShane, M. V. Pishko, and G. L. Coté, Implantable biosensors: analysis of fluorescent light propagation through skin, *SPIE Proceedings*. **4263**, 20-24 (2001).
84. B. L. Ibey, A. Meledeo, V. Gant, V. Yadavalli, and M. V. Pishko, In vivo monitoring of blood glocuse using poly(ethylene glycol) microspheres, *SPIE proceddings* **4965-01**, (2003).
85. B. L. Ibey, V. Vadavalli, *et al.*, Analysis of Longer Wavelength Alexa Fluor Dyes for Use in a Minimally Invasive Glucose Sensor, *EMBS Conference*, Mexico (2003).

FLUORESCENCE BIOSENSORS FOR CONTINUOUSLY MONITORING THE BLOOD GLUCOSE LEVEL OF DIABETIC PATIENTS

Sabato D'Auria, Giovanni Ghirlanda, Antonietta Parracino, Marcella de Champdoré, Viviana Scognamiglio, Maria Staiano, and Mosè Rossi ~

5.1. INTRODUCTION

Diabetes mellitus is increasing rapidly and will double in the next 15 years.[1, 2] In addition, diabetic patients have a mortality excess for cardiovascular disease up to 2.5-4 times more than non diabetic population.[3, 4] In the last years became evident a cluster of cardiovascular risk factors like hypertension, central obesity, dyslipidemia with low HDL-cholesterol and high triglycerides, impaired fibrinolysis, hypercoagulation and endothelial dysfunction that has been called metabolic syndrome.[5] The underlying defect that shares all these alterations is insulin resistance with compensatory hyperinsulinemia that is associated with increased cardiovascular events and mortality.[6, 7] Patients with diabetes and/or metabolic syndrome must be treated aggressively about every risk factors to minimize the cardiovascular events. Hyperglycemia is clearly related to microvascular complication of diabetes: retinopathy, nephropathy and neuropathy, while for macrovascular complication other coexisting risk factors are important. Many studies have demonstrated that an intensive treatment of diabetes reduces the macro and microvascular complications and the best results are obtained when every risk factor is aggressively treated.[8-10]

~ Sabato D'Auria, Antonietta Parracino, Marcella de Champdoré, Viviana Scognamiglio, Maria Staiano, and Mosè Rossi, Institute of Protein Biochemistry, Italian National Research Council, Naples, Italy.
Giovanni Ghirlanda, Servizio di Diabetologia, Policlinico A Gemelli, Rome, Italy.
Correspondence to: Sabato D'Auria, Institute of Protein Biochemistry, Italian National Research Council, Via P. Castellino, 111, 80131 Naples, Italy. Tel. +39-0816132250 Fax +39-0816132277 email: s.dauria@ibp.cnr.it

Glycated hemoglobin (HbA_{1c}), at present the gold standard of glycemic control, predicts diabetic complications and actions that lowered HbA_{1c} reduced also diabetic complications.[11, 12]

HbA_{1c} gives us the integrated mean of glycemia of the previous 6-10 weeks, but other parameters of glucose metabolism are also important to explain cardiovascular risk. The Verona Diabetes Study demonstrated that the coefficient of variation of fasting glucose best correlated to all causes mortality than mean fasting glucose. Diabetic patients with greater variation of fasting glucose had greater mortality irrespective of mean plasma glucose and of other known risk factors as diabetes duration, hypertension, gender and age.[13, 14]

Many epidemiological studies have shown that post prandial glycemia and glucose plasma levels 2 h after a glucose load are more powerful predictors of cardiovascular risk than HbA_{1c} and fasting glucose.[15]

A complete assessment of cardiovascular risk related to glycemia should comprise the HbA_{1c}, the daily glycemic excursions and the peak of postprandial hyperglycemia. Diabetic patients on intensive treatment adjust their therapy from data obtained by home blood glucose monitoring with reflectance or electrochemical meters. These measurement use capillary blood drawn from the fingertip through a microlance. Time needed to perform the measure, pain associated to the pinprick and the discomfort arising from having all the fingers pricked prevent the possibility that most of diabetic patients could have many blood glucose levels measured during the day.[16-18]

At present we have continuous glucose monitoring systems that are minimal invasive based on interstitial fluid sampling. Glucose is measured transdermally, after microdyalisis, or by a subcutaneous glucose sensor. Each of this method has a number of problems that till now have hampered a wide clinical use of these devices. Novel promising approaches are the optical glucose sensors that are totally non invasive and could overcome many problems.[19]

It is strongly desirable that in a reasonable time we can get a continous glucose monitoring system able to give to the diabetic patients the body of information required to better refine their therapy and ultimately improve their metabolic control.

The search of a suitable non-invasive glucose measurement method has resulted in decades of research with not great improvements. Included in this effort is the development of fluorescence probes specific for glucose, typically based on boronic acid chemistry.[20-25] An alternative approach to glucose sensing using fluorescence is based on proteins which bind glucose. Optical detection of glucose appears to have had its origin in the promising studies of Schultz and co-workers,[26-29] who developed a competitive glucose assay which does not require substrate and does not consume glucose. This assay used fluorescence resonance energy transfer (FRET) between a fluorescence donor and an acceptor, each covalently linked to concanavalin A (ConA) or to dextran. In the absence of glucose the binding between ConA and dextran resulted in a high FRET efficiency. The addition of glucose resulted in its competitive binding to ConA, displacement of ConA from the labeled dextran, and a decrease in the FRET efficiency. These early results generated considerable enthusiasm for fluorescence sensing of glucose.[30-32] The glucose-ConA system was also studied in the dr. Lakowicz's lab. The researchers, however, found that the system was only partially reversible upon addition of glucose, became less reversible with time, and showed aggregation.[33-35] For this reason, the use of other glucose-binding proteins as sensors was explored from several labs.[36-40]

However, there is a considerable interest in using enzymes or proteins as sensors for a wide variety of substances[41-45] and, if a reliable fluorescence assay for glucose could be developed, the robustness of lifetime-based sensing[46-48] could allow development of a minimally invasive implantable glucose sensor, or a sensor which uses extracted interstitial fluid. The lifetime sensor could be measured through the skin[49] using a red laser diode or light emitting diode (LED) device as the light source. These devices are easily powered with batteries and can be engineered into a portable device. An implantable sensor can be expected to report on blood glucose because tissue glucose closely tracks blood glucose with a 15 min time lag,[50, 51] and time delays as short as 2-4 minutes have been suggested.[52]

In this chapter we show the use of inactive form of enzyme probes for a non-consuming analyte stable biosensor, in particular we will show the data obtained on the glucose oxidase from *Aspergillus niger*, the glucose dehydrogenase from *Thermoplasma acidophilum*, and the glucokinase from *Bacillus stearothermophilius*.

5.2. GLUCOSE OXIDASE FROM ASPERGILLUS NIGER

Glucose oxidase (GO) (EC 1.1.3.4) from *Aspergillus niger* catalyzes the conversion of β-D-glucose and oxygen to D-glucono-1,5-lactone and hydrogen peroxide. It is a flavoprotein, highly specific for β-D-glucose, and is widely used to estimate glucose concentration in blood or urine samples through the formation of colored dyes.[53] Because glucose is consumed, this enzyme cannot be used as a reversible sensor. In a recent report[54] we extended the use of GO under conditions where no reaction occurs. In particular, in order to prevent glucose oxidation, we removed the FAD cofactor that is required for the reaction. The absorbance spectrum shows the characteristic shape of the coenzyme-free proteins, with a maximum of absorbance at 278 nm due to the aromatic amino acid residues. The absence of absorption at wavelengths above 300 nm indicates the FAD has been completely removed. The fluorescence emission spectrum of apo-GO at room temperature upon excitation at 298 nm displays an emission maximum at 340 nm, which is characteristic of partially shielded tryptophan residues. The addition of 20 mM glucose to the enzyme solution resulted in a quenching of the tryptophanyl fluorescence emission of about 18 % (data not shown). This result indicates that the apo-GO is still able to bind glucose. The observed fluorescence quenching may be mainly ascribed to the tryptophanyl residue 426.

In fact, as shown by X-ray analysis (Figure 5.1) and molecular dynamics simulations, the glucose-binding site of GO is formed by Asp 584, Tyr 515, His 559 and His 516.

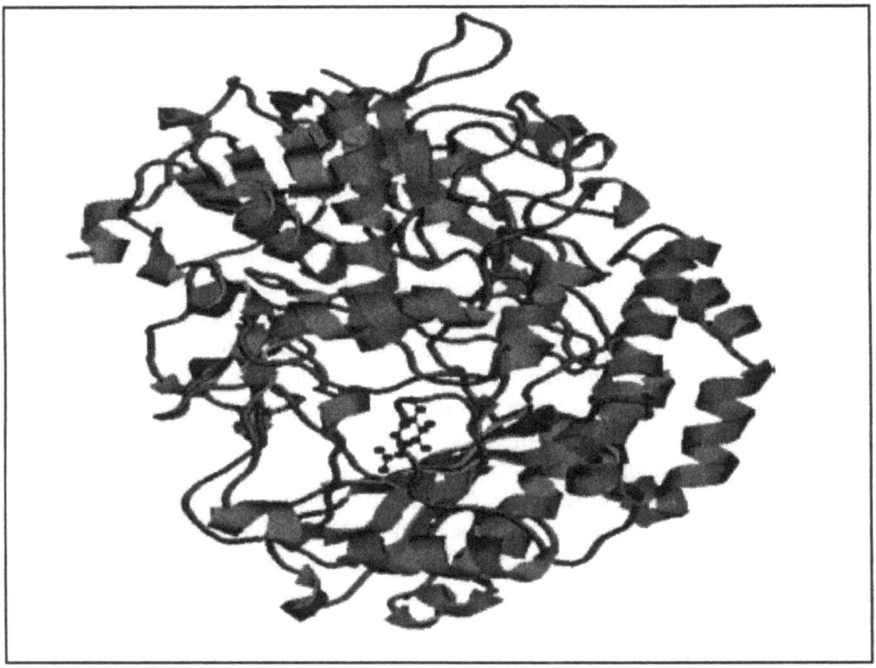

Figure 5.1. Structure of Glucose oxidase solved by X-ray.

Moreover, Phe 414, Trp 426 and Asn 514 are in locations where they might form additional contacts to the glucose. The intrinsic fluorescence from proteins is usually not useful for clinical sensing because of the need for complex or bulky light sources and the presence of numerous proteins in most biological samples. ANS is known to be a polarity-sensitive fluorophore which displays an increased quantum yield in low polarity environments.[54] Additionally, ANS frequently binds to proteins with an increase in intensity. We examined the effects of GO on the emission intensity of ANS. Addition of apo-GO to an ANS solution resulted in an approximate 30-fold increase in the ANS intensity.

Importantly, the intensity of the ANS emission was sensitive to glucose, decreasing of approximately 25 % upon glucose addition (Figure 5.2).

The ANS was not covalently bound to the protein Addition of glucose resulted in a progressive decrease in the ANS fluorescence intensity. This suggests that the ANS is being displaced into a more polar environment upon binding glucose. The decreased ANS intensity occurred with a glucose-binding constant near 10 mM, which is comparable to the K_D of the holoenzyme. Since the binding affinity has not changed significantly, one can suggest that the binding is still specific for glucose.

Intensity decays of ANS-labeled apo-GO in the presence of glucose were studied by frequency-domain fluorometer. Addition of glucose shifts the frequency responses to higher frequencies, which is due to a decreased ANS

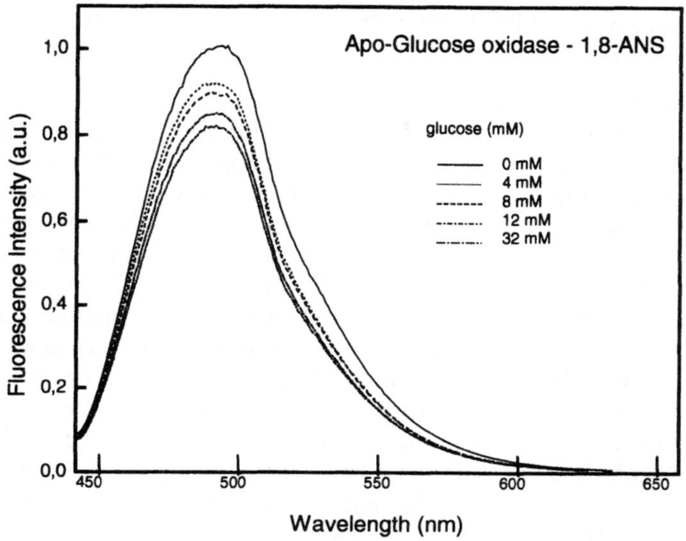

Figure 5.2. Emission spectra of 1,8-ANS-Glucose oxidase in the presence of glucose.

lifetime (data not shown). The shorter lifetimes of ANS apo-GO in the presence of glucose is consistent with the suggestion that glucose displaces the ANS to a more polar environment.[37]

The mean lifetime decreases by over 40 % upon addition of glucose. These results demonstrate that apo-glucose oxidase, when labeled with suitable fluorophores, can serve as a protein sensor for glucose.

5.3. THERMOSTABLE GLUCOSE DEHYDROGENASE FROM THERMOPLASMA ACIDOPHILUM

The widespread use of proteins as sensors depends on protocols to enhance their stability, such as introduction of changes in the amino acid composition leading to enhanced protein structural stability.[55] An alternative method is to use naturally thermostable enzymes or proteins isolated from thermophilic microorganisms: these macromolecules have intrinsically stabile structural features[56] and they can be considered as ideal biosensors.

Glucose dehydrogenase (GD) from the thermoacidophilic archaeon *Thermoplasma acidophilum* is a tetramer of about 160 kDa composed of four similar subunits of about 40 kDa each. The enzyme shows a K_D value of 10 mM for glucose, and it is resistant to high temperatures and organic solvents. At 55 °C, full activity is retained after 9 hours, and at 75 °C the half-life is approximately 3 hours. Moreover, incubation of the enzyme for up 6 hours at room temperature with 50 % (v/v) methanol, acetone or ethanol results without any appreciable loss of activity.[55] We examined the potential of this thermostable GD as a glucose sensor. The enzyme catalyzes the following reaction:

$$\text{Glucose} + \text{NAD(P)}^+ \longrightarrow \text{gluconate} + \text{NAD(P)}^+ + \text{H}^+$$

To prevent the glucose oxidation we used the apo-form of GD, that is the enzyme without the cofactor which is required for the reaction. We found that the apo-GD still binds glucose with an affinity comparable to the holo-enzyme.

However, GD from *Thermoplasma acidophilum* is a thermophilic protein and can be expected to be rigid under mesophilic conditions. We knew that thermophilic proteins often display increased activity at higher temperatures or in the presence of non-polar solvents,[57] which are conditions expected to increase the protein dynamics. Addition of acetone to the solution containing ANS and GD resulted in a dramatic increase of the ANS intensity as well as in a blue-shift of the emission maximum. Addition of similar amounts of acetone to ANS in the absence of the protein produced modest fluorescence increase. Hence the increase in the ANS intensity reflects a change in the local protein environment which is due to acetone. To be useful as a glucose sensor the ANS labeled GD must display usefully large spectral changes in the presence of glucose. Addition of glucose to ANS-GD in the presence of 3 % acetone resulted in an approximate 25% decrease in intensity (Figure 5.3).

This seemed to be the optimal acetone concentration because smaller spectra changes were seen at lower and higher acetone concentrations (data not shown). Apparently at higher acetone concentrations the ANS is already in an environment which results in much of the possible increase in quantum yield. At lower acetone concentrations the environment surrounding the ANS changes in response to glucose in a manner which increases the ANS intensity.

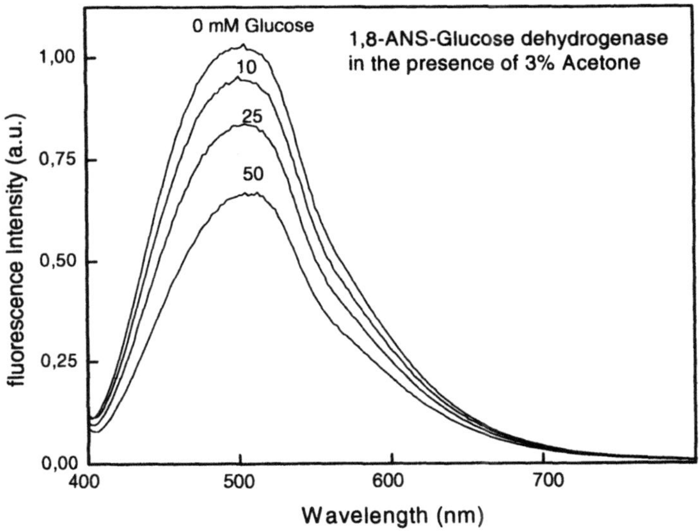

Figure 5.3. Emission spectra of 1,8-ANS-Glucose dehydrogenase in the presence of 3% Acetone.

In previous reports we described the value of fluorescence lifetimes as a basis for chemical sensing.[58] Hence we questioned whether the glucose-dependent decrease in intensity would be accompanied by a similar change in the ANS decay time. Glucose induces a modest shift in the response to higher frequencies, which indicate a decrease of the mean decay time. In the presence and in the absence of glucose the multi-exponential analysis (not shown) indicates that the decay is dominated by a sub-nanosecond component whose contribution is increased by glucose. However the changes in the intensity decay, or equivalent the phase and modulation, are not adequate for lifetime-based sensing. In the preceding discussion we interpreted the results in terms of a change in the protein environment caused by glucose. However, it is also possible that the changes are due to a difference in the amount of protein-bound ANS due to glucose. These preliminary data are not adequate to distinguish between these possibilities.

In a previous report we described the use of polarization sensing for systems which display changes in intensity, but not lifetime, in response to analytes.[59] Because the intensity changes of ANS-GD in response to glucose are modest, it is important to carefully select the best conditions.

Emission polarized spectra of ANS-GD at various concentrations of glucose were performed. The polarization decreases at higher glucose concentrations because the emission from this solution is observed through the horizontal polarizer. Moreover, the change in polarization is larger at shorter wavelengths, and this is due to the differences in the emission spectra of reference (ANS in buffer) and sample (ANS+GD) The wavelength dependent changes in polarization were used to create a calibration curve for glucose (data not shown). This curve shows that the present ANS-GD system can yield glucose concentrations accurate to about 2.5 mM, at a glucose concentration near 20 mM (data not shown).

5.4. A FLUORESCENCE COMPETITIVE ASSAY BY USING THE STABLE GLUCOKINASE

The structure of the hexokinase A from yeast has been determined by x-ray diffraction both in the absence and in the presence of glucose (Figure 5.4).[60, 61]

The polypeptide chain of 485 amino acid residues in the yeast protein is folded into 2 distinct domains, a smaller N-terminal domain and a larger C-terminal domain. From the high-resolution crystal structures of the enzyme is evident that in the absence of ligand, the two domains are separated by a deep cleft. This cleft represents the enzyme active site. It is in this region that the enzyme binds the substrate. In particular, the binding of glucose causes the small lobe of the molecule to rotate by 12° relative to the large lobe, moving the polypeptide backbone as much as 8°, closing the gap between the two domains. The domain rotation has two effects: the glucose molecule is buried into the interior of the protein and the side chains in the active site are rearranged.

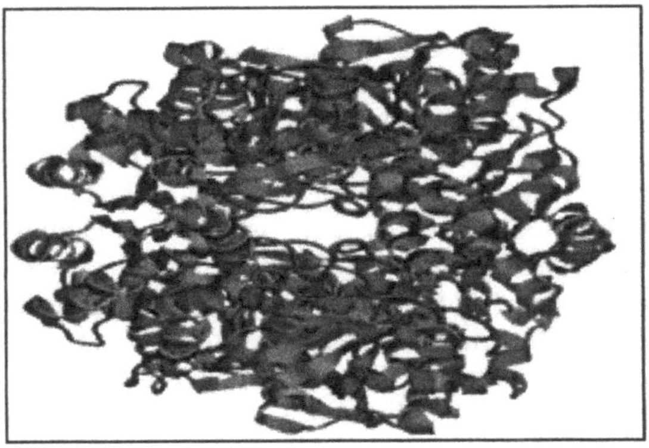

Figure 5.4. Structure of hexokinase solved by X-ray.

Fluorescence spectral data from literature suggest that hexokinase can be used as an optical glucose sensor. For instance, glucose binding to the native dimer and monomer hexokinase from *Saccharomyces cerevisiae* was monitored by following the concomitant quenching of the protein fluorescence.[62-64] This enzyme possesses four tryptophan residues that can be classified as two surface residues, one glucose-quenchable residue in the cleft, and one buried. The maximal quenching induced by glucose was about 25 % and the concentration of glucose at half-maximal quenching was 0.4 mM for the monomeric form and 3.4 mM for the dimeric one.[62-64]

For use as a probe for a sensor a protein should have long term stability. Unfortunately, yeast hexokinase and human hexokinase are unstable and quickly loose activity at room temperature. Thermophilic micro-organisms produce enzymes with unique characteristics such as high temperature-, chemical-, and pH-stability. These enzymes are already in use as bio-catalysts in industrial processes.[55] A glucokinase from the thermophilic organism *Bacillus stearothermophilus* has been characterized and it is known to display long term stability.[65] Hence we evaluated the use of this glucokinase (BSGK) in the absence of ATP as a glucose non-consuming glucose sensor. This protein has already been used as an active enzyme in glucose assays.[65]

In order to check the stability properties of BSGK, we incubated a solution of the enzyme (enzyme concentration 1.0 mg/mL) at room temperature. Enzyme aliquots were withdrawn and the enzyme activity as well as the fluorescence intensity were monitored.

Yeast hexokinase loses activity over several days and the fluorescence intensity simultaneously decreases. In contrast BSGK loses no activity over two weeks at room temperature and the fluorescence intensity remains constant (data not shown). Hence BSGK is a good candidate for a glucose sensing probe.

BSGK has a single cysteine residue located near the active site.[66] We labeled the residue with a sulfhydral-reactive fluorophore IA-ANS. The emission of the labeled protein was near 460 nm (data not shown). The intensity of the ANS-labeled protein decreased upon addition of glucose (data not shown). The decreased intensity is consistent with displacement of the water-sensitive ANS into the aqueous phase upon binding glucose. The change in intensity occurs at a glucose concentration near 3 mM, which is comparable to the concentration of glucose in blood. The important conclusions from these observations is that BSGK binds glucose in the absence of ATP and can thus serve as a non-consuming glucose sensor.

For highly accurate glucose measurements we were not satisfied of the magnitude of the intensity change. We examined the fluorescence lifetimes to determine if a change occurred upon glucose binding. Unfortunately, ANS-labeled BSGK displayed no change in lifetime upon glucose binding. Hence we considered alternative methods to use BSGK as a glucose sensor.

Resonance energy transfer (RET) reliably occurs whenever fluorescent donors and acceptors are in close proximity. We developed a method to use RET to develop a competitive glucose assay. To demonstrate the feasibility of a competitive glucose assay we used the unmodified protein and its intrinsic tryptophan emission as the donor. As the acceptor we used glucose containing the absorbing nitrophenyl group, ONPG. Addition of ONPG (3 μM) resulted in an approximate 80 % decrease in the tryptophan intensity. Addition of glucose resulted in the recovery of the fluorescence intensity. At about 6 mM glucose concentration fluorescence intensity returns to its initial value before addition of ONPG.[67] Further addition of glucose does not change the fluorescence signal. The fact that the intensity was sensitive to glucose demonstrates that the intensity changes are due to a binding event and not to trivial inner filter effects from ONPG.[67]

In recent publications we addressed the problem of obtaining reliable intensity measurements for sensing which could be used in the absence of useful changes in lifetimes. Polarization sensing is accomplished by constructing a sensor such that a stable intensity reference is observed through one polarizer and the sample is observed through a second orthogonal polarizer. In this case as reference we used a BSGK solution, which can be expected to display similar temperature, time or illumination dependent changes as the sample. To optimize the sensor response the reference intensity was about 65 % of the sample response, as calculated for the expected 2-3-fold intensity change. This reference is observed through a horizontally oriented polarizer.[67] The sample contains BSGK, ONPG and various concentrations of glucose, and is observed through a vertically oriented polarizer. The emission from both sides of the sensor is then observed through a vertically and horizontally oriented polarizer in order to measure polarization of the system. Figure 5.5 shows the observed polarization of the system for BSGK + ONPG and different glucose concentrations. An advantage of polarization measurements for sensing is that they are self-normalized and thus independent on the overall intensity of the sensor.

The results shown above demonstrate that a thermostable glucokinase can serve as a glucose sensor. Additional studies are needed to obtain a BSGK-based sensor which displays larger spectral changes. For example, we are hopeful that BSGK labeled with fluorophores other than IA-ANS will display larger intensity changes, spectral shifts or changes in lifetime. The results in the

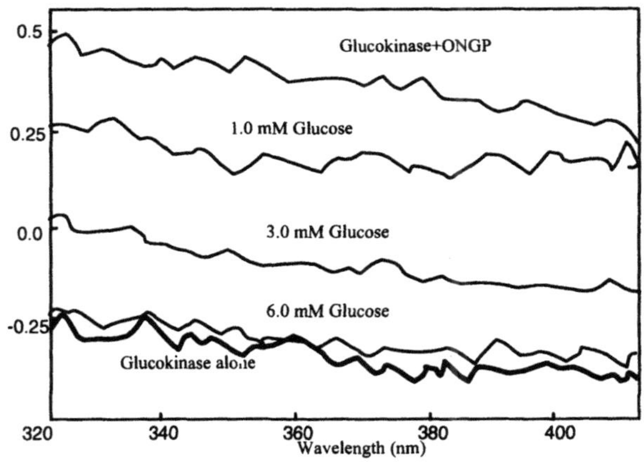

Figure 5.5. Effect of glucose on the polarization spectra of Glucokinase in the presence of ONGP

competitive RET are especially interesting because RET is a through-space interaction which occurs whenever the donor and acceptor are within the Forster distance (R_0), and does not require a conformational change and/or a change in the probe environment. For these reasons we are confident that BSGK can be used with longer wavelength donors and acceptors to develop practical glucose sensors for use in diabetes health care. Since the measurements through the skin can be easily performed by using a red laser diode or a light emitting diode (LED) as an excitation source, one may envision a polarization based device with an external calibrated standard that will allow noninvasive glucose determinations. The main advantage of using this method is the obtainment of ratiometric polarization measurements that are not influenced by light instability and sample perturbation.

5.5. CONCLUSIONS

The results described represent our first attempt to use inactive forms of enzymes as glucose sensor. Larger glucose-dependent spectral change would increase the accuracy of the glucose measurements. In spite of these difficulties we feel that our system demonstrates a useful approach to sensing. Our results suggest that the enzymes which use glucose as their substrate can be used as reversible and non-consuming glucose sensors in the absence of required co-factors. The possibility of using inactive apo-enzymes for a reversible sensor greatly expands the range of proteins which can be used as sensors, not only for glucose, but for a wide variety of biochemically relevant analytes. Hence one is no longer limited to use signaling proteins which bind the analyte without chemical reaction. The need for acetone may be eliminated by selecting proteins which are less thermophilic. The proteins can be engineered for covalent labeling by insertion of cysteine residues at appropriate locations in the sequence. The glucose induced spectral changes may be larger with other polarity sensitive probes or by the use of RET between two fluorophores on the protein. In summary, apo-enzymes appear to be a valuable source of protein sensors.

5.6. ACKNOWLEDGMENT

This project was realized in the frame of CRdC-ATIBB POR UE-Campania Mis 3.16 activities (S.D., M.R).

5.7. REFERENCES

1. H. King, R.E. Aubert, V.H. Herman, Global burden of diabetes, 1995-2025: prevalence, numerical estimates and projection, *Diabetes Care* **21**, 1414-1431 (1998).
2. J.P. Boyle, A.A. Honeycutt, K.M. Narayan, T.J. Hoerger, L.S. Geiss, H. Chen, T.J. Thompson, Projection of diabetes burden through 2050: impact of changing demography and disease prevalence in the U.S., *Diabetes Care* **24**, 1936-1940 (2001).
3. R.N. Anderson, Deaths: leading causes for 2000, *Natl. Vital. Stat. Rep.* **16**, 1-85 (2002).
4. L.S. Geiss, W.H. Herman, P.J. Smith, Mortality in non-insulin-dependent diabetes, in: *National Diabetes Data Group: Diabetes in America.* 2nd edition by NIH pub no. 95-1468 (Government Printing Office, Washington, DC 1995), pp. 233-257.
5. J.B. Meigs, Epidemiology of the metabolic syndrome. *Am. J. Manag. Care* **8**(11, Suppl 1), S283-S292 (2002).
6. S.I. McFarlane, M. Banerji, J.R. Sowers, Insulin resistance and cardiovascular disease. *J. Clin. Endocrinol. Metab.* **86**, 713-718 (2001).
7. Despres J.P., B. Lamarche, P. Mauriege, Cantin B., G.R. Dagenais, S. Moorjani, P.J. Lupien, Hyperinsulinemia as an independent risk factor for ischemic heart disease, *N. Engl. J. Med.* **334**(15), 952-957 (1996).
8. UK Prospective Diabetes Study (UKPDS) Group, Intensive blood-glucose control with sulphonylureas or insulin compared with conventional treatment and risk of complications in patients with type 2 diabetes (UKPDS 33), *Lancet* **352,** 837-53 (1998).
9. UK Prospective Diabetes Study Group, Tight blood pressure control and risk of macrovascular and microvascular complications in type 2 diabetes: UKPDS 38, *BMJ* **317**(7160), 703-13 (1998).
10. P. Gaede, P. Vedel, N. Larsen, G.V. Jensen, H.H. Parving, O. Pedersen, Multifactorial intervention and cardiovascular disease in patients with type 2 diabetes, *N. Engl. J. Med.* 348(5), 383-93 (2003).
11. The Diabetes Control and Complications Trial Research Group, The effect of intensive treatment of diabetes on the development and progression of long-term complications in insulin-dependent diabetes mellitus, *N. Engl. J. Med.* **329**(14), 977-86 (1993).
12. Y. Ohkubo, H. Kishikawa, E. Araki, T. Miyata, S. Isami, S. Motoyoshi, Y. Kojima, N. Furuyoshi, M. Shichiri, Intensive insulin therapy prevents the progression of diabetic microvascular complications in Japanese patients with non-insulin-dependent diabetes mellitus: a randomized prospective 6-year study, *Diabetes Res. Clin. Pract.*, **28**(2), 103-117 (1995).
13. M. Muggeo, G. Verlato, E. Bonora, F. Ciani, P. Moghetti, R. Eastman, G. Crepaldi, R. de Marco, Long-term instability of fasting plasma glucose predicts mortality in elderly NIDDM patients: the Verona Diabetes Study, *Diabetologia* **38**(6), 672-9 (1995).
14. M. Muggeo, G. Zoppini, E. Bonora, E. Brun, R.C. Bonadonna, P. Moghetti, G. Verlato, Fasting plasma glucose variability predicts 10-year survival of type 2 diabetic patients: the Verona Diabetes Study, *Diabetes Care* **23**(1), 45-50 (2000).
15. A. Ceriello, Postprandial hyperglycemia and diabetes complications: is it time to treat? *Diabetes* **54**(1), 1-7 (2005).

16. B.H. Ginsberg, An overview of minimally invasive technologies, *Clin. Chem.* **38**(9), 1596-1600 (1992). Review. Erratum in: *Clin. Chem.* **38**(11), 2360 (1992).
17. E. Boland, T. Monsod, M. Delucia, C.A. Brandt, S. Fernando, W.V. Tamborlane, Limitations of conventional methods of self-monitoring of blood glucose: lessons learned from 3 days of continuous glucose sensing in pediatric patients with type 1 diabetes, *Diabetes Care* **24**(11), 1858-62 (2001).
18. L. Heinemann, H. Overmann, I. Muhlhauser, How well do patients with type 1 diabetes measure their blood glucose in daily life, *Diabetes Care* **21**(3), 461-2 (1998).
19. L. Heinemann, G. Schmelzeisen-Redeker, Non-invasive continuous glucose monitoring in Type I diabetic patients with optical glucose sensors. Non-Invasive Task Force (NITF), *Diabetologia* **41**(7), 848-854 (1998).
20. W. Yang, H. He, and D.G. Drueckhammer,. Computer-guided design in molecular recognition: Design and synthesis of a glucopyranose receptor, *Angew. Chem. Int. Ed.* **40**(9), 1714-1718 (2001).
21. T.D. James, P. Limaane, and S. Shinkai, Fluorescent sacchariderecceptors: A sweet solution to the design, assembly, and evaluation of boronic acid derived PET sensors, *Chem. Commun.* 281-288 (1996).
22. C. R. Cooper, and T. D. James, Synthesis and evaluation of D-glucosamine-selective fluorescent sensors, *J. Chem. Soc., Perkin Trans.* **1**, 963-969 (2000).
23. A.J. Tong, A. Yamauchi, T. Hayashita, Z.-Y. Zhang, B. D. Smith, and N. Teramae, Boronic acid fluorophore/β-cyclodextrin complex sensors for selective sugar recognition in water, *Anal. Chem.* **73**,1530-1536 (2001).
24. N. Di Cesare, and J.R. Lakowicz, Wavelength-ratiometric probes for saccharides based on donor-acceptor diphenylpolyenes, *J. Photochem. Photobiol. A: Chem.* **143**(1), 39-47 (2001).
25. N. Di Cesare, and J.R. Lakowicz, Spectral properties of fluorophores combining the boronic acid group with electron donor or withdrawing groups. Implication in the development of fluorescence probes for saccharides, *J. Phys. Chem.* **105**, 6834-6840 (2001).
26. J.S. Schultz, and G. Sims, Affinity sensors for individual metabolites, *Biotech. Bioeng. Symp.* **9**, 65-71 (1979).
27. J. Schultz, S. Mansouri, and I.J Goldstein, Affinity sensor: A new technique for developing implantable sensors for glucose and other metabolites, *Diabetes Care* **5**(3), 245-253 (1982).
28. D. Meadows, and J.S. Schultz, Fiber-optic biosensors based on fluorescence energy transfer, *Talanta* **35**(2),145-150 (1988).
29. Ballerstadt, R., and Schultz, J. S. (2000). A galactose-specific affinity hollow fiber sensor based on fluorescence resonance energy transfer, *Methods Biotechnol.* 7:89-98.
30. R.J. Russell, and M.V. Pishko, A fluorescence-based glucose biosensor using concanavalin A and dextran encapsulated in a poly(ethylene glycol) hydrogel, *Anal. Chem.* **71,** 3126-3132 (1999).
31. R. Ballerstadt, and J.S. Schultz, A fluorescence affinity hollow fiber sensor for continuous transdermal glucose monitoring, *Anal. Chem.* 72, 4185-4192 (2000).
32. O.J. Rolinski, D.J.S. Birch, L.J. McCartney, and J.C. Pickup, Sensing metabolites using donor-acceptor nanodistributions in fluorescence resonance energy transfer, *App. Phys. Letts.* **78**(18), 2796-2798 (2001).
33. J.R. Lakowicz, and B.P. Maliwal, Optical sensing of glucose using phase-modulation fluorometry, *Anal. Chim. Acta* **271**, 155-164 (1993).
34. L. Tolosa, H. Malak, G. Rao, and J.R. Lakowicz, Optical assay for glucose based on the luminescence decay time of the long wavelength dye Cy5^{TM}, *Sensors and Actuators*, **45**, 93-99 (1997).
35. L. Tolosa, H. Szmacinski, H. Rao, and J.R. Lakowicz, Lifetime-based sensing of glucose using energy transfer with a long lifetime donor, *Anal. Biochem.* **250**, 102-108 (1997).
36. L. Tolosa, I. Gryczynski, L.R. Eichhorn, J.D. Dattelbaum, F.N. Castellano, G. Rao, and J.R. Lakowicz, Glucose sensor for low-cost lifetime-based sensing using a genetically engineered protein, *Anal. Biochem.* **267**, 114-120 (1999).
37. S. D'Auria, P. Herman, M. Rossi, and J.R. Lakowicz, The fluorescence emission of the apo-glucose oxidase from Aspergillus niger as probe to estimate glucose concentrations, *Biochem. Biophys. Res. Commun.* **263**, 550-553 (1999).

38. S. D'Auria, N. Di Cesare, Z. Gryczynski, I. Gryczynski, M. Rossi, and J.R. Lakowicz, A thermophilic apoglucose dehydrogenase as nonconsuming glucose sensor, *Biochem. Biophys. Res. Commun.* **274**, 727-731 (2000).
39. J.S. Marvin, and H.W. Hellinga, Engineering biosensors by introducing fluorescent allosteric signal transducers: Construction of a novel glucose sensor, *J. Am. Chem. Soc.* **120**, 7-10 (1998).
40. J.S. Marvin, E.E. Corcoran, N.A. Hattangadi, J.V. Zhang, S.A. Gere, and H.W. Hellinga, The rational design of allosteric interactions in a monomeric protein and its applications to the construction of biosensors, *Proc. Natl. Acad. Sci. USA* **94**, 4366-4371 (1997).
41. K.A. Giuliano, and D.L. Taylor, Fluorescent-protein biosensors: New tools for drug discovery, *TIB Tech.* **16**, 135-140 (1998).
42. K.A. Giuliano, and P.L. Post, Fluorescent protein biosensors: Measurement of molecular dynamics in living cells, *Annu. Rev. Biophys. Biomed. Struct.* **24**, 405-434 (1995).
43. G.S. Wilson, and Y. Hu, Enzyme-based biosensors for *in vivo* measurements, *Chem. Rev.* **100**, 2693-2704 (2000).
44. H.W. Hellinga, and J.S. Marvin, Protein engineering and the development of generic biosensors, *TIB Tech.* **16**, 183-189 (1998).
45. T. Schalkhammer, C. Lobmaier, B. Ecker, W. Wakolbinger, E. Kynclova, G. Hawa, and F. Pittner, Microfabricated glucose, lactate, glutamate and glutamine thin-film biosensors, *Sensors and Actuators B* **18-19**, 587-591 (1994).
46. H. Szmacinski, and J.R. Lakowicz, "Lifetime-based sensing," in *Topics in Fluorescence Spectroscopy,* (J. R. Lakowicz, ed.), Plenum Press, New York, Vol. 4, pp. 295-334 (1994).
47. H. Szmacinski, and J.R. Lakowicz, Fluorescence lifetime-based sensing and imaging, *Sensors and Actuators B* **29**, 15-24 (1995).
48. M.E. Lippitsch, S. Draxler, and D. Kieslinger, Luminescence lifetime-based sensing: new materials, new devices, *Sensors and Actuators B* **38-39**, 96-102 (1997).
49. S.B. Bambot, G. Rao, M. Romauld, G.M. Carter, and J.R. Lakowicz, Sensing oxygen through skin using a red diode laser and fluorescence lifetimes, *Biosensors and Bioelectronics*, **10**(6-7), 643-652 (1995).
50. K. Rebrin, T.V. Fishcher, P. Woedtke, and E. Brunstein, Automated feedback control of subcutaneous glucose concentration in diabetic dogs, *Diabetologia* **32**, 573-576 (1989).
51. G. Velho, P. Froguel, D.R. Thevenot, and G. Reach, In vivo calibration of a subcutaneous glucose kinetics, *Diab. Nutr. Metab.* **3**, 227-233 (1988).
52. Smith, A., Yang, D., Delcher, H., Eppstein, J., Williams, D., and Wilkes, S. (1999). Fluorescein kinetics in interstitital fluid harvested from diabetic skin during fluorescein angiography: Implications for glucose monitoring, *Diabetes Tech. & Therap.* 1(1):21-27.
53. D.J. Menstein, E.F. Pai, L.M. Schopfer, and V. Massey, Absolute stereochemistry of flavins in enzyme catalyzed reactions, *Biochemistry* **25**(22), 6807-6816 (1986).
54. S. D'Auria, P. Herman, M. Rossi, J.R. Lakowicz, The fluorescence emission of the apo-glucose oxidase from Aspergillus niger as probe to estimate glucose concentrations, *Bioch. Biophys. Res. Comm.* **263**, 550-553 (1999).
55. V.V. Mozhaev, I.V. Berezim, and K. Martinek, Structure-stability relationship in proteins: fundamental tasks and strategy for the development of stabilized enzyme catalysts for biotechnology, CRC *Crit. Rev. Biochem.* **25**, 235-281 (1998).
56. S. D'Auria, R. Barone, M. Rossi, R. Nucci, G. Barone, G. Fessas, E. Bertoli, and F. Tanfani, Effects of temperature and SDS on the structure of b-glycosidase from the thermophilic archaeon Sulfolobus solfataricus, *Biochem. J.* **323**, 833-840 (1997).
57. S. D'Auria, R. Nucci, M. Rossi, I. Gryczynski, Z. Gryczynski, and J.R. Lakowicz, The β-Glycosidase from the Hyperthermophilic Archaeon Sulfolobus Solfataricus: Enzyme Activity and Conformational Dynamics at Temperatures Above 100 °C, *Biophys. Chem.* **81**, 23-31 (1999).
58. J.R. Lakowicz, and H. Szmacinski, Fluorescence lifetime-based sensing of pH, Ca^{2+}, K^{+} and glucose, *Sensors Actuators B* **11**, 133-143 (1993).
59. S. D'Auria, N. Di Cesare, Z. Gryczynski, I. Gryczynski, M. Rossi, and J.R. Lakowicz, A thermophilic apoglucose dehydrogenase as nonconsuming glucose sensor, *Biochem. Biophys. Res. Comm.* **274**, 727-731 (2000).
60. T. Ureta, C. Medina, and A. Preller, The evolution of hexokinases, *Arch. Biol. Med. Exp.* **20**(3-4), 343-357 (1987).

61. W.S. J. Bennet, and T.A. Steitz, Structure of a complex between yeast hexokinase A and glucose: structure determination and refinement at 3.5 Å resolution, *J. Mol. Biol.* **140**(2), 183-209 (1980).
62. A.R. Woolfitt, G.L. Kellet, and J.G. Hogget, The binding of glucose and nucleotides to hexokinase from Saccharomyces cerevisiae, *Biochim. Biophys. Acta* **952**(2), 238-243 (1998).
63. I. Feldman, Ionic strength dependence of glucose binding by yeast hexokinase isoenzyme, *Biochem. J.* **217**(1), 335-337 (1984).
64. R.C. McDonald, T.A. Steitz, and D.M. Engelman, Yeast hexokinase in solution exhibits a large confomational change upon binding glucose or glucose 6-phosphate, *Biochemistry* **18**(2), 338-342 (1979).
65. C.R. Goward, M.D. Scawen, and T. Atkinson, The inhibition of glucokinase and glycerokinase from Bacillus stearothermophilus by the triazine dye procion blue MX-3G, *Biochem. J.* **246**, 83-88 (1987).
66. H. Ishikawa, T. Maeda, and H. Hikita, Initial-rate studies of a thermophilic glucokinase from Bacillus stearothermophilus, *Biochem. J.* **248**, 13-20 (1987).
67. S. D'Auria, N. Di Cesare, M. Staiano, Z. Gryczynski, M. Rossi, and J.R. Lakowicz, A novel fluorescence competitive assay for glucose determinations by using a thermostable glucokinase from the thermophilic microorganism Bacillus stearothermophilus, *Analitycal Biochem.* **303**, 138-144 (2001).

MICROCAPSULES AS "SMART TATTOO" GLUCOSE SENSORS: ENGINEERING SYSTEMS WITH ENZYMES AND GLUCOSE-BINDING SENSING ELEMENTS

Michael J. McShane[1]

6.1. THE "SMART TATTOO" CONCEPT

Though not obvious as a noninvasive tool due to the need for introduction of exogenous indicator materials, recent research has advanced concepts that may lead to application of fluorescence spectroscopy to noninvasive biosensing.[1-3] Key efforts have focused on developing carriers for glucose assay chemistry that physically restrict the transduction molecules yet allow sufficient transport of glucose. The ultimate goal is then to implant these devices (AKA "smart tattoos") in the skin, where they would be exposed to interstitial fluid, respond to local changes in glucose that correlate with blood levels,[4,5] and be "viewed" from the surface of the skin using optical instrumentation to excite and measure fluorescence (Figure 6.1).

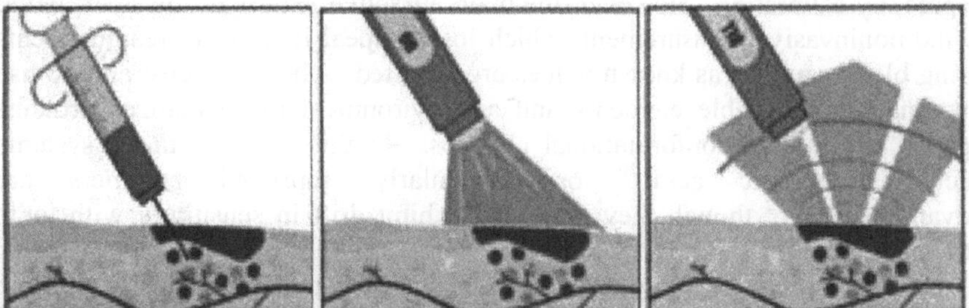

Figure 6.1. Illustration of "smart tattoo" concept: implantation, interrogation, and readout.

1 Biomedical Engineering, Institute for Micromanufacturing, PO Box 10137, 911 Hergot St.Ruston, LA 71272
E-mail: mcshane@latech.edu, Phone: 318-257-5100

A number of desirable features for "smart tattoos" can be identified.[6,7] First, and foremost, the materials used must be sufficiently *biocompatible* to avoid severe acute or sustained long-term host response to the implant. An acute inflammatory response to the implant is expected due to local tissue trauma during the procedure, though the steady-state response should be limited to minimal fibrous tissue formation, and activation of the immune response by antibodies recognizing foreign material must be avoided. Generically, this response is determined the surface properties of the implant[8] and, so long as the contents are not released or degraded over time, a stable interface will be formed between the implant surface and the host. The purity of the materials used in sensor fabrication can be a key factor in determining biocompatibility.

A second key property of tattoo systems is a *strong, sensitive, reversible, and fairly rapid response*. These terms are given together because the fluorescence signal from the implant must be measurable, must change significantly over the range of interest for the system, and must accurately reflect the current glucose levels to be useful. Neglecting the influence of the optical instrumentation employed, this essentially requires a combination of properties: adequate amounts of fluorophore with high quantum yield, use of wavelengths with sufficient penetration depth for tissue, and an assay that produces large percentage change from baseline measurements over the range of interest. Ideally, a linear response that produces a 2-10 times change in the measurand (e.g. intensity ratio, wavelength, lifetime) over the 0-600mg/dL range could be achieved with a sub-minute response time. However, even nonlinear profiles and response times on the order of a few minutes, sufficiently short to accurately reflect details of blood glucose excursions, would be sufficient if glucose could be measured with error <10%.[5]

A third desired characteristic of fluorescent glucose sensor implants is a *stable assay response*. Changes in sensitivity and signal levels due to denaturation of protein elements or photobleaching can be compensated, to a degree, by calibration. However, the main attractive feature of the smart tattoo is the noninvasive measurement, which loses appeal if frequent readjustments using blood samples as known values are required. Thus, it is desirable to use maximally photostable elements and an environment that stabilizes proteins against irreversible conformational changes. In this case, synthetic systems such as boronic acid[9,10] or molecularly imprinted polymers are advantageous,[11,12] though they may also exhibit drift in sensitivity with ionic strength and pH.[13]

Specificity is a key aspect of sensors in general, and this is a prime reason for great attention given to enzymatic transduction schemes for electrochemical and other sensor types. However, it can be generally stated that specificity need only be defined in terms of what will be encountered in extremes of normal operation. For example, specificity to glucose is desired in a glucose sensor, but since other sugars are not expected to be present in high levels within interstitial fluid, the response to glucose need only be such that the glucose sensitivity is minimally affected by the presence of potential interferents at their maximum expected physiological levels. Thus, while sensors that demonstrate high specificity for glucose over other sugars or potentially interfering species are attractive, sensors exhibiting low specificity should not be excluded from consideration in tattoo systems on that basis alone.

Another attribute desired in implants is *minimal or zero consumption of analytes and minimal or zero byproduct formation.* Enzymatic systems, while attractive because of specificity and potential for high sensitivity, generally consume glucose and co-substrates, decreasing the local levels, while producing other species that may have deleterious effects on the sensors or the surrounding tissue. For example, gluconic acid and hydrogen peroxide are typical products of glucose oxidation.[14] These can affect sensors through several means, including: 1) changing enzymatic activity due to reversible pH-induced conformational alterations; 2) changing fluorescence signals from slightly pH-sensitive fluorophores; 3) irreversible degradation of enzyme due to cleavage of peptide bonds by peroxide; and possibly other mechanisms. Damage to surrounding tissue caused by removal of key nutrients (oxygen, glucose) or production of toxic materials (acid, peroxide, reactive oxygen species, etc) must also be avoided. Thus, careful consideration must be given to consumption and production and appropriate design features used to reduce or eliminate their influence on the sensor and host. Tattoo materials based on non-consuming glucose-binding reactions (e.g. Concanavalin A,[15] boronic acid,[6,9,16] glucose/galactose binding protein, and apo-enzymes[17]) are advantageous in this regard.

Smart tattoos, because of their intended location in dermal tissue, must have *optical properties that enable interrogation with reasonable signal-to-noise ratio.* This essentially requires that the implant materials have excitation and emission in spectral regions that can be probed with low-cost instrumentation, while still allowing sufficient penetration of excitation light to elicit strong signals and collection of emitted photons at the skin surface. Given the reasonably well-understood optical properties of skin,[18,19] the ideal range for optical communication with smart tattoos is in the long wavelength visible and near-infrared region, for example 600-1000nm, due both to decreased scattering and the minimal absorption by tissue chromophores and water. However, it has been shown that even shorter wavelengths can be feasibly used with an appropriately-designed instrument, even for steady-state measurements;[1] improvement in SNR can be expected when modulated light or phase lifetime methods are employed.

The final desired feature for smart tattoos to be discussed here is *reliable, repeatable fabrication.* As with any device, particularly those for medical use, quality control is a key. This is particularly true for small-scale systems, and even more so for those for which function depends directly and sensitively on size, shape, and composition. Tattoo materials requiring delicate balance between diffusion and reaction will require careful control over assembly to achieve the necessary configuration and dimensions, as discussed in later sections.

Use of competitive-binding reactions requires free movement of molecules within a restricted range such as can be obtained with dialysis membranes.[15,20] However, using relatively large hollow membranes has drawbacks in production, fouling, and mechanical stability. Use of polymeric matrices where the movement of molecules is physically restricted has shown promise, but issues of response time and reversibility persist. Thus, while the concept of implantable fluorescent sensors is attractive, realizing this goal remains a challenge.

Microcapsules with precisely controlled properties offer promise to solve a number of existing problems in creating smart tattoo systems. The remainder of this chapter will focus on nanoengineered microcapsules as a platform for developing glucose-sensing smart tattoos, including both enzymatic and non-enzymatic transduction approaches. Before discussing the details of sensor construction and performance, an introduction to layer-by-layer self assembly, a versatile process used to produce microcapsules with finely tuned properties, is provided.

6.2. LBL NANOFILMS AND POLYELECTROLYTE MICROCAPSULES

Layer-by-layer self-assembly (LbL) of multilayer films involves the construction of complex composite materials with nanoscale precision in film thickness, one layer at a time, enabling the development of novel structures and devices with properties tailored by controlling the molecular makeup and arrangement.[21,22] Early fundamental studies of multilayer assemblies on planar substrates demonstrated the practicality and versatility of the approach,[23] and the process has been developed and extensively studied over the last decade for a wide variety of synthetic and natural materials.[24] This has included further investigation into the internal structure and composition of LbL films, including dynamic and long-term interactions between film components, solvents and solute, especially transport properties.[25]

LbL has recently been reviewed extensively,[25] and therefore will only be introduced here in the context of constructing glucose biosensors. The LbL process is general, relying primarily on the attractive force between oppositely charged molecules; the principal idea of the method consists of the resaturation of polyion adsorption, resulting in the alternation of the terminal charge after every subsequent layer deposition (see Figure 6.2). Materials employed in the assembly process range from charged polymers ("polyelectrolytes" or "polyions") to proteins, dyes, and semiconductor nanoparticles.[26,27] Each layer has a thickness on the order of a few nanometers, and this thickness is controllable through careful selection of materials and reaction parameters. Furthermore, the composition of films can be engineered through the sequential deposition of different materials; thus, complex film architecture may be achieved through a common process without need for complicated chemistry. Thus, the LbL method provides a practical procedure for building precisely engineered nanocomposite films in the range of 5-1000 nm, with a precision better than 1 nm and a definite knowledge of molecular composition. The procedure has also been elaborated to demonstrate the fabrication of three- dimensional layered structures by using charged substrates with micrometer and nano-meter dimensions.[28-30] Specifically, microspheres and nanoparticles have been extensively studied because of the wide availability and attractive surface properties of spherical particles. Functional nanocomposite films can be deposited on these carriers, which are attractive for controlling biocatalytic reactions and other interfacial phenomena due to the high surface-to-volume ratio.[31-34]

In the past several years, a technique for fabrication of hollow micro/nanocapsules has been developed using an extension of the polyelectrolyte layering process applied to colloids. Subsequent to deposition of coatings comprising charged molecules around the templates, the core material is removed using organic solvents or chemical etching to achieve hollow shells (Figure 6.2).[28,35-37] The method is very versatile, and shell properties can be controlled: inner diameters are dependent upon the size of templates used for assembly, wall thicknesses can be controlled precisely to within a few nanometers, and walls may be constructed using a diverse selection of charged molecules (polyelectrolytes, proteins, dyes, inorganic particles, copolymers, etc.).[28] The versatility in construction of these tiny capsules and control over their properties makes them attractive for use in sensor applications, specifically those in which encapsulation of active molecules and control over transport properties are critical to proper function.

Building upon the possibilities afforded by coating micro/nanoscale templates with functional materials such as enzymes and dyes, sensors for chemicals and biochemicals are being developed as tools for biological research, medical diagnostics and monitoring, and biodefense applications. Using nanoparticles coated with fluorescent materials and microcapsules loaded with indicator chemistry, ratiometric nanoscale probes have been developed for intracellular and extracellular measurements of ions and oxygen.[38-41] These micro/nanodevices have advantages over standard liquid-phase small molecule

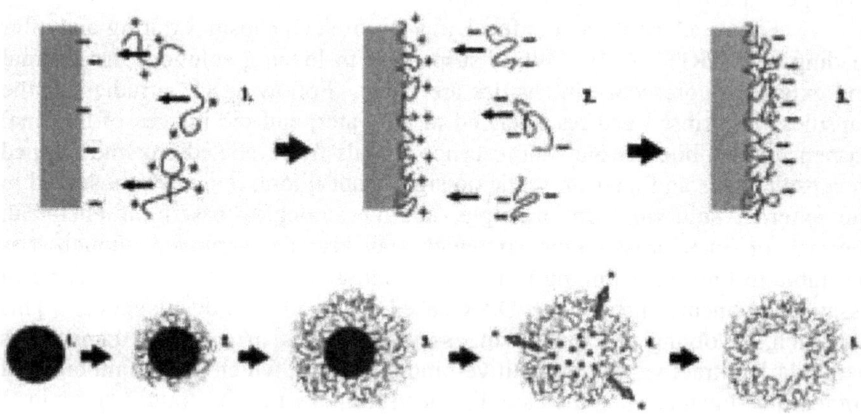

Figure 6.2. Schematic illustration of layer-by-layer nanofabrication process: (top) applied to planar templates and (bottom) applied to colloidal templates to create hollow microcapsules.

indicators in that they provide a protective package for the chemistry, separating the dyes from the biological environment and, in doing so, reducing nonspecific responses, dye-protein binding, and toxic effects. In addition, the immobilization of the indicators within the nanofilms provides physical linkage with reference fluorophores, allowing constant ratiometric monitoring without large shifts in calibrations otherwise seen. By coupling the chemical nanosensors with an enzyme nanoreactor,[33,34] a second class of sensors may be realized, using direct monitoring of co-reactants or products as an indirect means of determining the concentration of other substrates.

6.3. GLUCOSE SENSORS FROM NANOENGINEERED CAPSULES

Microcapsules based on the layer-by-layer assembly technique were recently discussed as potential candidates for glucose-sensing smart tattoo systems.[42] Such devices may take many forms, though in all cases, it can be assumed that the fluorescence assay chemistry must be stably entrapped within a carrier vehicle, typically constructed from a natural or synthetic polymer. Polyelectrolyte microcapsules present an attractive opportunity to construct these carriers, and offer a variety of options in achieving stable immobilization of the assay chemistry. Early investigations of the properties of polymer capsules revealed that permeability depends directly on the wall composition,[43,44] and some architectures allow for dynamic control over permeability to open and close pores.[45,46] For example, pH-induced formation of pores (~100nm) was observed for a combination of strong and weak polyelectrolytes.[47] Furthermore, changes in solvents have also been shown to induce encapsulation.[48-50]

A more recent development employs a photosensitive cationic resin (diazoresin, DAR)[51-53] to enable photoinduced crosslinking of hollow capsule walls by UV irradiation following diffusion loading of capsules,[54] providing a more permanent and stable encapsulation.

Figure 6.3 contains confocal images of such capsules during and after loading with TRITC-GOx. While suspended in loading solution, the internal and external fluorescence intensities are equal. Following UV-irradiation, the capsules were rinsed and resuspended in DI water, and the images of the final suspension exhibited strong fluorescence signals from labeled enzyme trapped in capsule walls and interior, while no significant fluorescence was observed in the external solution. In principle, similar strategies based on chemical, thermal, or other crosslinking strategies may also be employed, though it is desirable to limit crosslinking to the walls and avoid undesired crosslinking of assay components; hence, the DAR-based approach is advantageous. This approach, involving entrapment of assay molecules in a hollow capsule, is particularly attractive for competitive-binding assays, which require mobility of competitive ligands when they are displaced from the glucose binding protein.

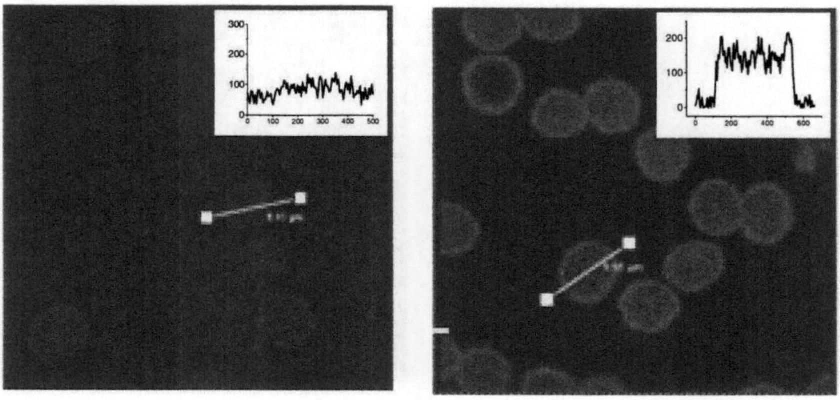

Figure 6.3. Encapsulation by DAR-induced crosslinking of capsule walls. Left: Capsules in TRITC-GOx loading solution, prior to irradiation. Right: Fluorescence from crosslinked DAR capsules following UV irradiation, rinsing, and resuspension in DI water.

Another approach for glucose assay entrapment relies on formation of hydrogel microspheres with the sensing chemistry included in the gel precursor. For example, formation of microspheres in a water-in-oil (w/o) emulsion has been used to trap glucose oxidase in calcium-crosslinked alginate microspheres (Figure 6.4, Figure 6.5). While alginate matrices have been studied for Gox encapsulation for some time, the primary limitation of these for long-term carriers is the leaching of enzyme from the gel. In this case, a nanofilm coating is easy to apply (Figure 6.5), serving as a simple and effective barrier to loss of encapsulated macromolecules.[55] A number of different material combinations are being compared to assess potential for stable immobilization of active enzyme.[56,57]

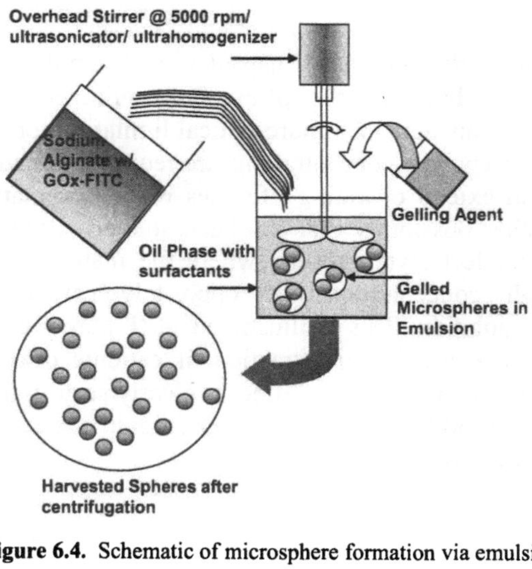

Figure 6.4. Schematic of microsphere formation via emulsion.

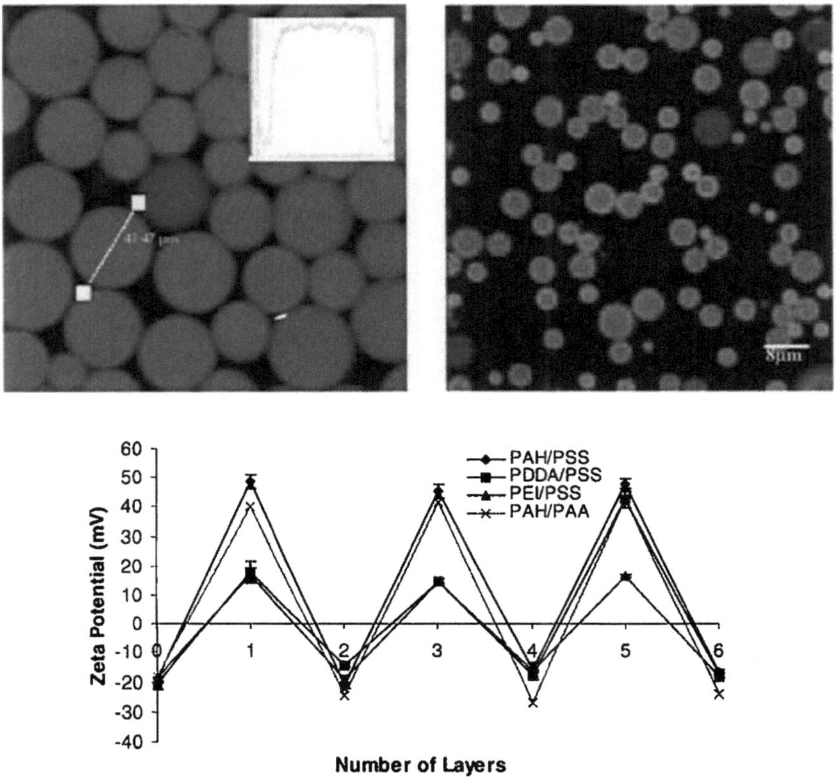

Figure 6.5. Top left: FITC-GOx loaded calcium-crosslinked microspheres formed by emulsion. Top right: Smaller alginate microspheres doped with FITC-GOx and coated with RITC-poly(allylamine) nanofilm. Bottom: Zeta potentials measured from alginate microspheres during formation of nanofilm coating. (see the color insert after p. 429.)

The diffusion-loaded capsules and emulsion-based systems have advantages in stability and ease of use, but are limited in concentrations that can be achieved. This may be a more critical limitation for enzyme-based systems, in which high enzyme concentrations are required to maintain diffusion-limited behavior and extend operating lifetimes when enzymatic activity is lost with time. Another possibility that has been studied is the use of charged matrix within a polyelectrolyte capsule, where the matrix serves to electrostatically adsorb high concentrations of oppositely-charged molecules from the surrounding solution. This is illustrated in (Figure 6.6), and demonstrated for anionic alginate matrix for attraction of cationic dextran (amino-dextran) in (Figure 6.7). The nature of the effect is clearly seen in the uptake of significant dextran-amino (500kDa), while smaller anionic dextran (77kDa) was excluded from the same particles.

These examples constitute only a limited view of many possibilities for microcapsule-based sensor construction; they show promise for building stable systems with entrapped glucose-sensing chemistry and therefore provide sufficient basis for discussion of the different sensing systems that can be achieved using them. The remainder of this chapter will focus on the assembly of systems with different assays in an effort to realize glucose-sensitive responses.

These systems can be divided for discussion based first on the *interaction with glucose* and second based on the *transduction employed.* For purposes of this discussion, interaction options for microcapsule-based smart tattoo sensors will be discussed in terms of enzymatic and binding systems, followed by examples of prototype sensors employing transduction through fluorescence quenching and energy transfer.

Enzymatic glucose sensing typically relies upon the oxidation of glucose driven by glucose oxidase. The reaction is discussed in more detail in the following section; here, it is sufficient to note that oxygen is a co-substrate with glucose, and both hydrogen peroxide and gluconic acid are produced. Fluorescence monitoring of oxygen, discussed below, is typically preferred, as it provides an indirect measure of glucose level. Alternatively, pH or peroxide could similarly be used. Drawbacks specific to using enzymes include: 1) changes in activity over time, leading to drifting calibration curves; 2) dependence on local oxygen levels; and 3) consumption of analyte and co-substrates, accompanied with production of byproducts. Even if these issues can be overcome, simply creating a system where diffusion and reaction are sufficiently balanced for a sensitive response is a difficult task. Because layer-by-layer self assembly allows deposition of ultrathin polyelectrolyte multilayer

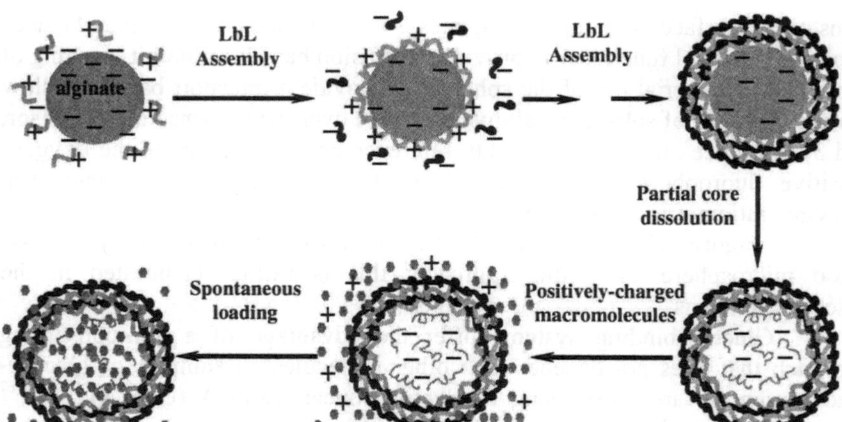

Figure 6.6. Electrostatic "sponge" encapsulation scheme employing charged matrix in polyelectrolyte shell.

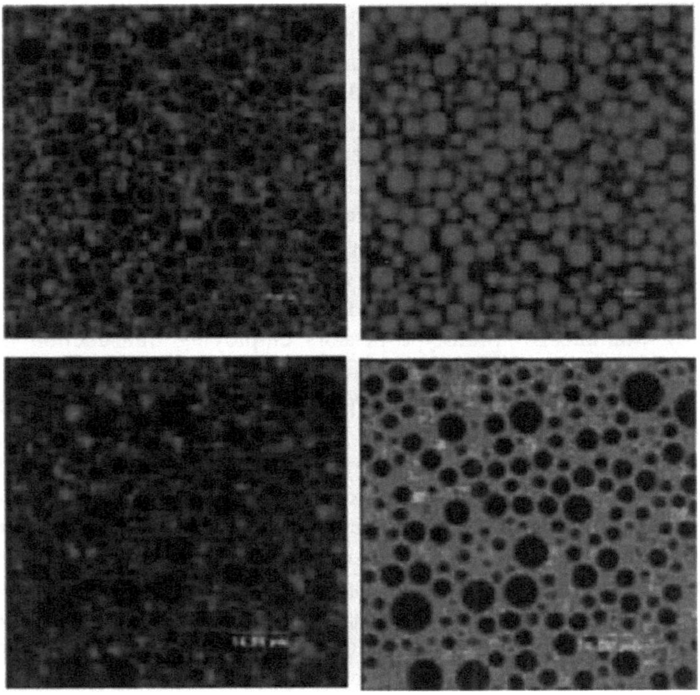

Figure 6.7. Electrostatic uptake of macromolecules. Left: alginate microspheres coated with TRITC-PAH/PSS nanofilms. Top right: same capsules following exposure to TRITC-POx. Bottom right: same capsules following exposure to anionic FITC-dextran 500kDa. (see the color insert after p. 429.)

films on the surface of colloidal templates, such a nanofilm coating may be used to perform several functions: 1) provide a diffusion barrier to inhibit leaching of encapsulated material out of the spheres, 2) provide a transport barrier to slow inward diffusion of substrates, allowing control over the response of the sensor, and 3) introduce an internal intensity reference complementary to the oxygen-sensitive fluorophore by the use of polyelectrolyte-fluorophore conjugates, allowing ratiometric measurements.

Figure 6.9 contains an illustration and image of a prototype smart tattoo microsphere using this approach; this is further elaborated in the following sections.

Glucose-binding systems offer the advantage of a non-consuming approach that does not depend upon other molecules. Examples of glucose-binding sensors are those using the lectin Concanavalin A (Con A),[15,20,58-62] glucose/galactose binding protein (GBP),[63,64] apo-enzymes (e.g. apo-GOx),[17] and synthetic systems using boronic acid (BA).[6,65] These approaches rely on either conformational changes or competitive-binding assays to transduce glucose information. In the former case, conformational changes can be transduced using environmentally-sensitive fluorophores attached directly to the protein, and fluorescence changes are observed as the dye is exposed to more polar environment. An alternative approach is to employ resonance energy transfer (RET)[66,67] by applying dual labels to the protein, such that in one conformation the fluorescent tags are close together (more energy transfer) and the conformational change induced by binding or dissociating from glucose displaces the fluorophores and results in decreased energy transfer.

Competitive binding assays also typically use RET; in these situations, the glucose-binding molecule is labeled with either the donor or acceptor, while a competitive ligand (glucose analog) is tagged with the other component of the RET pair. In the absence of glucose, the protein binds the ligand and energy transfer is increased; in the presence of glucose, the ligand and associated fluorophore are displaced and RET is decreased. As an example, Figure 6.8 illustrates a competitive-binding assay based on the differential affinity of TRITC-apo-GOx for glucose and FITC-dextran, where the energy transfer between FITC and TRITC is exploited.[68] It is noteworthy that RET approaches are extremely sensitive and inherently ratiometric, both of which are attractive for the proposed tattoo systems.

Smart tattoos using glucose binding approaches would require encapsulation of the recognition elements within a biocompatible, semi-permeable shell.

Figure 6.9 contains an illustration and images of a prototype glucose sensor employing apo-GOx with dextran in a competitive-binding assay. A typical disadvantage to these systems is the loss in specificity for glucose, for which enzymes are far superior. Furthermore, it is difficult to engineer proteins

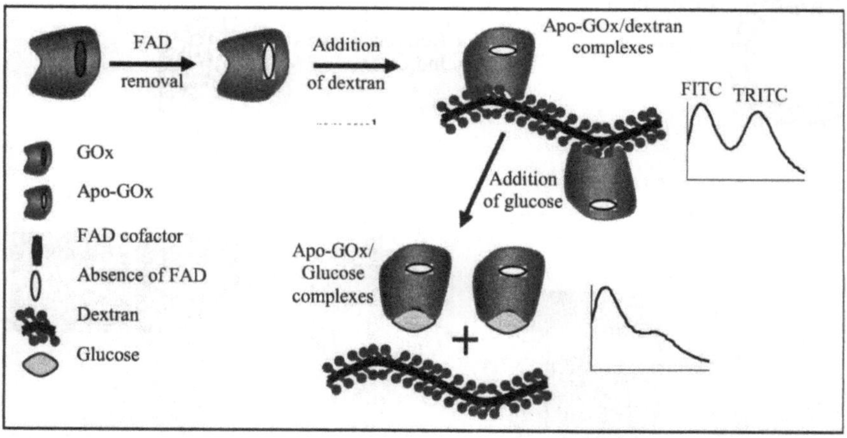

Figure 6.8. Schematic of a glucose assay based on competitive binding between dextran and glucose for binding sites on apo-GOx

to have sensitive response in the range of interest for physiological measurements, as glucose binding proteins possess glucose dissociation constants that fall well below the 1-10 mM range.[69]

6.4. ENZYME-BASED MICROCAPSULE SENSORS

While the layer-by-layer self assembly process coupled with other chemical production techniques such as emulsification enables the construction of a wide variety of microspheres from a seemingly infinite selection of materials, creating useful devices that function well at the microscale demands careful consideration of the system of interest. In the case of microsphere glucose sensors employing enzymes to drive a reaction that will be monitored with an oxygen indicator, this requires a balance between reaction (consumption) and diffusion (supply) of the co-substrates glucose and oxygen. This balance must be engineered to arrive at a measurable signal change for the expected glucose concentrations. This section focuses on a model-based approach to design of these devices, and concludes with an example of prototype sensors made using model output.

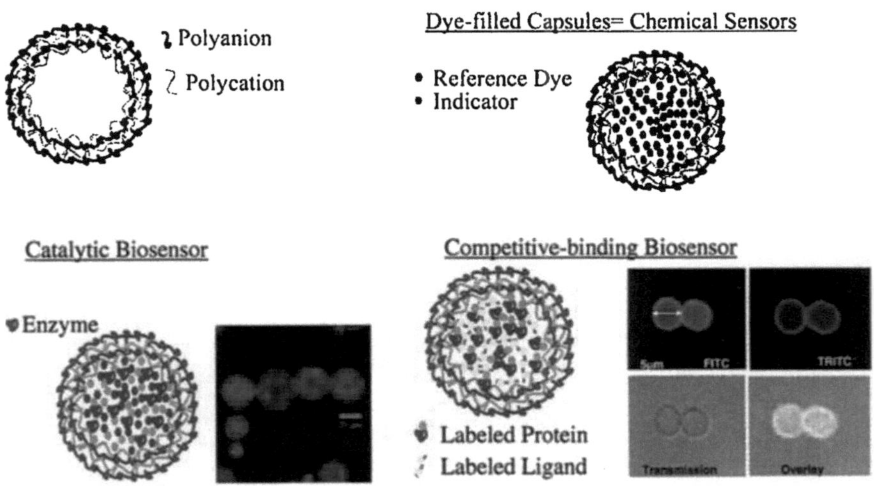

Figure 6.9. General description of microcapsule-based sensors. Top left: polyelectrolyte capsule, hollow or filled with polymer matrix; Top right: capsule filled with indicator and reference dye for ratiometric monitoring of a target analyte; Bottom left: enzyme-based microcapsule sensor containing enzyme and dyes; Bottom right: microcapsule sensor employing competitive-binding FRET assay, comprising glucose-binding protein labeled with acceptor and glucose analog labeled with donor. (see the color insert after p. 429.)

6.4.1. Design of Sensors through Mathematical Modeling

This section discusses the design of a functional glucose sensor using microcapsules containing an enzyme. While the presentation will focus on the glucose oxidase reaction, it is noteworthy that similar treatment may be given to other enzymatic reactions, so long as the appropriate rate constants and diffusion coefficients can be estimated. To begin with, a basic understanding of the reaction that occurs is necessary. The reaction scheme for a system involving glucose oxidase is

$$\begin{aligned} A + E &\underset{k_1, k_{-1}}{\longleftrightarrow} X_1 \xrightarrow{k_2} F + gluconate \\ F + C &\underset{k_3, k_{-3}}{\longleftrightarrow} X_3 \xrightarrow{k_4} E + H_2O_2 \end{aligned} \tag{1}$$

where k_1, k_2, k_3, k_4 and k_{-1}, k_{-3} are the forward and reverse reaction rate constants, respectively, A and C are primary and co-substrates (glucose and oxygen), F and E are the reduced and oxidized form of the enzyme, and X_1 and X_3 are the complexes formed in binding of enzyme with the substrates (glucose and oxygen, respectively). From this reaction scheme, a total of six reaction equations must be used to describe the entire system in 1-D space (z), where the supply of substrate is assumed to occur via diffusion with D_A and D_C being the coefficients for glucose and oxygen, respectively:

$$\begin{aligned} \frac{\partial A}{\partial t} &= D_A \frac{\partial^2 A}{\partial z^2} - k_1 AE + k_{-1} X_1 \\ \frac{\partial C}{\partial t} &= D_C \frac{\partial^2 C}{\partial z^2} - k_3 CF + k_{-3} X_2 \\ \frac{\partial E}{\partial t} &= -k_1 AE + k_{-1} X_1 + k_4 X_2 \\ \frac{\partial F}{\partial t} &= k_2 X_1 - k_3 CF + k_{-3} X_2 \\ \frac{\partial X_1}{\partial t} &= k_1 AE - (k_{-1} + k_2) X_1 \\ \frac{\partial X_2}{\partial t} &= k_3 CF - (k_{-3} + k_4) X_2 \end{aligned} \tag{2}$$

To simplify this description into a more manageable set of expressions, several steps must be taken. First, it can be noted that the total enzyme concentration in the system E_t is constant at $E_t = E + F + X_1 + X_2$. Then, assuming that the formation and dissociation of intermediate complexes (X_1, X_3) occur very rapidly, such that $\frac{\partial X_1}{\partial t} = \frac{\partial X_2}{\partial t} = \frac{\partial F}{\partial t} = 0$ (the pseudo-steady-state assumption),[70] the system is simplified to two coupled differential equations with intuitive nomenclature

$$\frac{\partial C_G}{\partial t} = D_G \frac{\partial^2 C_G}{\partial r^2} - R_G(C_G, C_O, E_t) \quad (3)$$
$$\frac{\partial C_O}{\partial t} = D_O \frac{\partial^2 C_O}{\partial r^2} - R_O(C_G, C_O, E_t)$$

where C_G and C_O are the concentrations of glucose and oxygen, D_G and D_O are the diffusion coefficients for glucose and oxygen, R_G and R_O are the forward reaction terms for glucose consumption $(k_1 C_G E - k_{-1} X_1)$ and oxygen consumption $(k_3 C_O F - k_{-3} X_2)$, respectively. In this equation, the 1-D coordinate z has been transformed to spherical coordinate r, the radial distance from the center of a sphere, to facilitate direct application to the geometries of interest (microspheres/capsules). The reaction terms can be described by the following reduced terms:

$$X_2 = \frac{k_3 C_O F}{k_{-3} + k_4}, \quad X_1 = \frac{k_1 C_G\left[(k_{-3} + k_4)(E_t - F) - (k_3 C_O F)\right]}{(k_1 C_G + k_{-1} + k_2)(k_{-3} + k_4)},$$
$$F = \frac{k_1 k_2 C_G E_t (k_{-3} + k_4)}{(k_{-3} + k_4 + k_3 C_O) k_1 k_2 C_G + \left\{\frac{\left[k_3 C_O (k_1 C_G + k_{-1} + k_2)\right]\left[k_{-3} + k_4 - k_3\right]}{k_{-3} + k_4}\right\}}. \quad (4)$$

Thus, the two differential equations can be solved simultaneously to compute the distribution of glucose and oxygen in time and space if appropriate initial conditions (IC) and boundary conditions (BC) can be applied. To use these mathematical models to predict the behavior of microsphere-based enzymatic sensors, the structure in Figure 6.10 was used to develop IC and BC assumptions.

Using this basic model, the boundary conditions and material properties may be varied to evaluate the potential of microcapsules of different architecture to produce a measurable change in oxygen levels in proportion to glucose concentration. The absolute dimensions and physico-chemical properties may be easily modified by changing sphere size, nanofilm thicknesses, enzyme concentration, and diffusion constants. A few exercises will demonstrate the potential of this approach. For each of the following cases, these assumptions will define the system properties: (1) enzyme is homogenously distributed throughout the matrix, but none is in the nanofilm; (2) the matrix and nanofilm are homogenous and have constant, but different diffusion coefficients for glucose and oxygen; (3) enzyme activity is constant, unaffected by spontaneous or H_2O_2-mediated deactivation or pH shifts; (4) the

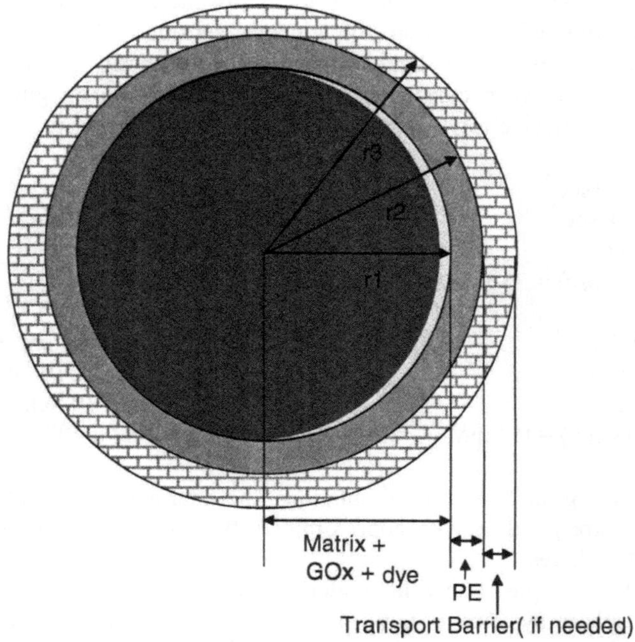

Figure 6.10. Schematic for simulations: spherical architecture with internal sphere containing homogenously distributed enzyme and fluorescent indicators; ultrathin films coating of variable composition and thickness. The fluorescent indicators are co-localized with the enzyme within the sphere, as this is where the greatest changes in oxygen concentration occur. The polyelectrolyte layers surrounding the microsphere act as a barrier to prevent outward diffusion of enzyme and fluorescent indicator, and to limit inward diffusion of substrate. A transport barrier to further limit substrate transport, or biocompatible coating, may be applied to the outer surface.

bulk concentration of oxygen remains constant at 250μM; (5) the bulk concentration of glucose does not change after the initial step increase at t=0; (6) a sample containing a single sphere will behave the same as one with many spheres (the model will simulate only one). For the boundary conditions, the following assumptions are also applied: (1) concentration of substrates is constant at boundaries; (2) flux at r=0 is zero.

This model can be used to predict sensor performance and determine appropriate combinations of materials and geometries that can yield useful devices. A typical analysis approach is to apply a step change in glucose concentration, then follow the depletion of oxygen until a steady-state situation is reached. In this case, it is assumed that oxygen is equilibrated within and outside of the sphere, and the change in glucose results in enzymatic catalysis, which involves consumption of both oxygen and glucose. At some point in time, a balance is reached between the diffusion and consumption of oxygen and glucose, and it is this steady-state situation in which glucose can be effectively monitored with oxygen probes.

A detailed explanation of the model is provided elsewhere;[71] here, the output of simulations using the model are used to describe the expected functional properties of enzyme-loaded microspheres with nanofilm coatings that may potentially be used for glucose sensing. Examples of system characteristics are given below using the following parameters, unless otherwise, noted, with constants derived from literature on glucose oxidase, alginate, and polyelectrolyte nanofilms:

$k_1 = 10^5$ M$^-$1sec^{-1}
$k_{-1} = 3 \times 10^3$ sec^{-1}
$k_2 = 300$ sec^{-1}
K_m (for glucose) = 33 mM

$C_0 = 250$ μM
$C_g = 0 - 30$ mM
$E_t = 10$ mM
r_1=20 μm

$k_3 = 10^6$ M^{-1}sec^{-1}
$k_{-3} = 150$ sec-1
$k_4 = 50$ sec^{-1}
K_m (for oxygen) = 0.2 mM

$D_{Gsphere} = 4 \times 10^{-6}$ cm^2 sec^{-1}
$D_{Osphere} = 2.45 \times 10^{-5}$ cm^2 sec^{-1}
$D_{Gfilm} = 6.8 \times 10^{-6}$ cm^2 sec^{-1}
$D_{Ofilm} = 9.87 \times 10^{-10}$ cm^2 sec^{-1}

Also, excepting the last set of simulations, the total nanofilm thickness was set to 16nm, corresponding with approximately two bilayers of PAH/PSS coating on alginate spheres.

An example of how the model is applied to simulating the properties of the microsphere systems is now given. First, Figure 6.11 contains plots of oxygen and glucose concentration profiles in the sphere, in time and space, after a 30mM step input of glucose. From these graphs, several important observations can be made. A steady-state distribution of both oxygen and glucose is achieved within a few seconds from the step change in glucose level. This rapidly-equilibrated distribution contains oxygen concentrations that decrease smoothly from the bulk value at the internal surface of the nanofilm coating to zero near the center of the sphere. The spatial concentration profile evolves very quickly to this curved shape due to rapid catalysis resulting from the large glucose concentration present immediately after the step. Between one and two seconds after the step is applied, the system reaches a steady-state situation where the internal oxygen profile possesses a similar shape, but does not completely drop to zero. This situation is due to the constant balance between diffusional supply and enzymatic consumption of oxygen and glucose. For glucose, a smooth profile is seen when looking at the time axis, whereas a very sharp drop is observed from the edge of the sphere inward for all points in time. This is due to the small glucose diffusion coefficient of the nanofilm coating, and the immediate reaction of glucose with the enzyme upon entering the matrix.

Several key observations may be made by consideration of these data. First, the step change in glucose is artificial and physically unrealistic; while a step change in ambient glucose concentration could be approximated by rapid addition and mixing, applying such a rapid increase throughout the interior of a sphere is not possible. Second, it is important to keep in mind that the shape and magnitude of these distributions are dependent upon the absolute size of the sphere as well as the transport coefficients.

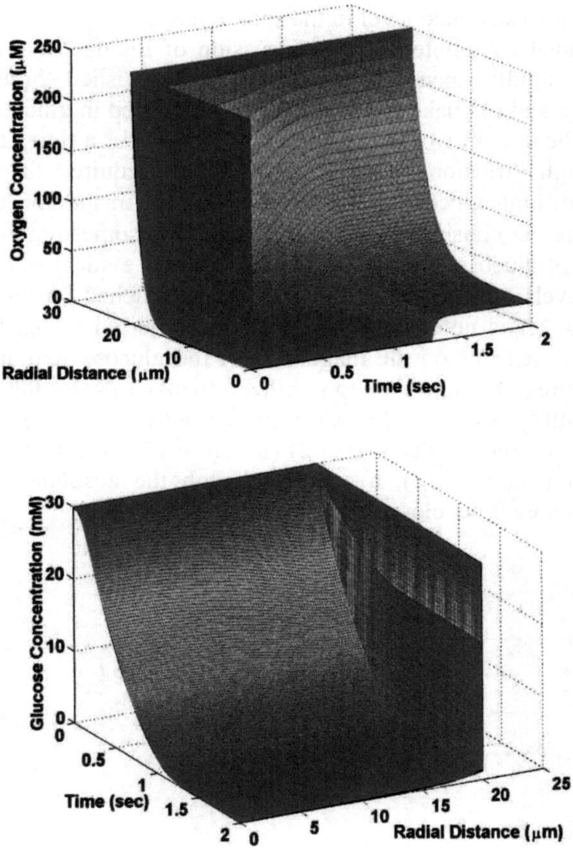

Figure 6.11. Time-dependent oxygen and glucose concentration distributions in microspheres due to glucose step changes. Note that the graph axes have different orientations to aid in observing important features.

It is also important to note that these plots versus time are not particularly useful for dynamic modeling at this stage, but simply provide insight into the time required to achieve steady-state balance for each step input, such that the simulations for steady-state conditions can be properly constructed. From the graphs in Figure 6.11, it appears that a 40μm diameter alginate sphere containing 10mM glucose oxidase and coated with a 16nm PSS/PAH nanofilm reaches steady-state within two seconds following application of the step. To more clearly define the temporal response of the system to steps of varying magnitude, additional simulations for glucose concentrations between 0 and 30 mM were performed.

Figure 6.12 contains graphs of the time course of oxygen and glucose concentrations at different points within the sphere, following each step addition of glucose. From these graphs, it can be seen that both oxygen and glucose reach a steady state level in the sphere interior within two seconds after the step is added. As noted in the discussion of Figure 6.11, oxygen levels initially drop rapidly, then rise as oxygen is replenished through diffusion. Figure 6.12 (a and c) makes it clear that, if such a step in glucose levels could be achieved, the system would be limited by oxygen for a time, and all oxygen arriving through diffusion would immediately be required for the oxidation reaction due to high glucose levels initially present in the sphere. Then, the excess glucose is consumed and the reaction becomes controlled by the diffusion rate of glucose into the sphere. This fact is evident from the different steady-state levels of glucose and oxygen that are reached for different glucose concentrations. It is noteworthy that the time required to reach steady-state conditions also depends on the magnitude of the glucose step applied; higher concentrations require more time to stabilize. In all cases, for the glucose range studied (0-30mM), steady-state inside of the sphere is achieved within 1.5 seconds. This is true for deep ($r=r_1/2$) (Figure 6.12), a and b) and superficial regions ($r=r_1$) (Figure 6.12), a and b), though the absolute values for the concentrations are higher closer to the surface.

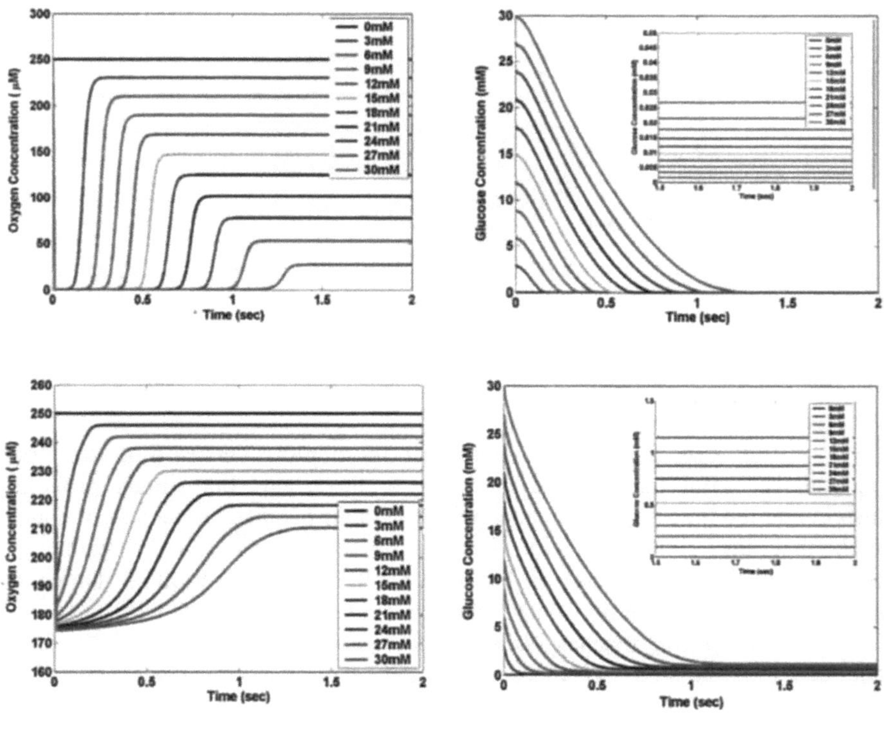

Figure 6.12. Oxygen (top left) and glucose (top right) temporal concentration profiles at $r=r_1/2$; oxygen (bottom left) and glucose (bottom right) temporal concentration profiles at $r=r_1$. The insets for glucose are exploded views of glucose levels 1.5-2 seconds following step addition. (see the color insert after p. 429.)

This leads to a consideration of the steady-state concentration profiles expected within the spheres. For the proposed enzymatic tattoo sensors, an indicator dye must be embedded within the sphere matrix to monitor oxygen, and therefore an understanding of the situation the molecular probe will encounter is essential. In particular, it is desirable to predict and optimize the difference in internal oxygen levels such that measurable differences can be achieved for the glucose range of interest. To assess the expected performance for the same microcapsule systems, the same simulations used to construct the graphs in Figure 6.12 were used to predict the spatial profiles for glucose and

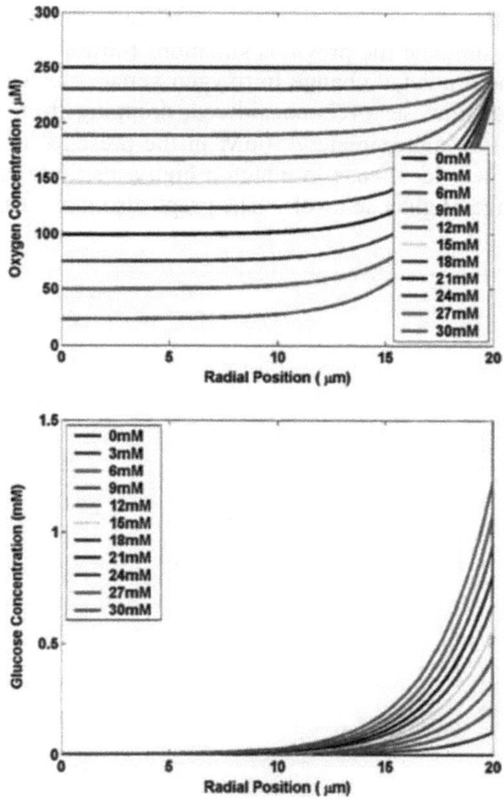

Figure 6.13. Steady-state spatial profiles for oxygen and glucose within spheres due to step inputs of glucose from 0-30 mM. (see the color insert after p. 429.)

oxygen under steady-state conditions (t>2 seconds). The plots confirm the expected glucose diffusion-limited behavior, where the concentration of both analytes decreases from the bulk level when moving from the surface to the center. Oxygen levels are maintained at intermediate concentrations, whereas glucose levels drop to zero as the diffusing molecules are oxidized. However, it is clear that higher bulk glucose concentrations result in deeper penetration and higher standing glucose profiles, due both to higher flux and saturation of surface enzyme, allowing glucose to diffuse to deeper regions to react with accessible glucose oxidase. These profiles represent a nearly ideal situation for monitoring the entire 0-30 mM glucose range due to the even spacing of oxygen levels (near-linear response).

Since monitoring with these devices would occur with an indicator distributed within the sphere, it is useful to plot the volume-averaged oxygen concentration versus glucose concentration as a predictor of the response of the system. Figure 6.14 contains such a graph, for which the average steady-state oxygen concentration for several situations has been calculated. First, for the structures used to simulate the previous situations (40μm diameter sphere with 16nm coating), the predicted change in oxygen versus glucose is highly linear across the range of glucose levels considered, dropping from the bulk oxygen level (no glucose) to approximately 50μM in the presence of 30mM glucose. This represents a near-ideal case in which a high sensitivity is nearly constant over the entire range of glucose levels. The graph also includes, for purposes of

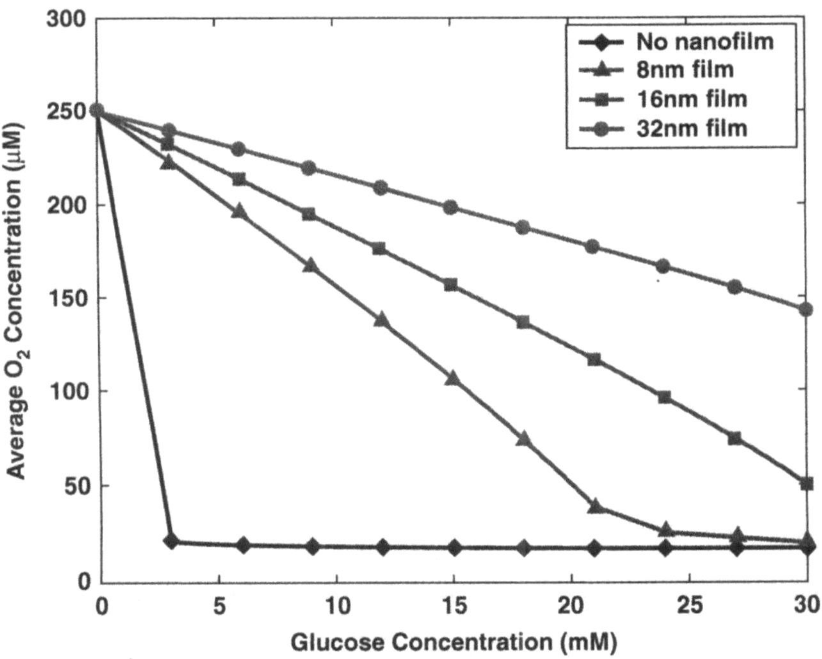

Figure 6.14. Steady-state sensitivity to glucose for microspheres with different nanofilm thickness. (see the color insert after p. 429.)

comparison, the expected response for uncoated microspheres, as well as spheres with thinner and thicker coatings of the same material.

This plot demonstrates clearly the utility of the nanofilm coating, which provides a critical diffusion barrier to limit the glucose delivery rate such that internal oxygen is not depleted too quickly, and so the system remains glucose diffusion-limited instead of oxygen- or reaction-limited. It is critical to note that the lines in the plot only connect the dots and therefore cannot be used as indications of behavior between points for which explicit calculations were performed. Thus, it can be stated that spheres without nanofilms exhibit almost zero sensitivity to glucose beyond very low levels.

Furthermore, the effect of nanofilm thickness can be observed from the changing slope and shape of the curves. Thicker nanofilms (twice the original thickness) further decrease glucose delivery to the enzyme, driving a balance between oxygen consumption and diffusion that results in a very linear response, but a higher steady-state oxygen level in all cases, and a lower overall sensitivity to glucose. In contrast, thinner nanofilms (half the original thickness) are predicted to provide higher sensitivity over a smaller linear range of glucose. This may, in fact, be advantageous for diabetic monitoring due to the greater clinical importance of accuracy at lower glucose concentrations.

From this discussion, the impact of nanofilms for controlling diffusion in enzymatic microsphere sensor applications should be obvious. It is worth reiterating here the ability to tailor nanofilms, both in terms of thickness and composition, such that desired properties may be used to tailor sensor response as needed. However, it is also important to again acknowledge the limitations of the simulations used to make these predictions. First, a number of assumptions have been made that may not be true in many cases. It is reasonable to believe that the microspheres can be produced with high uniformity and homogeneity in composition, but the assumptions regarding the physiological environment (constant bulk oxygen and glucose) are less likely to hold, though their validity is currently difficult to assess. Glucose, on one hand, will vary with blood glucose, as desired and by limiting glucose diffusion to slow rates, it is reasonable to anticipate that the microspheres will not remove significant glucose from the interstitial space. This can potentially be predicted from dynamic models in which the total glucose consumption is estimated for a given external supply, and from models in which bulk glucose is not required to be constant but is rather supplied to a "reservoir" of interstitial fluid from the blood according to a physiological transport model.

On the other hand, oxygen may vary significantly over time, and external oxygen levels are expected to significantly impact sensor function because they determine sensitivity as well as absolute readings. The most obvious means of compensation for this is the use of separate oxygen sensors that allow correction for different bulk oxygen. The ability to accurately model sensor response due to changing oxygen levels will be essential to understanding the impact of varying ambient levels of the co-substrate. As an example, Figure 6.15 is a graph similar to that presented in Figure 6.14, where the same simulations have been performed for different bulk oxygen levels.

The oxygen levels chosen were 80μM and 1400μM; the former is a value published for interstitial dissolved oxygen,[72] while the latter approximates the maximum dissolved oxygen concentration achievable in water. It is clear from this plot that different bulk oxygen levels drastically change the relationship between oxygen levels within the microspheres and the bulk glucose concentrations. While the sensitivity of the systems appears to be very stable, as the initial slopes of the lines are essentially unchanged, the different ambient oxygen produce an offset in the relationship. This can be easily compensated for cases where bulk oxygen can be measured and remains above a threshold; an example of what happens when bulk oxygen drops below the required value is seen in the curve for 80μM bulk O_2, where the curve flattens out at high glucose concentrations (zero oxygen) due to the inability of the bulk oxygen to diffuse quickly enough to replenish that being consumed in the reaction.

Enzymatic activity may also have a significant impact on sensor function, because the linear relationship between oxygen and glucose in the microsphere interior depends upon a system that is glucose diffusion-limited, not reaction-limited. To assess the impact of enzyme activity, which is effectively related to the concentration of active and accessible enzyme, more

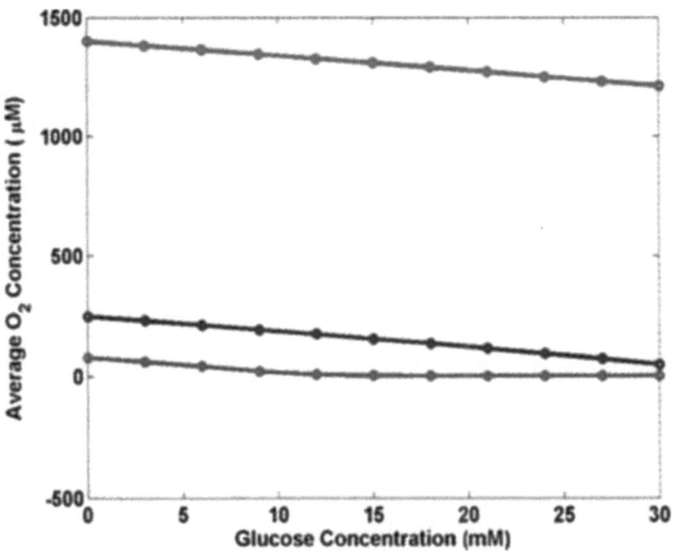

Figure 6.15. Average oxygen versus glucose for different bulk oxygen concentrations. The lines correspond to three different bulk oxygen levels. Top : 1400 μM; middle : 250 μM, bottom : 80 μM. (see the color insert after p. 429.)

simulations were performed with enzyme concentrations from 10mM (initial value) to 5, 1, and 0.1mM. The results of these modeling experiments are presented in Figure 6.16, from which it can be learned that the amount of active enzyme can be balanced to achieve maximum sensitivity and linearity. First, comparing the curves for 1 and 5mM enzyme to 10mM, it is observed that the decreased activity results in a higher sensitivity but smaller linear range. This behavior makes sense if one considers that the same amount of glucose can diffuse further into the spheres, depleting oxygen in deeper areas that cannot be replenished as quickly. This hypothesis is supported by observation of the spatial distributions of glucose and oxygen (Figure 6.16). Conversely, at very low enzyme concentrations (100 times less), the system becomes reaction-limited, a situation where oxygen cannot be removed any more rapidly, despite higher glucose levels, due to saturation of the enzyme.

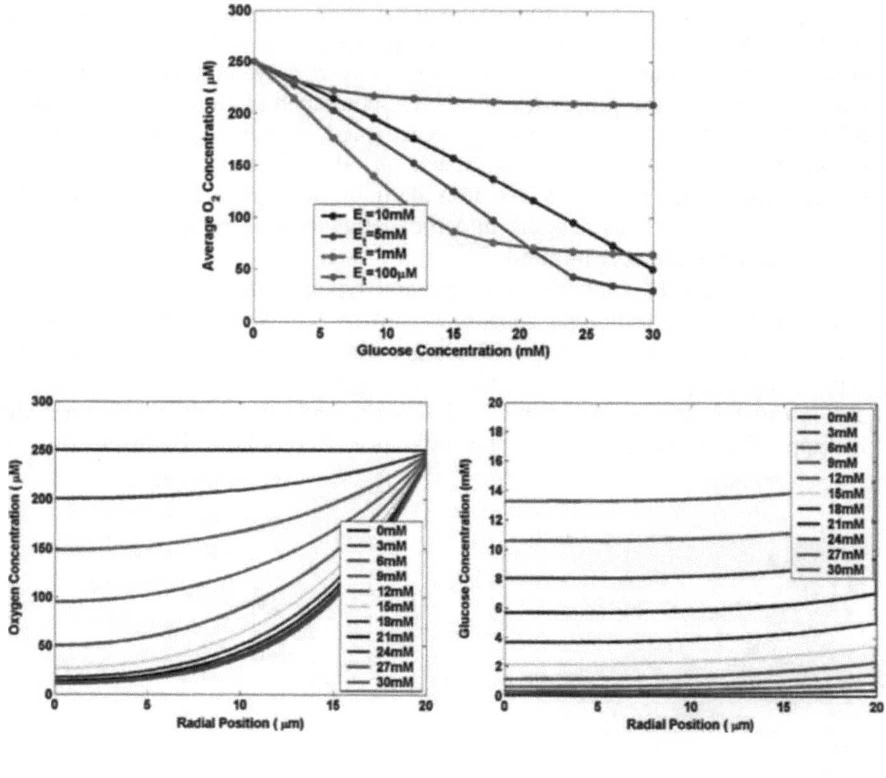

Figure 6.16. (Top) Average oxygen versus glucose for different enzyme concentrations and (bottom left) internal oxygen and (bottom right) glucose distributions for E_t=1mM. (see the color insert after p. 429.)

These simulations enable identification of composition and geometric combinations that can be used to construct sensitive systems for glucose sensing via oxygen monitoring using enzyme-doped microspheres. A similar approach could be taken for microspheres monitoring pH, or other local variables, that are changed as a result of reactions occurring within the confined environment of the microsphere/capsule.

6.4.2. Experimental Validation of Enzyme-Loaded Microcapsule Sensors

Following the discussion on sensor design above, prototype sensors have recently been developed and tested preliminarily for glucose response.[73] Calcium-crosslinked alginate gel microspheres were prepared in a water-in-oil emulsion, then loaded with ruthenium-tris(4,7-diphenyl-1,10-phenanthroline) dichloride, one of the more sensitive ruthenium-based metal-ligand complexes.[74,75] Poly(styrene sulfonate) (PSS, MW~70000) and poly(allylamine hydrochloride) (PAH, MW~70000), or PAH conjugated to Alexa Fluor® 488 (PAH-AF488, $\lambda ex = 488$ nm, $\lambda em = 515$ nm) were alternately assembled on the surfaces of the microspheres until two bilayer polyelectrolyte films were realized. The spectral properties of this dye combination closely matches those of the FITC/TRITC pair previously investigated, which suggests that *in vivo* monitoring with the current approach is also feasible,[1] though longer-wavelength oxygen indicators based on platinum porphyrins may potentially be used to extend the wavelengths into the yellow/orange region for excitation with emission in the near infrared.[76,77]

Figure 6.17 contains an image and typical fluorescence spectra of a suspension of the prototype sensors in buffers containing different glucose levels. The spectra were acquired using a single excitation wavelength (460 nm), while emission was collected by scanning from 500-650 nm. The peak at 520 nm is the contribution of the PAH-AF488, while the peak at 610 nm is the Ru(dpp) fluorescence. Since the reference fluorophore is insensitive to

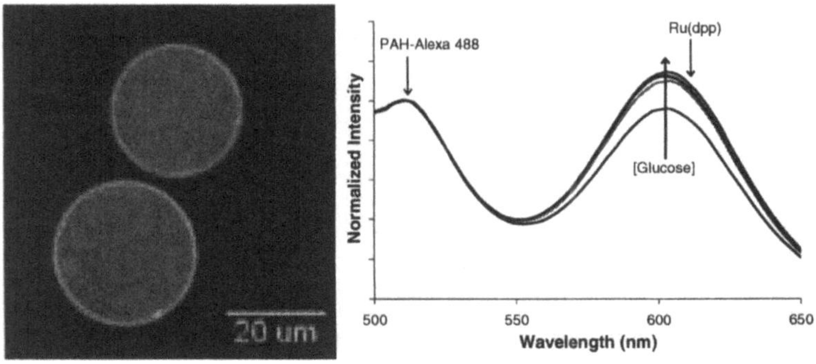

Figure 6.17. Left: Fluorescence micrograph of glucose sensors. Right: Fluorescence spectra of oxygen-sensitive, enzyme-loaded microspheres at different glucose levels.

fluctuations of oxygen or pH, differences in the intensity at 610 nm relative to 520 nm can be attributed to changes in local oxygen concentration. In experiments used to test oxygen sensitivity, the microspheres exhibited a typical quenching response with an overall change in the fluorescence peak ratio R_0/R of 0.73 between 0% to 100% dissolved oxygen (sensitivity=42%,) where R_0 denotes the ratio of Ru(dpp) to Alexa Fluor® 488 intensity at zero oxygen, and R is the intensity ratio at each respective oxygen concentration.

The sensitivity of the glucose sensors was assessed by adding aliquots of a glucose solution to a standard fluorimeter cuvette containing a suspension of sensors in pH 7.4 buffer (phosphate buffered saline, 0.01M). The fluorescence emission intensity at 520 and 610nm was measured in T-format mode using 460 nm excitation. Fluorescence emission was acquired every five seconds, with an integration time of two seconds, while the sample was continuously stirred except during measurements. During the experiments, oxygen was continuously bubbled to maintain the dissolved oxygen concentration at 1.3 mM (measured separately with an oxygen microelectrode) in order to ensure that changes in fluorescence intensity ratio were actually due to local oxygen changes internal to the microspheres, and not global oxygen depletion in the bulk solution.

Figure 6.18 contains a plot of the experimentally obtained intensity ratio versus glucose concentration, plotted as $R/R_{0\ glucose}$, where $R_{0\ glucose}$ denotes the intensity ratio of Ru(dpp) to Alexa Fluor® 488 ($I_{\lambda=610nm}/\ I_{\lambda=520nm}$) in O_2-saturated buffer without glucose, and R is the intensity ratio in O_2-saturated buffer at each respective glucose concentration. As glucose concentration

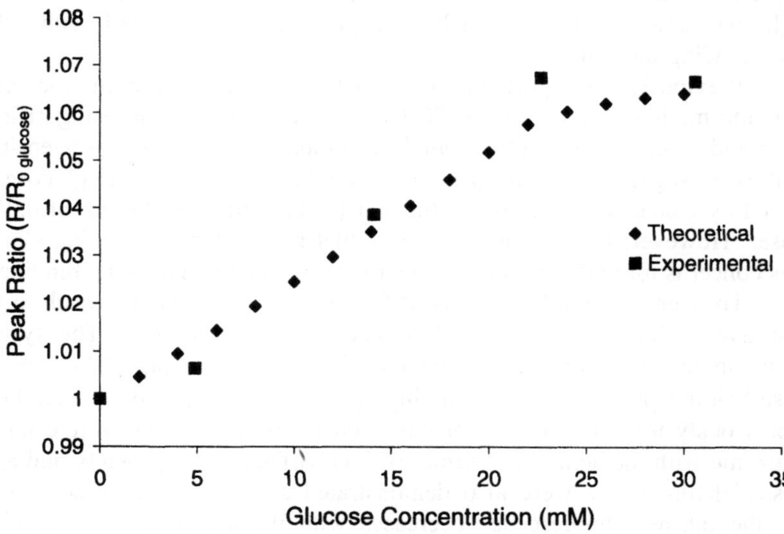

Figure 6.18. Experimental and theoretical data for glucose sensors in oxygen-saturated conditions.

increases, the ratio increases nearly linearly up to 20 mM glucose, where the response then saturates (at a change in $R/R_{0\ glucose}$ of about 7%.) On the same graph, predictions of the peak intensity ratio are given; these were calculated using the models described above, combined with the experimentally-determined oxygen sensitivity of the same microspheres. The results show excellent agreement, confirming that the mathematical models do produce realistic output, validating their accuracy and confirming their value for use in further sensor design.

These preliminary results are promising, as they suggest that further refinement of the sensor design (sphere size, enzyme concentration, permeability of the films to substrates, and film thickness) may significantly improve the response of the sensors. These findings show promise for the future development of similar microsensors with more optimized response. The microspheres are now being tested in skin phantoms immobilized in a flow-through apparatus that allows complete control over oxygen and glucose concentrations and enables time-dependent responses to be measured.

6.5. GLUCOSE-BINDING PROTEINS IN MICROCAPSULES

As noted above, a number of recognition elements for glucose have been identified: glucose-binding proteins, lectins, and synthetic receptors such as boronic acid. This section will discuss the possibility of encapsulating such materials within microcapsules for use as smart tattoo materials. The specific example of a competitive-binding assay using apo-glucose oxidase will be used, though the same general approach may potentially be applied with other glucose-binding molecules.

Previously, we reported on a nanoengineered polymeric microcapsule comprising multilayer films of TRITC-Con A and FITC-dextran using affinity binding and layer-by-layer self-assembly technique.[78] These glucose-sensitive thin films packaged in a microcapsule were shown to be sensitive to glucose, as observed by a change in the energy transfer (11%) with the addition of 0.05M glucose. However, it is important to note that any system employing Con A carries concerns for real applicability due to toxicity and non-specific binding.

To overcome the limitations of Con A, an alternative approach to the competitive-binding glucose assay has recently been reported. The system employs an inactive form of the enzyme glucose oxidase (apo-GOx) as the glucose-binding protein, which is highly specific to β-D-glucose. Apo-GOx was previously used as a biosensor based on its ability to reconstitute into a holoenzyme with the addition of semi-artificial cofactors.[79] Apo-GOx and apo-glucose dehydrogenese were also demonstrated as direct glucose sensors in which the intrinsic fluorescence decreases with the addition of glucose.[80,81] Apo-GOx was also tagged with an environmentally-sensitive fluorophore, which exhibited a decrease in intensity with the addition of glucose, due to partial shielding of tryptophan residues.[80] The disadvantages of these approaches include the short wavelengths required to excite intrinsic fluorescence, and the inability to correct for inner filter, dilution, or other non-specific effects due to non-ratiometric operation.

The design and operation of a novel RET assay using apo-GOx was discussed previously (illustrated in Figure 6.8). Briefly, apo-GOx is prepared from GOx by removing FAD cofactor.[80] When apo-GOx is labeled with TRITC and exposed to FITC-dextran, strong fluorescence peaks due to significant energy transfer between FITC and TRITC are observed. Because of the high affinity of apo-GOx for glucose, addition of glucose results in the displacement of dextran from apo-GOx. This change in physical proximity is manifested as a decrease in the energy transfer efficiency (Figure 6.8), as evidenced by a stronger FITC peak relative to TRITC. The glucose sensitivity of the TRITC-apo-GOx/FITC-dextran was demonstrated by measuring changes in RET between FITC and TRITC resulting from titration of glucose (figure 6.19).[17] This system retains the advantages of the competitive binding approach, including selectivity to the analyte of interest, elimination of reaction byproducts, and there is no consumption of the analyte during the sensing process.

Following demonstration of glucose sensitivity in solution phase, the RET assay components were encapsulated within a hollow polymeric shell. Polyelectrolyte microcapsules, templated on 5μm $MnCO_3$ cores with walls comprising multilayer nanofilms including a photosensitive diazoresin, were suspended in a solution of the RET assay and then irradiated with UV light. As illustrated in Figure 6.20, this encapsulation approach results in capsules loaded via diffusion, then the contents are locked within the capsule interior via crosslinking of the walls. A key advantage of this approach is the absence of polymer or other matrix in the capsul interior, maintaining a nonrestrictive environment for the molecular components of the competitive binding assay to

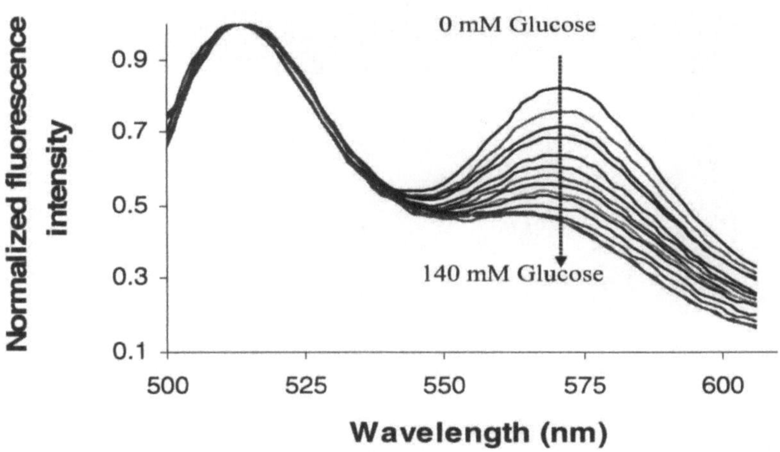

Figure 6.19. Fluorescence spectra recorded from RET competitive binding assay based on apo-GOx.

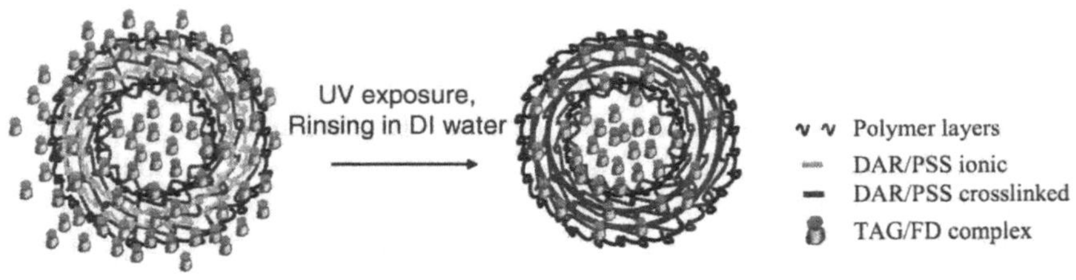

Figure 6.20. Schematic of encapsulation procedure for competitive-binding assay.

easily move relative to one another. The hollow shells act in a similar way to a dialysis membrane, as a semipermeable barrier that allows glucose to diffuse in and out while retaining the sensing components inside.

Confocal micrographs of loaded capsules are presented in Figure 6.21, from which it can be observed that the tagged molecules are trapped within the polyelectrolyte microcapsule wall and interior. The bright fluorescence from the walls suggests entrapment of sensing molecules in a more immobile manner. This could contribute a strong baseline signal to fluorescence measurements, and signals coming from molecules trapped in crosslinked walls are unlikely to be significantly affected by glucose due to limited mobility of the dextran.

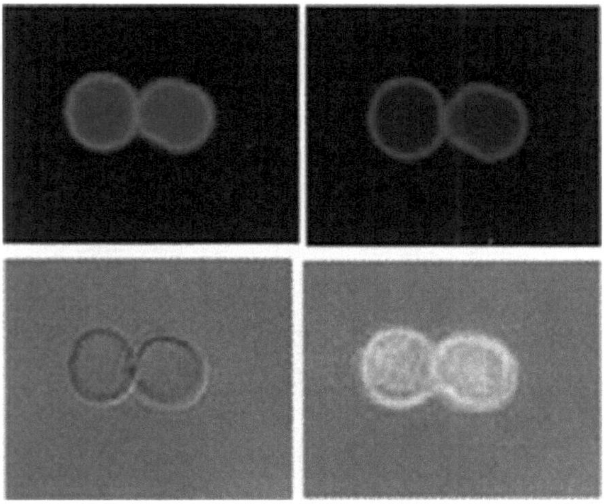

Figure 6.21. Confocal fluorescence (top, left: FITC, right: TRITC), phase (bottom left) and overlay (bottom right) images of polyelectrolyte capsules loaded with TRITC-apo-GOx and FITC-dextran. (see the color insert after p. 429.)

Fluorescence intensity ratio measurements performed at varying glucose levels showed a significant response to glucose over the physiologically-relevant range (Figure 6.22). This graph shows the change in the ratio of FITC (520nm) to TRITC (580nm) intensity with changing glucose for both 500kDa and 2MDa FITC-dextran competitive ligands. In the former case, the total change in peak ratio over the dynamic range of the assay was found to be ~52%, while a slightly lower value (47%) was calculated for the larger ligand. Calculating the linear slope over the lower concentration range (0-20mM), it was found that the sensitivity for the system using the 500kDa dextran is approximately twice that of the 2MDa dextran (~4%/mM versus ~2%/mM). The 2MDa dextran system exhibited a response that was smaller at low values but extended to higher glucose concentrations, with associated k_D values of 8mM and 20mM for 500kDa and 2MDa, respectively. These results corroborate those found for the solution-phase assay, and point to the possibility of tuning the sensitivity and range for the specific application by judicious choice of the ligand. These properties are advantageous compared to glucose/galactose binding proteins, which typically possess k_D values around 1 μM.[64,69]

While this apo-GOx competitive binding system is still under intense investigation, the preliminary results show a sensitive, reversible, and repeatable response that does not vary significantly over four weeks. It is also

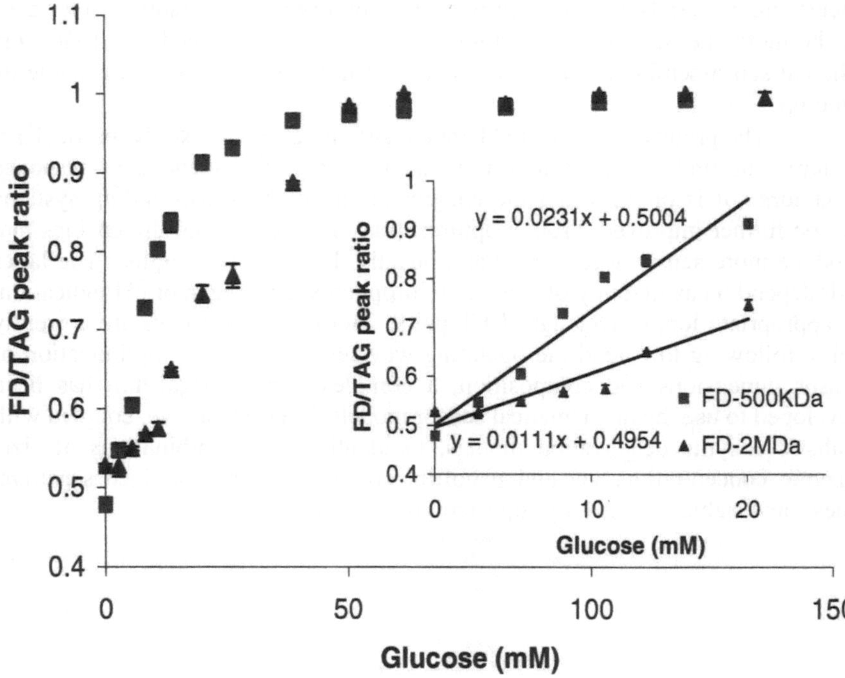

Figure 6.22. Glucose response for competitive-binding assays in microcapsules.

noteworthy that this same system can be easily extended to the near-infrared region by selection of an appropriate energy transfer pair.

6.6. CONCLUSIONS

This chapter highlighted some of the general needs for creating fluorescent smart tattoo glucose sensor systems, described structures and material combinations identified by mathematical modeling to be appropriate for microscale enzymatic systems, and presented some example sensors employing microcapsule carrier technology. Nanoengineered microcapsules based on polyelectrolyte multilayer nanofilms have enabled the flexible construction of several sensor types, and have helped to solve some technical obstacles in the areas of encapsulation of enzymes and other proteins, enzyme stabilization, and diffusion rates for glucose and oxygen. Early results of microcapsule-based sensors are promising, and the primary challenges for realizing long-term implantable sensors appear to be in the areas of biocompatibility and sensor stability. The advantages of these systems employing nanoengineered ultrathin films as capsule walls are the ability to precisely control dimensions and composition, which are critical to achieve desired function, especially with enzymatic-based sensors. Because of the flexibility in fabrication, it provides a facile approach to screening and integrating materials that will provide an appropriate bio/material interface. Furthermore, because the nanofabrication processes described are based on efficient self-assembly techniques, it is believed that production is amenable to scale-up.

The prototypes presented here are still under intense study, particularly in terms of understanding long-term stability and time-dependent response behaviors. It is obvious that the enzymatic and competitive-binding systems can be further improved through optimization of materials and dimensions that produce more sensitive responses and, ideally, longer wavelengths. The latter will depend on availability of suitable fluorophores for oxygen or pH indication, or appropriate long-wavelength RET pairs. Both of these routes are currently being following to extend the operating wavelengths. For the optimization of sensor dimensions and composition, a complex genetic algorithm has been developed to use the mathematical sensor models described above, coupled with realistic enzyme deactivation models, to identify best combinations of size, enzyme concentration, gel and nanofilm composition that produce sensitive, linear, and stable response for long-term operation.

6.7. ACKNOWLEDGEMENT

This work was funded in part by NIH (R01 EB000739), NSF (0210298), and the Louisiana Board of Regents (LEQSF(2001-04)-RD-A-18). J. Quincy Brown, Rohit Srivastava, Erich Stein, Swetha Chinnayelka, and Huiguang Zhu are gratefully acknowledged for their helpful technical and editorial feedback and assistance with illustrations.

6.8. REFERENCES

[1] McShane, M.J., Russell, R.J., Pishko, M.V., Coté, G.L., *IEEE-EMBS Mag*, **19**, 36, 2000.
[2] McNichols, R.J., Coté, G.L., *J. Biomed Opt*, **5**, 1, 2000.
[3] Wickramasinghe Y, Yang Y, and Spencer S.A., *J Fluor,* **14**, 513, 2004.
[4] Thennadil, S.N., Rennert, J.L., Wenzel, B.J., Hazen, K.H., Ruchti, T.L., Block, M.B., *Diab. Tech. Ther.*, **3**, 357, 2001.
[5] Thome-Duret, V.; Reach, G.; Gangnerau, M.N.; Lemonnier, F.; Klein, J.C.; Zhang, Y.; Hu, Y.; Wilson, G.S.; *Anal. Chem.* **68**; 3822, 1996.
[6] Moschou, E.A., Sharma, B.V., Deo, S.K., Daunert, S., *J Fluor*, **14**, 535, 2004.
[7] Koschinsky, T., Heinemann, L., *Diabetes Metab Res Rev* **17**, 113, 2001.
[8] Anderson, J.M., "Inflammation, Wound Healing, and the Foreign Body Response," in *Biomaterials Science*, Ratner, Hoffman, Schoen, Lemons, Eds. (Academic Press), pp. 165-173, 1996.
[9] Dicesare, N.; Lakowicz, J. R. *Anal. Biochem.* **294**, 154, 2001.
[10] Cao, H.; Diaz, D. I.; DiCesare, N.; Lakowicz, J. R.; Heagy, M. D.; *Org. Lett.;* **4**, 1503, 2002.
[11] Parmpi, P. Kofinas, P. *Biomat*, **25**, 1969, 2004.
[12] Byrne, M.E., Oral, E., Hilt, J.Z., Peppas, N.A. *Polym Adv Tech*, **13**, 798, 2002.
[13] Alexeev, V. L.; Sharma, A. C.; Goponenko, A. V.; Das, S.; Lednev, I. K.; Wilcox, C. S.; Finegold, D. N.; Asher, S. A.; *Anal. Chem.* **75**, 2316, 2003.
[14] Wilson, R. and Turner, A.P.F., *Biosens Bioelect*, vol. 7, pp.165-185, 1992.
[15] Mansouri, S., Schultz, J.S., *Biotechnol*, **2**, 885, 1984.
[16] Badugu, R.; Lakowicz, J. R.; Geddes, C. D.; *Anal. Chem.* **76**, 610, 2004.
[17] Chinnayelka S, McShane MJ, *Biomacromol,* **5**, 1657, 2004.
[18] Anderson, R.R., Parrish, J.A., *J Invest Dermatol*, **77**, 13, 1981.
[19] Van Gemert, M.J.C., Jacques, S.L., Sterenborg, H.J.C., Star, W.M., *IEEE Trans Biomed Eng*, **36**, 1146, 1989.
[20] Ballerstadt, R., Schultz, J.S., *Anal Chem*, **72**, 4185, 2000.
[21] Decher, G. *Science.* **227**, 1232, 1997.
[22] Lvov, Y; Decher, G.; Möhwald, H. *Langmuir* **9**, 481, 1993.
[23] Decher, G.; Hong J.D.; Schmitt J. *Thin Solid Films* **210**, 831, 1992.
[24] Keller, S.; Kim, H.; Mallouk, T. *J. Am. Chem. Soc.* **116**, 8817, 1994.
[25] McShane, M.J. and Lvov, Y.M., "Layer-by-Layer Electrostatic Self-Assembly," in *Encyclopedia of Nanoscience and Nanotechnology.*, Marcel Dekker, published online 12/24/2004.
[26] Lvov, Y.; Ariga, K.; Ichinose, I.; Kunitake, T. *J. Am. Chem. Soc.* **117**, 6117, 1995.
[27] Mamedov, A.; Kotov, N. *Langmuir* **16**, 5530, 2000.
[28] Sukhorukov, G.B.; Donath, E.; Davis, S.; Lichtenfeld, H.; Caruso, F.; Popov, V.I.; Möhwald, H. *Polym. Adv. Technol.* **9**, 759, 1998.

[29] Caruso, F.; Caruso, R.A; Möhwald, H. *Science* **262**, 1111, 1998.
[30] Lvov, Y.; Price, R.; Singh, A.; Selinger, J.; Spector, M.; Schnur, J. *Langmuir* **16**, 5932, 2000.
[31] Caruso, F., Schueler, C., *Langmuir*, **16**, 9595, 2000.
[32] Lvov, Y.; Caruso, F. *Anal. Chem.* **73**, 4212, 2001.
[33] Fang, M., Grant, P.S., McShane, M.J., Sukhorukov, G.S., Golub, V., Lvov, Y.M., *Langmuir*, **18**, 6338, 2002.
[34] Stein, E.W., McShane, M.J., *IEEE Trans Nanobiosci*, **3**, 133, 2003.
[35] Sukhorukov, G.B.; Brumen, M.; Donath, E.; Möhwald, H. *J. Phys. Chem. B.* **103**, 6434, 1999.
[36] Donath, E.; Sukhorukov, G.B.; Caruso, F.; Davis, S.A.; Möhwald, H. *Angew. Chem. Int. Ed.* **37**, 2201, 1998.
[37] Voigt, A.; Lichtenfeld, H.; Sukhorokov, G.B.; Zastrow, H.; Donath, E.; Bäumler, H.; Möhwald, H. *Ind. Eng. Chem. Res.* **38**, 4037, 1998.
[38] Grant, P.S., McShane, M.J., *IEEE Sensors*, **3**, 139, 2003.
[39] Brown, J.Q., McShane, M.J.; *IEEE-EMBS Mag.* **22**, 118, 2003.
[40] Duchesne, T.A., Brown, J.Q., Guice, K.B., Lvov, Y.M., McShane, M.J., *Sens Mat*, **14**, 293, 2002.
[41] McShane, M.J., Brown, J.Q, Guice, K.B., Lvov, Y.M.; *J Nanosci Nanotech*, **2**, 411, 2002.
[42] McShane, M.J., *Diabet Technol Ther*, **4**, 533, 2002.
[43] Sukhorukov, G.B.; Donath, E.; Moya, S.; Susha, A.; Voigt, A.; Hartmann, J.; Möhwald, H. *J. Microencapsulation* **17**, 177, 2002.
[44] Antipov, A.; Sukhorukov, G.; Donath, E.; Möhwald, H. *J. Phys. Chem.* **105**, 723, 2001.
[45] Antipov, A.A.; Sukhorukov, G.B.; Leporatti, S., Radtchenko, I. L.; Donath, E., Möhwald, H. *Coll Surf A: Physicochem Eng Aspects*, **198**, 535, 2002.
[46] Tiourina, O.P., Antipov, A.A., Sukhorukov, G.B., Larionova, N.I., Lvov, Y., Mohwald, H., *Macromol. Biosci.*, **1**, 209, 2001.
[47] Mendelson, J.; Barrett, C.; Chan, V.; Pal, A.; Mayes, A.; Rubner, M.F. *Langmuir* **16**, 5017, 2000.
[48] Moya, S.; Sukhorukov, G.B.; Auch, M.; Donath, E.; Möhwald, H. *J. Colloid Interface Sci.* **216**, 297, 1999.
[49] Lvov, Y.; Antipov, A.; Mamedov, A.; Möhwald, H.; Sukhorukov, G. *Nano Letters.* **1**, 125, 2001.
[50] Sukhorukov, G.B.; Dähne, L.; Hartman, J.; Donath, E.; Möhwald, H. *Adv Mater* **12**, 112, 2000.
[51] Zhong, H.; Wang, J.; Jia, X.; Li, Y.; Qin, Y.; Chen, J.; Zhao, X.; Cao, W.; Li, M.; Wei, Y. *Macromol. Rapid Commun.* **22**, 583, 2001.
[52] Zhang, Y.; Cao, W. *Macromol. Rapid Commun.* **22**, 842, 2001.
[53] Zhang, Y.; Yang, S.; Guan, Y.; Cao, W.; Xu, J. *Macromol* **36**, 4238, 2003.
[54] Zhu, H., McShane, M.J., *Langmuir*, **21**, 424, 2005.
[55] Srivastava, R.S., McShane, M.J., *J Microencaps*, (in press).
[56] Srivastava, R., Brown, J.Q., Zhu, H., McShane, M.J., *Biotechnol Bioeng*, (in press).
[57] Srivastava, R., Brown, J.Q., Zhu, H., McShane, M.J., *Macromol Biosci*, (submitted).'
[58] Schultz, J.S., Mansouri, S., Goldstein, I.J., *Diab Care*, **5**, 3, 1982.
[59] Meadows, D.L., Schultz, J.S., *Anal Chim Acta*, **280**, 21, 1993.
[60] Ballerstadt, R., Schultz, J.S., *Anal Chim Acta*, **345**, 203, 1997.
[61] Ballerstadt, R., Schultz, J.S., *Sens Actuat B: Chem*, **46**, 50, 1998.
[62] Ballerstadt, R., Polak, A., Beuhler, A., Frye, J., *Biosens Bioelect*, **19**, 905, 2004.
[63] Tolosa, L., Gryczynski, I., Eichhorn, L.R., Dattelbaum, J.D., Castellano, F.N., Rao, G., Lakowicz, J.R., *Analytical Biochemistry*, **267**, 114, 1999.
[64] Ge, X., Tolosa, L., Rao, G., *Analytical Chemistry*, **76**, 1403, 2004.
[65] DiCesare, N., Lakowicz, J.R., *Anal Biochem*, **294**, 1542001.
[66] Ullman, E.F., Schwarzberg, M., Rubenstein, K.E., *J Biol Chem*, **251**, 4172-4178, 1976.
[67] Stryer, L., *Ann Rev Biochem*, **47**, 819, 1978.
[68] Birch, D.J.S., Rolinski, O.J., *Res Chem Intermed*, **27**, 425, 2001.

[69] Boos, W., Gordon, A.S., Hall, R.E., Price, H.D., *J Biol Chem*, **247**, 917-924, 1972.
[70] Sakamoto, N., *J Membrane Sci*, **70**, 237, 1992.
[71] Brown, J.Q., McShane, M.J., submitted to *Annals of Biomedical Engineering*.
[72] Evans, N.T.S., Naylor, P.F.D., *Resp Phys*, **2**, 46, 1966/67.
[73] Brown, J.Q., Srivastava, R., McShane, M.J., *Biosens Bioelect*, (in press).
[74] Carraway, E.R.; Demas, J.N.; DeGraff, B.A.; Bacon, J.R. *Anal. Chem.* **63**, 337, 1991.
[75] Mills, A. *Sens Act B* **51**, 60, 1998.
[76] Mills, A.; Lepre, A. *Anal. Chem.* **69**, 4653, 1997.
[77] Koo, Y.-E. L., Cao, Y., Kopelman, R., Koo, S.M., Brasuel, M., Philbert, M.A., *Anal. Chem.*, **76**, 2498, 2004.
[78] Chinnayelka S, McShane MJ, *Journal of Fluorescence*, **14**, 585, 2004.
[79] Willner, I.; Heleg-Shabtai, V.; Blonder, R.; Katz, E.; Tao, G.; Bückmann, A.F.; and Heller, A. *J. Am. Chem. Soc.* **118**, 10321, 1996
[80] D'Auria, S.; Herman, P.; Rossi, M.; Lakowicz, J. R. *Biochem Biophys Res Comm* **263**, 550, 1999.
[81] Xiao, Y., Patolsky, F., Katz, E., Hainfeld, J.F., Willner, I., *Science* **299**, 1877, 2003.

NON-INVASIVE MONITORING OF DIABETES: Specificity, compartmentalization, and calibration issues

Omar S. Khalil[1]

7.1. INTRODUCTION

Non-invasive (NI) monitoring of glucose has attracted tremendous attention in the past two decades, mainly because diabetes is expected to be a major epidemic due to the increased overall obesity of the population. Non-Invasive monitoring of glucose decreases the pain associated with skin lancing used to sample blood for home glucose monitors. Reduction in pain can encourage more frequent testing and lead to tighter control of glucose levels, improve patient care, and delay the onset of diabetes complications and their associated health care costs. Patient-care needs and the commercial significance of NI glucose monitoring has led to a flurry of patenting activity and research on NI glucose detection methods. Several recent reviews discuss the importance of non-invasive glucose testing and report on attempts at its measurement (1-4).

The advent of several minimally-invasive techniques, alternate sample testing, and the decrease in size of the blood droplet required in new home glucose monitors has reduced the pain of obtaining a few micro liters of blood from the finger tip which is rich in nerve-endings. These methods include extracting the interstitial fluid (ISF) (5, 6), use of insertable electrodes (7, 8), testing at alternate sites other than the fingertip (9). These methods decreased the pain of blood sampling (8, 9), allowed continuous glucose monitoring for a limited time (7), and provided near real-time glucose trend data (7). They are defining new performance expectations for a non-invasive glucose monitor and raised the expectation from a truly non-invasive test.

---

[1] Abbott Laboratories, Diagnostics Division, Abbott Park, IL 60064

In a recent review we identified specificity, compartmentalization of glucose values, and calibration as the major issues for NI glucose measurements (1). Three questions were raised regarding NI- glucose determination (1). At least few questions need to be answered. The specificity question is what is being detected and determined? Is it an intrinsic property of the glucose molecule or is it the effect of change in glucose concentrations on the properties of the medium?

The compartmentalization question seeks answers as to in which body compartment is the glucose value determined, and how does the determined concentration relate to arterial blood glucose concentration? The calibration question deals with how to calibrate the NI testing device? Is it a single person calibration or multiple subject calibration? Is the testing device factory calibrated or can the user calibrate it?

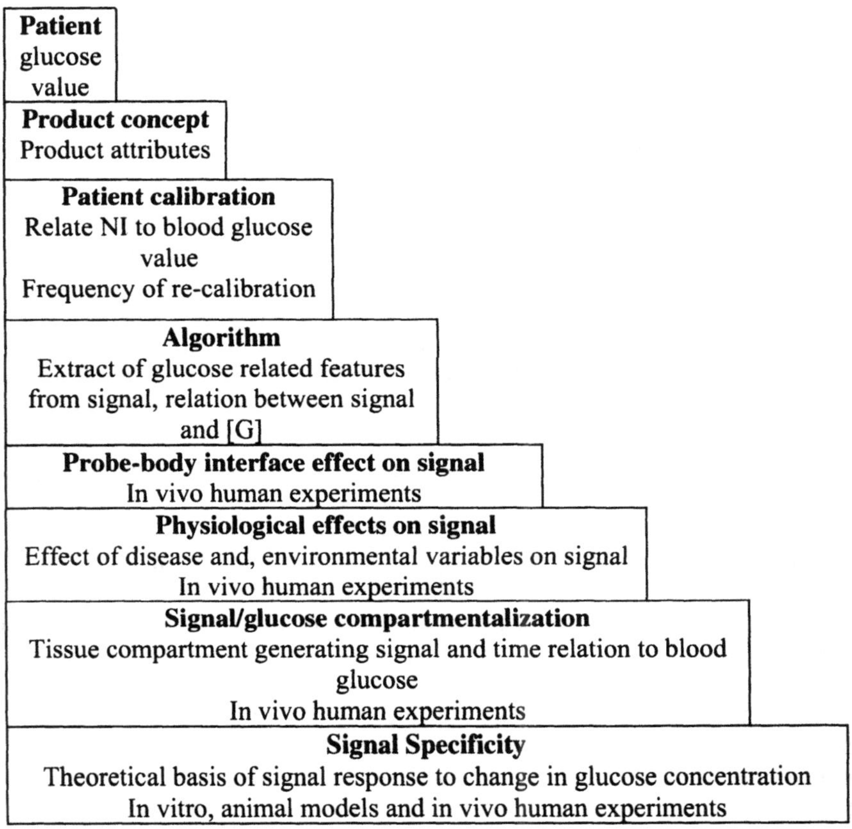

Figure 7.1. Laddered structure of the thought processed and steps towards a non-invasive glucose-measuring device.

The graph in Figure 7.1 shows a bottom-up hierarchal structure of the relationship between different steps towards a NI device from theoretical construct to a product. Presently, most of the NI methods for glucose determination lay in the bottom two cells of the graph.

The lack of understanding of the effect of changes in body physiology and the probe-tissue interaction on the measured signal leads to correlation of glucose concentration with spurious signals and algorithms that are contaminated with spurious information.

There have been attempts to develop methods to improve detection specificity and localize glucose detection to defined tissue compartment. The first set of methods involves the use of an external stimulus to assist in localizing the signal-generating tissue region. A second set of methods will include the use of specific capture agents, such as boronic acid derivatives, and detection using fluorescence or change in diffraction patterns. Specific capture methods will be discussed in several chapters of this volume. They involve the use of a chemically active agent and detection vehicles like a contact lens or a fluorescent tattoo.

This chapter expands discussion of the issues of specificity, compartmentalization and calibration that were presented recently (1). It will also discuss the work of our group on understanding the effect of interaction between the optical detection probe and the skin and the effect of temperature on skin optical properties. The temperature effect on optical properties of tissue was dubbed the thermo-optical response (THOR) of human tissue. We also discuss the work of our group on the use of temperature-modulated thermo-optical response of human skin, attempts to localize detection to changes in the vascular subsystem and tissue of the dermis layer.

7.2. SPECIFICITY OF NI GLUCOSE MEASUREMENTS

Methods used for the NI determination of glucose can be classified into two broad categories as: methods that track a glucose property, and methods that track the effect of glucose on tissue properties. The first category depends on measuring a signal that represents an intrinsic property of the glucose molecule such as near Infrared (NIR) absorption coefficient, mid-IR absorption coefficient, optical-rotation, Raman shifts, and the like. These methods assume the ability to detect glucose in tissue or blood independent of the presence of other body components and independent of the body physiological state. The second set of methods depends on measuring the effect of change in glucose concentration on the optical properties of tissue such as the scattering coefficient of tissue, refractive index of interstitial fluid (ISF) or sound propagation in tissue (1).

Obstacles to collecting reliable NI glucose data include the minute magnitude of the signal, repositioning error of the measuring probe with respect to the body part, temperature changes, and variations in the physical properties of the probe/body interface such as skin hydration, efficiency of optical and thermal coupling between the probe and tissue, and effect of probe/tissue interaction on signal magnitude (1). Both approaches face large obstacles.

Optical spectroscopy of human tissue is dominated by the infrared spectrum of water and fat, the visible absorption spectrum of hemoglobin, and absorption by skin pigments. In between, in the near IR spectral region, are the overtone and combination bands of water and lipids, as well as weak absorption features attributable to glucose and other tissue components that have similar molecular bonds. The earliest methods for the determination of glucose involved either near Infrared (NIR) transmission or reflectance measurements. Light penetrates deeper in tissues in the NIR spectral region and allows for optical sampling of tissue layers.

7.2.1. Tissue Spectroscopy

Light propagation in tissues is expressed by a set of spectroscopic parameters; the absorption coefficient μ_a, the scattering coefficient μ_s, and the anisotropy factor g (the average cosine of the angle at which a photon is scattered), and the reduced scattering coefficient μ'_s, where $\mu_s' = \mu_s[1\text{-}g]$. The absorption coefficient μ_a equals the absorbance per unit pathlength = 2.303 εC cm^{-1}, where ε is the molar extinction coefficient and C is the molar concentration. μ_a is the inverse of the average distance traveled by a scattered photon before it gets absorbed via interaction with the oscillating dipole of a tissue chromophore. The unit of μ_a is cm^{-1}. The larger the absorption coefficient, the shorter is the distance traveled by the photon in the medium; the lower is the reflected light intensity off the tissue. The scattering coefficient μ_s' is the inverse of the average distance traveled by a scattered photon before it looses memory of its initial propagation direction. The scattering coefficient μ_s' = $\sigma\rho$ where σ is the scattering cross section and ρ is the number density of the particles. It is equivalent to the product of an extinction coefficient due to scattering and the concentration of scattering centers and it has the unit ofcm^{-1}. The larger the absorption coefficient, the shorter is the distance traveled by the photon in the medium before scattering; the higher is the reflected light intensity off the tissue.

Equation 1 expresses attenuation of light intensity in tissue according to light transport theory. I_0 is the incident light intensity, I, is transmitted or reflected intensity, μ_{eff} is the effective attenuation coefficient and l is the pathlength.

$$I = I_0\, e^{-\mu_{eff} l} \qquad (1)$$

Equation 2 relates the effective attenuation coefficient to μ_a and μ'_s.

$$\mu_{eff} = (3\, \mu_a\, [\mu_a + \mu'_s])^{1/2} = (3\, \mu_a\, [\mu_a + \mu_s(1\text{-}g)])^{1/2} \qquad (2)$$

The reciprocal of μ_{eff} is the light penetration depth in tissue, δ (cm), is the distance light travels in tissue before its intensity is attenuated by 1/e of its initial incident value. Thus,

$$\delta = 1/(3\ \mu_a\ [\mu_a + \mu'_s])^{1/2} \qquad (3)$$

Optical measurements and theoretical light propagation models are used to extract μ_{eff}, μ_a and μ'_s. These measurements include transmission, diffuse and localized reflectance, frequency domain measurements, and optical coherence tomography (OCT). Glucose detection methods that track direct absorption of glucose generally utilize long wavelengths absorption features where μ_s' is low. On the other hand, methods that track effect of glucose on μ_s' generally utilize wavelength below 900 nm where μ_s' is large.

7.2.2. Infrared and Near Infrared Spectroscopy of Glucose

The fundamental IR absorption bands of glucose in the mid IR are in the spectral range between 2.5 microns (4000 cm^{-1}) and 10 microns (1000 cm^{-1}) (10-14). The mid-IR spectral bands of glucose and other carbohydrate have been assigned to by C-C, C-H, O-H stretching, and bending vibrations. The 800 cm^{-1} to 1200 cm^{-1} fingerprint region of the IR spectrum of glucose has bands at 836, 911, 1011, 1047, 1076, and 1250 cm^{-1} that have been assigned to C-H bending vibrations. The 1026 cm^{-1} band corresponds to C-O-H bend vibration. Spectral measurements in this frequency interval were used to determine glucose in serum and blood. There are several reports on the mid-IR spectra of skin and on attempts to use this range for NI glucose determination. These reports were recently reviewed (1). The fundamental vibrations of the Raman-active bands between 800 and 1200 cm^{-1} were used for determination of glucose in aqueous media and serum (1-3). Mid- IR emission was used for NI measurement of glucose and was discussed in a recent review (1).

The strongest fundamental bands that can generate intense combinations and overtones are the broad OH stretch at 3550 cm^{-1}, and the C-H stretch vibrations at 2961, and 2947 cm^{-1}. Possible combination bands are second OH overtones band at 939 nm (3νOH) and second harmonic CH overtone band at 1126 nm (3νCH). A first OH overtone band can be assigned at 1408 nm (2νOH). The 1536 nm band can be assigned as an OH and CH combination band (νOH +νCH). The 1688 nm is assigned as a CH overtone band (2νCH). The 2261 nm band is possibly a combination of a CH stretch and a CCH, OCH deformation (νCH + νCCH, OCH). The 2326 nm bang can be attributed to a (νCH + ν CCH, OCH) combination (2). Having the CCH, OCH ring deformation component confirms some glucose specificity on these bands (2). Table 7.1 shows these NIR bands used for NI glucose determinations at various body sites (15-17).

The calculated NIR overtone and combination spectra of glucose overlap with several (more intense) combinations and overtone bands of water and of tissue fat (2, 18, 19, 20-26)). They also overlap with hemoglobin electronic absorption bands as shown in Table 7.2.

Table 7.1: NIR glucose bands and NI transmittance and reflectance studies

Wavelength in nm	Possible assignment (2)	Body site,
939	3 ν O-H stretch	Oral mucosa (15)
1126	3 ν C-H stretch	Oral mucosa (15
1408	2 ν O-H	Arm (16)
1538	ν O-H + ν C-H	Arm (16)
1688	2 ν C-H	Arm (16), tongue (17)
2261	ν C-H + ν C-C-H, O-C-H	Arm (16), tongue (17)
2326	ν C-H, ν C-H, OCH	In vitro studies

Table 7.2: Near IR absorption features of glucose and human tissue components (2)

Glucose [a, b]	Water [c]	Hb [d]	Fat [e]	Protein [f]
714	749			
		760 (E)		
			770	
		805 (E)		
		820 (E)		
	880			
939		910 (E)	920	910
	980			
1126	1211	1020 (E)	1040	1020
1408	1450			
1536				
1689[b]				
	1787			
	1934			
2270[b]				2174
2293[b]			2299	2288

a) Calculated from glucose fundamental vibrations reference 2, b) reference 26, c) reference 18, 19, d) "E" indicates an electronic absorption band, others are vibration overtone and combination bands, e) reference 20, f) reference 21

Most of these bands are more intense than glucose NIR absorption bands due to the much larger relative concentration of the absorbing components. Variation in the intensity of some of these bands will affect the precision of measuring a NIR signal due to changes in glucose only. Combination bands of water, fat, proteins and hemoglobin will cause major interferences with the NI determination of glucose. Several studies showed the ability to determine glucose in aqueous media, serum and blood, in the NIR (2000 - 2500 nm) (18-24). Arnold and Small contributed to understanding the critical fitting conditions for these determinations. An example of the magnitude of NIR glucose intrinsic absorption signals is illustrated by the recently determined values of the molar extinction coefficient, ε, of glucose in water in Table 7.3 (23, 24).

Table 7.3: Summary of glucose NIR absorption bands and extinction coefficients

Uncorrected for water displacement [a]		**Corrected for water displacement** [b]	
Wavelength nm	Absorptivity $molar^{-1}cm^{-1}$	Wavelength nm	Absorptivity $molar^{-1}cm^{-1}$
1689	0.23	1689	0.463
2257	0.4	2270	0.129
		2293	0.113

a) From reference 23, b) from reference 24

The absorptivity values in Table 7.3 are far smaller than the ε value of NADH at 340 nm of $6.2x10^{+3}$ $molar^{-1}cm^{-1}$, which is usually used for the determination of serum glucose values on automated blood analyzers. Using a 1 mm pathlength, a 10-mmolar glucose solution will have $2.3\ x10^{-4}$ absorbance units at 1686 nm and $4x10^{-4}$ absorbance units at 2257 nm (uncorrected for water displacement). A 1mm pathlength is longer than the pathlength encountered in diffuse reflectance measurements, and is of comparable magnitude to the pathlength in some spatially resolved and diffuse reflectance measurements (15, 16, 25). The intrinsic extinction coefficient of glucose will be much lower at the higher overtone bands between 800 nm and 1300 nm. Quantitative interpretation of the data in this spectral range requires extremely sensitive detection system and elimination of sources of biological and measurement background noise. The very low values of the molar extinction coefficient of glucose necessitate experimental measurement methods with extremely high signal to noise ratio and tight temperature control. Although IR-absorption measurement of glucose has reasonable specificity in aqueous solutions, it faces higher hurdles when attempted at human body sites.

Raman spectroscopy is another example of a detection method that track a specific molecular property of glucose. Advances in Raman studies using NIR wavelength lasers to minimize the effect of tissue and serum fluorescence were achieved by Feld's group (26, 27, 28).

The low values of NIR glucose molecular absorptivity were the impetus for seeking indirect methods that track the effect of glucose on tissue optical properties. Although these methods lack the molecular specificity of absorption measurements, they offer larger signal changes (1, 29, 33-45). These are mainly based on scattering or on including both absorption and scattering information.

7.3. COMPARTMENTALIZATION OF GLUCOSE VALUES

Glucose in human body is distributed in various body fluids such as blood, interstitial fluid, eye vitreous fluid and sweat in sweat glands. Other fluids include tears and saliva. In-vitro invasive testing methods determine glucose in samples of venous blood or arterialized venous blood, which uniquely defines the body compartment in which glucose is determined. A NI glucose-monitoring device will process optical signals transmitted through, or reflected by, multiple body compartments such as the stratum corneum, dermis and epidermis layers, subcutaneous tissue, interstitial fluid, arteries, veins, and capillaries. A pictorial presentation of the skin layers and their optical properties is given in Tuchin's monograph on tissue optics (29). Each of these compartments has different optical properties and may have different concentration of glucose. A reported NI glucose concentration might represent an average of glucose values in several compartments. Correlation between a NI-determined glucose measurement and blood glucose values may vary from one body site to another depending on differences in tissue and vascular properties of each site (1, 2).

Several NI methods propose determination of glucose in ISF. There is a reported lag time between change in blood glucose value and the concentration of glucose in other body fluids. Lag times between blood and ISF glucose levels were reported for implantable glucose sensors (6). Lag time between ISF and blood glucose values varied from < 5 minutes to > 30 minutes depending on the measurement technology (10). Delays in ISF or any other body fluid glucose equilibration, if predictable, could be corrected for in the algorithm (1).

7.4. CALIBRATION MODELS AND PATIENT-SPECIFIC CALIBRATION

7.4.1. Establishing a Standard Response Relationship Calibration Models

Calibration of NI glucose devices requires establishing a correlation between invasively- determined blood glucose concentrations and the corresponding NI signals. One-way is to induce a change in glucose concentration that spans a range over which the NI measured signal is monitored. Glucose clamp experiments, oral glucose tolerance tests (OGTT), and meal tolerance tests (MTT) are example of methods for inducing change in blood glucose concentration. A reference (in vitro) glucose measurement is simultaneously performed and a calibration model is established using a regression procedure. The calibration model that is generated during the test period is then used to predict glucose concentration from subsequent NI measurements.

7.4.2. Prediction Parameters

Multivariate analysis is generally used for determination of the concentration of a component in a complex mixture (30-32). The quality of the predicted NI glucose data is judged by the magnitude of the standard error of prediction (SEP), and the prediction correlation coefficient r_p. A low SEP, when associated with a high r_p, is indicative of a successful prediction. One must guard against the possibility of chance correlation with other time–dependent events, and of overfitting the experimental data. Extreme care must be taken to avoid overfitting. Arnold et al first discussed the issue of chance instrument correlation (-). The issue of physiological chance correlation was discussed in a recent review (1). The training set and the prediction set need to be separated in time. The number of input NI data points must be larger than the number of terms in the fitting equations.

A high SEP can be interpreted in one of three ways. First, the calibration model is void of glucose specific information. A second interpretation is that glucose predictions are limited by measurement noise. A third reason for a high SEP is that glucose predictions are limited by biological background noise in the body part and repositioning errors of the probe with respect to the body site. Noise sources such as circulation parameters, skin surface condition, skin water content, circadian rhythm effects, and temperature changes are examples of biological variables that haves not been discussed in sufficient details yet. Time-dependent artifacts can influence the results from multivariate calibrations when randomized sampling over time cannot be performed. In addition to instrument-related time variables, the circadian rhythm of the human body can lead to spurious time-dependent biological background that will be superimposed on the sequential MTT data points (1).

Although glucose clamp experiments, OGTT, and MTT data are necessary to prove that a particular NI signal varies with induced change in glucose concentration, these experimental data are not sufficient to establish a reliable calibration of signal versus blood glucose values over a long time period. As a NI signal may embody non-glucose-related physiological effects, relying on calibration based on correlating OGTT or MTT data may lead to a calibration model that is unique to the individual tested. Calibration will need to be periodically updated. If frequent recalibration is required, it will present an acceptance problem for NI testing device (1).

7.5. THERMO-OPTICAL RESPONSE OF HUMAN SKIN

In order to improve specificity, one must find a measurement method where the signal is related to glucose through some quantitative relationship between glucose concentration and its effect on the signal. Another method is to use external stimulus to induce a signal change that may enhance that is due to glucose of minimized some of the experimental variables during the measurement. Improving compartmentalization is achieved by directing the optical measurement to a localized section of the body site that has homogeneous physical and structural properties.

There are several attempts to improve specificity, compartmentalization, and calibration. One method is to use external stimuli such as temperature and defined probe design (33-37). These studies attempt to improve on μ_s' measurements for tracking glucose (38-42). Another method is occlusion spectroscopy, which attempts to measure scattering due to erythrocytes (43), and OCT measurements that localizes detection to the upper dermis region (44, 45).

Limiting light penetration depth to a defined tissue region was achieved by using a localized reflectance optical probe with short source-detector distance, which limits the optical depth in tissue to about 2 mm (33). This depth spans the epidermis and the dermis that include the cutaneous vascular system and blood capillaries. It lies just above the adipose tissue (29). The use of the interplay between temperature and light wavelengths and the special probe design was used to localize the optical signal within a confined range of cutaneous tissue (36). The motivation to study of the thermo-optical response of human skin was predicated upon:

a. It is possible to sample a limited tissue compartment (epidermis and the dermis) by proper design of the optical probe.
b. There are structural differences between human diabetic skin and non-diabetic skin
c. Glucose causes physical changes and has a physical manifestation on human tissue optical properties as shown by the scattering coefficient.
d. Glucose and insulin invoke physiological response in human tissue that is manifested by change in blood flow. Temperature change has a physical effect on blood flow.
e. Skin vascular system plays an important role in body temperature regulation and consequently in body response to glucose change.
f. In order to measure glucose: it is important to study the probe skin interaction, temperature effect, probe skin repositioning.

In this chapter we will discuss the work of our group on the use of temperature-modulated thermo-optical response (THOR) of human skin to localize compartmentalization to the dermis layer and to study the effects of probe-skin interaction (33-37). The studies involved both diabetic and non-diabetic volunteers.

7.5.1: Thermo-Optical Response (THOR) Localized Reflectance Measurements

Our group studied the effect of the interaction between the detection probe and tissue during a non-invasive optical measurement (33-37). The effect of probe–tissue interaction and the effect of temperature changes on the measured non-invasive optical signal were not well studied. Glucose affects μ_s' of tissue-simulating phantoms and of human tissue. However, this effect is non-specific.

Figure 7.2: Photograph of the temperature-controlled THOR probe.

Other soluble analytes decrease μ_s' as the concentration of the analyte (or glucose) is increased (38-41).

We measured localized reflectance signals over a set of small source-detector distances, using a temperature-controlled detection probe. Figure 7.2 shows a picture of the probe. The source-detector distances were 0.4 mm to 2 mm. A 400-micrometer fiber illuminated the tissue, which was the dorsal side of human forearm skin, and light was collected using 4-6 fibers of the same diameter. The wavelengths used were between 550 nm and 980 nm (33-37).

The center ring in Figure 7.2 is the end of the fiber bundle having the illumination fiber and light collection fibers at set distances. The larger middle ring is a 2-cm diameter temperature-controlled aluminum disk. Its temperature is controlled by a thermoelectric element glued to its and a thermistor embedded it. The thermistor provided feedback signal to the temperature controller powering the thermoelectric element. The outer black ring is a plastic insulating ring.

Modeling the temperature distribution in human skin showed that it was controlled to a depth of 2 mm (33). This depth is close to the to the light penetration depth in skin at the wavelengths used in the experiments. This depth includes the epidermis, dermis and the vasculature therein. Light penetration and temperature change interact with the upper and lower plexus and skin nutritive capillary structure. Monte Carlo simulations and use of tissue simulating phantoms of known optical properties were used to calibrate the system and to extract μ_a and μ_s' from the measured signals (33). The temperature controlled probe in Figure 7.2 was part of a body interface module, where the probe was brought in contact with the dorsal side of a human arm under a constant pressure of approximately 100 gm/cm2. The body interface is shown in Figure 7.3.

The temperature-controlled localized reflectance optical system was used to study the effect of temperature on the μ_a and μ_s' of human skin. Localized reflectance of the skin of 5 light-skin subjects was measured at 590 nm with the

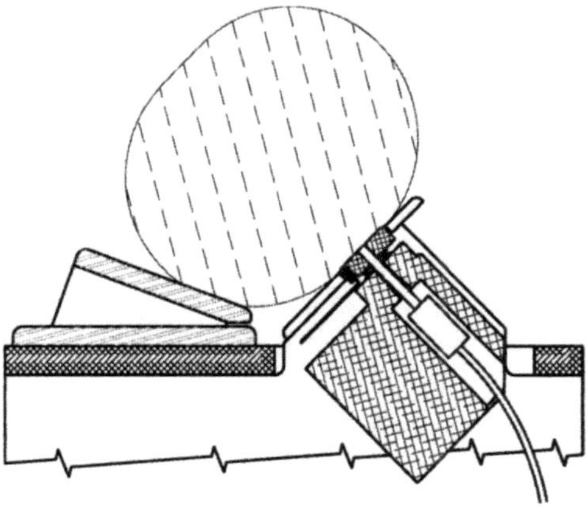

Figure 7.3: A cross section of the arm is shown in contact with the optical probe mounted in the body interface module.

dorsal side of the forearm in constant contact with the temperature-controlled probe. Temperature was switched back and forth among three values between 25 and 41 °C. Plots of the regression lines for μ_s' vs. probe temperature are shown in Figure 7.4.

Plots of μ_s' vs. temperature fit straight lines. Data are labeled by subject's designation A through F. The fitted slope ($\partial\mu_s'/\partial T$) for individual subjects was 0.053±0.0094 cm^{-1}/°C. The intercept and the slope of the regression lines (of μ_s' vs. probe temperature plots) varied from subject to subject and for different measurements on the same subject. These differences were attributed as probably due to differences in repositioning the probe on skin (33).

The linear temperature dependence of μ_s' is reversible and can extend over several cycles of cooling and heating the skin. It is also observed at several wavelengths (33). This reversible dependence on temperature was interpreted based on the effect of temperature on the refractive index of the ISF, and hence the refractive index mismatch between the ISF and skin scattering centers. The effect of temperature on μ_s' of human skin is opposite to the observed effect of glucose on μ_s' of scattering media, including human tissue.

The behavior of μ_a of intact human skin as a function of temperature differed from that of the μ_s' (33). Figure 7.5 shows a plot of μ_a at different temperatures, determined at 590 nm and using the same temperature controlling steps used to generate the data in Figure 7.4. Measurements are designated A through F in both Figures 7.4 and 7.5. Dependence of μ_a data of intact human skin in Figure 7.5 first shows the short-term irreversibility of μ_a values upon heating to or above 38 °C. Thus μ_a generally increased upon heating and decreased upon cooling. Secondly, the values of consecutively-determined μ_a zigzagged upwards. There is a cumulative effect of multiple times of heating and cooling on the value of μ_a suggesting that it depends on the thermal history of the skin. The upward zigzagging drift of μ_a, was explained as due to pooling of blood in the cutaneous capillaries (33).

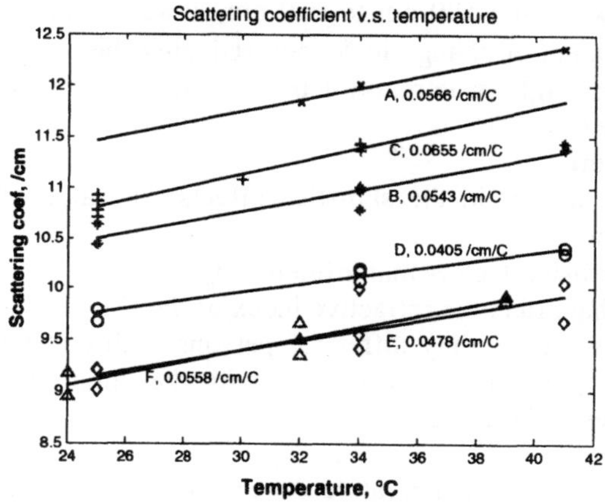

Figure 7.4: Effect of temperature on the scattering coefficient of human skin. Each letter denotes a volunteer; the numbers beside each letter is the slope of the fitted ms vs. temperature line

Changes in μ_a and μ_s', as a function of temperature, will inversely affect light penetration depth in tissue as is defined by Equation 3. Decrease in blood flow to the subsurface capillaries, i.e. a decrease in μ_a of the top layers, or a decrease in μ_s' as temperature is lowered leads to an increase in δ, which allows the sampling deeper dermis layers.

Table 7.4 shows the mean δ values of the forearm skin of seven light skin non-diabetic volunteers (36, 37). Light penetration depth, δ, increases as the

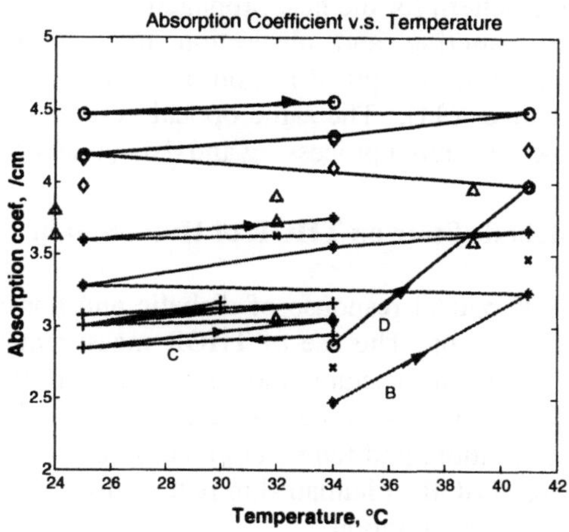

Figure 7.5: Effect of temperature on the absorption coefficient of human skin. The letters B, C, and D refer to the temperature response of μ_a of the same subjects whose scattering coefficient data are plotted in Figure 7.4. Figures 7.4 and 7.5 are reproduced from Khalil et al, *J. Biomed. Opt.* **8**:191-205 (2003), with permission from the American Institute of Physics

wavelength increased from 590 nm to 950 nm, at skin temperatures of 38 °C and 22 °C. The extent of change in δ upon reducing the temperature depends on the wavelength. Light penetration depth in human skin depends on several factors that include (33, 34):

a. Temperature
b. Structural factors in the skin that affects the value of the scattering coefficient
c. Blood perfusion factors that influence μ_a
d. Factors that affect the refractive index of the ISF

The effect of these factors may differ for persons with diabetes as compared with non-diabetics (34). It may also vary with the duration of diabetes and other co-morbidities and medications.

Table 7.4: Wavelength and temperature dependence of Light penetration depth in skin (36)

Wavelength of Incident light ⇒ Temperature ⇓	Light penetration depth δ in skin in micrometers		
	590 nm	750 nm	950 nm
38 °C	970± 80	1800±180	2020±200
22 °C	1110±90	2060±180	2210±180
Mean Δδ in microns	130	260	190
% Change in Δδ	14%	14.4%	9.9%

After establishing a pattern for the heat propagation in cutaneous tissue and the response of the scattering and absorption to thermal stimulus, we investigated the difference in the optical response of diabetic and non-diabetic skin to temperature changes (34). The same optical system was used in these studies with improvement in signal processing and data collection (33).

7.5.2. Effect of the Diabetic State on THOR of Human Skin

We studied the NIR optical response of diabetic and non-diabetic human skin to temperature modulation. The use of THOR measurements is based on the premises that due to known structural and circulatory differences in the skin of diabetics, the THOR of human skin may be sensitive to the diabetic state. As diabetes, glucose concentration, and temperature cause cutaneous vascular and refractive index changes, THOR of human skin potentially may be used to track changes in tissue glucose concentration.

Cooling and heating have both physiological and optical effects on skin tissue. Cooling induces vasoconstriction accompanied by decreased blood perfusion, which decreases μ_a. Heating induces vasodilatation, increases blood perfusion, and increases μ_a at the hemoglobin wavelengths. The optical effects are decrease in μ_s' by cooling and its increase upon heating (33, 34).

Measurements were performed at two sites, designated A_1 and A_2, on the dorsal side of the arm. The two were not overlapping and were selected to be morphologically similar in such aspects as the absence of hair, closeness to bone, and appearance of veins. Each of the two skin sites was independently subjected to temperature perturbation in the form of programmed temperature/time events during the optical measurements.

THOR measurement is performed at skin site A_1 where a temperature perturbation involved cooling the skin at a defined rate and for a defined duration. THOR measurement is performed at skin site A_2 involved heating the skin at a defined rate and for a defined duration.

THOR signals at A1; $R_{A1}(\lambda_i, r_j, T_k)$ and $R_{A1}(\lambda_i, r_j, T_l)$ were recorded at two time points corresponding to two temperature values, where λ_i is the wavelength of light and r_i is the distance between the illumination fiber and a light collection fiber in contact with the skin (source-detector distance). T_k and T_l are cutaneous temperatures attained after *k* seconds and *l* seconds of contact between the temperature-controlled optical probe and the skin. From the measured localized reflectance we calculated the natural log of the ratio of the reflectance at the first (cooling) site, for each wavelength, each source detector distance, and different temperatures; $Ln\{R_{A1}(\lambda_i, r_j, T_l)/ R_{A1}(\lambda_i, r_j, T_k)\}$, where T_l is at a time point later than T_k.

Simultaneously or sequentially, we performed a measurement at A_2 where the temperature program involves heating the skin to determine $R_{A2}(\lambda_i, r_j,T_p)$ and $R_{A2}(\lambda_i, r_j,T_q)$ at two time points corresponding to two temperature values. We then calculated the value: $Ln\{R_{A2}(\lambda_i, r_j,T_q)/ R_{A2}(\lambda_i, r_j,T_p)\}$, where T_q is at a later time than T_p.

Since the change in temperature induces a small change in the reflectance over the range studied, the value of the natural logarithm can be expressed using the expansion:

$$Ln(1 \pm x) = \pm x \, , \; x \ll 1 \tag{4}$$

Thus it is possible to express $Ln\{R_{A1}(\lambda_i, r_j, T_l)/ R_{A1}(\lambda_i, r_j, T_k)\}$as;

$$Ln\{R_{A1}(\lambda_i, r_j, T_l)/ R_{A1}(\lambda_i, r_j, T_k)\} = Ln\,(R \pm \Delta R) = \pm \Delta R_T(A_1) \tag{5}$$

ΔR is the temperature-induced fractional change in $R(\lambda_i, r_j)$ at site A_1. Similarly, $Ln\{R_{A2}(\lambda_i, r_j,T_q)/ R_{A2}\, (\lambda_i, r_j,T_p)\}$ is the temperature-induced change in $R(\lambda_i, r_j)$ at site A_2. The quantities are also proportional to optical density changes induced by temperature.

THOR was measured at two sites on the dorsal side of the arm of test subjects. The time for the two sequential measurements was 240 seconds. The time the skin was kept at a constant temperature was 30 seconds and the transition time for temperature change was 180 seconds. Three different temperature limits were attempted and were described in detail (34). A THOR at two skin sites, $\Delta R_T(A_1, A_2)$, is expressed as:

$$\Delta R_T(A_1, A_2) = \mathrm{Ln}\{R_{A1}(\lambda_i, r_j, T_l)/ R_{A1}(\lambda_i, r_j, T_k)\} - \mathrm{Ln}\{R_{A2}(\lambda_i, r_j, T_q)/ R_{A2}(\lambda_i, r_j, T_p)\} \quad (6)$$

Figure 7.7 shows a difference plot of the THOR function $\Delta R_T(A_1, A_2) = [\mathrm{Ln}\ (R_{A1Tm}/R_{A1T30}) - \mathrm{Ln}\ (R_{A2Tm}/R_{A2T30})]$ plotted against probe-skin contact time. A qualitative separation of diabetic and non-diabetic skin responses after temperature change at 360 seconds from probe-skin contact is quite noticeable.

The qualitative separation of the THOR function plots (temperature-induced fractional change in the localized reflectance versus time) for each body mass index group was also noticeable at 890 nm and 0.9 mm source-detector distance. The separation was not as noticeable under the same-programmed temperature change at other source-detector distances. It is then apparent that the use of THOR at a single wavelength and at a single source-detector distance is not sufficient to overcome person-to-person differences in skin properties.

7.5.3. Qualitative Assessment of the Diabetic State from THOR of Human Skin Using Nonlinear Discriminant Functions for Data Analysis

We then moved to expand the analysis of THOR data to multiple wavelengths and multiple source-detector distances. Yeh et al combined four wavelengths and four source-detector distances and used a nonlinear

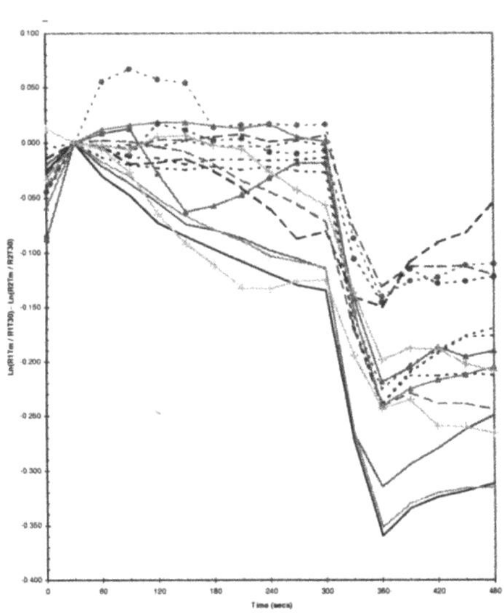

Figure 7.6: THOR at 890 nm and a source-detector distance of 0.9 mm. The solid lines represent diabetic's data; the dashed lines represent no-diabetic's data points.

discriminant function to overcome the person-to-person differences in body mass index and skin properties (34). Non-diabetic subjects and Type 2 diabetic subjects with diabetes duration 2-15 years volunteered for the experiment. The subject's diabetic state was determined by previous diagnoses as diabetic or non-diabetic. Two data sets were collected, a training set and a prediction set. The training set consisted of four diabetic and four non-diabetic subjects, each tested six times (total of 48 data points). Measurements for the training set were performed at the time slots that were selected to allow including several blood glucose concentrations, physiological, and circadian rhythm conditions in the training set. The independent prediction set consisted of six diabetic and six non-diabetic subjects, each tested twice (total of 24 data points). Measurement times for the prediction set were randomly chosen within a week.

Two sequential measurements were performed, each lasting 240 seconds. The time the skin was kept at a constant temperature was 30 seconds and the transition time for temperature change was 180 seconds (34). Three different temperature limits were attempted:

a. In the first experiment the probe at skin site 1 was maintained at 34 °C for 30 seconds and then cooled over 180 seconds to 22 °C at 4 °C.min-^{1}.
The probe at skin site 2 was maintained at 34 °C for the 240 seconds duration of the measurement.
b. In the second experiment skin site 1 was maintained at 38 °C for 30 seconds and cooled over 180 seconds period to 22 °C at 5.33 °C.min^{-1}. Skin site 2 was maintained at 22 °C for 30 seconds and then raised to 38 °C at 5.33 °C.min^{-1}.
c. In the third experiment skin site 1 was maintained at 30 °C for 30 seconds and then cooled over 180 seconds period to 22 °C at 2. °C.min^{-1}. Site 2 was treated in the opposite way; it was maintained at 30 °C for 30 seconds and then raised to 38 °C at 2.67 °C.min^{-1}.

An example of a temperature perturbation program at two sites is shown in Figure 7.7.

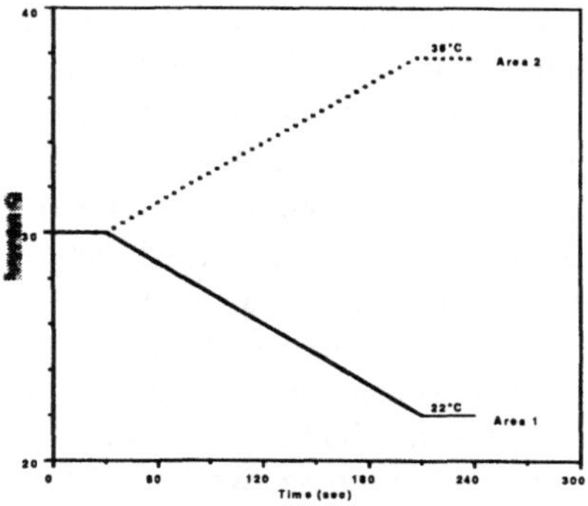

Figure 7.7: Temperature program reproduced from Khalil et al, *J. Biomed. Opt.* **8**:191-205 (2003), with permission from the American Institute of Physics

The difference THOR function at the two skin areas. $\Delta R_T(A_1, A_2)$ is expressed as a function *f*:

$$f(R_{A1t1}, R_{A1t2}, R_{A2t1}, R_{A2t2}) = [\mathrm{Ln}\,(R_{A1t1} / R_{A1t2})] - [\mathrm{Ln}\,(R_{A2t1} / R_{A2t2})] \quad (7)$$

Where R_{A1t1} is the localized reflectance within skin site A_1 at probe-skin contact time t_1, R_{A1t2} is the localized reflectance within site A_1 at time t_2; R_{A2t1} is a localized reflectance within skin site A_2 at time t_1, and R_{A2t2} is the same localized reflectance within site A_2 at t_2. In this expression the contact time between the probe and the skin is used instead of the contact temperature. The temperature perturbation program employed at each site determines the cutaneous temperature at that site at any time point. The function *f* is applied to the optical signals at all source-detector distances and for all wavelengths.

The function(s) $f(R_{A1t1}, R_{A1t2}, R_{A2t1}, R_{A2t2})$, determined at each of the four source-detector distances and at four wavelengths, together with the known diabetic or non-diabetic state of each of subjects in the training set, are used to generate a discriminant function D. A subject is classified as diabetic if $D > 0$, and non-diabetic if $D < 0$. D is a quadratic expression comprising multiple functions of the type expressed in Equation 7, and has the form (34):

$$D = \sum_i \sum_j a_{ij} (\delta_i f_i)(\delta_j f_j) + \sum_i a_i \delta_i f_i + a_o \quad (8)$$

Where

$$\delta_i = 1 \; or \; 0; \quad \text{and} \sum_i \delta_i = K \quad (9a)$$

$$\delta_j = 1 \; or \; 0; \quad \text{and} \sum_j \delta_j = K \quad (9b)$$

a_{ij}, a_i, and a_0 are constants determined from the training set, and *i* or *j* are indices to specific combinations of wavelength and source-detector distance. The parameter d in equations 8 through 9b is not the light penetration depth in tissue. The number *K* limits the total number of wavelength/source-detector distance combinations used in D to avoid overfitting, and δ_i and δ_j were determined from the training set through a leave-one-out cross validation procedure that minimizes the number of false classification.

The true diabetic state of a subject was represented by S_i where; $S_i = +1$ for a diabetic subject, and $S_i = -1$ for a non-diabetic subject. D was calculated for each subject i as D_i. The subject is categorized as concordant if D_i and S_i had the same sign and was discordant if they had different signs. The coefficients of the quadratic function D of the calibration set were used to calculate the value of the function D_i for the prediction set. If $D_i > 0$, the subject was classified as diabetic. On the other hand, if $D_i < 0$, the subject was classified as non-diabetic.

The optical data points collected were divided into two classes; the class of diabetic and the class of non-diabetic subjects. A 2x2 prediction matrix $\begin{bmatrix} \alpha & \beta \\ \chi & \gamma \end{bmatrix}$ was established, where the elements of the matrix are the number of data points representing a true or false classification: α is the number of true diabetic data points identified as diabetic, γ is the number of true non-diabetic identified as non-diabetics, χ is the number of diabetics identified as non-diabetic (false negative), and β is the number of non-diabetic identified as diabetic data points (false positive). We then calculated several prediction parameters using the formulas in Table 7.5.

Table 7.5: Diabetic state prediction parameters

Parameter	Sensitivity	Specificity	Positive predictive value (PPV)	Negative predictive value (NPV)
Formula	$100\alpha/(\alpha+\chi)$	$100\gamma/(\gamma+\beta)$	$100\alpha/(\alpha+\beta)$	$100\gamma/(\chi+\gamma)$

Classification was then attempted utilizing equations 8, 9a and 9b, and using optical signals at eight wavelength-distance combinations (i.e. K=8) at t_1 = 30 seconds and at any of these time points t_2: 120, 150, 180, and 210 seconds from the probe-skin contact. The prediction parameters are given in Table 7.6.

Table 7.6: Performance parameters at different probe–skin contact time [a]

Contact time (second)	Probe temperature (°C)						
	Site 1	Site 2	p	Sensitivity %	Specificity %	PPV	NPV
120	30	30	0.0001	92	90	92	100
150	27.3	32.7	0.0016	73	100	100	77
180	24.7	35.3	0.0004	90	90	90	90
210	22	38	0.0004	82	92	100	83

a) Using the temperature program shown in Fig. 7

Reproduced from Yeh et al, *J. Biomedical Optics.* **8**:534 544 (2003), with permission of the American Institute of Physics

Prediction of the diabetic state, using the three different temperature perturbations conditions and using THOR function at 120 seconds from the onset of probe-skin contact is given in Table 7.6. Also given is the probability p of overlap between the two populations. The data in the table indicates that it was possible to achieve a separation between the diabetics and non-diabetics using the nonlinear discriminant function and eight distance-wavelength combinations, which was not consistently achievable at a single wavelength and a single source detector distance.

Differentiation between diabetic and non-diabetic subjects was not sensitive to the contact time between the skin and the probe beyond 120 seconds, as shown for the first temperature perturbation program in Table 7.6.

Similar data were obtained for the two other temperature perturbation programs, indicating that prediction of diabetic status was achievable at 120 seconds of probe-skin contact for all the three temperature perturbation programs. Increasing the probe-skin contact times did not appreciably enhance the quality of the classification data. Further, the classification was not limited to one unique programmed temperature perturbation as shown in Table 7.7.

The thermo-optical response (THOR) of the skin of diabetic subjects appeared to differ from that of non-diabetic subjects. Measuring cutaneous localized reflectance under specific thermal perturbation, and the use of a nonlinear discriminant function as a classifier allowed assessment of the diabetic state of human subjects in the small studied population (34). The method involved two sequential localized reflectance measurements under different temperature perturbation conditions, each measurement lasting less than 300 seconds. Acceptable classification was achieved after 120 seconds contact time for each measurement.

Table 7.7: Prediction of the diabetic status at 120 seconds after probe-skin contact

Temperature program #	Optical Test Result	True Diabetic State	
		Diabetic	Non-diabetic
1	Diabetic	11	3
	Non-diabetic	1	9
2	Diabetic	11*	1*
	Non-diabetic	0*	9*
3	Diabetic	10	0
	Non-diabetic	2	12

*Total number of data points is 12, three data points were rejected as they resulted in outlier optical signals. Reproduced from Yeh et al, J. Biomedical Optics 8:534-544 (2003) with permission of the American Institute of Physics

It is possible to explain the observed differentiation in THOR between diabetics and non-diabetics as due to a combination of two or more of:

1. Temperature dependence of absorption and scattering due to red blood cell structural changes, difference in cell aggregation, and the refractive index.

2. Temperature dependence in tissue scattering due to the effect of diabetes on cutaneous collagen fibers and subsequent difference in refractive index mismatch.
3. Difference in microcirculation response to temperature change. Laser Doppler Flowmetry accompanied by iontophoreses of vasodilators showed dependence of the vasodilator induced blood flow and the diabetic state of the subject (46).

Monitoring a THOR function exemplified by the temperature-induced fractional change in the localized reflectance showed the ability to differentiate between diabetic and non-diabetic subjects in the studied population. The use of a nonlinear discriminant function as a classifier did overcome person-to-person differences in skin optical properties and the variable glucose concentrations at the time of measurement.

The preliminary study discussed shows the potential for the non- invasive screening for diabetes using THOR of human skin (34). Current methods for screening for diabetes include a fasting glucose test, followed by a glycated hemoglobin test, and in some instances a meal tolerance test is also performed. This preliminary study is interesting, but further studies are required. Larger population studies are needed to include subjects with impaired glucose tolerance, with early-diagnosed diabetes, and with advanced diabetes, to improve the statistical predictability and to relate changes in signal to skin pathology and disease progress and to concentration of glycated hemoglobin (HbA1c) (47, 48). The thermo-optical response of human skin has the potential for studies on vascular diseases or the effect of disease on the peripheral vascular system, in a way similar to the use of laser Doppler flowmetry.

The observation of a difference in thermo-optical properties of human skin between diabetics and non-diabetics has an important ramification in comparing data between the two groups, as any interaction between a probe and skin will affect temperature equilibration. It is thus advisable to analyze the results studies on diabetics and non-diabetics separately without mixing the data from the two groups.

7.5.4. Glucose Calibration Models from Thermo-Optical Response of Human Skin

Yeh at al studied the effect of changes in glucose concentration on the THOR of human kin in the spectral range between 550 nm to 980 nm (35, 36). The approach is based on the premises that measuring the skin's THOR signal, which was sensitive to that diabetic state, may be used to track effect of glucose on optical properties of skin (35). Glucose has no intrinsic absorption bands in this spectral range. Reflected light intensity corresponds to blood absorption and scattering by cutaneous tissue. It relates to hemodynamic changes in cutaneous tissue and the effect of glucose on refractive index mismatch between the ISF and tissue connective fibers.

Glucose concentrations were correlated with THOR signals at wavelengths between 590 and 935 nm in a series of MTT experiments (35, 36). Skin temperature was modulated between 22 °C and 38 °C over 2 hours to cause periodic changes in skin reflectance corresponding to changes in cutaneous refractive index and cutaneous vascular response. Blood glucose concentrations were fitted the THOR signals using the 4-tem linear least squares fitting equation:

$$[\text{Glucose}] = a_0 + \Sigma_i\, a_i \bullet \text{Ln}_e R_i\,(r, \lambda, T) = a_0 + \Sigma_i\, a_i \bullet R'_i\,(r, \lambda, T) \quad (11)$$

In this equation, $R_i\,(r, \lambda, T)$ is the THOR signal at a source-detector distance r (mm), wavelength λ (nm), an temperature T (°C); a_0 and a_i are the regression coefficients.

7.5.5. Calibration Models at Fixed Source-Detector Distance

The results of the MTT experiments were analyzed to determine the terms that contributed to the model at a single S-D distance. Table 7.8 shows the calibration data obtained at a single S-D distance of 0.92 mm for a diabetic subject.

The data (r_c and SEC) in Table 7.8 show the ability to establish models that correlate blood glucose level with THOR signal at a S-D distance of 0.92 mm at four wavelengths, as T was modulated between 22 °C and 38 °C. SEC was < 16.9 mg/dL and r_c was > 0.92 and. Similar data were obtained for the three subjects at S-D distances of 0.44 mm and 1.84.

The regression models of glucose concentration at a single S-D distance had the form:

$$\begin{aligned}[G] = \; & a_0 + a_1\text{LnR}\,(660\text{ nm}/\,22\text{ °C}) + a_2\text{LnR}\,(590\text{ nm}/\,22\text{ °C}) + \\ & a_3\text{LnR}\,(935\text{ nm}/\,38\text{ °C}) - a_4\text{LnR}\,(890\text{ nm}/\,38\text{ °C}) \qquad (12)\end{aligned}$$

Thee short λs of 590 nm and 660 nm are paired with the low temperature of 22 °C, and of the long λ of 890 nm and 935 nm, are paired with the higher temperature of 38 °C (36). LSQ regression models in Table 7.8 suggest that the interaction between light and tissue may occur in a specified cutaneous region. This region is confined between the depth specified by light penetration at shorter λ and lower temperature and the depth specified by light penetration expected at longer λ and higher temperature. In effect, the selection of temperature and wavelength can specify a cutaneous volume in the skin that can provide a satisfactory calibration relationship between THOR signal and blood glucose concentration.

Table 7.8: CLSQ regression 4-term fitting equation at a sampling distance of0.92 mm

Test	Calibration model	r_c	SEC (mg/dL)
Run 1 control	[G] = [5.65 - 1.32R'(660 nm/ 22 °C) + 0.278 R'(590 nm/ 22 °C) +5.15R'(935 nm/ 38 °C) −1.77R'(890 nm/ 38 °C)]x1000	0.95	9.22
Run 2 control	[G] = [-9.61 + 2.10R'(660 nm/ 22 °C) – 1.03 R'(590 nm/ 22 °C) - 7.71 R'(935 nm/ 38 °C) + 2.30 R'(890 nm/ 38 °C)] x 1000	0.96	13.2
Run 3 meal	[G] = [-4.25 + 2.49R'(660 nm/ 22 °C) – 4.73 R'(590 nm/ 22 °C) - 9.42R'(935 nm/ 38 °C) – 3.07 R'(890 nm/ 38 °C)] x 1000	0.93	14.9
Run 4 meal	[G] = [-2.72 - 2.16R'(660 nm/ 22 °C) – 2.51 R'(590 nm/ 22 °C) - 2.99 R'(935 nm/ 38 °C) – 7.96R'(890 nm/ 38 °C)] x 1000	0.95	11.3
Run 5 meal	[G] = [-3.12 + 4.82R'(660 nm/ 22 °C) – 8.30 R'(590 nm/ 22 °C) + 2.48 R'(935 nm/ 38 °C) – 5.43 R'(890 nm/ 38 °C)] x 1000	0.94	16.9
Run 6 meal	[G] = [-4.01 - 3.79R'(660 nm/ 22 °C) – 1.76 R'(590 nm/ 22 °C) +9.10 R'(935 nm/ 38 °C) – 1.08 R'(890 nm/ 38 °C)] x 1000	0.92	10.1
Run 7 meal	[G] = [-5.60 - 3.10R'(660 nm/ 22 °C) – 1.83 R'(590 nm/ 22 °C) - 1.52R'(935 nm/ 38 °C) + 1.80 R'(890 nm/ 38 °C)] x 1000	0.94	15.7

7.5.6. Glucose Prediction Models from Thermo-Optical Response of Human Skin

THOR parameters R'(r, λ, T) = Ln (Measured Localized Reflectance) were used in predicting glucose concentrations. Thirty-two sequences of

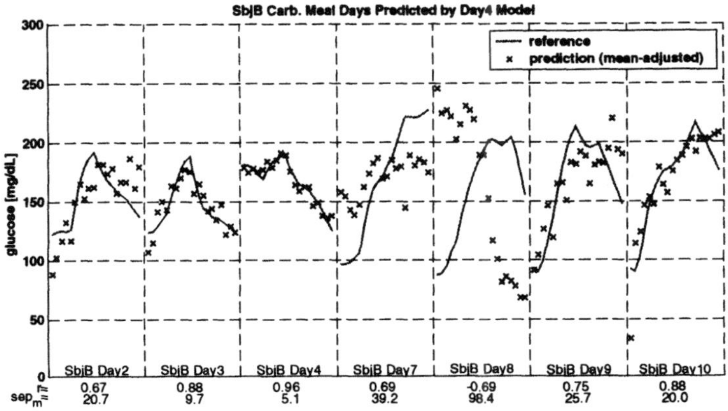

Figure 7.8: Reproduced with permission from Reference 35. Yeh et al, *Clin. Chem.* **49**: 924-934 (2003), with permission of the American Association of Clinical Chemistry

THOR data at temperatures $T_{22\,C}$ and $T_{38\,°C}$ were obtained. These were R' (at $T_{22\,°C}$) = Ln R(r, λ, $T_{22\,°C}$) and R' (at $T_{38\,°C}$) = Ln R(r, λ, $T_{38\,°C}$). Changes in glucose concentrations were predicted with a model based on MTT calibration for diabetic volunteers, with a SEP <1.5 mmol/L and r_p = 0.73 in 80% of the experiments. There were run-to-run shift in the whole response curve, which was attributed to site-to-site structural differences. Figure 7.8 shows an example of the MTT results for a Type 2 diabetic volunteer (35).

The correlation between glucose concentration and optical signals was attributed to effect of glucose on cutaneous hemodynamic and refractive index change. Temperature modulation was suggested to drive one or both of the vascular response and cutaneous light scattering in a way that enhances the physiological or physical effect of changes in glucose concentration on the dermis optical properties (35). Table 7.9 lists the mean–adjusted 4-term linear least squares prediction parameters for MTT experiments on the two diabetic subjects in the study (35).

Using a meal tolerance (MTT) calibration model predicted changes in glucose concentration in 80% of the NI-MTT runs on diabetic subjects. The day-to-day shift in the response curve (bias in the signal) was attributed to probe positioning, cutaneous structural effects, or other physiological changes. Maruo et al observed a similar bias in the predicted glucose values in a fiber optics based NIR absorption experiment (27, 28). There was a fortuitous observation for volunteer B on run #8 in Table 7.9. The subject had fever due to flue. The predicted glucose values based on day 3 model were in the opposite direction to the invasively measured values, with a negative r_p value of - 0.69, and SEP > 98.36 mg/dL. The effect of fever in this case suggests that testing a NI glucose detection method should be performed under different disease conditions (1). Glucose concentrations in non-carbohydrate meal runs showed considerable scatter when predicted by a MTT calibration model having a wide glucose swing. For the case of protein meal runs, cutaneous hemodynamic changes due to digestion affected the optical signal (35). Yeh et al postulated that the extent and rate and direction of cutaneous hemodynamic response to protein meals may be different from that caused by the change in glucose concentration (35).

Table 7.9: Mean–adjusted 4-term prediction parameters for MTT experiments (35)

	Volunteer A			Volunteer B		
Day	[G] Range mmol/L	r_p*	SEP* mmol/L	[G] range mmol/L	r_p*	SEP* mmol/L
2	6.93	0.84	1.51	3.84	0.67	1.15
3	6.67	note#	#	3.62	0.88	0.54
4	6.43	0.86	1.17	3.49	note#	#
7	7.04	0.94	0.96	7.27	0.69	2.18
8	4.45	0.73	1.10	6.46	-0.69	5.46
9	7.22	0.85	1.34	6.84	0.75	1.43
Day 8 repeat	--	--	--	7.03	0.88	1.11

Reproduced with permission from reference 35, Yeh et al, *Clin. Chem.* **49**: 924-934 (2003), with permission of the American Association of Clinical Chemistry

* SEP and r_p are the mean-adjusted standard error of prediction and prediction correlation coefficient

\# Data set used as a calibration model

Table 7.10. Human data of methods based on light scattering measurements

Method	Specificity	Compartmentalization	Calibration
THOR [a]	μ_s' and μ_a Glucose effect on n_{ISF} and perfusion	Forearm, Dermis 0.4 to <2 mm depth,	OGTT, calibration model, Prediction
Localized reflectance of abdomen [b]	μ_s' of tissue Glucose effect on n_{ISF}	Abdomen, Dermis, subcutaneous fat and muscles 0.8 to 10 mm depth	Glucose clamp or OGTT, calibration model
Occlusion spectroscopy [c]	μ_s' of blood Erythrocyte scattering Glucose effect on n_{plasma}	Finger, Blood vessels	Glucose Clamp, calibration model
OCT[d]	μ_s' of tissue Glucose effect on n_{ISF}	Forearm, Upper dermis 200-600 μm depth	OGTT, calibration model

a) Reference 35, b) References 39 and 40, c) References 42, and d) references 43-44

Measurement of a THOR function of human skin offers compartmentalization advantages over localized reflectance measurements that use large source-detector separation (39-43), as it limits sampling depth to the dermis due to probe design. The detection method incorporates temperature effect on both the absorption and scattering processes in the near-IR. Measurement of Optical Coherence Tomography (OCT) of human skin tracks the change in μ_s' of the dermal layer as a function of change in glucose concentration (43-45). It was possible to establish calibration models in animal and human experiments.

As summarized in Table 7.10, it was possible to establish calibration models for THOR, occlusion spectroscopy and OCT, but not for the scattering experiments on the abdomen. It was possible to achieve prediction under some conditions using THOR. The prediction ability of the other methods is not reported yet. Thor, occlusion spectroscopy and OCT have compartmentalization advantages over localized reflectance (with large source-detector separation) and frequency domain methods.

7.6. ENHANCING SPECIFICITY BY AFFINITY CAPTURE AGENTS

In the first sections we discussed the issue of specificity using the example of direct NIR absorption measurements. In section 5, we discussed the use of glucose effect on a tissue property, with the thermo-optical response of human skin as an example. In this section we discuss molecular recognition as a method of enhancing specificity.

This recent set of methods is predicated upon the specific capture of a glucose molecule by a boronic acid derivative or a glucose-specific enzyme. As a result of this interaction a change in an optical property take place and can be used to track the concentration of glucose. Molecular capture or molecular displacement imparts a level of specificity not found in direct spectral measurements.

Two sets of detection methods were investigated. Fluorescence measurements, where the glucose capture step affects an energy-transfer process in a fluorescent system. A change in fluorescence intensity or lifetime indicates the effect of glucose capture or displacement.

The second set of methods has the capture agent as part of a macro-molecular lattice that acts as a diffraction grating for visible light. Interaction with glucose and capture with the embedded agent leads to a different charge distribution and modifies the grating layer spacing. The change leads to a spectral shift in light falling and detected at a certain glancing angle.

Both of the fluorescence detection method and the light diffraction method involve introduction of a reagent, usually immobilized in a microcapsule or a contact-lens type vehicle to measure glucose. Glucose can thus be tracked in the ISF, like the case of using implantable electrodes, or in tears, which leads to interesting compartmentalization issues, especially for the case of glucose measurements in tears.

As several chapters in this volume will discuss various aspects of fluorescence measurements of glucose, we will briefly discuss fluorescence measurements as it relates to other technologies.

7.6.1. Fluorescence Measurements:

Fluorescence is known to have more detection sensitivity than absorption measurements. Glucose does not have electronic absorption bands in visible or NIR spectral range. When glucose solutions were excited with an excimer laser line at 308 nm, fluorescence was detected in the 340 and 400 nm, with fluorescence maximum at 380 nm (49). Fluorescence intensity was reported to change with change in glucose concentration in an aqueous medium. There is no explanation as to the molecular origin of this fluorescence.

In vitro florescence assays for glucose have been studied as steps towards its in vivo determination. Several methods are based on fluorescent resonance energy transfer (FRET) and on competition between glucose and dextran for concanavalin-A (con-A) binding sites. The assay components are con-A labeled with an energy acceptor or an energy donor and dextran labeled with the complimentary molecule for FRET (50-52).

A fluorescence biosensor incorporates rhodamine labeled con-A (TRITC-con-A) as energy acceptor and fluorescein-dextarn (FITC-dextran) as energy donor. Both molecules are chemically conjugated into a hydrogel network (50). In the absence of glucose, TRITC-con-A binds with FITC-dextran, and the FITC fluorescence is quenched through FRET. Competitive glucose binding to TRITC-con-A liberates FITC-dextran, resulting in increased fluorescence intensity proportional to the glucose concentration. In vitro fluorescence response was linear over a glucose range between 0 and 33 mmol/l (50). Another glucose assay uses FRET between con-A labeled with fluorescent protein allophycocyanin (APC) as donor, and dextran-malachite green as an acceptor. Glucose displaces dextran- malachite green, increasing APC fluorescence intensity. The in-vitro assay had glucose dynamic range extending from 2.5 to 30 mmoles/liter (51, 52).

Conceptually there are two configurations for applying fluorescence measurements for glucose detection. One configuration is use of implanted encapsulated reagents under the skin, in the form of a tattoo. This implanted tattoo can be used for the measurement of glucose in the ISF. A second configuration is incorporating fluorescence-generating reagents is in the form of contact lens for glucose measurement in eye tears. In both configurations fluorescence-generating reagents are encapsulated in polymer matrices. Encapsulation of florescent agents in polymer matrices was described (53-55), and advances in encapsulation have been made over the past few year.

Lakowicz's group developed a sensing technology, using a daily use, disposable contact lens embedded with sugar-sensing boronic acid containing fluorophores. Boronic acid fluorophores embedded in commercial contact lens and immersed in glucose solutions undergo change in fluorescence intensity and wavelength upon binding to glucose (56, 57). The fluorescent contact lens film has the highest sensitivity at low glucose concentrations, which correspond to that in tear fluid. In in-vitro experiments, they showed that this approach is indeed suitable for the continuous monitoring of changes in glucose levels in the concentration range (50-500 micro molar). The polymeric hydrogel embedded fluorophore approach allowed sensing low glucose concentrations similar to the concentration range in tears.

March et al polymerized a fluorescent complex within a hydrogel to make an intraocular lens that responds well to glucose concentration (58, 59). The patient wears the lens, which changes its fluorescence depending on glucose concentration. A TRITC-con-A / FITC-dextran, system was incorporated in the contact lens material. FITC fluorescence is quenched through FRET, displacement of FITC-dextran by glucose molecules decrease the extent of quenching and allows observation of FITC fluorescence. The contact lens was excited at 488 nm and detected by a handheld device. An OGTT was performed on five fasting patients with type-2 diabetes. Contact lens fluorescence and venous blood glucose were measured over a three hours period. Contact lens fluorescence appeared to track blood glucose with a time delay. There is no details of the data analysis, but predicted glucose values were plotted against true glucose values and showed reasonable distribution. The predicted glucose values are, most probably, the results of cross-validation prediction process. The fluorescent contact lenses were comfortable and were tolerated well during the three hours experiment, even in patients who had not previously worn contact lenses (59).

A novel fluorescence technique was described by Strano' group, who demonstrated new opportunities for nanoparticle optical sensors that fluoresce in strongly absorbing media of relevance to medicine or biology (60). These single-wall carbon nano-tube sensors can be excited and do emit in the NIR. Molecular detection using near-infrared light because of greater tissue penetration and reduced auto-fluorescent background in thick tissue or whole-blood media.

Carbon nanotubes have a tunable near-infrared emission that responds to changes in the local dielectric function but remains stable to permanent photobleaching. They reported the synthesis of single-walled carbon nanotubes that modulate their near-infrared emission in response to the adsorption of glucose. Derivatized single-wall carbon nanotubes can be implanted under the skin. Fluorescence can be excited and detected in the same way the FRET polymer-encapsulated tattoos can be excited.

7.6.2 Use Light Diffraction Elements

Asher's group developed a novel photonic sensing material, which responds to glucose concentrations via diffraction of visible light. Polymerized crystalline colloidal arrays (PCCAs) are periodic crystalline lattices of polystyrene micro spheres polymerized within thin hydrogel films (61-64). These were called photonic crystals. The arrays are brightly colored and act as diffraction gratings for white light following Bragg diffraction equation (62,63):

$$\Delta\lambda = n.d/\sin\Theta \tag{13}$$

In equation 13, n is the refractive index of the system (medium, hydrogel and colloids), d is the spacing between the diffracting planes, λ, is the diffracted wavelength and, Θ is the glancing angle between the incident light and the diffracting planes. Change in electric charge in the PCCAs by binding molecular or ionic species causes changes the spacing, d, and there is a subsequent wavelength shift, Δλ of the light reflected off the array.

Bragg diffraction depends on the refractive index of the system (solvent, hydrogel and colloids), the spacing d between the diffracting planes. The array will act a s a diffraction grating for white light allowing a specific diffracted wavelength λ, to be detected at a specific glancing angle between the incident light propagation direction and the diffracting planes Incorporation of charged species, or change in electric charge in the PCCAs, causes the arrays to expand, changing the spacing d. The diffraction pattern then changes causing a wavelength shift in the light reflected off the array.

Several studies show the ability of the PCCA films to detect metal ions, creatinine and glucose (61, 63, 64). Asher et al constructed a glucose photonic sensor in the form of a thin acrylamide diffracting PCCA hydrogel film that contains glucose oxidase or phenyl boronic acid crystals as the molecular recognition elements. Attachment of glucose causes change in charge distribution. Glucose capture by glucose oxidase, or phenyl boronic acid, results in change in the spacing d in the Bragg equation, causing shifts in wavelengths of diffracted light. For example, 0.1 mmol/l glucose concentration causes the diffracted light to shift from yellow at 550 nm to red at 600 nm (64).

A glucose photonic sensor in the form of thin acrylamide PCCA hydrogel films that contained glucose molecular recognition elements was constructed (61- 64). Phenyl boronic acid derivatives in the lattice bind glucose, causing a change in charge distribution and a blue shift, -Δλ of diffracted light (61). In-vitro experiments showed that Δλ responded to change in glucose concentration with highest sensitivity < 10 mmol/l glucose (61, 64). The magnitude of the blue shift decreased as a function of glucose concentration, when glucose approached 20 mmol/l (61). This made the PCCAs suitable only for detection of tear glucose concentration, which is considerably lower than that of blood. The PCCA lattice shrinks as glucose concentration increases (61), and possibly reaches a lower limit where minimal Δλ is observed. The PCCA polymer film sensor is conceived to be used as a contact lens, which changes color according to glucose concentration in tears.

Lowe's group developed a different way of making molecular diffractive sensing elements Polymeric diffraction gratings that can be used for detecting glucose and other small analytes (65-68). A method of replication was developed as opposed to the synthetic method of making the PCCAs (61). The fabrication method has the possibility of large scale manufacturing of these disposable diffraction gratings. Molecular recognition elements are grafted into the polymeric structure and they act in the same way the photonic crystals, they change the angle of diffraction as glucose is captured. A shift in the wavelength, of the diffracted light results and change and is perceived as a change in the color of the polymer diffractive element. They can be used as contact lens in the same the photonic crystal polymer film can.

7.6.3. Specificity and Compartmentalization in Contact Lens Sensors

Use of boronic-acid containing capture elements depends on interaction between glucose and a specific binding molecule. It will have a specificity advantage over near infrared absorption and scattering methods, but successful application in vivo is awaited. The dynamic range is limited to the glucose concentration in tears (61). Reversibility of $\Delta\lambda$ change upon reversing the change in glucose concentration must be demonstrated in animal model and in human volunteers (69).

The contact lens sensor construct utilizes tear as a body fluid, and so the relative concentration and lag time between glucose concentrations in tears and in the vascular system will require detailed studies.

There are limited reports on the relationship between tear and blood glucose concentrations. The relationship seems depend on the sampling tears method and extent of eye irritation during sampling (70). Tears are generated in the lachrymal glands. External stimuli affect the rate of tear generation. Citing earlier references, Van Haeringen suggested that tissue fluid, and not the lachrymal gland fluid, contributed to the "tear glucose" after mechanically stimulated methods of collection, making a relationship between tear glucose and blood glucose concentrations as that between blood glucose and tissue fluid (70).

Earlier tear glucose studies were semi quantitative using color-strips and did not include glucose surge experiments. Tear glucose concentration tracked blood glucose with a time lag in an OGTT (71). A time lag of about 20 minutes was calculated from graph (69). It generally tracked blood glucose during the day (72, 73). Tear glucose levels can be elevated by mild abrasion of the conjunctival epithelium and exposure to hypotonic solutions. Non-mechanical stimuli causing reflux tears, such as light flashes and noxious vapors decrease tear glucose levels (69). However, March et al report the use of a fluorescent contact lens sensors in an OGTT experiments the fluorescent contact lenses were comfortable and were tolerated well during the three hours experiment. Examination of the corneas immediately at the end of the three-hour experiment showed no signs of abrasion, cloudiness, or damage (59). Longer term wear experiments are till needed.

The in-vitro, artificial tear fluid data suggests a simple in vitro calibration of $\Delta\lambda$ versus tear glucose concentration in artificial tear fluid (61). In vivo calibration must correlate $\Delta\lambda$ with blood glucose concentration. If lag times or other confounding factors are found, algorithms that account for these factors will be needed (69). Effect of eye irritation, lachrymating agents and environment must be studied. It is important to decide whether the readout will be visual semi-quantitative using color charts or quantitative using a spectrometric device. In vivo, animal and human experiments are needed to address these issues (69).

The polymer film sensor is conceived to be used as a contact lens, which changes color according to the glucose concentration in tears. The use of glucose oxidase or phenyl boronic acid to capture glucose from tears has specificity advantages. The polymer film methods measure glucose in tears, the relative concentration and lag time between glucose concentration in and in tears requires detailed studies (69).

7.7. CONCLUSIONS

There is no NI method for the determination glucose that offers clinically acceptable results that will allow it to replace present invasive methods. There are several recent attempts at improving the understanding the factors affecting a non-invasive glucose measurement.

Methods used for the NI determination of glucose can be classified into two broad categories as: methods that track a glucose property, and methods that track the effect of glucose on tissue properties. Specificity, compartmentalization and calibration were identified as the main issues with the NI glucose measurements.

We examined NIR absorption as an example if the first type of methods, showing the extremely small of magnitude of specific absorptivity of glucose in this spectral range and the need to develop methods for enhancing the signal/noise ratio in order to improve the chance of glucose quantitation.

We examined NIR thermo-optical response of human skin as an example of the second type, of detection methods, where the effect of glucose on the optical properties of the skin may be correlated with the change in glucose concentration.

There are some general conclusions that can be derived from the studies on the thermo- optical response of human skin. There are mainly:

a. A measuring probe and skin changes temperature distribution in tissue.
b. Changes in cutaneous temperatures caused changes in the measured μ_s', μ_a, and light penetration depth in tissue.
c. The thermo-optical response, THOR function, of human skin varies with the diabetic state of the subject.
d. It was possible to establish a classification of the diabetic state based on a THOR function and discriminant analysis.
e. THOR response correlated with the change in glucose concentration in a set of meal tolerance tests in diabetic individuals. It was possible to establish calibration models for all the cases and to predict changes in glucose concentration for most of the cases.

We examined the use of specific capture agents as an example of a way to enhance specificity. The use of these methods involves incorporating a fluorescent reagent in an encapsulated format under the skin, insertion of a derivatized carbon nano-tube under the skin, or use of a contact lens configuration. Recent clinical studies on the use of fluorescent contact lens showed the feasibility of specific displacement in a human experiment. The use of encapsulated reagents, insertable nano-wires may not be construed as a truly non-invasive and may be classified as minimally invasive. A similar argument can be used for the case of the contact lens configuration. At this stage in development, it may be useful to forgo the classification attempt and dwell on resolving the issues of specificity, compartmentalization and calibration.

Attempts to improve specificity include methods to minimize biological noise, use external stimuli, and controlling measurement condition. Methods that define specific compartments include limiting tissue volume targeted for the measurement and selecting body site with homogeneous tissue structure. Enhancing calibration will involve methods to assure absence of overfitting and testing for overfitting, methods to avoid chance physiological correlations and methods to simplify the calibration process. The use of a limited number of variables for the fitting equation and spreading the measurement time over multiple sessions covering various physiological conditions can decrease the chance of overfitting.

Methods to improve calibration include decreasing the number of terms/factors in the fitting equation, simplifying the variable, and relating the variables to physically and physiologically meaningful effects of glucose and diabetes.

Although our understanding of non-invasive glucose measurements has advanced in the past few years and novel dew detection technologies keep appearing on the horizon, improvements in minimally invasive electrochemical detection methods using insertable electrodes advances too and raise the bar for any future truly non-invasive detection technique. This will undoubtedly result in improvements in diabetes care.

7.8. REFERENCES

1. Khalil O. S., Noninvasive glucose measurements at the dawn of the new millennium, *Diabetes Technol. Ther.* **6**:660-697 (2004).
2. Khalil O. S., Spectroscopic and clinical aspects of noninvasive glucose measurements, *Clin. Chem.* **45**:165-177 (1999).
3. McNichols R. J., Cote G.L., 2000, Optical glucose sensing in biological fluids: an overview, *J. Biomed. Opt.* **5**:5-16 (2000).
4. Tamada J. A., Garg S., Jovanovic L., Pitzer K. R., Fermi S., Potts R. O., Noninvasive glucose monitoring: comprehensive clinical results Cygnus Research Team. *JAMA*. **282**:1839-1844 (1999).
5. Mitragotri S., Coleman M., Kost J., Langer R., Analysis of ultrasonically extracted interstitial fluid as a predictor of blood glucose levels, *J Appl Physiol* **89**:961-966 (2000).
6. Boyne M., Silver D., Kaplan J., and Saudek D., Timing of changes in interstitial and venous blood glucose measured with a continuous subcutaneous glucose sensor, *Diabetes* **52**:2790-5974 (2003).
7. Feldman B., Brazg R., Schwartz S., Weinstein R., A continuous glucose sensor based on wired enzyme technology -- results from a 3-day trial in patients with type 1 diabetes. *Diabetes Technol. Ther.* **5**:769-779 (2003).

8. Yum S.I., and Roe J., Capillary blood sampling for self monitoring of blood glucose, *Diabetes Tecnol Therap* **1**;29-37 (1999)
9. Cunningham D. D., Henning T. P., Shain E. B., Young D. F., Elstrom T. A., Taylor E. J., Schroder S. M., Gatcomb P. M., Tamborlane W.V., Vacuum-assisted lancing of the forearm: an effective and less painful approach to blood glucose monitoring, *Diabetes Technol Ther.* **2**:541-548 (2000).
10. Rebrin K., Steil G. M., van Antwerp W. P., Mastrototaro J. J., Subcutaneous glucose predicts plasma glucose independent of insulin: implications for continuous monitoring, *Am. J. Physiol.* **277**:E561-571 (1999).
11. Vasko P. D. , Blackwell J., and Koenig J. L., Infrared and Raman spectroscopy of carbohydrates, I: Identification of O-H and C-H vibrational modes for D-glucose, malose, cellobiose, and dextran by deuterium substitution methods, *Carbohyd Res.* **19**:297-310 (1971).
12. Vasko P. D., Blackwell J., and Koenig J. L., Infrared and Raman spectroscopy of carbohydrates, II Normal coordinate analysis of α-D-glucose, *Carbohyd Res.* **23**:407-417 (1972).
13. Cael J. J, Koenig J. L., and Blackwell J., Infrared and Raman spectroscopy of carbohydrates. IV Identification of configuration and conformation sensitive modes for D-glucose by normal coordinate analysis, *Carbohyd Res.* **32**:79-91 (1974).
14. Shen Y. C., Davies A. G., Linfield E. H., Elsey T. S., Taday P. F., and Arnone A. D., The use of Fourier-transform infrared spectroscopy for the quantitative determination of glucose concentration in whole blood, *Phys. Med. Biol* **48**:2023-2032 (2003).
15. Marbach R., Koschinsky T. H., Gries F.A., and Heise H.M., Non-invasive glucose assay by near-Infrared diffuse reflectance spectroscopy of the human inner lip, *Appl Spect* **47**:875-881 (1993).
16. Malin S., Ruchti T. L., Blank T. B., Thennadil S. N., Monfre S., Noninvasive prediction of glucose by near-infrared diffuse reflectance spectroscopy, *Clin Chem* **45**:1651-1658 (1999).
17. Burmeister J. J., Arnold M. A., and Small G. W., Noninvasive blood glucose measurements by near-infrared transmission spectroscopy across human tongues, *Diabetes Technol. Ther.* **2**:5-16 (2000).
18. Bayly J. J., Kartha V. B., and Stevens W. H., The absorption spectra of liquid phase H_2O, HOD and D_2O from 0.7 μm to 10 μm, *Infrared Phys.* **3**:211-223 (1963).
19. Hall J. W., Quaresima V., and Ferrari M., Can we get more tissue biochemistry information from in vivo near-Infrared spectra? *SPIE Proceedings* **2387**:225-231 (1995).
20. Burmeister J. J., Chung H., and Arnold M. A., Phantoms for noninvasive blood sensing with near infrared transmission spectroscopy, *Phochem. Photobiol.* **67**:50-55 (1998).
21. Pan H., Chung H., Arnold M. A., and Small G. W., Near-Infrared spectroscopic measurement of physiological glucose levels in variable matrices of protein and triglycerides, *Anal. Chem.* **68**:1124-1135 (1996).
22. Chen J., Arnold M. A., Small G. W., Comparison of combination and first overtone spectral regions for near-infrared calibration models for glucose and other biomolecules in aqueous solutions, *Anal. Chem.* **76**:5405-5413 (2004).
23. Saptari V., Youcef-Toumi K., Design of a mechanical-tunable filter spectrometer for noninvasive glucose measurement, *Appl. Opt.* **43**:2680-2688 (2004).
24. Amerov A. K., Chen J., Arnold M. A., Molar absorptivities of glucose and other biological molecules in aqueous solutions over the first overtone and combination regions of the near-infrared spectrum, *Appl. Spectrosc.* **58**:1195-1204 (2004).
25. Maruo K., Tsurugi M., Tamura M., and Ozaki Y., In Vivo noninvasive measurement of blood glucose by near-infrared diffuse-reflectance spectroscopy, *Appl. Spectrosc.* **57**:1236-1244 (2003).
26. Koo T. W., Berger A. J., Itzkan I., Horowitz G., Feld M.S., Reagentless blood analysis by near-infrared Raman spectroscopy, *Diabetes Technol Ther.* **1**:153-7 (1999).
27. Hanlon E. B., Manoharan R., Koo T.W., Shafer K. E., Motz J. T., Fitzmaurice M., Kramer J. R., Itzkan I., Dasari R. R., Feld M.S., Prospects for in vivo Raman spectroscopy, *Phys Med Biol.* **45**:R1-59 (2000).
28. Berger A. J., Itzkan I., Feld M.S., Feasibility of measuring blood glucose concentration by near-infrared Raman spectroscopy, *Spectrochim Acta A* **53A**:287-292 (1997).
29. Tuchin, V., *Tissue Optics, Light scattering Methods and Instruments for Medical Diagnostics*, SPIE Press, Billington, WA, 2000, chapter 1.

30. Martens H., and NaesT.1989, *Multivariate Calibration*, John Wiley and Sons, New York,
31. Haaland D. M., and Easterling R. G., Application of new least squares analysis methods for the quantitative analysis of multicomponent samples, *Appl. Spectosc.* **36**: 665-673 (1982).
32. Haaland D. M., Easterling R. G., Vopicka D. A., Multivariate least-squares methods applied to quantitative spectral analysis of multicomponent samples, *Appl. Spectrosc.* **39**:73-84 (1985).
33. Khalil O. S., Yeh S-j., Lowery M. G., Wu X., Hanna C. F., Kantor S., Jeng T-W., Kanger J., Bolt R. A., and de Mul F. F., Temperature modulation of optical properties of human skin, *J. Biomed. Opt.* **8**:191-205 (2003).
34. Yeh S-j, Khalil O. S., Hanna C. F., and Kantor S., Near Infrared thermo-optical response of the localized reflectance of intact diabetic and non-diabetic human skin, *J. Biomed. Opt.* **8**:534-544 (2003).
35. Yeh S-j., Hanna C. F., and Khalil O. S., Monitoring blood glucose changes in cutaneous tissue by temperature-modulated localized reflectance measurements, *Clin. Chem.* **49**: 924-934 (2003).
36. Yeh S-j., Kantor S., Hanna C. H., Hohs R., Lindberg J., Khalil O. S., 2005, Calculated calibration models for glucose in cutaneous tissue from temperature modulation of localized reflectance measurements, *Saratov Fall Meeting, September 2004, Proc. SPIE* **5771**: in press
37. Khalil O. S., Yeh S-j, Hanna C. F., Stalder A., Wu X., Lowery M., Kanger J., Bolt R. A., and de Mul F. F. M., A Non-invasive sensor having controllable temperature feature,*United States Patent* 6,662,030 (2003)
38. Kohl M., Essenpreis M., Cope M.., The influence of glucose concentration upon the transport of light in tissue-simulating phantoms, *Phys Med Biol.* **40**:1267-1287 (1995)
39. Bruulsema J. T., Hayward J. E., Farrell T., Patterson M., Heinemann L., and Berber M., Correlation between blood glucose concentration in diabetics and noninvasively measured tissue optical scattering coefficient, *Opt. Lett.* **22**:190-192 (1997).
40. Heinemann L., Kramer U., Klotzer H. M., Hein M., Volz D., Hermann M., Heise T., Rave K.; Non-Invasive Task Force, Noninvasive glucose measurement by monitoring of scattering coefficient during oral glucose tolerance tests. *Diabes Technol. Ther.*, **2**:211-220 (2000).
41. Heinemann L, Schmelzeisen-Redeker G., et al, Non-invasive continuous glucose monitoring in Type I diabetic patients with optical glucose sensors. *Diabetologia* **41**:848-854 (1998).
42. Maier J., Walker S., Fantini S., Franceschini M., and Gratton E., Non-invasive glucose determination by measuring variations of the reduced scattering coefficient of tissues in the near-infrared, *Opt. Lett.* **19**:2062-2064 (1994).
43. Cohen O., Fine I., Monashkin E., and Karasik A., Glucose correlation with light scattering patterns--a novel method for non-invasive glucose measurements, *Diabes Technol. Ther.* **5**:11-17 (2003).
44 Larin K. V., Eledrisi M. S., Motamedi M., and Esenaliev R. O., Noninvasive blood glucose monitoring with optical coherence tomography: a pilot study in human subjects, *Diabetes Care.* **25**:2263-2267 (2002).
45. Larin K. V., Motamedi M., Ashitkov T. V., and Esenaliev R. O., Specificity of noninvasive blood glucose sensing using optical coherence tomography technique: a pilot study. *Phys. Med. Biol.* **48**:1371-1390 (2003).
46. Caballero A. E., Arora S., Saouaf R., Lim S. C., Smakowski P., Park J. Y., King G. L., LoGerfo F. W., Horton E. S., and Veves A., Microvascular and macrovascular reactivity is reduced in subjects at risk for type-2 diabetes, *Diabetes* **48**:1856-1862 (1999).
47. Khan F., Elhadd T. A., Greene S. A., Belch J. J., Impaired skin microvascular function in children, adolescents, and young adults with type 1 diabetes, *Diabetes Care.* **23**:215-220 (2000).
48. Smulders R. A., Stehouwer C. D., Schalkwijk C. G., Donker A. J., van Hinsbergh V. W., TeKoppele J. M., Distinct associations of HbA1c and the urinary excretion of pentosidine, an advanced glycosylation end-product, with , markers of endothelial function in insulin-dependent diabetes mellitus, *Thromb. Haemost.* **80**:52-57 (1998).
49. Snyder W. J., and Grundfest W. S., Glucose monitoring apparatus and method using laser-induced emission spectroscopy. *US patent* 6,232,609B1 (2001)
50. Russell R. J., Pishko M. V, Gefrides CC, McShane MJ, Cote GL: A fluorescence-based glucose biosensor using concanavalin A and dextran encapsulated in a poly(ethylene glycol) hydrogel. *Anal Chem.* **71**:3126-32 (1999)

51. McCartney, Pickup JC, Rolinski OJ, Birch DJS: Near-infrared fluorescence lifetime assay for serum glucose based on allpophycocyanin-labeled concanavalin-A. *Anal Biochem.* **292**:216-21(2001).
52. Rolinski OJ, Birch DJS, McCartney LJ, and Pickup JC: Fluorescence nanotomography using resonance energy transfer: demonstration with a protein-sugar complex. *Phys. Med. Biol.* **46**:N221-6 (2001).
53. McShane M.J., Russell R.J., Pishko M.V., Cote G.L., Glucose monitoring using implanted fluorescent microspheres. *IEEE Eng Med Biol Mag.* **19**:36-45 (2000).
54. Chinnayelka S., McShane M.J., Glucose-sensitive nanoassemblies comprising affinity-binding complexes trapped in fuzzy microshells. *J. Fluoresc.* **14**:585-95 (2004).
55. McShane M.J., Potential for glucose monitoring with nanoengineered fluorescent biosensors. *Diabetes Technol Ther.* **4**:533-8 (2002).
56. Badugu R., Lakowicz J.R., Geddes C.D., Noninvasive continuous monitoring of physiological glucose using a monosaccharide-sensing contact lens. *Anal. Chem.* **76:** 610 –618 (2004).
57. Badugu R., Lakowicz J. R., Geddes C. D., Ophthalmic glucose sensing: a novel monosaccharide sensing disposable and colorless contact lens, *Analyst.* **129:**516-21 (2004).
58. March W. F., Ochsner K., and Horna J., Intraocular lens glucose sensor. *Diabetes Technol Ther.* **2**:27-30 (2000).
59. March W. F., Mueller A., Herbrechtsmeier P., Clinical trial of a noninvasive contact lens glucose sensor, *Diabetes Technol Ther.* **6**:782-9 (2004).
60. Barone P. W., Baik S, Heller D. A., M. S., Near-infrared optical sensors based on single-walled carbon nanotubes. *Nat Mater.* **4:**86-92 (2005).
61. Alexeev V. L., Das S., Sharma A. C., Finegold D. N., and Asher S. A., Noninvasive photonic crystal tear glucose sensing material. *Clin Chem* **50:**2353-60 (2004).
62. Holtz J. H., and Asher S. A.: Polymerized colloidal crystal hydrogel films as intelligent chemical sensing materials. *Nature* **389:** 829-32 (1997).
63. Alexeev V. L., Sharma A. C., Goponenko A. V., Das S., Lednev I. K., Wilcox C. S., Finegold D. N., Asher S. A., High ionic strength glucose-sensing photonic crystal. *Anal Chem* **75:** 2316–2323 (2003).
64. Asher S. A., Alexeev V. L., Goponenko A. V., Sharma A. C., Lednev I. K., Wilcox C. S., Finegold D. N., Photonic crystal carbohydrate sensors: low ionic strength sugar sensing. *J Am Chem Soc* **125**:3322–3329 (2003).
65. Kabilan S., Blyth J., Lee M. C., Marshall A. J., Hussain A., Yang X. P., Lowe C. R., Glucose-sensitive holographic sensors. *J Mol Recognit.* **17:**162-6 (2004).
66. Marshall A. J., Young D. S., Blyth J., Kabilan S., Lowe C. R., Metabolite-sensitive holographic biosensors. *Anal Chem.* **76:**1518-23 (2004).
67. Kabilan S., Marshall A. J., Sartain F. K., Lee M. C., Hussain A., Yang X., Blyth J., Karangu N., James K., Zeng J., Smith D., Domschke A., Lowe C. R., Holographic glucose sensors. *Biosens Bioelectron.* **20:**1602-10 (2005).
68. Lee M. C., Kabilan S., Hussain A., Yang X., Blyth J., Lowe C.R., Glucose-sensitive holographic sensors for monitoring bacterial growth., *Anal Chem.* **76:**5748-55 (2004).
69. Khalil O. S., Noninvasive photonic-crystal material for sensing glucose in tears, Editorial, *Clin Chem.* **50** :2236-7 (2004).
70. Van Haeringen N. J., Clinical biochemistry of tears. *Survey of Ophthalmol.* **26:**84-96 (1981).
71. Gasset A. R., Braverman L. E., Fleming M. C., Arky R. A., and Alter B. R., Tear glucose detection in hyperglycemia, *Am. Opthalmol. J.* **65**:414-420 (1968).
72. Daun K. M., and Hill R. M., Human tear glucose, *Invest. Ophthalmol. Vis. Sci.* **22**:509-514 (1982).
73. Chatterjee P. R., De S., Datta H., Chatterjee S., Biswas M. C., Sarkar K, and Mandal L. K., Estimation of tear glucose level and its role as a prompt indicator of blood sugar level, *J. Indian Med. Assoc.* **101**:481-483 (2003).

OPTICAL ENZYME-BASED GLUCOSE BIOSENSORS

Xiao Jun Wu and Martin M.F. Choi*

8.1. ABSTRACT

The development of glucose enzyme biosensor and its associated techniques are succinct reviewed, in the framework of optical transducer. The coverage of this review mainly includes a concise discussion on the characteristics of common optical transducer, enzyme immobilization techniques, enzyme stabilization and mass transfer efficiency, followed by a general description of format and performance of glucose biosensors, and finally a brief commentary on analytical features and applications. Very recent achievements of our group in the technique of glucose biosensor construction, particularly in the use of mixed water-miscible organic solvents and micelle are also covered in this review.

8.2. INTRODUCTION

Biosensors are widely applied to many fields such as biocatalytic process, medical care, food, environment, industries, security and defense.[1-4] Enzyme-based biosensors are analytical devices that utilize immobilized enzyme(s) as biological recognition component(s) and a transducer for biosensing signal. Accordingly, the biosensing process of a biosensor is regarded as a heterogeneous phase reaction with the immobilized enzyme functioning as a biocatalyst, and the biosensing signal is captured on the interfacial area between the immobilized enzyme and transducer. The development of techniques for glucose determination has attracted considerable

* Xiao Jun Wu, Genomic & Protein Engineering Laboratory, Tian Kang-Yuan Biotechnology Company, Yingkou, Liaoning, China. Martin M.F. Choi, Department of Chemistry, Hong Kong Baptist University, Kowloon Tong, Hong Kong SAR. On sabbatical leave at The University of North Carolina at Chapel Hill from July 2004 to July 2005.

interest over past several decades[5-8] since glucose determination is essential for diabetes diagnosis, fermentation process and food industry. Numerous techniques for glucose detection such as enzyme-based biosensor, optical rotation, near-infrared absorbance, Raman scattering, and resonance energy transfer have been developed successfully. Among them, biosensors based on glucose oxidase (GOx) are most widely used as it is robust and offers high sensitivity. Since Clark and Lyons reported the first enzyme glucose electrode in 1962,[9] the development of glucose biosensors has been growing extensively. Glucose oxidase plays a very important role in the construction of glucose biosensor as GOx is a very efficient biocatalyst for the oxidation of glucose[10] as presented in the following reactions:

$$\text{D-glucose} + \text{GOx(FAD)} \longrightarrow \text{D-gluconolactone} + \text{GOx(FADH)}_2 \quad (1)$$

$$\text{GOx(FADH}_2) + O_2 \longrightarrow \text{GOx(FAD)} + H_2O_2 \quad (2)$$

During the reaction, flavin adenine dinucleotide (FAD) is a prosthetic group, functioning as an electron mediator between glucose and oxygen; sequentially a glucose molecule is enzymatically oxidized by an oxygen (O_2) molecule, releasing stoichiometrically equivalent hydrogen peroxide (H_2O_2) and D-gluconolactone. Furthermore, D-gluconolactone is quickly converted into gluconic acid according to the following hydrolysis reaction:

$$\text{D-gluconolactone} + H_2O \longrightarrow \text{D-gluconic acid} \quad (3)$$

Thus, based on these reactions, numerous enzyme glucose biosensors have been developed. The optical measurement of glucose relies mainly on detection of O_2,[11, 12] H_2O_2,[13-16] gluconic acid,[17,18] or GOx[19,20] as schematically illustrated in Figure 8.1. Glucose oxidase from *Aspergillus niger* is very stable and relatively cheap; therefore, it is not only used for development of glucose biosensors, but also in other fundamental studies.

Up till now, a vast number of glucose biosensors in literature are amperometric electrodes that rely on the electrochemical properties of the biocatalytic reactions. These biosensors are based on the consumption of O_2, oxidation of H_2O_2 with/without redox mediators, and direct electron transfer between the immobilized GOx and the electrode surface.[21-24] On the other hand, optical biosensors have been developing very rapidly since 1970s.[25-31] The promising features of an optical biosensor are generally related to simple design, ease of operation, free from electric and magnetic interference, and suitability for in situ or on-line remote monitoring. To date, the optical method can, in principle, image single molecule with extremely high spatial resolution.[32] This latest advancement has catalyzed the expansion of optical glucose biosensors. In developing optical glucose enzyme-based biosensors, the most important technology is to merge an immobilized enzyme with a transducer since the techniques in the interfacial design determine the analytical characteristics of

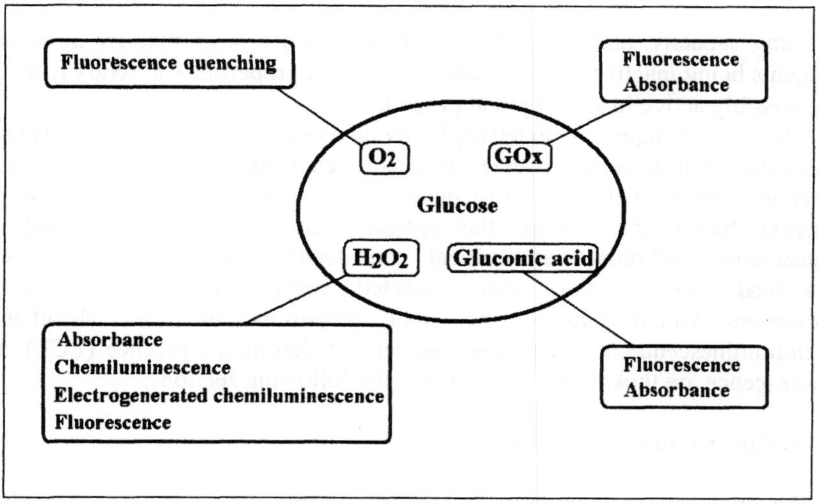

Figure 8.1. Optical biosensing strategies based on enzymatic reaction of glucose oxidase.

the resulting biosensor. As light or electromagnetic signal is able to transfer freely over space, the performance of an optical device is less dominated by the spatial combination of these two component systems.

Since many excellent reviews have appeared in recent years,[7, 25-31, 33-42] in which and also elsewhere the development of optical techniques and glucose biosensor are extensively summarized, the present review does not intend to cover the exhaustive research domain of glucose sensor, but selectively describes and evaluates the progress of optical glucose biosensor, and its recent advances together with our research. The primary focus is mainly on optical enzyme-based glucose biosensor in the aspects of enzyme immobilization, assembling, and operation mode with a slight touch on other techniques.

8.3. OPTICAL TRANSDUCTION USING COMMON OPTICAL TRANSDUCERS

Over the past years, various transduction methods have been developed. Among them, amperometric electrodes have attracted the most attention because of their well-known fundamental concepts and high sensitivity in performance.[22] On the other hand, optical transducers have been playing a more and more significant role in the fabrication of biosensor for more than two decades.[43] The foremost reasons are probably that: (1) the fundamental physical and chemical principles of the optical techniques are better understood since 1970s; (2) numerous useful opto/electronic components are commercially available; (3) optical fibers offer superior advantages such as remote in-situ or on-site detection, high degree of wavelength selectivity, high information carrying capacity, and multi-channel analysis; (4) a large number of sensing reagents being inactive or less active in electrical properties e.g. redox reactions are possibly active in some optical properties.

Optical techniques are extensively reviewed by many authors. For instance, Kuswandi et al. technically reviewed sensing element for coupling with optical fiber with detailed illustration of numerous measurement examples including glucose biosensors.[31] Since the optical measurement methods and the transduction techniques are critical considerations in achieving desirable analytical characteristics, some selected enzyme-based optical glucose biosensors which employ the light properties such as absorbance, chemiluminescence (CL) and electrogenerated chemiluminescence (ECL), and fluorescence are thus briefly described in the following sections.

8.3.1. Absorbance Measurement

Absorbance measurement is still one of the most useful approaches in both fundamental research and analytical application for glucose biosensors because the technical procedure is quite simple and readily obtainable either in literature or from any biological technique company. In addition, the glucose biosensor can even achieve high and satisfactory sensitivity with a set of cheap visible colorimeter/photometer. As a matter of fact, numerous commercial glucose test kits, test papers, and biosensing systems are based on the knowledge of absorbance. In general, one reagent or reaction has to be involved in the biosensing scheme to react with H_2O_2 or proton ion to develop color products. For example, colorimetric assay H_2O_2 can be carried out precisely at 460 nm by measuring absorbance change of oxidized *o*-dianisidine; thereby, in conjugation with this assay, glucose determination can be performed by using an immobilized GOx.[44] This similar method is not only widely employed to measure glucose concentration, but also frequently use to assay the biocatalytic activity of immobilized GOx such as GOx entrapped in the sol-gel matrix.[45]

In addition to colorimetric assay, there are several other significant absorbance techniques for glucose determination without using peroxidase and chromogen. For example, the color of polyaniline membrane is strongly pH dependent; thus, by measuring the absorbance of the polyaniline membrane in which GOx is entrapped, glucose determination can be induced.[18] The merit of this optical transducing technique is that the sensitive membrane is easy to prepare and the biosensor does not need other pH indicator dyes. In other design, the glucose biosensor can also be designed to measure the absorbance change of GOx complex due to its conformational change during the reaction. For example, the glucose measurement is realized by detecting the absorbance change of the reversible complex of GOx and meso-tetra(4-carboxyphenyl)porphine at peak wavelength of 427 nm.[46] Furthermore, the absorption change in FAD of the immobilized GOx can even be employed directly to detect glucose.[20, 47] The initial rates of reduction of the FAD is measured in the presence of glucose; thus, the decrease in the absorbance is related to the glucose concentration.

In general, there are two distinct advantages with absorbance-based biosensing techniques. Firstly, the number of available light wavelengths can be selectively employed. Secondly, they are generally considered as high precise methods for the measurements of glucose when the absorption is taken at the complete conversion of glucose to gluconic acid. However, the key disadvantage is that it requires a transparent sample solution and color or turbid samples hamper the detection.

8.3.2. Chemiluminescence or Electrogenerated Chemiluminescence Measurement

Measurement of CL or ECL is widely used for detecting H_2O_2 formed during the enzymatic reaction of glucose and O_2. The typical luminescence response mechanism is based on the biocatalytic or electrochemical schemes as follow:[37,48]

$$\text{luminol} + H_2O_2 + OH^- \xrightarrow{\text{horseradish peroxidase}} \text{3-aminophthalate} + 2H_2O + N_2 + h\upsilon \ (\text{maximum} \sim 430\text{ nm}) \quad (4)$$

or

$$\text{luminol} + H_2O_2 + OH^- \xrightarrow{\text{electric potential}} \text{3-aminophthalate} + 2H_2O + N_2 + h\upsilon \ (\text{maximum} \sim 430\text{ nm}) \quad (5)$$

Accordingly, the sensitivity of the luminescence measurement is mainly determined by the amount of H_2O_2 generated from the enzymatic reaction, photon yield efficiency of the luminescent reaction, and sensitivity of the optical instrumentation. Since excitation light source is not required, this biosensing system is generally simpler than that of traditional absorbance-, fluorescence-, and reflectance-based optical techniques. As such, these methods have distinct advantages of higher selectivity and sensitivity because the background light is greatly reduced.

In the past two decades, CL biosensing devices composed of co-immobilized GOx and peroxidase, and luminol have been reported to determine glucose. The majority of these biosensing systems are carried out under flow mode and slight alkaline conditions.[14, 49] By using cetyltrimethylammonium bromide, an enzyme CL glucose biosensor demonstrates the feasibility of conducting both biocatalytic reaction and CL at neutral pH of around 7.5–8.5.[50] In conjugation with optical fiber, an enzyme CL glucose biosensor can remotely measure glucose.[51] Recently, several microchip CL biosensors have been reported. For example, one microchip CL biosensor is developed by co-immobilizimg GOx and horseradish peroxidase (HRP) on porous silicon flow through microchips for monitoring glucose.[52] The measurement is carried out under microfluidic sequential injection analysis and the sensitivity of luminescence generated from oxidation of luminol can be enhanced in the presence of *p*-iodophenol. This microfluidic CL system is facile for monitoring glucose released from living cells.[53] The fabrication technique of the microfluidic CL system is also demonstrated by composing poly(dimethylsiloxane)/glass microfluidic system with an enzymatic bed-reactor packed with immobilized GOx particle.[54] Another microfluidic CL system is reported for determination of glucose in human serum.[55] The ECL techniques are very similar to that of CL, several fiber optic ECLs[56-58] and microfluidic ECL[15] systems based on glassy carbon electrode and luminol have also been reported in recent years. However, two distinct weaknesses of the CL or ECL method are that luminescent reagent is, so far, required, which makes the biosensing system fairly complicated; and that the luminescent reactions must be conduced under fairly high pH condition since the sensitivity of the biosensor is significantly affected by variations in ambient pH. This may limit their possible practical applications.

8.3.3. Fluorescence Measurement

For many years, fluorescence techniques have been widely applied to be the transduction signal of biocatalytic glucose reaction, resulting in considerable achievement in the development of glucose biosensor. Although each fluorescence method has its own strengths and weaknesses, the fluorescence techniques have steadily become the favorable choice in biosensing applications. When fluorescence techniques and devices are coupled with immobilized GOx, the resulting glucose biosensors generally offer high selectivity and sensitivity.

The direct measurement of the fluorescence intensity is sometimes employed in biosensing field because this technique is relatively simple and straightforward. For example, one optical fiber glucose biosensor is constructed by immobilizing a pH sensitive dye onto a solid support to monitor the increase in fluorescence intensity *via* pH decrease of the biosensing system with the oxidation of glucose to gluconic acid by the biocatalytic action of an immobilized GOx.[17] Another example used immobilized Eu(III)-tetracycline complex as a fluorescent probe for H_2O_2.[16] The largest signal changes with H_2O_2 are observed at pH levels between 6.5 and 7.5; thus, measurement of glucose is done in desirable pH media by detecting the fluorescence at 616 nm. The performance is reversible in terms of analytical fluorescence signal. The measurement of fluorescence of FAD of GOx or a complex of GOx with a reagent is also useful technique for detecting glucose.[59] For example, the change in intrinsic fluorescence of GOx with the change of glucose content can be exploited as an optical biosensing scheme for glucose determination.[19] The glucose determination can also be induced by detecting fluorescence of the chemical labeled GOx.[60]

On the other hand, a large proportion of the fluorescence techniques to date have been concerned with the depletion of O_2 concentration by oxidation of glucose under the biocatalytic action of immobilized GOx. Optical O_2 transducers based on O_2 dynamic quenching of fluorescence intensity emitted from fluorescence dye has received much attention recently.[61] So far, there are many enzyme glucose biosensors based on employment of generic optical O_2

transducers,[11, 12, 62, 63] and the working principle and characteristics of these transducers are comprehensively discussed in a review article with a few typical examples.[31] This biosensing scheme, in conjugation with optical fibers, was successfully employed to monitor glucose concentration[11] and free cholesterol concentration[64] by Wolfbeis' group. Figure 8.2 illustrates a typical optical O_2 transducer (OOT) which is attached to the window of a flow-though cell where immobilized enzyme is loaded. Oxygen sensitive particles (OSP) are prepared from tris(4,7-diphenyl-1,10-phenanthroline) ruthenium(II) complex absorbed onto silica gel particles.[65, 66] An OOT is composed of an OSP layered behind a piece of Teflon membrane as shown in Figure 8.2, or a silicon rubber membrane entrapped with some OSP. The response mechanism of OOT is based on collisional quenching by O_2 of the fluorescence of the ruthenium complex. As such, the biosensing measurement can be carried out by detecting the decrease in the luminescence intensity during enzymatic oxidation under the rear-detection mode. Optical transducer with a black carbon paste membrane separating the sample solution from the optical elements can easily reduce any ambient or stray light interference on the bioreactor.[11, 64] The attractive features of OOT are high sensitivity and good specificity, no consumption of O_2, immune to electric interference from electro-active species, and not easily poisoned by other interferants. Moreover, a superior advantage of OOT-based biosensor over amperometric electrode is that a slight change in O_2 concentration during the biocatalytic oxidation can be continuously accumulated making it amenable to rather slow biosensing process under the stop-flow mode; hence it can be a more sensitive method. Employing such an analogous OOT, Choi et al. [65-68] have developed several oxidase-based biosensors including four GOx-based biosensors. This transducing method is proved to be very useful for assaying glucose concentration. In addition, in order to compensate the variation of initial dissolved O_2 concentration in the glucose solution, two OOTs have also been used to construct an optical glucose biosensor.[12] On the other hand, the luminescence lifetime measurement can be employed to overcome the variation of initial signal intensity during biosensing.[69] For example, a lifetime-based optical sensing device is reported to determine glucose by using energy-transfer mechanism.[70] More recently, time-resolved luminescence lifetime imaging is employed to visualize glucose by using GOx and Eu(III)-tetracycline complex as the fluorescent probe for H_2O_2. The attractive feature of this technique is that the sensing system does not involve the use of HRP, and can perform at neutral pH.[71]

8.4. IMMOBILIZATION OF GLUCOSE OXIDASE

It is essential to immobilize or co-immobilize enzyme for enzyme biosensors. The primary purpose is to confine the enzyme within a bioreactor to maximize its lifetime while maintaining the enzymatic activity.[28, 31, 72, 73] However, once enzyme has been immobilized, the apparent enzyme activity is usually reduced due to the diffusional barriers between the enzyme and its substrate. The most common techniques of enzyme immobilization are

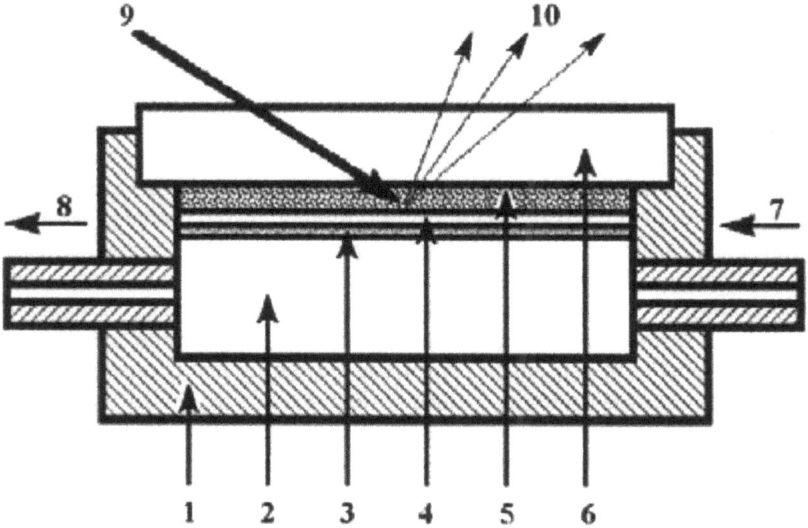

Figure 8.2. Schematic diagram of an optical glucose biosensor embroidered with a flow-through cell and an optical oxygen transducer: (1) stainless steel cell body, (2) chamber for loading immobilized GOx, (3) carbon black paste, (4) Teflon membrane, (5) oxygen sensitive reagent phase, (6) transparent window, (7) sample inlet, (8) sample outlet, (9) excitation light beam, and (10) emission light beam.

entrapment, micro-encapsulation, cross-linking, covalent linkage, chemical or physical adsorption, and electrochemical deposition or entrapment, to name but just a few. Accordingly, the ideal requirements of immobilized enzyme(s) are larger amounts of enzyme loading, higher retention of enzyme activity, better enzyme accessibility to substrate, longer operation and storage stability, higher reactant mass/electron transfer efficiency in the supporting matrix and/or on the interfacial area, but no leaching and contamination problems, no inter-interference between solution phase, enzyme phase and transducer element. Over the years, numerous immobilization techniques have been developed. Although the approaches are different, the main consideration should engender the immobilized enzyme with good stability and high apparent biocatalytic activity, as well as offering high mass transfer efficiency for reactant transport. In this review, the underlining immobilization strategies are described in three basic categories: (1) enzyme to be immobilized onto surface of transducer or solid supporting materials such as beads, film or membrane; (2) enzyme to be entrapped beneath/in supporting materials such as permselective membranes and polymer gels; and (3) enzyme to be immobilized with bulk polymer matrix such as bulk sol-gel encapsulation and blend composite. Types of immobilized GOx and the general comparison of the common configurations of them are listed in Table 8.1

Table 8.1. Comparison of common immobilization formats for glucose oxidase

Format of immobilized glucose oxidase (GOx)	Supporting material	Enzyme loading	Surface area to volume	Diffusion efficiency	Operation stability	Storage stability
GOx immobilized on the surface of support materials	Porous glass beads	Very high	Very large	Very high	Low	Low
	Carbon black powder	High	Large	High	Low	Low
	Nylon membrane	Fair	Fair	High	Low	Low
GOx mixed within support materials	Bovine serum albumin network	High	Moderate	Moderate	Fair	Fair
	Carbon paste with mineral oil	High	Large	Fair	Fair	Good
	Rigid matrix of graphite powder and Teflon powder	High	Very large	Moderate	Fair	Fair
	Gold nanoparticle in silicate network	Moderate	Very large for gold nano-particle	Fair	Fair	Fair
GOx beneath/ behind membrane	Dialysis membrane	High	Small	Low	Low	Low
	Ultrafiltration membrane	High	Small	Low	Low	Low
GOx entrapped in thin film(s)	Sol-gel silicate film	Moderate	Small	Fair	Low	Fair
	Polyacryl-amide film	Moderate	Small	Fair	Good	Good
	Sol-gel silicate film	Moderate	Small	Fair	Good	Good
	Poly(*p*-chlo-rophenyl-amide)	Fair	Small	Fair	Good	Good
	Poly(*p*-chlo-rophenyl-amide)/Nafion	Fair	Small	Low	Better	Better
GOx entrapped in bulky matrix	Sol-gel monolith	High	Very small	Very low	Good	Excellent
	Poly(vinyl alcohol) gel	Moderate	Very small	Extremely low	Good	Excellent
GOx entrapped in the particles of support matrix	Polysacch-aride polymer bead	Moderate	Moderate	Fair	Fair	Good
	Sol-gel silicate particle	High	Very high	High	Good	Excellent
	Nano-particle	High	Very high	Fair	Fair	Fair

8.4.1. Immobilizing Enzyme on Surface

The techniques of immobilizing enzyme on surface are useful because it is usually quite simple and the immobilized enzyme can achieve rather high apparent biocatalytic efficiency. When the supporting material is particles e.g. Sepharose beads,[14] or large surface membrane e.g. rough nylon,[74] it possesses a large surface area to volume ratio which appears more favorable for enzyme immobilization. In this approach, the role of the supporting material mainly provides large surface area with desirable chemical and physical properties; thereby the enzyme can be immobilized on sites for biocatalytic purpose. Because the enzyme is not covered by a film, the enzymatic reaction is quite fast due to high mass transfer efficiency of the substrate. A variety of synthetic and naturally occurring materials, as partially listed in Table 8.1, have been used for GOx. In general, materials with large surface area and porous structure such as controlled porous glass beads (CPG) can provide large surface area for GOx immobilization, resulting in high apparent enzyme activity. However, the operation stability of GOx directly exposed to solution is generally unsatisfactory because the immobilized GOx is easily interfered by external factors.

Recently, the attachment of enzyme onto solid support e.g. surface of electrode is considered to significantly influence the apparent enzyme activity of the immobilized enzyme because of both electrical and mechanical means of confinement.[21] Fortunately, the optical transducing process is not significantly limited by the electronic detention. Enzyme may adsorb on supporting surface in monolayer or multilayer in various mechanisms. Lately, the mechanisms of direct electrodeposition of GOx on a platinum electrode are fundamentally investigated by using GOx micelle solution in order to precisely control the layers and properties of deposited GOx on a surface.[75] This technique is helpful to explore and optimize the apparent enzyme activity.

In essence, three attractive advantages of this biosensing system are that: (1) the diffusional distance of the substrate from interface to enzyme is the shortest among any other immobilization systems; (2) there is no barrage region between the enzyme and the substrate; and (3) it is possible to achieve high enzyme loading on huge surface area of the supporting material. Consequently, the biosensing system can sometimes show quite high sensitive response to analyte. On the other hand, the means of immobilizing GOx onto supporting surface has two major shortcomings: (1) the immobilized enzyme is difficult to be accommodated on the interfacial area; and thus, (2) a wide variety of interferants from biological fluids can disable the performance on line in few hours. Fortunately, these weaknesses are not so significant for high degree conversion of glucose when using larger amount of immobilized GOx since the conversion is less sensitive to the variation of enzyme activity. For instance, a number of glucose biosensors based on absorbance,[76] CL,[14, 54, 55] and ECL[15] rather prefer to immobilize GOx onto small particles such as CPG. in order to provide high enzyme loading.

8.4.2. Enzyme Entrapped In or Buried Behind Thin Homogenous Film

For many years, the techniques of the film/membrane masking/entrapping enzyme have become attractive means for developing enzyme-based biosensor for real sample analysis.[77-79] The most important feature of these techniques and configurations is that the enzyme immobilization membrane can be ready to couple to the transducer surface while the massively immobilized enzyme can be accommodated much better by the functional environment than that onto the uncontaminated surface of solid support. Accordingly, this approach can help effectively protect enzymes from denaturing and avoid unfavorable micro-environments; as a result, the operation stability of the immobilized enzyme is profitably improved. A variety of polymers, both synthetic and naturally occurring as shown in Table 8.1, can be employed to construct membrane configuration glucose biosensor. For optical biosensor, the polymer membrane and the structure of the immobilization enzyme is designed to hold the enzyme for biocatalytic reaction, create the interface for transducing the biosensing signal, and sometimes control the diffusion of the substrate to the enzyme in order to optimize the sensitivity, reproducibility, selectivity, and response time. Hence, the design of the membrane is crucial for serving these multiple purposes.

Several membrane techniques have been explored to fabricate membrane glucose biosensors. .Some membranes can separate the larger molecule of the enzyme from test solution while the smaller substrate molecules can easily diffuse though the membrane. Thus, the simplest way is to preserve the enzyme behind a semi-permeable membrane e.g. dialysis membrane.[19] Although this technique is even simpler than immobilizing the enzyme on surface, it is a less promising method in view of sensitivity as this skill uses a large amount of enzyme that creates serious mass transfer barrier for both substrates and products. This is because the massive enzyme behind the membrane tends to aggregate, leading to low apparent biocatalytic activity. So far this mean of immobilization enzyme has become less attractive.

Coating or dropping enzyme cocktail solution onto the surface of transducer or support is more often employed because this procedure is also quite simple, and the enzyme can be evenly dispersed in the resulting homogenous membrane or film. However, the apparent enzyme activity can be reduced because the enzyme movement is significantly limited and the substrate diffusion efficiency is much declined by the polymer network. Therefore, the main consideration of the membrane design is that the pore size must allow rapid diffusion of glucose and O_2 and yet retain the enzyme. Many kinds of hydrophilic polymers listed in Table 8.1 have desirable densities and compositions similar to those of common natural tissue; thereby, they are frequently employed to entrap GOx. The amount of polymer matrix and enzyme is also crucial in designing immobilized GOx membrane. Biosensor with the thin film technique can attain very fast optical response time, but may not achieve high sensitivity. This is because the conversion of the substrate is largely restricted by the available enzyme when the rate of response is limited by the enzymatic reaction, but not the substrate diffusion due to very limited interfacial area, and difficulty in high enzyme loading. On the other hand, although a biosensor employing a thicker membrane for loading more enzymes can generate more stable response signal, it is very sluggish in response to analyte due to slow diffusion of the substrate.

Comparing to monolayer film, multilayer films technique can effectively optimize the most important issues such as loading and movement of enzyme, substrate diffusion, and interferants exclusion by carefully configuring the multiple films and sensing strategy. For sandwich configuration of immobilization, the outside film is generally a thin film to restrict the enzyme, enlarge the surface of the membrane for enhancing substrate flux, and repel interferants whereas the base layer between enzyme and transducer is arranged to favorably support the enzyme and increase efficiency in transducing the biosensing signals.[12, 44] However, with the middle slushy layer mainly composed of large amount of high freedom enzyme, the biosensor shows low mechanical strength. Moreover, such a biosensor still exhibits unsatisfactory substance flux due to small interfacial area.

The most effective film immobilized GOx technique is one that is able to in situ polymerize electrically and chemically to create some immobilization architectures.[22] The key advantage is that the immobilized enzyme in electro-polymerized films provides the possibility of precisely electro-generating a polymer film on conductive micro-surface of complex geometry. In addition, the film or membrane can be well adhered onto or contacted with the surface of transducer without trapping any gap between them, which is very important in facilitating high mass transfer or electron transfer efficiency. Although it seems to be a promising technique, it is not yet the preferred approach probably due to its complicated procedures and the difficulty in increasing enzyme loading in an electro-polymerized membrane.

8.4.3. Enzyme Entrapped Within Sol-Gel Matrix

Recently, more complex film and three dimensional immobilization techniques have been growing very rapidly. Among them, the entrapment of GOx with sol-gel matrix is the most fashionable; as a result, numerous sol-gel based glucose biosensors have emerged.[80] In general, GOx can be well entrapped within the cages of sol-gel network, without much losing its intrinsic biocatalytic activity. The porous nature of the sol-gel structure allows small glucose to diffuse through the matrix to reach the encapsulated GOx; but the pore size may restrict the GOx's biocatalytic function in the micro-environment structure of this silicate matrix. At the same time, the thermal stability of the encapsulated GOx can even been much enhanced.[45] Thus, it is recognized that the entrapment of GOx cannot be simply immobilized by this technique. In fact, an essential role of the silica cage is to protect the enzyme to preserve its biocatalytic activity for longer term and more stable use.[67] Unfortunately, the sol-gel approach has two distinctive drawbacks. First, despite the significant increase in enzyme loading together with large amounts of supporting matrix, the resulting glucose biosensor may still present fairly low sensitivity. This is because the interfacial area between solution and the silicate matrix is still rather small and the thicker silicate matrix also creates longer diffusional pathway for the substrate; consequently, the entrapped GOx shows low mass transfer efficiency for the substrate. Under the serious substrate flux limitation with small interface of the thicker silicate matrix, the response of the glucose biosensor can only reach a certain constant level even though the glucose concentration is higher. Therefore, such biosensor is practically useless for the determination of high glucose concentration when its response is governed by serious mass transfer process. Second, such entrapped GOx membrane or monolith in conjugation with an optical transducer, especially with an optical fiber, has only a small interfacial transducing area between rather limited available surfaces of the sol-gel immobilization monolith or membrane and the optical transducer or optical fiber tip. As a result, the biosensor can only attain very weak signals, and also small transducing signal with low signal-to-noise ratio.

To facilitate high mass transfer efficiency, the matrix of entrapped GOx can be ground into small particles as they not only offer larger surface area to volume ratio, but also shorten the diffusional distance for the substrate. As such, the substrate flux is much enhanced and internal diffusion of the substrate is also greatly improved. Moreover, the huge surface area of the particles provide potential for coupling with several types of transducing devises such as OOT[66] and absorbance.[45]

8.4.4. Enzyme Immobilized By Nano-Technique and With Nanoparticles

In recent years, nano-technique and nanomaterials have attracted considerable interest in the field of biosensor research, especially in amperometric transducer. The applications of nanomaterials in biosensors are mainly for: (1) improving the micro-environment of the immobilized enzyme by hybridizing conductive nanoparticles such as gold,[81, 82] platinum,[83] and carbon,[84] (2) creating a huge interfacial area and/or a shortened pathway for electron transfer between the enzyme and electrode,[85] and (3) fabricating the immobilized enzyme in nanobiosensors[63] which has high potential for microscale bioanalysis.[86] Nano-technique is also exploited to create nano-composite of enzyme immobilization in order to achieve extremely high enzyme loading, thus enhancing sensitivity.[87] Certainly, there is potential to integrate nanotechnology with biocatalytic element such as GOx in view of the recent development in nanotechnology.[88] Actually, two optical glucose nanobiosensors have been reported recently. One is developed by using a microemulsion polymerization process.[89] By ratiometric fluorescence intensity technique, the nanobiosensor attempts to *in vivo* image glucose in biological cells. Another optical glucose nanobiosensor is composed of GOx coated with magnetite nanoparticles in conjugation with a ruthenium phenanthroline fluorescent complex.[63] This nanobiosensor can determine glucose in solution at concentration up to 20 mM.

In general, the major merits of glucose nanobiosensors are high sensitivity and fast response because the configuration of immobilized nano-size enzyme provides extremely high surface area to volume ratio so that the glucose mass transfer efficiency between the solution phase and the enzyme phase is enhanced.

8.4.5. Blend Composites of Immobilization Enzyme

Blend composites prepared by hybridizing enzyme, conductive particles, and binding component for developing rigid, semi-rigid, flexible, or paste configuration of bulk mixed bioelectrodes have been reported. Most blend composite electrodes draw their advantages from the enhancement of signal-to-noise ratio by the edge effect of a microelectrode and high enzyme loading with co-immobilizing components.[84, 90-92] Since the most important characteristics such as structure format, interfacial property, hydrophilicity of micro-environment, solvent penetration and substrate diffusion are largely determined by the blend composite and the nature of each individual phase, the applications of this technique of enzyme immobilization are vast. For example, a glucose biosensor is reported to immobilize GOx within a blending matrix of regenerated silk fibroin and poly(vinyl alcohol) (PVA), and then coupled to a Clark O_2 electrode.[93] The water uptake within the matrix can be controlled by regulating the proportion of the two polymers. The blend technique of enzyme immobilization has also high potential in optical biosensors. For example, an alcohol biosensor is fabricated by blending silica gel particles adsorbed with alcohol oxidase with silicon rubber to monitor ethanol.[94]

In general, the major merit of the blend technique is the feasibility in bulk modification of the enzyme immobilization architecture. Optical biosensors incorporating the blend technique possess the advantages of high enzyme loading on extremely large surface area of micro particles, flexibility in designing the characteristics of the blend matrix, and fast response time (in seconds). The main shortcomings are that the structure of the blend composite cannot effectively restrict the enzyme from leaching and color matrix may seriously obstruct the optical measurement. A brief summary of various selected configurations of the immobilization GOx is displayed in Table 8.2.

Table 8.2. Some types of immobilized glucose oxidase

Format of immobilized glucose oxidase (GOx)	Supporting material	Reference
GOx immobilized on surface of support materials	Porous glass beads	54, 55, 76, 95-97
	Agarose particles (Sepharose)	14
	Carbon black powder	11, 98, 99
	Iron oxide (Fe_3O_4) magnetic nanoparticles	63
	Nylon	20, 74, 100, 101
	Eggshell membrane	65, 68, 102
	Platinum electrode	75

	Polyethylenimine/chip surface	52
	Silicone rubber membrane	103
	Polypropylene or polyethylene	104
GOx mixed within support materials	Bovine serum albumin	105, 106
	Graphite powder impregnated in sol-gel	56
	Graphite powder/epoxy, methacrylate, silicone, or polyester	90, 92
	Carbon nanotube/Teflon	84
	Gold nanoparticle/silicate network	82
	Silk fibroin/poly(vinyl alcohol)	93, 107
	Polyallylamine/polystyrene sulfonate	87
GOx behind membrane	Dialysis membrane	19
	Ultrafiltration membrane	40
	Sol-gel silicate film	47
	Sol-gel/sol-gel sandwich conformation	12
	Polyurethane/Silastic sandwich conformation	108
GOx entrapped in thin film(s)	Cellulose acetate film	109
	Polyacrylamide film	110
	Poly(vinyl chloride)	111, 112
	Sol-gel silicate films	44, 113
	Hybrid of sol-gel/poly(vinyl alcohol)-(4-vinylpyridine)	114
	poly(ester-sulfonic acid)	115
	Poly(*p*-chlorophenylamide)	116
	Poly(*p*-chlorophenylamide)/Nafion	116
	Prussian Blue film	117
GOx entrapped in bulky matrix	Sol-gel monolith	118, 119
	Poly(vinyl alcohol) polymeric network	120
GOx entrapped in particles of support matrix	Silicate gel	45, 66, 121
	Hybrid sol-gel	12, 67
	Polyacrylamide	89

8.5. CONSTRUCTION OF GLUCOSE BIOSENSOR

8.5.1. Bulky Bi-phase Bioreactor

It is well known that the performance of enzyme biosensors depends significantly not only on the character of immobilized enzyme but also on the structure and operation mode of biosensor.[122, 123] Until now the majority of biosensors have been constructed as two bulky phase bioreactors: a solution phase and an enzyme immobilization phase so that the solution can pass through the sensor chamber fairly freely. Most enzymatic optical sensing devices including optical fiber have an immobilized membrane mounted on the window of a reactor or on the top of a transducer as this is a straightforward method.[28, 31, 38, 41] In general, the volume of the chamber is in the order of micro- to nano-liters,[29, 86, 110] and the response sensitivity depends very much on the volume taken by the solution phase in the biosensor. Smaller volume can much enhance both sensitivity and rate of the response signal. A key advantage of this biosensing format is that the immobilization membrane can flexibly adapt to most types of optical transducers for continuous flow measurement, or batch analysis.[124] Obviously, in order to achieve high sensitivity, it is essential to arrange the immobilization membrane or film in conjunction with a high sensitive transducer e.g. fluorescence detector, or with a mass assay technique based on absorbance spectroscopy *via* the accumulation of H_2O_2 within a short reaction time. This is because on one hand it is difficult to immobilize a large amount of enzyme using thin film techniques; on the other hand only a portion of glucose molecules in a relatively large volume of the solvent phase is catalyzed by the immobilized GOx resulting in low sensitivity.

An archetypal optical technique using optical fiber can compare well with Clark O_2 electrode.[38] Wolfbeis' group has developed two enzyme-based glucose biosensors with two bulky phase structure by fixing a GOx immobilized membrane on the tip of optical fibers coupled with a flow-through cell.[11, 74] The GOx is covalently immobilized onto the surface of an OOT membrane. Because the enzyme membranes show good porosity and possess relatively large surface area, and the biosensors have rather small chamber volumes of approximately 20 μl, both biosensors exhibit adequate biocatalytic activities resulting in large output signals even in flow mode conditions. Their results prove that the biosensing systems can achieve fast dynamic balance of glucose and O_2 from the solution phase to the enzyme membrane.

Recently, Choi et al.[65] reported a glucose biosensor based on OOT and an immobilized GOx membrane for determination of glucose in beverage. Glucose oxidase is directly covalently immobilized onto a piece of eggshell membrane that is in close vicinity to an OOT. Thus, the biosensor is constructed into two separate bulky phases and the biosensor structure is very similar to that in Figure 8.2. The glucose determination is realized by measuring the increase in fluorescence intensity which is directly correlated to the decrease in dissolved O_2 which is consumed by the biocatalyst. Figure 8.3 displays typical response curves of such a glucose biosensor on exposure to various glucose solutions. These results, together with the observations from a similar GOx based biosensor,[68] suggest that the biosensor response is still governed by a dynamic process in the diffusion of glucose and O_2, and the rate of enzymatic oxidation whereas the conversion of the overall reaction is fairly slow with a relative large solution phase. This response mechanism is further verified by the result of another glucose biosensor which has a similar GOx immobilized eggshell

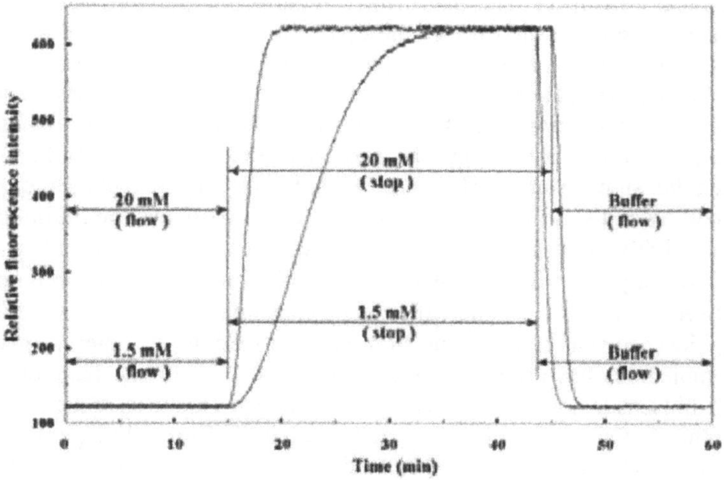

Figure 8.3. Response curves of a glucose biosensor fabricated from a GOx immobilized eggshell membrane and an optical oxygen transducer under the stop-flow mode. Air-saturated 1.5 and 20.0 mM glucose in aqueous phosphate buffer solutions were pumped into the biosensor.

membrane directly coupled with a Clark O_2 electrode.[102] When the concentration of the dissolved O_2 in a stirred test solution is reinforced from open air, the biosensor can generate a steady-state response signal, indicating a steady gradient of the O_2 concentration established over the eggshell membrane when the balance between the substrate diffusion and the enzymatic reaction is reached.

For optical glucose biosensors using OOT, the attractive advantage of the bulky biphase structure is that, by enlarging the volume of the solvent phase, the analytical working range can be extended under both flow and stop-flow operation modes though it is habitually considered as a disadvantage. This is because a larger amount of dissolved O_2 in the larger volume of the solution phase will only cause a smaller percentage conversion of O_2; thus, the rate of increase in the biosensor response becomes slower. The fluorescence signal will then need longer time to reach its highest value when the rate of enzymatic oxidation is slower than that of the O_2 diffusion. As such, the biosensor can respond to higher concentrations of glucose in samples.

8.5.2. Particle Packed Bioreactor

Multiple tiny biphase is very useful for biosensing configuration as it can be constructed as a flow-through cell or a column in which enzyme immobilized particles are closely packed. Thus, heterogeneous biphase biosensor is composed of numerous immobilized phases, the separated solution phases and the optical transducer. The biosensor can carry out batch or flow analysis with a system to control the flow of the carrier to the biosensor or the reactor. The innate feature of this biosensing scheme is that the enzyme immobilization phases have huge interfacial areas and the solvent can randomly pass through the gaps between the tiny particles. Consequently, the mass transfer of substrate across the phase boundary and the diffusion between the enzyme and solution phases can be significantly enhanced. Armed with the advantages of high enzyme loading and short diffusion path of the small particles, the immobilized enzyme is highly accessible to substrate, and thereby shows rather high apparent biocatalytic activity. Moreover, the solution phase must fill up the space between the particles; thereafter high substrate conversion is achieved when the biosensor has a dense and high particle loading. In other words, the substrate molecules have more chances to undergo biocatalytic reaction; as a result, the biosensor shows high sensitivity and fast response, which is very similar to the characteristics of small size biosensor.[125] Bioreactors packed with particles, as a mature technique of biosensor construction, provide very high efficient enzymatic reaction.[54,55,97] This technique has been frequently employed to construct calorimetric micro-biosensor over many years.[126] For example, two glucose biosensors have employed GOx and catalase co-immobilized solid particles and packed them in micro-columns as bioreactors.[127, 128] Both biosensors can determine glucose in whole blood by using a flow injection manifold. The biosensors display high sensitivity to glucose in terms of thermal output. In general, the distinctive features of the column bioreactor are: (1) the solid particle has very high surface area to volume ratio so that more enzymes can be immobilized; as such, the internal mass transfer resistance of the immobilized enzyme is reduced and larger amount of immobilized GOx is available to glucose; (2) the amount of sample solution in the column is small since spaces between the gaps of the solid particles is small; thus the external mass transfer resistance is also much reduced; and (3) the heat capacity of the solid particles such as glass beads is much lower than that of aqueous buffer; as a result, the change in temperature of the column is larger and the biosensor is more sensitive.

The heterogeneous biphase bioreactor also provides more flexibility in the design of optical biosensor. For example, the immobilized GOx column can be employed as an efficient bioreactor to convert glucose into H_2O_2 which is subsequently subjected to CL detection.[14, 129] An optical glucose biosensor is fabricated by loading the immobilized GOx particles on the upside of a flow cell and the immobilized HRP particles with H_2O_2 sensitive resin on the down-side.[76] The biosensor can determine H_2O_2 and glucose in serum. Recently, Wu et al.[66, 67] reported two glucose biosensors based on OOT and sol-gel based immobilized GOx for determination of glucose in beverage and urine samples. Both glucose biosensors are constructed by densely packing the particles of the immobilized GOx into a flow-through cell equipped with an OOT similar to Figure 8.2. Since GOx in the sol-gel pores or cages is easily accessed by glucose, the biosensors show high apparent biocatalytic activity toward glucose. Again, as the solution phase is largely minimized and there is no risk of overloading the biosensor, a relative high mass conversion of the substrate for a high concentration solution can be realized at ambient temperatures. Consequently, these two glucose biosensors can continuously detect glucose in flow mode and their steady-state fluorescence intensities in response to glucose are obtained within several minutes. The high response sensitivity of the glucose biosensors is also attributed by two aspects. (1) A relative large flow-through cell can entrap larger amount of immobilized GOx while significantly reduces the sample loading; thus enhances the percentage of conversion of substrate to product.[130] (2) Slower flow rate of the sample solution can increase the consumption of O_2; consequently lowers the O_2 concentration in the system with a concomitant effect of enhancing the response signal. In addition, since the enzymatic activity is usually more stable at lower concentrations of the substrate, smaller sample solution with high concentration and high enzyme loading are helpful ways to ensure long-term stable operation lifetime.

8.6. PERFORMANCE OF OPTICAL GLUCOSE BIOSENSOR

8.6.1. In Batch Operation

For measurement of glucose, enzyme glucose biosensor can either operate in batch condition where the bioassay is taken in a cuvette/bioreactor, or in continuous condition where the bioassay is done in a flow-through cell/bioreactor under flow mode. In general, the speed of bioconversion of the biocatalytic processes within the immobilized enzyme depends on the operation mode markedly, while good stability is crucial for the development of reliable glucose biosensors capable of quantitative analysis under flow mode.

In most cases, the conversion of glucose in test solution is almost complete and the detection is not affected by the ways of delivery of solution. Taking this advantage, as well as partially avoiding instability, numerous commercial glucose test kits are based on batch mode operation in order to achieve high precise results. These full-grown approaches are very simple and reliable because glucose is almost completely reacted to generate a sensitive and stable steady-response signal. Moreover, it only requires relatively inexpensive instrumentation such as colorimeter/photometer or fluorometer, and is easier and cheaper to design for clinical and sports purposes. For example, the colorimetric assay of the biocatalytic activity of immobilized GOx is by measuring the absorbance of oxidized *o*-dianisidine at 460 nm in the presence of HRP.[44] The absorbance measurement is preferably carried out under batch mode to achieve not only high sensitivity, but also precise results. Another example of glucose assay done on fluorescence detection where GOx and HRP are co-immobilized on a cylindrical magnetic stirrer.[131] The biocatalytic processes are quite quick under the stirring mode, thereby the biosensor shows very fast response to glucose in plasma. However, two main shortcomings come with the batch operation mode. First, the measurement based on H_2O_2 may take a long time for precise determination since the biosensing system has to almost completely convert O_2 to H_2O_2. Second, the measurement of light based on the generation of H_2O_2 is not suitable to assay color samples such as whole blood, red wine, and chocolate.

8.6.2. In Stop-Flow Mode

When the immobilized GOx is placed in a flow-through cell or bioreactor, the resulting biosensor can perform in stop-flow mode with accessories such as pump, valve, and injection device. The major difference between stop-flow and batch modes is that the bioassay in stop-mode can be sequentially carried out to detect glucose on line. In most cases, the solvent or sample solution functions as a carrier which is delivered through the flow-through cell by a syringe or peristaltic pump. The sample solution is then kept in the flow-through cell for a succinct period of time while the response signal is recorded against time within this duration. In this way, the overall biosensing reactions and the transport of reactant are accomplished. Thus, the reaction of glucose is much enhanced by the stop-flow mode as compared to flow mode. In principle, the glucose assay is based on measurement of the magnitude of signal either at a fixed time interval even before the complete reaction of the substrate or at the thermodynamic equilibrium of the biocatalytic reaction. Of course, the best sensitivity is obtained at the saturated response signal. With this advantage, glucose biosensor can employ low apparent enzyme activity of immobilized enzyme but still can achieve rather high sensitivity by taking longer incubation time under the stop-flow mode. For instance, Figure 8.3 displays typical response curves recorded by an optical glucose biosensor in the stop-flow mode.[65] Coupled with an OOT, the response rate is mainly governed by the flux and diffusion of glucose and O_2 into the enzyme immobilized membrane, and the rate of the enzymatic reaction. Therefore, when an aqueous buffer is pumped through the biosensor, a blank signal of fluorescence intensity is recorded as a baseline. When a glucose sample solution is delivered through the biosensor and then stopped, the fluorescence intensity is continuously increased with time. Finally, the fluorescence intensity saturated at a steady-state value when the kinetic equilibrium of the biocatalytic reaction is reached. The rate of the fluorescence intensity reaching the saturation point from the baseline can reflect the overall rate of O_2 diffusion from the solvent, into the immobilization phase, and then onto OOT. The difference between the saturation point and baseline is approximately the operation window. The other advantages of stop-flow mode operation are that only a relatively small sample solution is required and there is no need to precisely control the flow rate of the solution. However, the obvious drawback is that full continuously monitoring on line is not possible.

8.6.3. In Flow Mode

In order to continuously monitor glucose on line and in time, glucose biosensors are fabricated to operate under flow mode with flow rates of sample solution from micro- to nano-liters per min.[11, 86, 132] The flow sample solution is monitored until a steady-state signal is reached in a brief period of time as a result of a dynamic balance in substrate diffusion into the biosensor, substrate consumption and product generation, and flow rate of the solution. Numerous optical glucose biosensors can perform in flow mode. For example, Figure 8.4 is a typical example of the performance of an optical glucose biosensor based on OOT in response to glucose solution in flow mode.[66] When the buffer solution flows through the biosensor, the initial part of the curve is recorded as a baseline. As a stream of glucose solution continuously flows through the biosensor, an increase in the response is generated due to the forward process of the enzyme biocatalytic reaction, and subsequently the response reaches a steady value when the dynamic equilibrium is attained. Once the flow solution is switched to a buffer, the response signal starts to decline and returns to the baseline as glucose is fully washed out from the biosensor by the buffer. In general, the higher the concentration of the substrate, the higher rate of increase and the larger signal is observed. If the biocatalytic reaction takes place in a significantly slow pace, the magnitude of the steady-state signal will be related to the flow rate of the solution. As such, a larger signal response can be achieved only under lower flow rate of the sample solution since the biocatalytic reaction is closer to the thermodynamic equilibrium. Hence, in order to attain high sensitivity and short response time, it is vital to assemble a biosensor by maximizing its enzyme loading but minimizing the flow cell volume.[86,110]

The other advantages of the flow mode are that the response signal may be continuously and stably recorded in real-time and the working range can be extended under high flow rate of the solution. Unfortunately, the major problems are that a relatively larger sample solution is required and the biosensor shows lower sensitivity compared with that in the stop-flow mode.

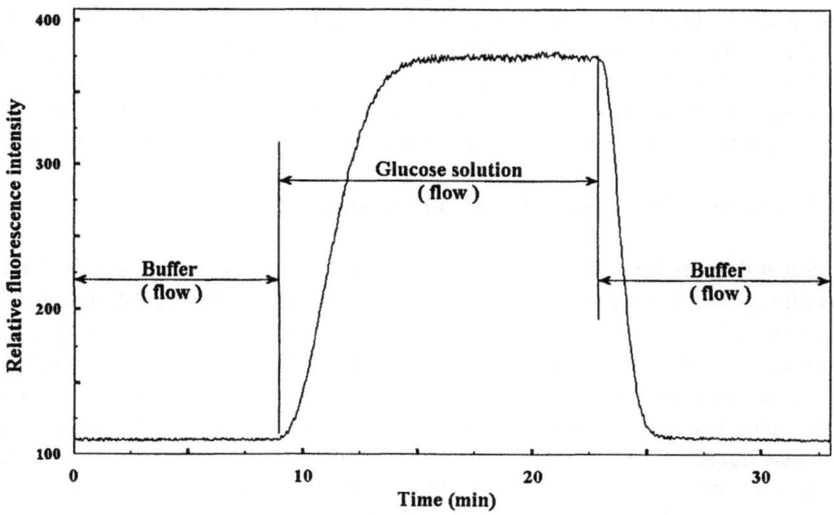

Figure 8.4. Response curves of a glucose biosensor composed of sol-gel entrapped GOx particles and optical oxygen transducer in flow mode. An air-saturated 3.0 mM glucose aqueous phosphate buffer solution continuously flew through the biosensor.

8.6.4. Flow Injection Analysis

Flow injection analysis (FIA) offers a simple automation system, the possibility of real-time monitoring, on-line sampling, and sample pre-treatment. A number of commercial devices such as pump, sample injection valve, one-choice-multi-channel valve, flow cell, and tubes are available, which enable users to setup precise FIA system with flexibly. Accordingly, the immobilized enzyme is mostly located within a flow-through cell; the solvent carrier is continuously pumped through the flow-through cell; and the glucose solution is intermittently loaded into the analysis system by an injection valve; thereby avoiding the procedures of manual changing of solvents and samples. Flow injection analyses are easier to couple with the CL or ECL techniques in order to precisely control their luminescence on spot and in time.[48, 133, 134] For FIA measurement, the peak signal mainly depends on the substrate concentration, the mass transfer efficiency, the rate of the enzymatic reaction, and the duration of stay of sample in the flow cell. Hence, in order to enhance sensitivity, wherever possible, the FIA biosensing system should utilize a small flow cell packed with large amount of high apparent active enzymes since the substrate can only contact the immobilized enzyme in a very brief period of time. In general, the flow rate has significant effect on both sample throughput and detection limit. When the flow rate is slower, although the substrate conversion is better complete, the peak signal will become wider but not necessarily higher due to the longitudinal diffusion of the substrate in the FIA tube system. At the same time the sample throughput declines. Therefore, the flow rate of a FIA system has to be optimized for reasonable sample throughput, sensitivity and detection limit. Nevertheless, the FIA technique has its unique advantages of automatic analysis at high speed for a vast number of samples and is reproducible. Over the past years, both research and application of glucose biosensors have been devoted to merge the FIA technique with various types of immobilized GOx and optical transduction techniques including absorbance, fluorescence, fluorescence O_2 quenching, CL and ECL. In fact, optical glucose biosensors have been applied to various fields with the merit of FIA technique.

8.6.5. Mixture of Organic Solvent and Aqueous Micelle Carriers

Several immobilized GOx systems can function well in organic solvents containing water such as acetonitrile, acetone, tetrahydrofuran, 2-butanol, and 2-propanol.[135-137] So far, there is no report on optical organic phase glucose biosensor based on immobilized GOx. Very recently the application of optical sol-gel glucose biosensor operating in aqueous buffer,[66] a mixture solvent of 2-propanol and micelle buffer has been explored.[138] Figure 8.5 displays the response curves of a glucose biosensor on exposure to successive step changes of various concentrations of glucose in a mixture solvent of 2-propanol, Triton X-100 and phosphate buffer. Relatively short response times and rapid return to baseline are observed. This indicates the facile transport of the substrate and the

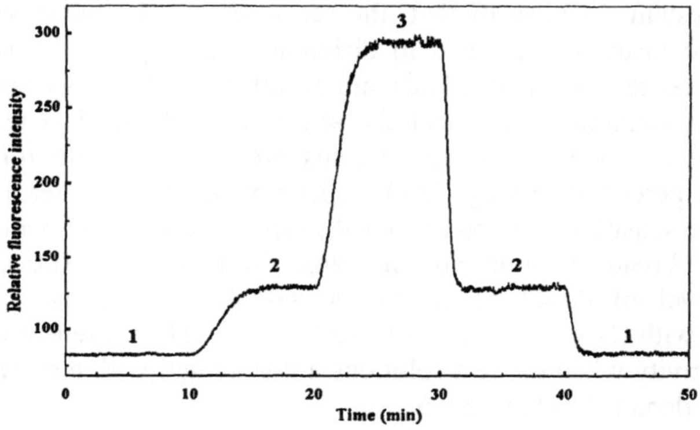

Figure 8.5. Response curves of a glucose biosensor constructed from sol-gel entrapped GOx particles and optical oxygen transducer in response to various concentrations of glucose in mixture solvents of 2-propanol/Triton X-100/phosphate buffer in flow mode: (1) 0.0, (2) 3.0, and (3) 10.0 mM glucose.

removal of the reaction product in the biosensing system. This result also strongly suggests that glucose in the mixture solvent can readily access the enzyme entrapped in the sol-gel pores/cages whilst the sol-gel matrix provides GOx with sufficient conformational flexibility for catalysis. The sensitivity is lower than that in aqueous buffer medium probably because the immobilized enzyme, as well as the biocatalytic reaction is significantly affected by the mixture solvent. However, this preliminary investigation can offer potential to develop organic phase optical glucose biosensor. A prospective advantage of using mixture solvent is that it may offer multi-analyses of glucose and cholesterol in whole blood or other biological fluids with only one single sample pretreatment.

8.7. IMPLICATION OF THE DISSOLVED OXYGEN CONCENTRATION

The dissolved oxygen concentration in the test solution plays a significant role in adjusting the sensitivity of the glucose biosensors in order to widen their applications. For example, for quick and entire generation of H_2O_2, glucose biosensors based on absorbance, CL, ECL, or pH measurements prefer a higher concentration of dissolved O_2 in the glucose solution. As such, the biosensor can achieve not only high sensitivity, but also precise result if the conversion of glucose nearly completes. On the other hand, the presence of O_2 in glucose solution is sometimes rather troublesome in optical glucose bioassay. For example, when glucose is measured by the decrease in absorption of the FAD of immobilized GOx, the dissolved O_2 can significantly affect the detection limit.[20] The removal of dissolved O_2 by purging nitrogen can improve the detection limit from 2.0 to 0.5 mM. Therefore, a key issue for dissolved O_2 is to effectively control its supplement and concentration within the glucose biosensing system. In view of that, the sensitivity of the biosensor based on OOT can, for instance, be tuned to different levels by simply adjusting the dissolved O_2 content in the standards and samples.[65, 67] The lower the dissolved O_2 in the test solutions, the higher the sensitivity will be. The reason is that although low dissolved O_2 concentrations are not favorable for enzymatic reactions; the percentage change in O_2 consumption can be increased thereafter enhancing the sensitivity. A sol-gel based glucose biosensor is demonstrated to have a rather broad dynamic working range from 0.009 to 100 mM glucose when the level of dissolved O_2 in the sample solution is controlled by equilibrating with O_2 gases from 0.1 to 100 % v/v.[67] The biosensor achieves the highest sensitivity when the test solutions are saturated with a stream of 0.1 % v/v O_2 gas as depicted in Figure 8.6.

8.8. ENHANCEMENT OF ENZYME STABILITY

Although glucose biosensors have been officially accepted as analytical instruments in the clinical domain for several decades and the native stability of GOx is quite suitable for biosensor application, the operation and shelf lifetime of the immobilized GOx still requires further improvement in order to maintain its biocatalytic activity in any environment over a longer period. In reality, the stability and the lifetime of the immobilized GOx determine mainly the success of a glucose biosensor system in real application. The stability of an enzyme primarily relates to its active protein conformation. In general, the change in the active conformation of an enzyme can be induced by thermal

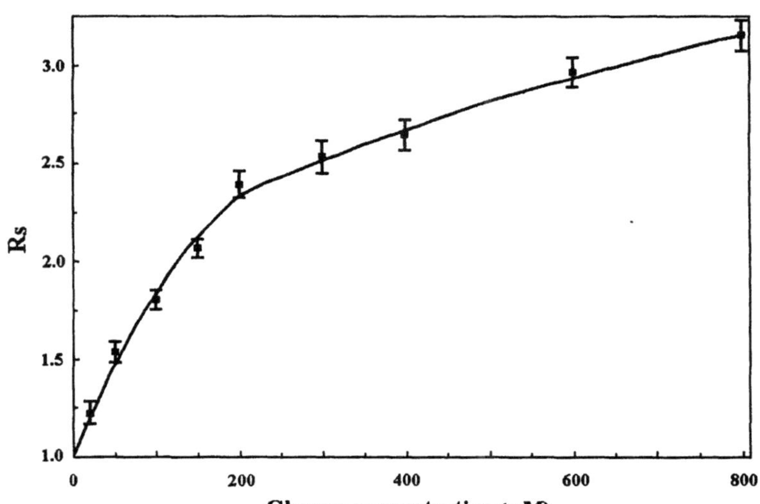

Figure 8.6. Calibration curve of a glucose biosensor assembled from hybrid sol-gel GOx particles and optical oxygen transducer at various concentrations of glucose in phosphate buffer saturated with 0.1 % v/v oxygen gas in flow mode (from reference 67).

denaturation, proteolytic degradation, and non-specific metal-catalyzed oxidation. For many years, efforts to make enzyme more stable have been realized by using novel immobilization techniques such that the chemical and physical micro-environment surrounding the enzyme is maintained;[139-141] as a result, reproducible and accurate analytical systems are available on the market.

Carbon materials are very useful for stabilizing GOx in the enzyme immobilization process. For example, the immobilization of GOx within carbon-paste matrices can enhance the thermal stability of the immobilized GOx because the hydrophobic-binding environment of an enzyme contributes to its conformational rigidity.[142] When diethylaminoethyl-dextran GOx is massively adsorbed on porous active carbon, the resulting glucose biosensor can exhibit excellent operation stability.[98, 143]

The entrapment of an enzyme is a very valuable non-covalent technique for enzyme immobilization to stabilize the protein structure against denaturing. For instance, sol-gel matrix can provide effective protective cages that precisely fit the structure of GOx; thereby the thermal stability of the entrapped GOx can gain 200 folds compared with free GOx.[45] Hence, the operation lifetime of the biosensor can be much increased when such sol-gel entrapped GOx is used as the biocatalytic element. The approach of entrapping GOx within synthetic polymer is also an effective means for improving enzyme stability. For example, a film of entrapped GOx on an electrode can be electrochemically induced from a polymer and GOx suspension.[144] The resulting amperometric biosensor shows extraordinary operation stability and demonstrates 5000 measurements of 2.5 mM glucose solution with sequential injection analyzer. In addition, Nafion polymer can improve the stability of entrapped GOx with the formation of a polyelectrolyte membrane.[145] It is found that a *p*-chlorophenylamine GOx membrane with Nafion is more stable than that of a neat *p*-chlorophenylamine GOx membrane.[116] Furthermore, when GOx is entrapped within PVA, the immobilized GOx is more stabilized to some extent.[146] Recently, an optical glucose biosensor using a novel entrapment technique of hybrid sol-gel matrix has been reported.[67] This work demonstrates that the enzyme activity and the accessibility of the immobilized enzyme are reasonably preserved in storage for as long as three years. The long shelf lifetime is mainly attributed by the attenuation in the shrinkage process of the silicate gel system as a result of hybridization of the silicate gel and hydroxyethyl carboxymethyl cellulose polymer.

Finally, GOx can also be stabilized in a favorable polymeric or polyelectrolyte micro-environment.[141] The formation of enzyme-polyelectrolyte complex can stabilize the enzyme due likely to an electrostatic cage that helps sustain the functional conformation of the enzyme over time.[140] In other cases, when the surface hydrophobicity of glass beads is created by a silanization process, the thermal stability of immobilized GOx is largely enhanced at 75 °C.[147] The stability of immobilized GOx can also be enhanced by thiols reagent[148] or sucrose[149] which chemically modify the enzyme. Moreover, the technique of immobilizing GOx with polystyrene latex beads can be applied to improve operation stability.[121]

8.9. ANALYTICAL FEATURE AND APPLICATION

The analytical applications of glucose biosensors have been increasing mainly in health care, food industry, and biological investigation. However, their analytical applications are vitally dictated by their analytical features that are commonly described in terms of specificity or selectivity, sensitivity including dynamic working range and detection limit, shelf lifetime, operation lifetime, response time, repeatability and reproducibility, etc.

Typical characteristics and applications of some selected optical glucose biosensors in literature are summarized in Table 8.3.

Table 8.3. Various optical glucose biosensors and their analytical characteristics

Detection technique	Immobilization format	Operation mode	Analytical range	Response time or speed	Sample	Ref.
Absorbance (GOx-meso-tetra (4-carboxyphenyl)-porphine)	On microscope slide	Batch	1.1 - 11.1 mM	6 s	No real sample	46
Absorbance (FAD)	In sol-gel	Batch	1 - 100 mM	~2 min	No real sample	47
Absorbance (H_2O_2)	In sandwich of sol-gel	Batch	5 - 35 mM	4 - 12 min	No real sample	44
Absorbance (H_2O_2)	On Amberlite IRA-743 resin	Flow injection	10 - 100 μM	~2 min	Blood	13
Absorbance (H_2O_2)	On film developed with Prussian blue film and 4-(Pyrrol-1-yl)-benzoic acid	Flow injection	0.05 - 2 mM	~3 min	Serum, urine, drinks, powder milk, etc.	117
Fluorescence (GOx)	On tip of fiber covered by a semipermeable membrane	Flow	1.5 - 2 mM	2 - 30 min	No real sample	20
Fluorescence (FAD)	In sol-gel membrane	Batch	Up to 10 mM	1 - 3 min	No real sample	118
Fluorescence (GOx-7-hydroxyl-coumarin-4-acetic acid)	In solution	Batch	0.5 - 6 mM	~3 min	Fruit juices	60

Fluorescence (H_2O_2)	On polyacryl-onitrile-co-polyacrylamide	Flow	0.1 - 5 mM	~10 min	No real sample	16
Fluorescence (pH)	Mixed with 1-hydroxy-pyrene-3,6,8-tri-sulfonate	Flow	0.1 - 2 mM	8 - 12 min	No real sample	17
CL (Glucose oxidase/ H_2O_2/ luminol)	On porous silicon of microchips modified with polyethylene-imine	Micro-fluidic flow injection	10 μM - 1 mM	12 sample/h	Yeast cells	53
	Immobilized on nylon tube	Sequential injection	30 - 600 μM	54 sample/h	Fermen-tation broth	101
	On aminopropyl-(controlled porous glass beads)	Flow injection	Up to ~1 mM	20 sample/h	Beverage	129
	On Sepharose beads	Flow injection	0.01-100 μM	25 sample/h	Blood	14
	In sol-gel matrix	Flow injection	0.05-12 mM	15 sample/h	Rabbit blood	150
	On polyamide membrane	Flow injection	detection limit 2.5 nM	25 sample/h	Soft drink	151
	On controlled porous glass beads	Flow injection	0.01 - 50 mM	20 sample/h	Cell culture	152
	On controlled porous glass beads	Flow injection	0.2 μM - 1 mM	30 sample/h	No real sample	153
	On Teflon membrane or between poly-carbonate and cellulose acetate membranes	Batch	2 to 18 mM	60 sample/h	Blood	154
CL (Glucose dehydro-genase/ H_2O_2/ luminol)	On hydrophilic vinyl polymer beads	Flow injection	0.1 - 30 μM	30 sample/h	Human serum	49
CL (Pyranose oxidase/ /H_2O_2/ luminol)	On controlled-pore glass beads	Flow injection	0.2 to 500 μM	60 sample/h	Serum	155
ECL (Glucose oxidase/H_2O_2/luminol)	In sol-gel/ graphite powder electrode	Flow injection	0.01-10 mM	20 s	No real sample	56
	In sol-gel deposited on glassy carbon electrode	Flow injection	0.050 - 10 mM	30 sample/h	Soft drink	134
	On polyamide membrane	Flow injection	150 pM - 1 mM	5 sample/h	Serum	58
	On imidodi-acetic acid chelating beads	Batch	20 μM - 2 mM	3 min	Human serum	15
Fluorescence (OOT)	On controlled porous glass	Flow injection	Up to 30 mM	~3 min	Cell culture matrix	95
	On carbon black	Flow	0.06 - 1 mM	2 – 6 min	No real sample	11
	On carbon black	Flow	0.01 - 200 mM	8 s -1 min	No real sample	156

(continued)

Table 8.3. (cont.)

Detection technique	Immobilization format	Operation mode	Analytical range	Response time or speed	Sample	Ref.
	On carbon black	Flow injection	0.1 - 500 mM	~1 min	Fruit juice, wine	99
	On nylon membrane	Flow	0.1 - 20 mM	1 – 6 min	No real sample	74
	On eggshell membrane	Stop-flow	0.3 - 4 mM	5 min	Beverage	65
	In sandwich of hybrid sol-gel	Flow	0.1 - 15 mM	50 s	No real sample	12
	In polyacryl-amide	Batch	0.7 - 10 mM	2 s	No real sample	157
	In sol-gel particles	Flow	0.06 - 30 mM	5 – 8 min	Beverage	66
	In hybrid sol-gel particles	Flow	0.009 - 100 mM	6 – 9 min	Urine	67
	In nano-spherical polyacryl-amide micro-emulsion system	Batch	~0.3 - 8 mM	3.5 min	No real sample	89
	On (Fe_3O_4) nano-particles	Batch	1-20 mM	~2 min	No real sample	63

The specificity is largely determined by the specificity of the transducer, rather than GOx. In general, fluorescence assays offer enough specificity for glucose measurement. The sensitivity and response time are mainly restricted by the apparent enzyme activity of the immobilized GOx and the size of the biosensing system, but lesser by the transducer. Operation and shelf lifetimes are a major problem that originates from denaturation of enzyme during operation or on storage. In real determination of glucose, reproducibility is a very important issue; unfortunately, it is significantly influenced by many factors including operation conditions such as pH medium, composition of the solution, temperature, and interferant or inhibitor to the enzyme. Therefore, a standard technique should be developed to evaluate the analytical properties of glucose enzyme biosensor over a reasonably long period of time. In our view it is not fair to only compare the analytical characteristics of the biosensors with each other.

8.10. CONCLUSION

This review has briefly summarized the development of optical enzyme-based glucose biosensors in regard to some fundamental aspects such as enzyme immobilization and stabilization, biosensor structure, biosensing scheme and operation mode, etc. The marriage of immobilized GOx with optical transducing techniques has also been gently discussed. The prime considerations are how the achievements of these studies in literature can stimulate the development of a more sensitive and useful glucose biosensing system, but not to deal with all the issues that glucose biosensor design faces at this time. Yet, we would like to make a final remark that although the development of glucose biosensor has been steadily progressing, the reported characteristics of these biosensors differ greatly even though they are based on the same type of GOx using similar immobilization techniques and same type of transducers. Biosensor personnel should bear in mind that the sensitivity of glucose biosensors can also be considerably influenced by their structure and operation mode.

To date the determination of glucose in blood, urine, and other biological fluids is very important for assessment, diagnosis or treatment of diseases, and in food samples for processing and quality control. Each assay depends indeed much on technical skill rather than extensive knowledge. Therefore, it seems that we have reached the point where identified works should be progressed in the direction of producing user-friendly glucose devices that can be mass-produced, robust and cheaper. In reality many criteria are required to take into account when developing a successful glucose sensor, but end-user feedback and input can play a vital role in the design during the development and even launch of a glucose sensor. As chemists come to expect glucose sensors suitable for schools, research institutes and quality control laboratories alike, the future will continue to hold challenges for the scientists who develop them. In our belief, we have to make sensor at a price that most people can afford it; otherwise, glucose sensors will not bring much benefit to our society but only to commercial firm.

8.11. ACKNOWLEDGMENT

The book chapter was supported by a grant from the Research Grants Council of the Hong Kong Special Administrative Region, China (project No. HKBU 2058/98P).

8.12. REFERENCES

1. J. Castillo, S. Gaspar, S. Leth, M. Niculescu, A. Mortari, I. Bontidean, V. Soukharev, S. A. Dorneanu, A. D. Ryabov, and E. Csoregi, Biosensors for life quality: Design, development and applications, *Sens. Actuators B* **102**, 179-194 (2004).
2. H. Nakamura, and I. Karube, Current research activity in biosensors, *Anal. Bioanal. Chem.* **377**, 446-468 (2003).
3. G. L. Cote, R. M. Lec, and M. V. Pishko, Emerging biomedical sensing technologies and their applications, *IEEE Sensor. J.* **3**(3), 251-266 (2003).
4. J. E. Pearson, A. Gill, and P. Vadgama, Analytical aspects of biosensors, *Ann. Clin. Biochem.* **37**, 119-145 (2000).

5. E. Wilkins, and P. Atanasov, Glucose monitoring: state of the art and future possibilities, *Med. Eng. Phys.* **18**(4), 273-288 (1996).
6. J. D. Newman, and A. P. F. Turner, Home blood glucose biosensors: a commercial perspective, *Biosens. Bioelectron.* **20**(12), 2435-2453 (2005).
7. R. McNichols, and G. L. Cote, Optical glucose sensing in biological fluids: an overview, *J. Biomed. Opt.* **5**(1), 5-16 (2000).
8. R. Badugu, J. R. Lakowicz, and C. D. Geddes, Ophthalmic glucose monitoring using disposable contact lenses - a review, *J. Fluoresc.* **14**(5), 617-633 (2004).
9. C. L. Clark, and C. Lyons, Electrode systems for continuous monitoring in cardiovascular surgery, *Ann. N.Y. Acad. Sci.* **102**, 29-45 (1962).
10. J. Raba, and H. A. Mottola, Glucose-oxidase as an analytical reagent, *Crit. Rev. Anal. Chem.* **25**, 1-42 (1995).
11. M. C. Moreno-Bondi, O. S. Wolfbeis, M. J. P. Leiner, and B. P. H. Schaffar, Oxygen Optrode for use in a fiber-optic glucose biosensor, *Anal. Chem.* **62**(21), 2377-2380 (1990).
12. O. S. Wolfbeis, I. Oehme, N. Papkovskaya, and I. Klimant, Sol-gel based glucose biosensors employing optical oxygen transducers, and a method for compensating for variable oxygen background, *Biosens. Bioelectron.* **15**(1-2), 69-76 (2000).
13. A. C. A. de Oliveira, V. C. Assis, M. A. C. Matos, and R. C. Matos, Flow-injection system with glucose oxidase immobilized on a tubular reactor for determination of glucose in blood samples, *Anal. Chim. Acta* **535**(1-2), 213-217 (2005).
14. D. T. Bostick, and D. M. Hercules, Quantitative determination of blood glucose using enzyme induced chemiluminescence of luminol, *Anal. Chem.* **47**(3), 447-452 (1975).
15. C. A. Marquette, A. Degiuli, and L. J. Blum, Electrochemiluminescent biosensors array for the concomitant detection of choline, glucose, glutamate, lactate, lysine and urate, *Biosens. Bioelectron.* **19**(5), 433-439 (2003).
16. O. S. Wolfbeis, M. Schaeferling, and A. Duerkop, Reversible optical sensor membrane for hydrogen peroxide using an immobilized fluorescent probe, and its application to a glucose biosensor, *Microchim. Acta* **143**, 221-227 (2003).
17. W. Trettnak, M. J. P. Leiner, and O. S. Wolfbeis, Fibre-optic glucose sensor with a pH optrode as the transducer, *Biosensors* **4**(6), 15-26 (1989).
18. S. A. Piletsky, T. L. Panasyuk, E. V. Piletskaya, T. A. Sergeeva, A. V. Elkaya, E. Pringsheim, and O. S. Wolfbeis, Polyaniline-coated microtiter plates for use in longwave optical bioassays, *Fresenius J. Anal. Chem.* **366**(8), 807-810 (2000).
19. W. Trettnak, and O. S. Wolfbeis, Fully reversible fiberoptic glucose biosensor based on the intrinsic fluorescence of glucose-oxidase, *Anal. Chim. Acta* **221**, 195-203 (1989).
20. I. Chudobova, E. Vrbova, M. Kodicek, J. Janovcova, and J. Ka, Fibre optic biosensor for the determination of glucose based on absorption changes of immobilized glucose oxidase, *Anal. Chim. Acta* **319**(1-2), 103-110 (1996).
21. W. Schuhmann, Amperometric enzyme biosensors based on optimized electron-transfer pathways and non-manual immobilisation procedures, *Rev. Mol. Biotechnol.* **82**, 425-441(2002).
22. E. Bakker, Electrochemical sensors, *Anal. Chem.* **76**(12), 3285-3298 (2004).
23. E. Magner, Trends in electrochemical biosensors, *Analyst* **123**, 1967-1970 (1998).
24. M. Mehrvar, and M. Abdi, Recent developments, characteristics, and potential applications of electrochemical biosensors, *Anal. Sci.* **20**, 1113-1126 (2004).
25. O. S. Wolfbeis, Fiber-optic chemical sensors and biosensors, *Anal. Chem.* **76**(12), 3269-3284 (2004).
26. J. C. Pickup, F. Hussain, N. D. Evans, O. J. Rolinski, and D. J. S. Birch, Fluorescence-based glucose sensors, *Biosens. Bioelectron.* **20**(12), 2455-2465 (2005).
27. J. R. Epstein, and D. R. Walt, Fluorescence-based fibre optic arrays: a universal platform for sensing, *Chem. Soc. Rev.* **32**(4), 203-214 (2003).
28. M. M. F. Choi, Progress in enzyme-based biosensors using optical transducers, *Microchim. Acta* **148**(3-4), 107-132 (2004).
29. W. Tan, Z. Shi, S. Smith, D. Birnbaum, and R. Kopelman, Submicrometer intracellular chemical

optical fiber sensors, *Science* **258**, 778-781 (1992).
30. B. Lee, Review of the present status of optical fiber sensors, *Opt. Fiber Technol.* **9**, 57-79 (2003).
31. B. Kuswandi, R. Andres, and R. Narayanaswamy, Optical fibre biosensors based on immobilised enzymes, *Analyst* **126**, 1469-1491(2001).
32. M. A. Medina, and P. Schwille, Fluorescence correlation spectroscopy for the detection and study of single molecules in biology, *Bioessays* **24**, 758-764 (2002).
33. O. S. Wolfbeis, Fiber-optic chemical sensors and biosensors, *Anal. Chem.* **74**(12), 2663-2678 (2002).
34. O. S. Wolfbeis, Fiber-Optic chemical sensors and biosensors, *Anal. Chem.* **72**(12), 81R-89R (2000).
35. D. J. Monk, and D. R. Walt, Optical fiber-based biosensors, *Anal. Bioanal. Chem.* **379**, 931-945 (2004).
36. M. Mehrvar, C. Bis, J. M. Scharer, and M. Moo-Young, Fiber-optic biosensors - trends and advances, *Anal. Sci.* **16**, 677-692 (2000).
37. H. Y. Aboul-Enein, R. Stefan, J. F. van Staden, X. R. Zhang, A. M. Garcia-Campana, and W. R. G. Baeyens, Recent developments and applications of chemiluminescence sensors, *Crit. Rev. Anal. Chem.* **30**(4), 271-289 (2000).
38. O. S. Wolfbeis, Fiber optic biosensing based on molecular recognition, *Sens. Actuators B* **5**(1-4), 1-6 (1991).
39. R. Narayanaswamy, Current developments in optical biochemical sensors, *Biosens. Bioelectron.* **6**(6), 467-475(1991).
40. T. Scheper, C. Muller, K. D. Anders, F. Eberhardt, F. Plotz, C. Schelp, O. Thordsen, and K. Schugerl, Optical sensors for biotechnological applications, *Biosens. Bioelectron.* **9**, 73-82 (1994).
41. M. Marazuela, and M. Moreno-Bondi, Fiber-optic biosensors - an overview, *Anal. Bioanal. Chem.* **372**, 664-682 (2002).
42. M. P. Byfield, and R. A. Abuknesha, Biochemical aspects of biosensors, *Biosens. Bioelectron.* **9**, 373-400 (1994).
43. R. Narayanaswamy, and O. S. Wolfbeis, Optical Sensors for Industrial, Environmental and Clinical Applications, *Springer Series on Chemical Sensors and Biosensors*, edited by O. S. Wolfbeis, Vol. 1 (Springer Verlag Press, 2003).
44. U. Narang, P. N. Prasad, and F. V. Bright, Glucose biosensor based on a sol-gel-derived platform, *Anal. Chem.* **66**(19), 3139-3144 (1994).
45. Q. Chen, G. L. Kenausis, and A. Heller, Stability of oxidases immobilized in silica gels, *J. Am. Chem. Soc.* **120**, 4582-4585 (1998).
46. B. J. White, and H. J. Harmon, Novel optical solid-state glucose sensor using immobilized glucose oxidase, *Biochem. Biophys. Res. Commun.* **296**(5), 1069-1071 (2002).
47. S. Shtelzer, and S. Braun, An optical biosensor based upon glucose oxidase immobilized in sol-gel silicate matrix, *Biotechnol. Appl. Biochem.* **19**, 293-305 (1994).
48. P. Fletcher, K. N. Andrew, A. C. Calokerinos, S. Forbes, and P. J. Worsfold, Analytical applications of flow injection with chemiluminescence detection - a review, *Luminescence,* **16**(1), 1-23 (2001).
49. N. Kiba, Y. Inoue, M. Tachibana, K. Tani, and H. Koizumi, Simultaneous determination of D-glucose and 3-hydroxybutyrate by chemiluminescence detection with immobilized enzymes in a flow injection system, *Anal. Sci.* **19**(8), 1203-1206 (2003).
50. M. S. Abdel-Latif, and G. G. Guilbault, Peroxide optrode based on micellar-mediated chemiluminescence reaction of luminol, *Anal. Chim. Acta* **221**, 11-17 (1989).
51. M. V. Cattaneo, and J. H. Luong, On-line chemiluminescence assay using FIA and fiber optics for urinary and blood glucose, *Enzyme Microb. Technol.* **15**(5), 424-428 (1993).
52. R. Davidsson, F. Genin, M. Bengtsson, T. Laurell, and J. Emneus, Microfluidic biosensing systems. Part I. Development and optimisation of enzymatic chemiluminescent micro-biosensors based on silicon microchips, *Lab. Chip.* **4**(5), 481-487 (2004).
53. R. Davidsson, B. Johansson, V. Passoth, M. Bengtsson, T. Laurell, and J. Emneus, Microfluidic biosensing systems. Part II. Monitoring the dynamic production of glucose and ethanol from microchip-immobilised yeast cells using enzymatic chemiluminescent micro-biosensors, *Lab. Chip.* **4**(5), 488-494 (2004).
54. Z. Xu, and Z. Fang, Composite poly(dimethylsiloxane)/glass microfluidic system with an

immobilized enzymatic particle-bed reactor and sequential sample injection for chemiluminescence determinations, *Anal. Chim. Acta* **507**(1), 129-135 (2004).

55. Y. Lv, Z. Zhang, and F. Chen, Chemiluminescence microfluidic system sensor on a chip for determination of glucose in human serum with immobilized reagents, *Talanta* **59**(3), 571-576 (2003).
56. L. Zhu, Y. Li, F. Tian, B. Xu, and G. Zhu, Electrochemiluminescent determination of glucose with a sol–gel derived ceramic–carbon composite electrode as a renewable optical fiber biosensor, *Sens. Actuators B* **84**(2-3), 265-270 (2002).
57. C. A. Marquette, A. Degiuli, and L. J. Blum, Fiberoptic biosensors based on chemiluminescent reactions, *Appl. Biochem. Biotechnol.* **89**(2-3), 107-115 (2000).
58. C. A. Marquette, and L. J. Blum, Luminol electrochemiluminescence-based fibre optic biosensors for flow injection analysis of glucose and lactate in natural samples, *Anal. Chim. Acta* **381**(1), 1-10 (1999).
59. S. D'Auria, and J. R. Lakowicz, Enzyme fluorescence as a sensing tool: new perspectives in biotechnology, *Curr Opin Biotechnol.* **12**(1), 99-104 (2001).
60. J. F. Sierra, J. Galban, S. de Marcos, and J. R. Castillo, Fluorimetric-enzymatic determination of glucose based on labelled glucose oxidase, *Anal. Chim. Acta* **368**(1-2), 97-104 (1998).
61. J. N. Demas, and B. A. DeGraff, Design and applications of highly luminescent transition metal complexes, *Anal. Chem.* **63**(17), 829A-837A (1991).
62. L. B. McGown, B. J. Kirst, and G. L. LaRowe, A new fluorimetric method for measurement of oxidase-catalyzed reactions, *Anal. Chim. Acta* **117**, 363-365 (1980).
63. L. M. Rossi, A. D. Quach, and Z. Rosenzweig, Glucose oxidase-magnetite nanoparticle bioconjugate for glucose sensing, *Anal. Bioanal. Chem.* **380**(4), 606-613 (2004).
64. W. Trettnak, and O. S. Wolfbeis, Fiber optic cholesterol biosensor with an oxygen optrode as the transducer, *Anal. Biochem.* **184**, 124-127 (1990).
65. M. M. F. Choi, W. S. H. Pang, D. Xiao, and X. Wu, An optical glucose biosensor with eggshell membrane as an enzyme immobilisation platform, *Analyst* **126**, 1558-1563 (2001).
66. X. Wu, M. M. F. Choi, and D. Xiao, A glucose biosensor with enzyme-entrapped sol-gel and an oxygen-sensitive optode membrane, *Analyst* **125**, 157-162 (2000).
67. X. J. Wu, and M. M. F. Choi, An optical glucose biosensor based on entrapped-glucose oxidase in silicate xerogel hybridised with hydroxyethyl carboxymethyl cellulose, *Anal. Chim. Acta* **514**, 219-226 (2004).
68. M. M. F. Choi, M. M. K. Liang, and A. W. M. Lee, A biosensing method with enzyme-immobilized eggshell membranes for determination of total glucosinolates in vegetables, *Enzyme Microb. Technol.* **36**, 91-99 (2005).
69. H. Szmacinski, and J. R. Lakowicz, Fluorescence lifetime-based sensing and imaging, *Sens. Actuators B* **29**(1-3), 16-24 (1995).
70. L. Tolosa, H. Szmacinski, G. Rao, and J. R. Lakowicz, Lifetimebased sensing of glucose using energy transfer with a long lifetime donor, *Anal. Biochem.* **250**, 102-108 (1997).
71. M. Schaeferling, M. Wu, and O. S. Wolfbeis, Time-resolved fluorescent imaging of glucose, *J. Fluoresc.* **14**(5), 561-568 (2004).
72. J. F. Liang, Y. T. Li, and V. C. Yang, Biomedical application of immobilized enzymes, *J. Pharm. Sci.* **89**(8), 979-990 (2000).
73. S. Zhang, G. Wright, and Y. Yang, Materials and techniques for electrochemical biosensor design and construction, *Biosens. Bioelectron.* **15**, 273-282 (2000).
74. W. Trettnak, M. J. P. Leiner, and O. S. Wolfbeis, Fibre optic glucose biosensor with an oxygen optrode as the transducer, *Analyst* **113**, 1519-1523 (1988).
75. N. Matsumoto, X. Chen, and G. S. Wilson, Fundamental Studies of Glucose Oxidase Deposition on a Pt Electrode, *Anal. Chem.* **74**(2), 362-367 (2002).
76. J. M. Fernandez-Romero, and M. D. L. de Castro, Flow-through optical biosensor based on the permanent immobilization of an enzyme and transient retention of a reaction product, *Anal. Chem.* **65**(21), 3048-3052 (1993).
77. M. Albareda-Sirvent, A. Merkoci, and S. Alegret, Configurations used in the design of screen-printed enzymatic biosensors: A review, *Sens. Actuators B* **69**(1-2), 153-163 (2000).
78. F. Davis, and S. P. J. Higson, Structured thin films as functional components within biosensors, *Biosens. Bioelectron.* in press.
79. H. Muguruma, and I. Karube, Plasma-polymerized films for biosensors, *Trends Anal. Chem.* **18**(1), 62-68 (1999).

80. W. Jin, and J. D. Brennan, Properties and applications of proteins encapsulated within sol-gel derived materials, *Anal. Chim. Acta* **461**, 1-36 (2002).
81. X. Zhong, R. Yuan, Y. Chai, Y. Liu, J. Dai, and D. Tang, Glucose biosensor based on self-assembled gold nanoparticles and double-layer 2d-network (3-mercaptopropyl)-trimethoxysilane polymer onto gold substrate, *Sens. Actuators B* **104**(2), 191-198 (2005).
82. S. Bharathi, and M. Nogami, A glucose biosensor based on electrodeposited biocomposites of gold nanoparticles and glucose oxidase enzyme, *Analyst* **126**(11), 1919-1922 (2001).
83. S. Hrapovic, Y. Liu, K. B. Male, and J. H. T. Luong, Electrochemical biosensing platforms using platinum nanoparticles and carbon nanotubes, *Anal. Chem.* **76**(4), 1083-1088 (2004).
84. J. Wang, and M. Musameh, Carbon nanotube/Teflon composite electrochemical sensors and biosensors, *Anal. Chem.* **75**(9), 2075-2079 (2003).
85. M. L. Mena, P. Yanez-Sedeno, and J. M. Pingarron, A comparison of different strategies for the construction of amperometric enzyme biosensors using gold nanoparticle-modified electrodes, *Anal. Biochem.* **336**(1), 20-27 (2005).
86. D. R. Meldrum, and M. R. Holl, Microscale bioanalytical system, *Science* **297**, 1197-1198 (2002).
87. D. Trau, and R. Renneberg, Encapsulation of glucose oxidase microparticles within a nanoscale layer-by-layer film: immobilization and biosensor applications, *Biosens. Bioelectron.* **18**(12), 1491-1499 (2003).
88. J. Chen, Y. Miao, N. He, X. Wu, and S. Li, Nanotechnology and biosensors, *Biotechnol. Adv.* **22**, 505-518 (2004).
89. H. Xu, J. W. Aylott, and R. Kopelman, Fluorescent nano-PEBBLE sensors designed for intracellular glucose imaging, *Analyst* **127**, 1471-1477 (2002).
90. S. Alegret, Rigid carbon-polymer biocomposites for electrochemical sensing, *Analyst* **121**, 1751-1758 (1996).
91. J. Wang, A. J. Reviejo, and L. Angnes, Graphite-Teflon enzyme electrode, *Electroanalysis* **5**, 575-579 (1993).
92. S. Alegret, F. Cespedes, E. Martinez-Fabregas, D. Martorell, A. Morale, E. Centelles, and J. Munoz, Carbon-polymer biocomposites for amperometric sensing, *Biosens. Bioelectron.* **11**, 35-44 (1996).
93. Y. Liu, X. Zhang, H. Liu, T. Yu, and J. Deng, Immobilization of glucose oxidase onto the blend membrane of poly(vinyl alcohol) and regenerated silk fibroin: morphology and application to glucose biosensor, *J. Biotechnol.* **46**(2), 131-138 (1996).
94. O. S. Wolfbeis, and H. E. Posch, A fibre optic ethanol biosensor, *Fresenius Z. Anal. Chem.* **332**, 255-257 (1988).
95. B. A. Dremel, S. Y. Li, and R. D. Schmid, On-line determination of glucose and lactate concentrations in animal cell culture based on fibre optic detection of oxygen in flow-injection analysis, *Biosens Bioelectron.* **7**(2), 133-139 (1992).
96. K. B. Male, and J. H. Luong, Determination of urinary glucose by a flow injection analysis amperometric biosensor and ion-exchange chromatography, *Appl. Biochem. Biotechnol.* **37**(3), 243-254 (1992).
97. M. Masoom, and A. Townshend, Determination of glucose in blood by flow injection analysis and an immobilized glucose oxidase column, *Anal. Chim. Acta* **166**, 111-118 (1985).
98. V. G. Gavalas, and N. A. Chaniotakis, Polyelectrolyte stabilized oxidase based biosensors: effect of diethylaminoethyl-dextran on the stabilization of glucose and lactate oxidases into porous conductive carbon, *Anal. Chim. Acta* **404**(1), 67-73 (2000).
99. B. A. A. Dremel, B. P. H. Schaffar, and R. D. Schmid, Determination of glucose in wine and fruit juice based on a fibre-optic glucose biosensor and flow-injection analysis, *Anal. Chim. Acta* **225**, 293-301 (1989).
100. M. J. Valencia-Gonzalez, Y. M. Liu, M. E. Diaz-Garcia, and A. Sanz-Medel, Optosensing of glucose with an immobilized glucose oxidase minireactor and an oxygen room-temperature phosphorescence transducer, *Anal. Chim. Acta* **283(1)**, 439-446 (1993).
101. X. Liu, and E. H. Hansen, Sequential injection determination of D-glucose by chemiluminescence using an open tubular immobilised enzyme reactor, *Anal. Chim. Acta* **326**(1-3), 1-12 (1996).
102. B. Wu, G. Zhang, S. Shuang, and M. M. F. Choi, Biosensors for determination of glucose with glucose oxidase immobilized on an eggshell membrane, *Talanta* **62**(2), 546-553 (2004).

103. R. Schuler, M. Wittkampf, and G. C. Chemnitius, Modified gas-permeable silicone rubber membranes for covalent immobilisation of enzymes and their use in biosensor development, *Analyst* **124**(8), 1181-1184 (1999).
104. A. Naqvi, P. Nahar, and R. P. Gandhi, Introduction of functional groups onto polypropylene and polyethylene surfaces for immobilization of enzymes, *Anal. Biochem.* **306**, 74-78 (2002).
105. N. Opitz, and D. W. Lubbers, Electrochromic dyes, enzyme reactions and hormone-protein interactions in fluorescence optic sensor (optode) technology, *Talanta* **35**, 123-127 (1988).
106. M. Shimohigoshi, K. Yokoyama, and I. Karube, Development of a bio-thermochip and its application for the detection of glucose in urine, *Anal. Chim. Acta* **303**(2-3), 295-299 (1995).
107. Y. Liu, X. Chen, J. Qian, H. Liu, Z. Shao, J. Deng, and T. Yu, Immobilization of glucose oxidase with the blend of regenerated silk fibroin and poly(vinyl alcohol) and its application to a 1,1'-dimethylferrocene-mediating glucose sensor, *Appl. Biochem. Biotechnol.* **62**(2-3), 105-117 (1997).
108. S. Yang, P. Atanasov, and E. Wilkins, Glucose biosensors based on oxygen electrode with sandwich-type membranes, *Ann. Biomed. Eng.* **23**(6), 833-839 (1995).
109. D. Jiang, E. Liu, and J. Huang, Novel fiber optic glucose biosensor based on fluorescence quenching, *Proc. SPIE-Int. Soc. Opt. Eng.* **4920**, 205-212 (2002).
110. Z. Rosenzweig, and R. Kopelman, Analytical properties and sensor size effects of a micrometer-sized optical fiber glucose biosensor, *Anal Chem.* **68**(8), 1408-1413 (1996).
111. S. M. Reddy, and P. Vadgama, Entrapment of glucose oxidase in non-porous poly(vinyl chloride), *Anal. Chim. Acta* **461**, 57-64 (2002).
112. Q. Deng, B. Li, and S. Dong, Self-gelatinizable copolymer immobilized glucose biosensor based on prussian blue modified graphite electrode, *Analyst* **123**, 1995-1999 (1998).
113. S. Yang, Y. Lu, P. Atanossov, E. Wilkins, and X. Long, Microfabricated glucose biosensor with glucose oxidase entrapped in sol-gel matrix, *Talanta* **47**, 735-743 (1998).
114. B. Wang, B. Li, Q. Deng, and S. Dong, Amperometric glucose biosensor based on sol-gel organic-inorganic hybrid material, *Anal. Chem.* **70**(15), 3170-3174 (1998).
115. J. Wang, D. Leech, M. Ozsoz, and S. M. R. Smyth, One-step fabrication of glucose sensors based on entrapment of glucose oxidase within poly(ester-sulfonic acid) coatings, *Anal. Chim. Acta* **245**, 139-143 (1991).
116. J. Xu, Z. Yu, and H. Chen, Glucose biosensor prepared by electropolymerization of *p*-chlorophenylamine with and without Nafion, *Anal. Chim. Acta* **463**, 239-247 (2002).
117. T. Lenarczuk, D. Wencel, S. Glab, and R. Koncki, Prussian blue-based optical glucose biosensor in flow-injection analysis, *Anal. Chim. Acta* **447**(1-2), 23-32 (2001).
118. A. M. Hartnett, C. M. Ingersoll, G. A. Baker, and F. V. Bright, Kinetics and thermodynamics of free flavins and the flavin-based redox active site within glucose oxidase dissolved in solution or sequestered within a sol-gel-derived glass, *Anal. Chem.* **71**, 1215-1224 (1999).
119. R. A. Shimkus, V. Laurinavichius, L. Boguslavsky, S. W. Tanenbaum, D. J. Slomczynski, and J. P. Nakas, Laccase containing sol-gel based optical biosensors, *Anal. Lett.* **29**, 1907-1913 (1996).
120. W. R. Heineman, Biosensors based on polymer networks formed by gamma irradiation crosslinking, *Appl. Biochem. Biotechnol.* **41**(1-2), 87-97 (1993).
121. C. G. Koopal, and R. J. Nolte, Highly stable first-generation biosensor for glucose utilizing latex particles as the enzyme-immobilizing matrix, *Enzyme Microb. Technol.* **16**(5), 402-408 (1994).
122. W. Zhan, G. H. Seong, and R. M. Crooks, Hydrogel-based microreactors as a functional component of microfluidic systems, *Anal. Chem.* **74**(18), 4647-4652 (2002).
123. R. Q. Thompson, H. Kim, and C. E. Miller, Comparison of immobilized enzyme reactors for flow-injection systems, *Anal. Chim. Acta* **198**, 165-172 (1987).
124. L. D. Ward, and D. J. Winzor, Relative merits of optical biosensors based on flow-cell and cuvette designs, *Anal. Biochem.* **285**, 179-193 (2000).
125. G. H. Seong, J. Heo, and R. M. Crooks, Measurement of enzyme kinetics using a continuous-flow microfluidic system, *Anal. Chem.* **75**(13), 3161-3167 (2003).
126. R. N. Goldberg, E. J. Prosen, B. R. Staples, R. N. Boyd, G. T. Armstrong, R. L. Berger, and D. S. Young, Heat measurements applied to biocheical analysis: glucose in human serum, *Anal. Biochem.* **64**(1), 68-73 (1975).
127. B. Xie, U. Hedberg, M. Mecklenburg, and B. Danielsson, Fast determination of whole blood glucose with a calorimetric micro-biosensor, *Sens. Actuators B* **15**(1-3), 141-144 (1993).

128. U. Harborn, B. Xie, R. Venkatesh, and B. Danielsson, Evaluation of a miniaturized thermal biosensor for the determination of glucose in whole blood, *Clin. Chim. Acta* **267**(2), 225-237 (1997).
129. K. Matsumoto, and K. Waki, Simultaneous biosensing of ethanol and glucose with combined use of a rotating bioreactor and a stationary column reactor, *Anal. Chim. Acta* **380**(1), 1-6 (1999).
130. S. Banu, G. M. Greenway, T. McCreedy, and R. Shaddick, Microfabricated bioreactor chips for immobilized enzyme assays, *Anal Chim. Acta* **486**(2), 149-157 (2003).
131. S. W. Kiang, J. W. Kuan, S. S. Kuan, and G. G. Guilbault, Measurement of glucose in plasma, with use of immobilized glucose oxidase and peroxidase, *Clin. Chem.* **22**(8), 1378-1382 (1976).
132. J. Krenkova, and F. Foret, Immobilized microfluidic enzymatic reactors, *Electrophoresis* **25**(21-22), 3550-3563 (2004).
133. U. Spohn, F. Preuschoff, G. Blankenstein, D. Janasek, M. R. Kula, and A. Hacker, Chemiluminometric enzyme sensors for flow-injection analysis, *Anal. Chim. Acta* **303**(1), 109-120 (1995).
134. L. Zhu, Y. Li, and G. Zhu, A novel flow through optical fiber biosensor for glucose based on luminol electrochemiluminescence, *Sens. Actuators B* **86**(2-3), 209-214 (2002).
135. E. I. Iwuoha, M. R. Smyth, and M. E. G. Lyons, Solvent effects on the reactivities of an amperometric glucose sensor, *J. Electroanal. Chem.* **390**, 35-45 (1995).
136. L. Campanella, G. Favero, L. Persi, M. P. Sammartino, M. Tomassetti, and G. Visco, Organic phase enzyme electrodes: applications and theoretical studies, *Anal. Chim. Acta* **426**, 235-247 (2001).
137. N. Adanyi, M. Toth-Markus, E. E. Szabo, M. Varadi, M. P. Sammartino, M. Tomassetti, and L. Campanella, Investigation of organic phase biosensor for measuring glucose in flow injection analysis system, *Anal. Chim. Acta* **501**, 219-225 (2004).
138. X. J. Wu, and M. M. F. Choi, Hydrogel network entrapping cholesterol oxidase and octadecylsilica for optical biosensing in hydrophobic organic or aqueous micelle solvents, *Anal. Chem.* **75**, 4019-4027 (2003).
139. J. J. Roy, and T. E. Abraham, Strategies in making cross-linked enzyme crystals, *Chem. Rev.* **104**(9), 3705-3721 (2004).
140. B. Appleton, T. D. Gibson, and J. R. Woodward, High temperature stabilisation of immobilised glucose oxidase: potential applications in biosensors, *Sens. Actuators B* **43**, 65-69 (1997).
141. N. A. Chaniotakis, Enzyme stabilization strategies based on electrolytes and polyelectrolytes for biosensor applications, *Anal. Bioanal. Chem.* **378**, 89-95 (2004).
142. J. Liu, and J. Wang, Remarkable thermostability of bioelectrodes based on enzymes immobilized within hydrophobic semi-solid matrices, *Biotechnol. Appl. Biochem.* **30**, 177-183 (1999).
143. V. G. Gavalas, N. A. Chaniotakis, and T. D. Gibson, Improved operational stability of biosensors based on enzyme-polyelectrolyte complex adsorbed into a porous carbon electrode, *Biosens. Bioelectron.* **13**(11), 1205-1211 (1998).
144. C. Kurzawa, A. Hengstenberg, and W. Schuhmann, Immobilization method for the preparation of biosensors based on pH shift-induced deposition of biomolecule-containing polymer films, *Anal. Chem.* **74**, 355-361 (2002).
145. A. A. Karyakin, E. A. Kotel'nikova, L.V. Lukachova, E. E. Karyakina, and J. Wang, Optimal environment for glucose oxidase in perfluorosulfonated ionomer membranes: improvement of first-generation biosensors, *Anal. Chem.* **74**(7), 1597-1603 (2002).
146. P. A. D'Orazio, T. C. Maley, R. R. McCaffrey, A. C. Chan, D. Orvedahl, J. Foos, D. Blake, S. Degnan, J. Benco, C. Murphy, P. G. Edelman, and H. Ludi, Planar (bio)sensors for critical care diagnostics, *Clin. Chem.* **43**, 1804-1845 (1997).
147. V. R. S. Babu, M. A. Kumar, N. G. Karanth, and M. S. Thakur, Stabilization of immobilized glucose oxidase against thermal inactivation by silanization for biosensor applications, *Biosens. Bioelectron.* **19**(10), 1337-1341 (2004).
148. K. C. Gulla, M. D. Gouda , M. S. Thakur, and N. G. Karanth, Enhancement of stability of immobilized glucose oxidase by modification of free thiols generated by reducing disulfide bonds and using additives, *Biosens. Bioelectron.* **19**(6), 621-625 (2004).
149. M. D. Gouda, M. A. Kumar, M. S. Thakur, and N. G. Karanth, Enhancement of operational stability of an enzyme biosensor for glucose and sucrose using protein based stabilizing

agents, *Biosens. Bioelectron.* **17**(6-7), 503-507 (2002).
150. B. Li, Z. Zhang, and Y. Jin, Chemiluminescence flow sensor for in vivo on-line monitoring of glucose in awake rabbit by microdialysis sampling, *Anal. Chim. Acta* **432**(1), 95-100 (2001).
151. L. J. Blum, Chemiluminescent flow injection analysis of glucose in drinks with a bienzyme fiberoptic biosensor, *Enzyme Microb. Technol.* **15**(5), 407-411 (1993).
152. Y. L. Huang, S. Y. Li, B. A. A. Dremel, U. Bilitewski, and R. D. Schmid, On-line determination of glucose concentration throughout animal cell cultures based on chemiluminescent detection of hydrogen peroxide coupled with flow-injection analysis, *J. Biotechnol.* **18**(1-2), 161-172 (1991).
153. C. A. K. Swindlehurst, and T. A. Nieman, Flow-injection determination of sugars with immobilized enzyme reactions and chemiluminescence detection, *Anal. Chim. Acta* **205**, 195-205 (1988).
154. X. Zhou, and M. A. Arnold, Internal enzyme fiber-optic biosensors for hydrogen peroxide and glucose, *Anal. Chim. Acta* **304**(2), 147-156 (1995).
155. N. Kiba, F. Ueda, and M. F. T. Yamane, Flow-injection determination of glucose in serum with an immobilized pyranose oxidase reactor, *Anal. Chim. Acta* **269**(2), 187-191 (1992).
156. B. P. H. Schaffar, and O. S. Wolfbeis, A fast responding fibre optic glucose biosensor based on an oxygen optrode, *Biosens. Bioelectron.* **5**(2), 137-148 (1990).
157. Z. Rosenzweig, and R. Kopelman, Analytical properties of miniaturized oxygen and glucose fiber optic sensors, *Sens. Actuators B* **35-36**, 475-483 (1996).

SACCHARIDE RECOGNITION BY BORONIC ACID FLUOROPHORE/CYCLODEXTRIN COMPLEXES IN WATER

Akiyo Yamauchi, Iwao Suzuki , and Takashi Hayashita*

9.1. ADVANCES IN SYNTHETIC RECEPTORS FOR SACCHARIDES

Saccharides are important biological molecules because they are essential in such processes as nutrition and metabolism, and cell structure maintenance.[1] Saccharides are also physiologically active and are involved in controlling an individual's birth, differentiation and immunity. Because of these important properties, methods for *in situ* saccharide sensing in aqueous solution are very much required.[1] In particular, continuous noninvasive monitoring of blood glucose is essential to the management of diabetes. Although biosensors using biological receptors (proteins) have been widely studied as tools for detecting saccharides in biological fluids, their poor stability limits their wide application in actual monitoring. Moreover, expected signals for biosensors are generally associated with changes in electrical current/potential, mass, and refractive index. These signals are not compatible with the imaging of intracellular events. To overcome the drawbacks of existing biosensors, an alternative method to detect saccharides must be developed without using biological receptors. For this purpose, two major strategies have been considered.[1, 2] One is based on the use of multiple hydrogen bonding interactions mimicking the carbohydrate recognition by biological receptors.[1-6] However, these synthetic receptors work only in organic solvents, since water interferes with the hydrogen-bonding interaction. Recently, Davis and coworkers have synthesized a water-soluble tricyclic polyamide receptor **1**, which binds with saccharides in water, but whose affinity is low.[7]

Another approach is the use of boronic acid that reversibly forms cyclic esters with various saccharides in water. For example, Czarnik and coworkers

* Akiyo Yamauchi and Iwao Suzuki, Graduate School of Pharmaceutical Sciences, Tohoku University, Aramaki, Aoba-ku, Sendai 980-8578, JAPAN. Phone: +81-22-795-6802; Fax: +81-22-795-5916; e-mail: yamauchi@mail.pharm.tohoku.ac.jp. Takashi Hayashita, Department of Chemistry, Sophia University, 7-1 Kioicho, Chiyoda-ku, Tokyo 102-8554, JAPAN. Phone: +81-3-3238-3372; Fax: +81-3-3238-3361; e-mail: ta-hayas@sophia.ac.jp.

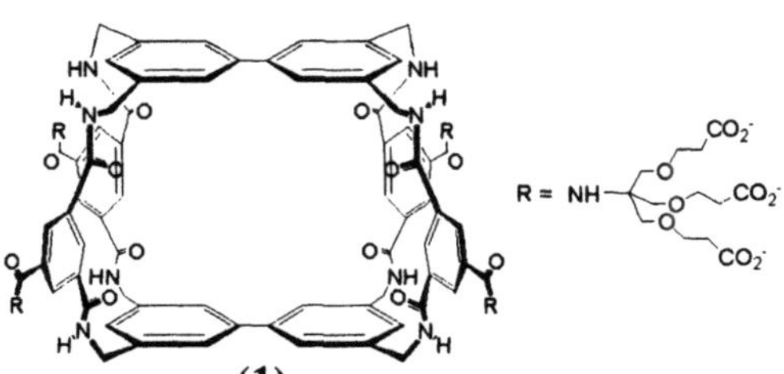

(1)

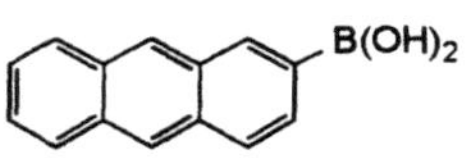

(2)

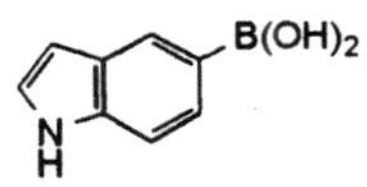

(3)

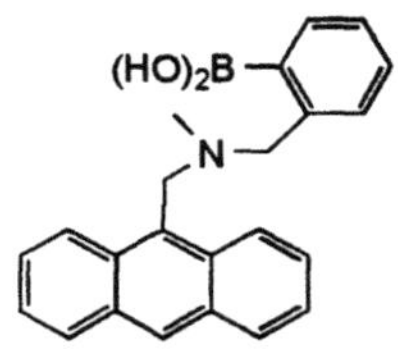

(4)

reported that anthrylboronic acid **2** senses saccharides in neutral aqueous solution *via* a fluorescence quenching process. The fluorescence quenching is based on electron transfer from the boronate ester anion to the anthryl moiety (Fig. 9.1).[8,9] Similarly, indolylboronic acid **3** synthesized by Aoyama and coworkers recognizes fructose and oligosaccharides at pH 9 with chain length selectivity due to the CH-π interaction in addition to boronate ester formation.[10]

Shinkai *et al.* proposed a notable design of a fluorescent probe for precise and selective saccharide recognition.[11-16] They connected a fluorescent tertiary amine to the boronic acid unit as shown in Fig. 9.2.[16] The fluorescence of **4** is quenched by intramolecular photoinduced electron transfer (PET). Binding with saccharide causes fluorescence recovery due to the strengthening of the boron-nitrogen Lewis acid-base interaction or by the recently proposed hydrolysis mechanism[17] which suppresses electron transfer quenching. This fluorescent probe can sense saccharides in neutral aqueous methanol solution with an increase in fluorescence intensity.

It is known that monoboronic acids display higher selectivity for fructose over other monosaccharides.[18] Shinkai and coworkers synthesized a glucose-selective sensor **5**[19] by introducing a second boronic acid group. Since then, various diboronic acid sensors have been developed.[20-24]

There are a number of non-PET sensors for saccharides. James and coworkers synthesized a fluorescent resonance energy transfer (FRET) saccharide sensor **6** with phenanthrene as the donor and pyrene as the acceptor.[25, 26]

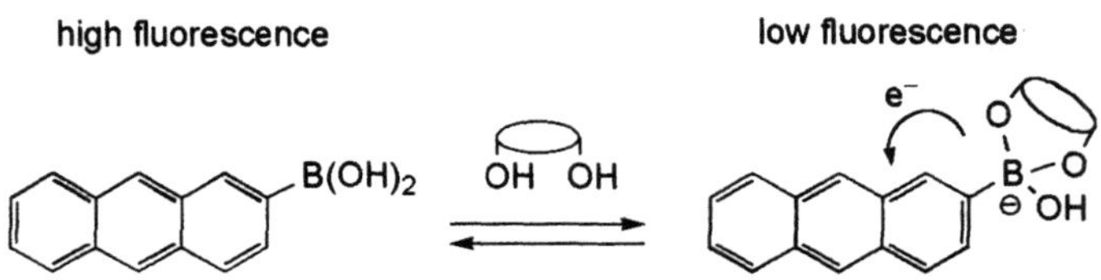

(2) **2** · saccharide complex

Figure 9.1. Fluorescence quenching mechanism by anionic ester formation of **2** with saccharides.

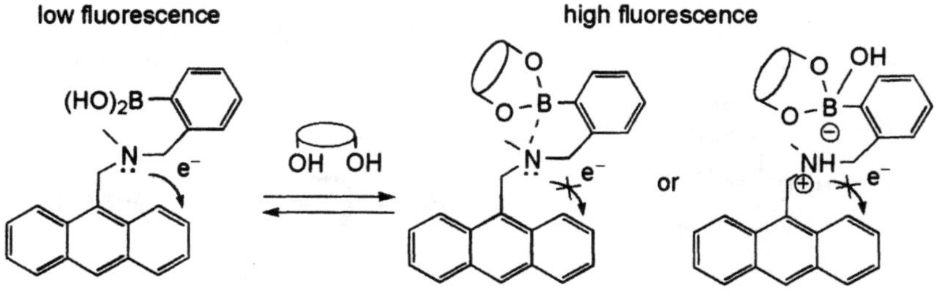

(4) 4 · saccharide complex

Figure 9.2. Fluorescence emission mechanism by ester formation of **4** with saccharides.

The 6-methoxyquinolinium-based fluorophore **7** reported by Geddes and coworkers is more fluorescent in acidic media because of the quaternary nitrogen. In the presence of saccharides, the electronic density at the boron atom is increased and this facilitates the partial neutralization of the positive charge on the nitrogen so that the fluorescence intensity decreases.[27]

DiCesare *et al.* utilized intramolecular charge transfer (ICT) which is very sensitive to small perturbations that can result in spectral shift and intensity changes.[28-30] They synthesized a substituted stilbene **8** which showed a large redshifted fluorescence due to ICT from the dimethylamino group to the boronic acid group.[28] Binding with fructose inhibited the ICT, resulting in the blue shift and increased fluorescence. Diphenyloxazole substituted with a dimethylamino and a boronic acid group, **9**[29], also responds to fructose based on the ICT mechanism. They also synthesized **10** and **11** whose boronic acid groups are not directly involved in the charge transfer.[30] In these receptors, a carbonyl group was introduced as an electron-withdrawing group. Binding with saccharide produced a spectral change, since the anionic form of the boronic acid group became an electron donor, which competes with dimethylamino groups for the ICT. Arimori *et al.* reported **12** that showed both a change in the intensity and wavelength of the emission maxima upon addition of saccharides due to a change in the environment of the anilinic nitrogen.[31] Wang and coworkers developed a waster-soluble sensor **13** following the ICT mechanism.[32]

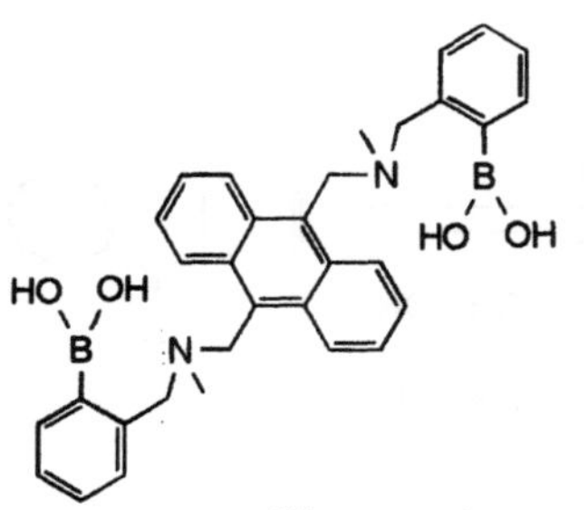

(5)

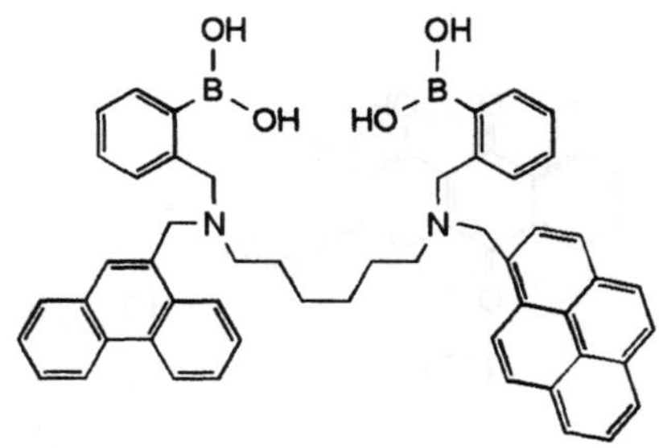

(6)

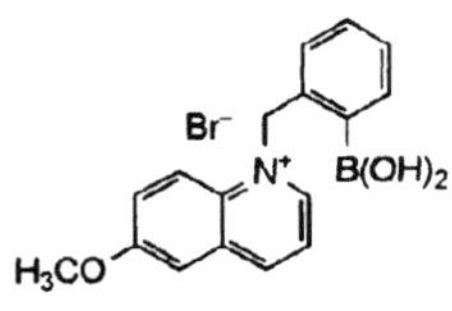

(7)

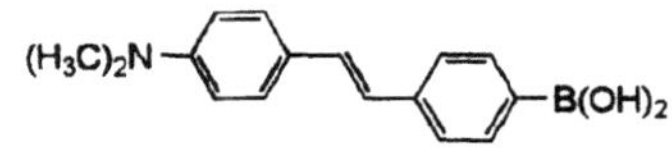

(8)

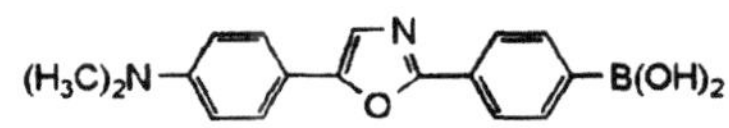

(9)

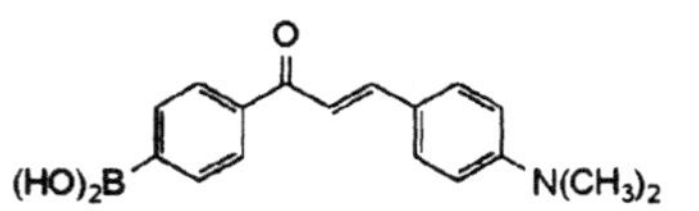

(10)

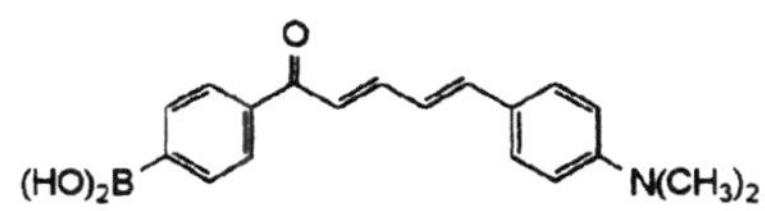

(11)

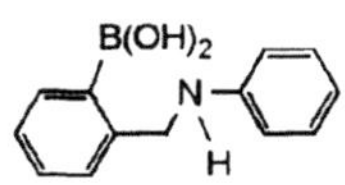

(12)

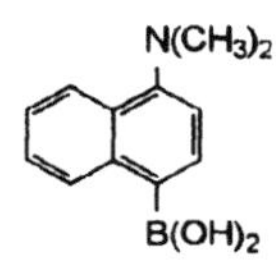

(13)

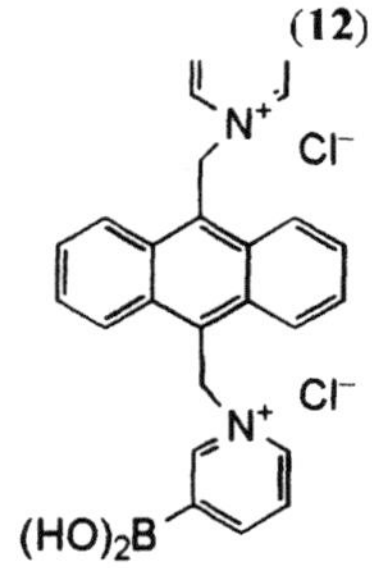

(14)

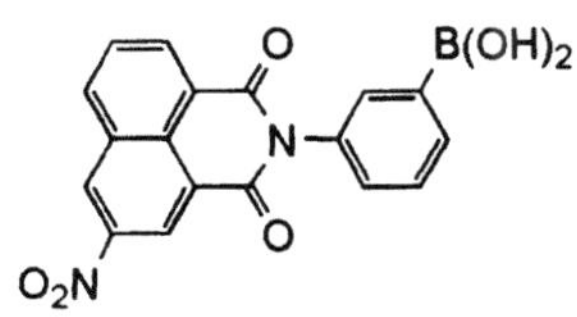

(15)

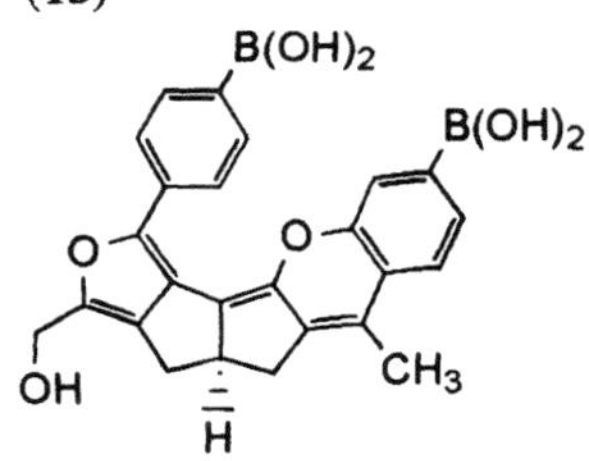

(16)

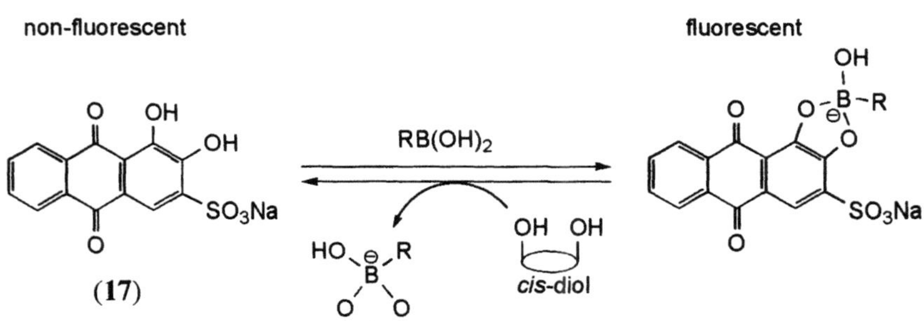

Figure 9.3. Competitive binding of a boronic acid with **17** (ARS) and a *cis*-diol.

Water solubility is one of the requirements of saccharide sensors. Eggert *et al.* synthesized a novel glucose selective receptor **14**, which is highly water soluble due to the presence of cationic pyridinium moieties.[33] Its fluorescence intensity increases in the presence of glucose, probably because of increased rigidity. Heagy and coworkers developed the naphthalic anhydride sensor **15**[34, 35], which is soluble in aqueous solution. It displayed the most visible fluorescence intensity change with glucose, in which its dissociation constant for glucose was higher than that for fructose.

Drueckhammer and coworkers developed the glucose-selective receptor **16** based on computer-guided design.[36] The polycyclic scaffold that permits proper orientation of two phenylboronic acid groups for binding with 1,2 and 4,6 positions of glucopyranose was chosen. The fluorescence intensity of **16** decreased with the addition of monosaccharide, and the selectivity for glucose over fructose was 43-fold. This indicated that the rigidity of the linker between two boronic acid groups is important for determining saccharide selectivity.

Although synthetic receptors possessing both a saccharide-binding motif and a signal transducer have been introduced, there are unique sensing systems that function with the cooperation of more than two molecules. Springsteen and Wang indirectly detected the binding of phenylboronic acid and fructose by monitoring the fluorescence change of Alizarin Red S. (ARS, **17**) in aqueous phosphate buffer as illustrated in Fig. 9.3.[37] Binding of a boronic acid to the catechol moiety of ARS causes a fluorescence increase. Singaram and coworkers prepared a novel boronic acid-substituted 4,7-phenanthrolinium viologen **18**, which quenched the fluorescence of 8-hydroxypyrene-1,3,6-trisulfonic acid trisodium salt (pyranine, **19**).[39] When monosaccharide binds with the viologen, the complex of viologen and pyranine dissociates, resulting in strong fluorescence (Fig. 9.4).

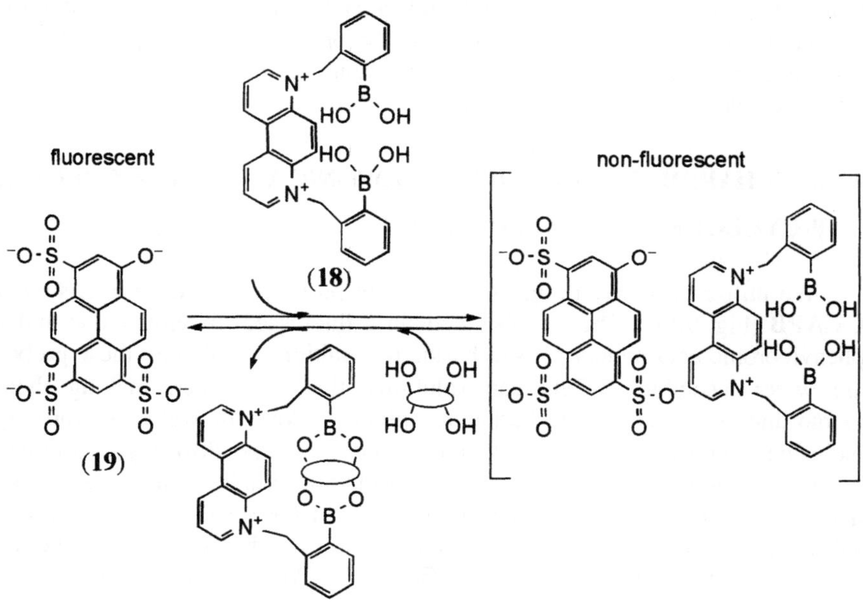

Figure 9.4. Equilibria between pyranine (**19**), viologen **18** and monosaccharides.

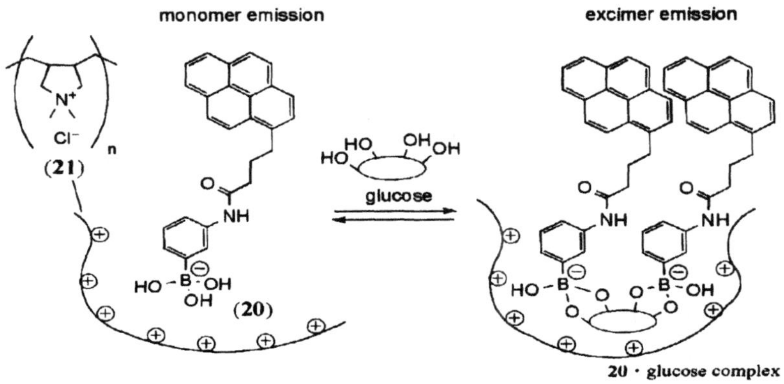

Figure 9.5. Excimer emission based on 1:2 complex formation between glucose and fluorescent boronic acid **20**.

There is another interesting example of a supramolecular sensing system for glucose.[38] The pyrene-appended boronic acid **20** forms 2:1 complex with glucose in the presence of polycation **21**, and displays excimer emission in aqueous solution. It is considered that **20** is enriched along the polycation chain *via* electrostatic interaction, which facilitates the 2:1 complex formation between **20** and glucose (Fig. 9.5).

For alkali-metal ion sensing, we have shown a similar pyrene dimer formation using crown ether fluoroionophore/γ-cyclodextrin (γ-CyD) complexes in water.[40-42] Cyclodextrins (CyDs) are water-soluble host compounds having nanosize hydrophobic cavities that enable them to incorporate various organic molecules in water. In addition, optically inert CyDs can be efficiently combined with various types of chromo- and fluoroionophores. We have recently developed a novel sensing system for saccharide recognition by constructing boronic acid fluorophore/β-CyD complexes in water.[43,44] The detailed functions of the supramolecular CyD sensors are described in the next section.

9.2. SACCHARIDE RECOGNITION BY BORONIC ACID FLUOROPHORE / β-CYCLODEXTRIN COMPLEXES

We utilized a newly designed fluorescent probe, **Cn-CPB** (n = 1, 4) and **C1-APB** (Fig. 9.6). These probes contain a fluorescent pyrenyl group and a phenylboronic acid moiety for saccharide recognition. Although these probes are water insoluble, β-CyD can solubilize them in water, making them extraordinarily effective saccharide receptors working in water by forming inclusion complexes. We first found that the **C4-CPB**/β-CyD complex exhibited fluorescence increase upon saccharide binding in water. We determined the response mechanism of the **C4-CPB**/β-CyD complex, and subsequently examined the structural effect of the boronic acid probe on saccharide sensing by comparing **C1-CPB** and **C1-APB** with **C4-CPB**.

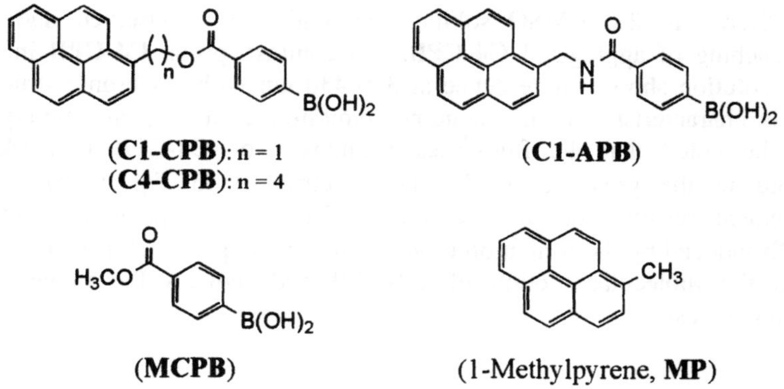

Figure 9.6. Fluorescent probes and model compounds used in this section.

9.2.1. Saccharine recognition by the C4-CPB/β-CyD Complex

9.2.1.1. Spectral Properties of C4-CPB in Aqueous DMSO Solution

Figure 9.7 shows UV-Vis and fluorescence spectra of **C4-CPB** recorded in various aqueous DMSO solutions. In 2% DMSO, the absorption peaks of **C4-CPB** are broader than those in 25% DMSO solution, which indicates the aggregation of **C4-CPB** in water. The absorption and fluorescence spectra of **C4-CPB** in 2% DMSO solution are markedly changed by the addition of β-CyD. The absorption spectrum of **C4-CPB** in 2% DMSO solution with 5.0 mM β-CyD (spectrum 3 in Fig. 9.7a) is similar to that in 25% DMSO solution. This indicates the presence of monomeric **C4-CPB** in a β-CyD cavity. The fluorescence spectra of **C4-CPB** in the same solvent mixtures are shown in

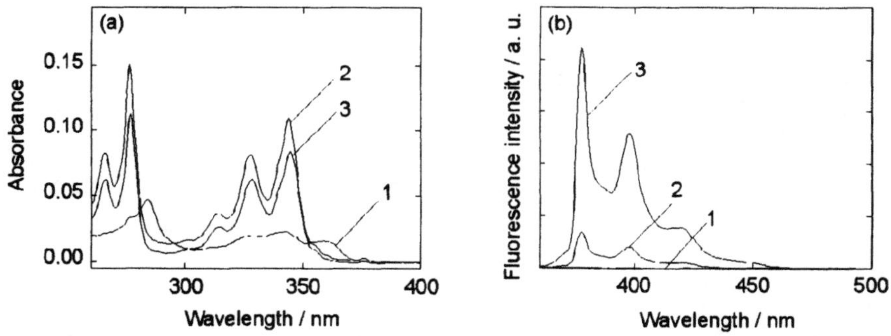

Figure 9.7. Spectral properties of **C4-CPB** in aqueous DMSO solution. (a) UV-Vis spectra: optical path length, 5 cm. (b) Fluorescence spectra: excitation wavelength, 328.0 nm; excitation bandwidth, 5 nm; emission bandwidth, 3 nm. [**C4-CPB**] = 1.05 μM in (1) 2% DMSO-98% water (v/v), (2) 25% DMSO-75% water (v/v), and (3) 2% DMSO-98% water (v/v) containing 5.0 mM β-CyD. pH = 7.5 adjusted by 0.015 M phosphate buffer (I = 0.08 M with NaCl). Reproduced by permission of American Chemical Society.

Figure 9.7b. In 2% DMSO solution, no emission is observed due to a self-quenching of aggregated **C4-CPB**. In contrast, probe **C4-CPB** in 25% DMSO solution shows fluorescence at 370-430 nm with a vibronic structure, which is characteristic of the monomer emission of the pyrene moiety. It should be noted that the fluorescence intensity of **C4-CPB** is markedly enhanced in the presence of β-CyD (spectrum 3 in Fig. 9.7b). This enhancement results from an increase in fluorescence quantum yield for **C4-CPB** induced by the formation of an inclusion complex with β-CyD, which restricts the molecular motion of **C4-CPB** and reduces the radiationless transition process.

*9.2.1.2. Fructose Recognition by the **C4-CPB**/β-CyD Complex in Water*

The response to fructose of **C4-CPB** was examined at pH 7.5 in 2% DMSO solution, 25% DMSO solution, and 2% DMSO solution containing 5.0 mM β-CyD by monitoring fluorescence emission. Figure 9.8a shows a typical fluorescence response for the **C4-CPB**/β-CyD complex upon fructose addition. It is noted that the fluorescence of the **C4-CPB**/β-CyD complex in 2% DMSO solution increases as the fructose concentration is increased from 0 to 30 mM (plots 3 in Fig. 9.8 b), while no fluorescence response is observed for **C4-CPB** in 2% DMSO solution (plots 1 in Fig. 9.8b). In 25% DMSO solution, fluorescence intensity of probe **C4-CPB** exhibits a reasonable increase (plots 2

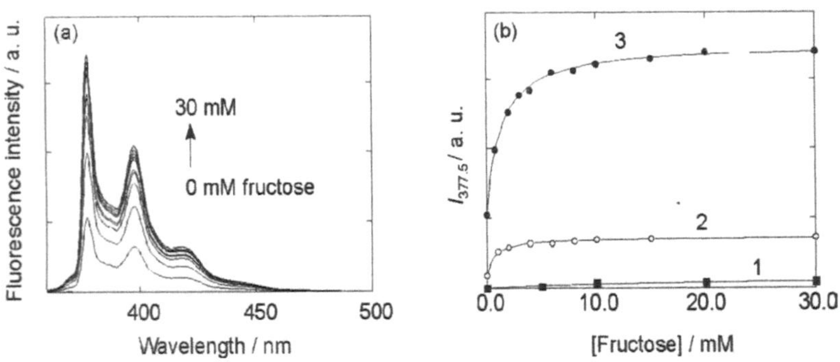

Figure 9.8. (a) Fluorescence spectra of the **C4-CPB**/β-CyD complex upon addition of fructose: [**C4-CPB**] = 1.05 μM and [fructose] = 0-30.0 mM in 2% DMSO-98% water (v/v) containing 5.0 mM β-CyD; excitation wavelength, 328.0 nm; excitation bandwidth, 5 nm; emission bandwidth, 3 nm; pH = 7.5 adjusted by 0.015 M phosphate buffer (I = 0.08 M with NaCl). (b) Dependence of $I_{377.5}$ on the concentration of fructose: [**C4-CPB**] = 1.05 μM in (1) 2% DMSO-98% water (v/v), (2) 25% DMSO-75% water (v/v), and (3) 2% DMSO-98% water (v/v) containing 5.0 mM β-CyD. pH = 7.5 adjusted by 0.015 M phosphate buffer (I = 0.08 M with NaCl). Reproduced by permission of American Chemical Society.

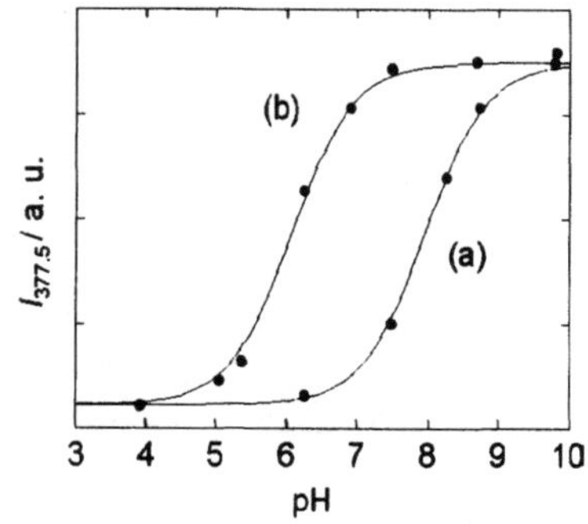

Figure 9.9. Dependence of $I_{377.5}$ as a function of pH: [**C4-CPB**] = 1.05 μM in 2% DMSO-98% water (v/v) containing 5.0 mM β-CyD; (a) [fructose] = 0.0 mM, (b) [fructose] = 30.0 mM. Reproduced by permission of American Chemical Society.

in Fig. 9.8b). These results demonstrate that the **C4-CPB**/β-CyD complex sensor exhibits fructose recognition in water with high sensitivity.

To clarify the response mechanism, the dependence of fluorescence intensity of the **C4-CPB**/β-CyD complex on pH was analyzed. Figure 9.9 shows the titration curves of the **C4-CPB**/β-CyD complex in the absence (a) and presence of 30 mM fructose (b). The apparent pK_a values of **C4-CPB** were determined by nonlinear curve fitting with eq. (1),

$$I = \frac{\beta[\mathrm{L}]_0\left(\phi_{\mathrm{HL}} + \phi_{\mathrm{L}}\dfrac{K_a}{[\mathrm{H}^+]}\right)}{1+\dfrac{K_a}{[\mathrm{H}^+]}} \qquad (1)$$

where ϕ_{HL} and ϕ_L are the fluorescence quantum yields for HL and L^- species of **C4-CPB** (see Fig. 9.10), respectively, and β is a constant which is proportional to the intensity of the excitation light and the molar extinction coefficient of **C4-CPB**. The results reveal that the apparent pK_a value of **C4-CPB** decreases from 7.95 ± 0.03 to 6.06 ± 0.03 resulting in efficient fructose recognition at neutral pH. Figure 9.10 shows proton dissociation and saccharide binding equilibria of **C4-CPB** in water. It is evident that the fluorescence quantum yield for **C4-CPB** increases when **C4-CPB** is converted into its tetrahedral boronate form. Thus, saccharide binding causes an increase in the amount of tetrahedral boronate, resulting in a fluorescence enhancement. It is notable that the fluorescent recovery of **C4-CPB** (ϕ_L/ϕ_{HL}) reaches 15.2. The emission intensity of the fluorescence quenching probes[8-10] decreases with increasing pH, which is different from that of the **C4-CPB**/β-CyD complex. Similarly, the fluorescent probes that exhibit increased emission response to saccharides have

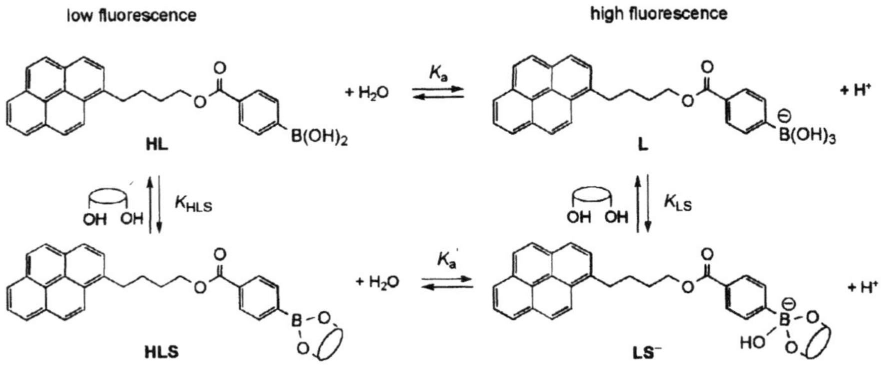

Figure 9.10. Proton dissociation and saccharide binding equilibria of **C4-CPB** in aqueous solution.

been developed by Shinkai's group, but their probes' fluorescence intensities decrease with increasing pH, which is different from the profile shown in Fig. 9.9.[16] Therefore, the response mechanism of the **C4-CPB**/β-CyD complex is clearly different from those of the previously reported probes.

*9.2.1.3. Response Mechanism of the **C4-CPB**/β-CyD Complex in Water*

The fluorescence intensity for the neutral form of **C4-CPB** is low, indicating that the arylboronic acid can act as an electron acceptor from the excited-state pyrene donor.[45] To clarify this hypothesis, we examined the effect of proton dissociation on the UV-Vis and fluorescence spectra of **C4-CPB** in 95% methanol-5% water (v/v). The addition of 1.0 mM benzyltrimethylammonium hydroxide markedly enhanced the fluorescence of **C4-CPB** (Fig. 9.11). It is evident that there is no ground state interaction, because no changes are observed in the UV-Vis spectra (dotted lines in Fig. 9.11). This result strongly supports the existence of photoinduced electron transfer (PET) from the pyrene donor to the neutral form of the arylboronic acid acceptor.[45] Since 95% methanol is relatively hydrophobic and no UV-Vis spectral change is noted, another sensing mechanism based on changes in solubility or aggregate formation[46] can be excluded.

To obtain further evidence, fluorescence quenching using model compounds was examined. We used 1-methylpyrene (**MP**) and 4-methoxycarbonylphenylboronic acid (**MCPB**) as donor and acceptor models of the intramolecular pyrene/arylboronic acid system in **C4-CPB**. The UV-Vis and fluorescence spectra of **MP** (50 μM) and increasing amounts of the neutral form of **MCPB** (0-20 mM) in 95% methanol-5% water (v/v) are shown in Fig. 9.12a. Similar to the result in Fig. 9.11, the addition of the neutral form of **MCPB** significantly quenched the fluorescence of **MP** without changing its UV-Vis spectra. In contrast, when 40 mM benzyltrimethylammonium

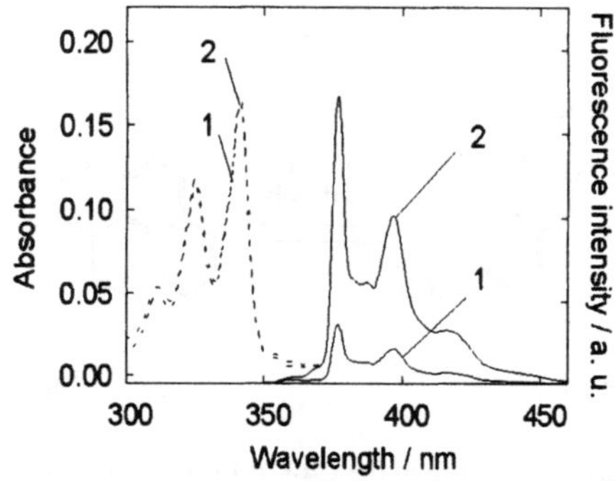

Figure 9.11. UV-Vis and fluorescence spectra of **C4-CPB** in 95% methanol-5% water (v/v): [**C4-CPB**] = 1.05 μM with (1) [BTA] = 0.0 mM and (2) [BTA] = 1.0 mM. BTA: Benzyltrimethylammonium hydroxide. Reproduced by permission of American Chemical

hydroxide is added, which converts **MCPB** to its tetrahedral boronate form, no fluorescence quenching is observed. Figure 9.12b shows the Stern-Volmer plot for this system in the absence (condition 1) and presence of the base (condition 2). From the slope analysis, the Stern-Volmer constant (K_{SV}) in condition 1 is determined to be 110 M^{-1}. The quenching rate constant is estimated to be around 10^9 $M^{-1}s^{-1}$ assuming that the fluorescence lifetime of **MP** is ca. 100 ns.[47] This rate constant is in the expected range for a diffusion-controlled quenching by **MCPB**. Thus, we conclude that the main mechanism for fluorescence quenching is electron transfer from the excited **MP** to the neutral form of **MCPB**.

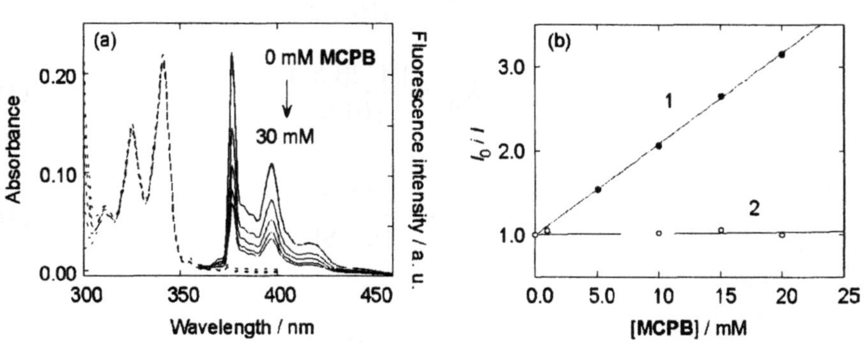

Figure 9.12. (a) UV-Vis and fluorescence spectra of **MP** upon addition of **MCPB** in 95% methanol-5% water (v/v): [**MP**] = 50.0 μM and [**MCPB**] = 0-20 mM. (b) Stern-Volmer plots: (1) [BTA] = 0.0 mM and (2) [BTA] = 40.0 mM. Reproduced by permission of American Chemical Society.

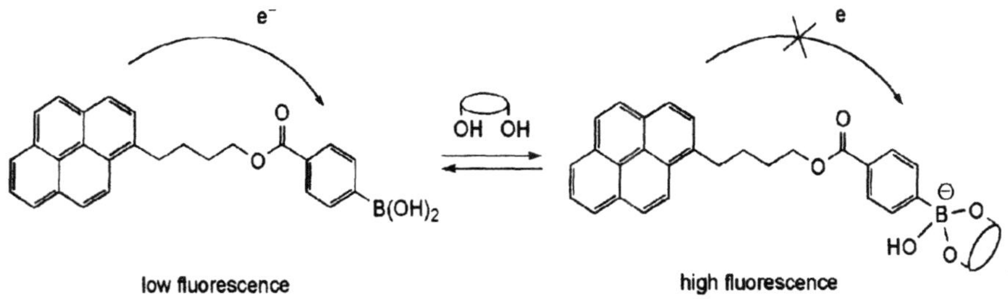

Figure 9.13. Response mechanism of the **C4-CPB**/β-CyD complex for saccharide binding.

These results strongly support the finding that the fluorescence of **C4-CPB** is quenched due to an intramolecular PET from the pyrene to the trigonal form of arylboronic acid. Saccharide binding converts the boronic acid to the tetrahedral boronate, which inhibits PET quenching and increases the monomer fluorescence intensity (Fig. 9.13). It is evidently a new mechanism for saccharide sensing in water.

*9.2.1.4. Binding Equilibrium and Recognition Selectivity of **C4-CPB/β-CyD** Complex for Saccharides*

Four equilibrium constants shown in Fig. 9.10 are expressed by the following equations:[48]

$$K_a = \frac{[H^+][L^-]}{[HL]} \quad (2)$$

$$K_a' = \frac{[H^+][LS^-]}{[HLS]} \quad (3)$$

$$K_{LS} = \frac{[LS^-]}{[L^-][S]} \quad (4)$$

$$K_{HLS} = \frac{[HLS]}{[HL][S]} \quad (5)$$

For an equilibrium analysis, three main equilibrium constants must be taken into consideration, since K_a' is equal to K_aK_{LS}/K_{HLS}. From eqs. (2)-(5), the fluorescence intensity of **C4-CPB**/β-CyD is expressed as a function of concentrations for proton and saccharide as follows,

$$I = \frac{\beta[\mathrm{L}]_0\left[\phi'_{\mathrm{HL}} + \phi'_{\mathrm{L}}\dfrac{K_a(1+K_{\mathrm{LS}}[\mathrm{S}])}{[\mathrm{H}^+](1+K_{\mathrm{HLS}}[\mathrm{S}])}\right]}{1+\dfrac{K_a(1+K_{\mathrm{LS}}[\mathrm{S}])}{[\mathrm{H}^+](1+K_{\mathrm{HLS}}[\mathrm{S}])}} \tag{6}$$

$$\phi'_{\mathrm{HL}} = \frac{\phi_{\mathrm{HL}} + \phi_{\mathrm{HLS}}K_{\mathrm{HLS}}[\mathrm{S}]}{1+K_{\mathrm{HLS}}[\mathrm{S}]} \tag{7}$$

$$\phi'_{\mathrm{L}} = \frac{\phi_{\mathrm{L}} + \phi_{\mathrm{LS}}K_{\mathrm{LS}}[\mathrm{S}]}{1+K_{\mathrm{LS}}[\mathrm{S}]} \tag{8}$$

where ϕ_{HLS} and ϕ_{LS} are the fluorescence quantum yields for HLS and LS^- species of the **C4-CPB**/β-CyD complex, respectively. By comparing eqs. (1) and (6), the apparent acid dissociation constant (K_a^{app}) is defined as:

$$K_{\mathrm{a}}^{\mathrm{app}} = \frac{1+K_{\mathrm{LS}}[\mathrm{S}]}{1+K_{\mathrm{HLS}}[\mathrm{S}]}K_{\mathrm{a}} \tag{9}$$

$$\Delta \mathrm{p}K_{\mathrm{a}} = -\log\frac{K_{\mathrm{a}}}{K_{\mathrm{a}}^{\mathrm{app}}} = \log\frac{1+K_{\mathrm{LS}}[\mathrm{S}]}{1+K_{\mathrm{HLS}}[\mathrm{S}]} \tag{10}$$

Equations (9) and (10) imply that no pK_a shift can be observed upon addition of saccharides if the K_{HLS} is comparable with the K_{LS}. However, it is known that only anionic boronate complexes (LS^-) are formed to a detectable extent, since the trigonal boronate esters (HLS) are very unstable at least in dilute aqueous solution.[18, 49] Thus, in Fig. 9.10, the formation of HLS complex can be neglected in water ($1 \gg K_{\mathrm{HLS}}[\mathrm{S}]$). In addition, the fluorescence intensity of the **C4-CPB**/β-CyD complex with 30 mM fructose reaches the same level of intensity at pH 10 without saccharide. Thus, ϕ_{L} is mostly equal to ϕ_{LS}, and eq. (6) is more simply expressed as:

$$I = \frac{\beta[\mathrm{L}]_0\left[\phi_{\mathrm{HL}} + \phi_{\mathrm{L}}\dfrac{K_{\mathrm{a}}(1+K_{\mathrm{LS}}[\mathrm{S}])}{[\mathrm{H}^+]}\right]}{1+\dfrac{K_{\mathrm{a}}(1+K_{\mathrm{LS}}[\mathrm{S}])}{[\mathrm{H}^+]}} \tag{11}$$

Therefore, the binding constant (K_{LS}) can be determined by nonlinear curve fitting analysis of fluorescence intensity (I) as a function of saccharide concentration ([S]) at the fixed pH condition. Alternatively, the K_{LS} can be evaluated by determining the pK_a shift (ΔpK_a) on the basis of eq. (12).

$$\Delta pK_a = \log(1 + K_{LS}[S]) \qquad (12)$$

It should be noted that the Benesi-Hildebrand plot[50] is only applicable at constant pH under the condition of $1 \ll K_{LS}[S]$. In such a case, the apparent experimental binding constant (K_{LS}^{app}) is simply expressed as eq. (13).

$$K_{LS}^{app} = K_{LS} \frac{K_a}{[H^+]} \qquad (13)$$

On the basis of the above equations, we examined the saccharide binding of the **C4-CPB**/β-CyD complex in water containing 2% DMSO. Figure 9.14 shows the fluorescence intensity changes of the **C4-CPB**/β-CyD complex upon addition of L-arabinose, D-galactose, D-glucose, and D-fructose at pH 7.5. The plots are fitted well with eq. (11) (solid lines), and the 1:1 binding constants calculated from the nonlinear regression analyses are summarized in Table 9.1. The binding constants are compared with the reported values of phenylboronic acid,[18] indolylboronic acid **3**[10] and the Shinkai's fluorescent probe bearing tertiary amine unit, **4**[16]. The selectivity of the **C4-CPB**/β-CyD complex for monosaccharide binding is essentially the same as that of the phenylboronic acid, and thus, the binding constants decrease in the following order: D-fructose >> L-arabinose > D-galactose > D-glucose. The binding constant of the **C4-CPB**/β-CyD complex for D-fructose calculated from the ΔpK_a [see eq. (12)] is 2600 M^{-1}, which corresponds well with the K_{LS} of 2500 M^{-1} in Table 9.1.

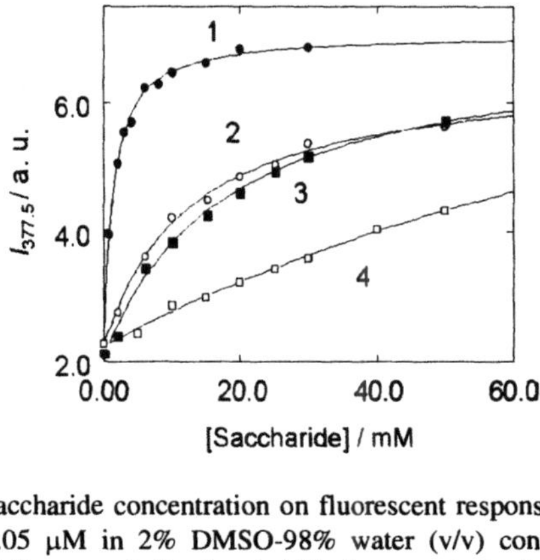

Figure 9.14. Effect of saccharide concentration on fluorescent response of the **C4-CPB**/β-CyD complex: [**C4-CPB**] = 1.05 μM in 2% DMSO-98% water (v/v) containing 5.0 mM β-CyD; excitation wavelength, 328.0 nm; excitation bandwidth, 5 nm; emission bandwidth, 3 nm; pH = 7.5 adjusted by 0.015 M phosphate buffer (I = 0.08 M with NaCl). (1) D-fructose, (2) L-arabinose, (3) D-galactose, and (4) D-glucose. Reproduced by permission of American Chemical Society.

Table 9.1. Binding constants of various fluorescent probes for monosaccharides

	Binding constants (K_{LS} / M^{-1})			
	D-fructose	L-arabinose	D-galactose	D-glucose
Phenylboronic acid	4370	391	276	110
3	6300	–	–	71
4 (in 33% MeOH)	1000	158	–	63
C4-CPB/β-CyD complex	2500 ± 130	270 ± 30	200 ± 30	80 ± 30

9.2.2. Effect of Phenylboronic Acid Probe Spacer

As shown above, the response mechanism of the **C4-CPB**/β-CyD complex is based on the PET pathway from pyrene to phenylboronic acid. This suggests that the spacer length between pyrene and phenylboronic acid strongly affects the response efficiency of the probe to saccharides. Thus, the response behavior of **C1-CPB** having a methylene spacer was examined as a β-CyD complex in water. As an additional spacer variation, **C1-APB** possessing an amide spacer was also investigated, since the amide bond is more stable than the ester bond.

The UV-Vis and fluorescence spectra of **C1-CPB** and **C1-APB** in various aqueous DMSO solutions are shown in Figs. 9.15 and 9.16, respectively. The absorption spectrum of **C1-CPB** shown in Fig. 9.15 a (spectrum 1) reveals that **C1-CPB** aggregates in 2%

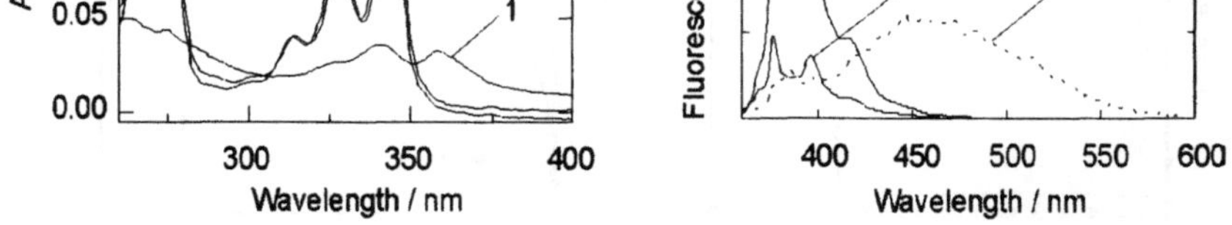

Figure 9.15. Spectral properties of **C1-CPB** in aqueous DMSO solution. (a) UV-Vis spectra: optical path length, 5 cm. (b) Fluorescence spectra: excitation wavelength, 328.0 nm; excitation bandwidth, 5 nm; emission bandwidth, 3 nm. [**C1-CPB**] = 1.05 μM in (1) 2% DMSO-98% water (v/v), (2) 25% DMSO-75% water (v/v), and (3) 2% DMSO-98% water (v/v) containing 5.0 mM β-CyD. pH = 7.5 adjusted by 0.015 M phosphate buffer (I = 0.08 M with NaCl).

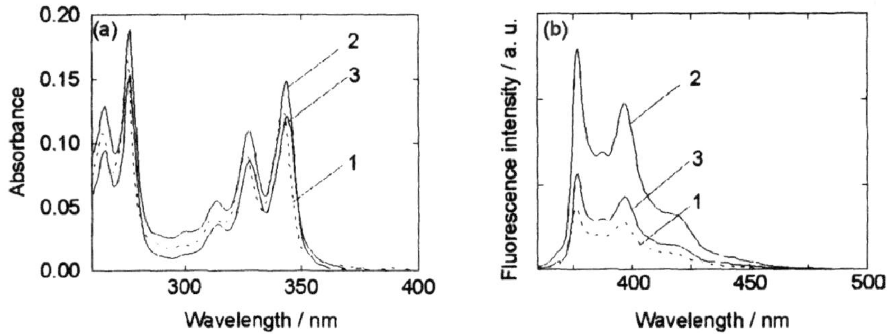

Figure 9.16. Spectral properties of **C1-APB** in aqueous DMSO solution. (a) UV-Vis spectra: optical path length, 5 cm. (b) Fluorescence spectra: excitation wavelength, 328.0 nm; excitation bandwidth, 5 nm; emission bandwidth, 3 nm. [**C1-APB**] = 1.05 μM in (1) 2% DMSO-98% water (v/v), (2) 25% DMSO-75% water (v/v), and (3) 2% DMSO-98% water (v/v) containing 5.0 mM β-CyD. pH = 7.5 adjusted by 0.015 M phosphate buffer (I = 0.08 M with NaCl).

DMSO solution, but it is solubilized in water by forming an inclusion complex with β-CyD (spectrum 3). The aggregated **C1-CPB** exhibits weak and broad emission, as shown in Fig. 9.15b (spectrum 1). The pyrene monomer fluorescence becomes significant in the presence of β-CyD (spectrum 3). The fluorescence intensity in the β-CyD solution is higher than that in 25% DMSO solution (spectrum 2), indicating that **C1-CPB** is incorporated inside the β-CyD cavity. In contrast, the UV-Vis spectrum of **C1-APB** in 2% DMSO is similar to that in 25% DMSO solution, indicating that **C1-APB** is soluble even in 2% DMSO (Fig. 9.16a). The pyrene fluorescence in 2% DMSO also supports the

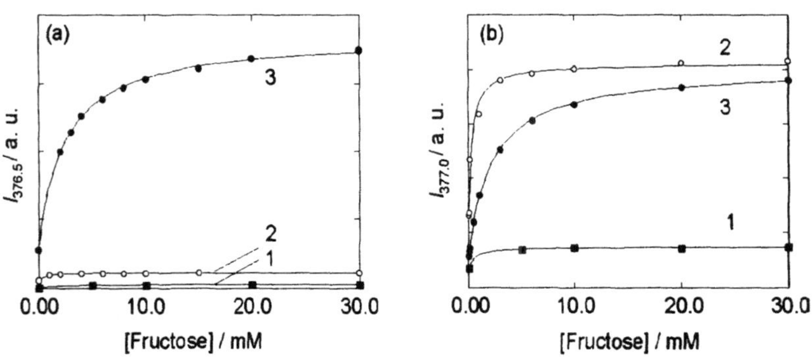

Figure 9.17. Effect of D-fructose concentration on fluorescent response of the (a) **C1-CPB**/β-CyD and (b) **C1-APB**/β-CyD complexes: [**C1-CPB**] or [**C1-APB**] = 1.05 μM in (1) 2% DMSO-98% water (v/v), (2) 25% DMSO-75% water (v/v), and (3) 2% DMSO-98% water (v/v) containing 5.0 mM β-CyD; excitation wavelength, 328.0 nm; excitation bandwidth, 5 nm; emission bandwidth, 3 nm; pH = 7.5 adjusted by 0.015 M phosphate buffer (I = 0.08 M with NaCl).

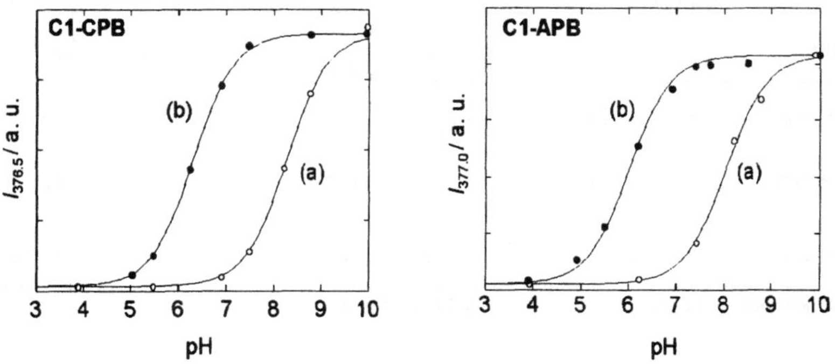

Figure 9.18. Dependence of fluorescence intensity as a function of pH: [**C1-CPB**] or [**C1-APB**] = 1.05 μM in 2% DMSO-98% water (v/v) containing 5.0 mM β-CyD; (a) [fructose] = 0.0 mM, (b) [fructose] = 30.0 mM.

high water solubility of **C1-APB** (Fig. 9.16b). The addition of 5 mM β-CyD caused an increase in fluorescence intensity of **C1-APB** in 2% DMSO (spectrum 3). However, the intensity is smaller than that in 25% DMSO (spectrum 2). This is probably because **C1-APB** has relatively high water solubility and its binding ability with β-CyD is low to be included inside the β-CyD cavity.

The effects of fructose concentration on the fluorescence spectra of **C1-CPB** and **C1-APB** are evaluated at pH 7.5 in 2% DMSO solution, 25% DMSO solution, and 2% DMSO solution containing 5.0 mM β-CyD. Fluorescence intensities both of the **C1-CPB**/β-CyD and **C1-APB**/β-CyD complexes increase as the fructose concentration is increased. The dependence of fluorescence intensity on fructose concentration, as shown in Fig. 9.17, indicates that the **C1-CPB**/β-CyD and the **C1-APB**/β-CyD complexes efficiently recognize fructose in water. Figure 9.18 shows the fluorescence intensity dependence of the **C1-CPB**/β-CyD and **C1-APB**/β-CyD complexes on pH in the absence and presence of 30.0 mM fructose. From curve fitting analysis with eq (1), the pK_as of **C1-CPB** and **C1-APB** are calculated to be 8.26 ± 0.02 and 8.04 ± 0.03, respectively. The apparent pK_a decreases to 6.30 ± 0.01 and 6.03 ± 0.02, respectively, by the addition of 30 mM fructose. It is interesting to note that the ratio of quantum yield between the neutral and basic forms of **C1-CPB** (ϕ_L/ϕ_{HL}) reaches 57.4. This fluorescence enhancement with acid dissociation is much higher than that of **C4-CPB** (ϕ_L/ϕ_{HL} = 15.2). This demonstrates that a more efficient PET process between pyrene and phenylboronic acid is taking place in the case of **C1-CPB** due to its shorter spacer length. On the other hand, the fluorescence recovery of the **C1-APB**/β-CyD complex (ϕ_L/ϕ_{HL}) is 34.5. This may be due to the difference in the conformation of the boronic acid probe inside the β-CyD

Table 9.2. Comparison of characteristics of boronic acid probe/β-CyD complex

probe	pK_a	pK_a (with 30 mM fructose)	ϕ_L/ϕ_{HL}	K_{LS} / M^{-1}*
C1-CPB	8.26 ± 0.02	6.30 ± 0.01	57.4	2800 ± 70
C1-APB	8.04 ± 0.03	6.03 ± 0.02	34.5	3900 ± 120
C4-CPB	7.95 ± 0.03	6.06 ± 0.03	15.2	2500 ± 130

*K_{LS} is the binding constant of the probe/β-CyD complex for D-fructose.

cavity, since the calculated reduction potential of carboxyphenylboronic acid is mostly equal to that of the amidophenylboronic acid. It is plausible that the amide group of **C1-APB** interacts with the hydroxyl groups of the β-CyD rim *via* hydrogen bonding to stabilize the inclusion complex, which may affect the efficiency of electron transfer. The pK_a, ϕ_L/ϕ_{HL}, and the 1:1 binding constants for fructose as calculated by nonlinear regression analyses of three boronic acid probe/β -CyD complexes are summarized in Table 9.2. The **C1-APB**/β-CyD complex has the highest binding constant among the three probes, and there is no correlation between the apparent pK_a and the binding constant for fructose. This result indicates that the affinity for saccharides of the boronic acid probe/β-CyD complex depends not only on the nature of the probe itself, but also on other factors such as the conformation of the inclusion complex.

In conclusion, the **C1-APB**/β-CyD complex is a desirable fluorescent probe for saccharide recognition in water because of its high affinity to saccharides as well as the relatively high fluorescent recovery upon saccharide binding. Since **C1-APB** is highly water soluble, it is possible to dissolve the probe in water without the addition of DMSO. This is an additional advantage of **C1-APB** as the β-CyD complex sensor for use in water.

9.3. FUTURE PERSPECTIVE OF SUPRAMOLECULAR CYCLODEXTRIN COMPLEX SENSORS

It should be noted that the Shinkai group has been providing efficient glucose sensors, as well as probes for chiral saccharide recognition, by incorporating two boronic acid units as a saccharide recognition site with a tertiary amine PET system.[11-16, 19, 51] In comparison, the advantages of the self-assembled boronic acid probe/β-CyD complex sensors shown in section 2 are (1) the high probe solubility in water and (2) the high fluorescent quantum yield due to inclusion complex formation. In addition, the combination of boronic acid probes with the CyD derivatives bearing various functional

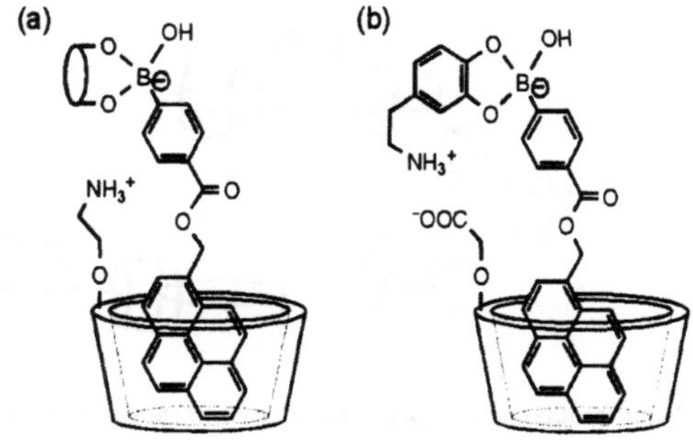

Figure 9.19. Molecular recognition by combination of boronic acid probe with modified CyD. (a) Anion-exchange CyD. (b) Cation-exchange CyD.

groups[52] may provide multipoint saccharide recognition. For example, the anion exchange CyD bearing the ammonium group should stabilize the anionic boronate complex with saccharide, as shown in Fig. 9.19a. On the other hand, the cation exchange CyD possessing a carboxyl group should provide smultipoint recognition for dopamine in combination with a boronic acid probe (Fig. 9.19b).

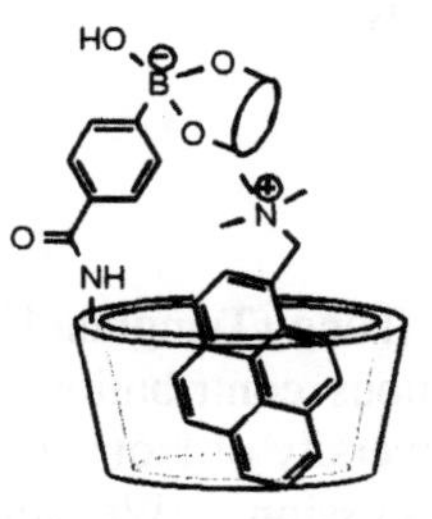

Figure 9.20. Saccharide recognition by boronic acid modified CyD.

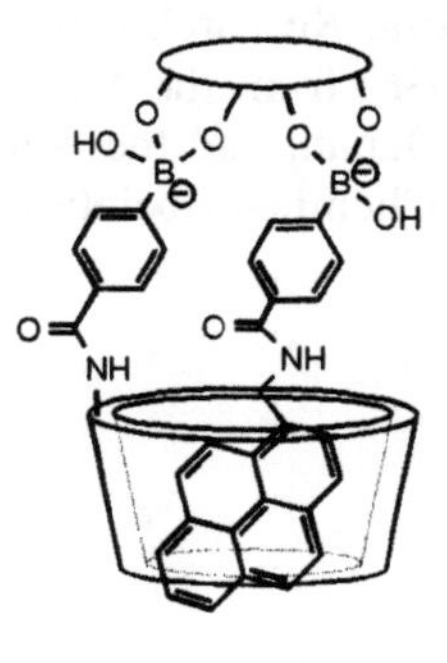

Figure 9.21. Glucose recognition by cooperative binding.

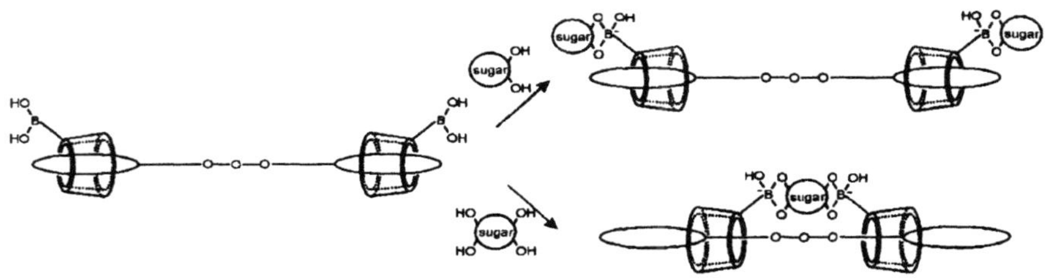

Figure 9.22. Psuedorotaxane composed of phenylboronic acid modified CyDs.

Phenylboronic acid can also be attached to CyD. In this case, binding with saccharides may be analyzed by measuring the spectral change of the included fluorescent molecule (Fig. 9.20). Since CyD has hydroxyl groups that can bind with saccharides *via* hydrogen bonding on its rims, it is expected to display a novel selectivity for saccharides in water. When a phenylboronic acid probe is incorporated inside the cavity of modified CyD, we may design the glucose sensing system by utilizing the cooperative binding of two phenylboronic acid groups (Fig. 9.21). A pseudorotaxane composed of phenylboronic acid-modified CyDs is another approach to glucose sensor design, as shown in Fig. 9.22. The evolutional development of such probe/CyD complex sensors is being actively pursued by our research group as a future prospect of the present study.

9.4. ACKNOWLEDGEMENT

We are grateful to Prof. A.-J. Tong (Tsinghua University) and Mrs. Ayako Takahashi (Kato) for their enormous contribution to the present study. We also thank Prof. B. D. Smith (University of Notre Dame) for providing boronic acid probes and for fruitful discussion. We acknowledge Prof. Teramae (Tohoku University) for his guidance. This work was supported by Toray Science and Technology Grant, the Japan Society for the Promotion of Science (JSPS), a Grant-in-Aid for Scientific Research (16550066) and a Grant-in-Aid for Young Scientists (B) (16790036) from the Ministry of Education, Culture, Sports, Science and Technology, Japan, and the National Science Foundation, USA. We also thank American Chemical Society for permission to reproduce copyrighted materials.

9.5. REFERENCES

1. A. P. Davis and R. S. Wareham, *Angew. Chem. Int. Ed.* **38**, 2978-2996 (1999).
2. A. Lützen, in: *Encyclopedia of Supramolecular Chemistry*, edited by J. L. Atwood and J. W. Steed (Marcell Dekker , New York, 2004), pp. 169-177.
3. G. Lecollinet, A. P. Dominey, T. Velasco, and A. P. Davis, *Angew. Chem. Int. Ed.* **41**, 4093-4096 (2002).
4. O. Rusin, K. Lang, and V. Král, *Chem. Eur. J.* **8**, 655-663 (2002).
5. J.-M. Fang, S. Selvi, J.-H. Liao, Z. Slanina, C.-T. Chen, and P.-T. Chou, *J. Am. Chem. Soc.* **126**, 3559-3566 (2004).
6. J.-L. Hou, X.-B. Shao, G.-J. Chen, Y.-X. Zhou, X.-K. Jiang, and Z.-T. Li, *J. Am. Chem. Soc.* **126**, 12386-12394 (2004).
7. E. Klein, M. P. Crump, and A. P. Davis, *Angew. Chem. Int. Ed.* **44**, 298-302 (2005).
8. J. Yoon and A. W. Czarnik, *J. Am. Chem. Soc.* **114**, 5874-5875 (1992).
9. A. W. Czarnik, in: *Fluorescent Chemosensors for Ion and Molecule Recognition*, edited by A. W. Czarnik (ACS Symposium Series 538; ACS Books: Washington, DC, 1992), pp. 104-129.
10. Y. Nagai, K. Kobayashi, H. Toi, and Y. Aoyama, *Bull. Chem. Soc. Jpn.* **66**, 2965-2971 (1993).
11. S. Shinkai, in *Chemosensors of Ion and Molecule Recognition* (NATO ASI Series C492, Kluwer, London, 1996), pp. 37-59.
12. T. D. James, K. R. A. S. Sandanayake, and S. Shinkai, *Supramol. Chem.* **6**, 141-157 (1995).
13. T. D. James, P. Linnane, and S. Shinkai, *Chem. Commun.*, 281-288 (1996).
14. T. D. James, K. R. A. S. Sandanayake, and S. Shinkai, *Angew. Chem. Int. Ed. Engl.* **35**, 1910-1922 (1996).
15. K. R. A. S. Sandanayake, T. D. James, and S. Shinkai, *Pure. Appl. Chem.* **68**, 1207-1212 (1996).
16. T. D. James, K. R. A. S. Sandanayake, and S. Shinkai, *J. Chem. Soc., Chem. Commun.*, 477-478 (1994).
17. W. Ni, G. Kaur, G. Springsteen, B. Wang, and S. Franzen, *Bioorg. Chem.* **32**, 571-581 (2004).
18. J. P. Lorand and J. O. Edwards, *J. Org. Chem.* **24**, 769-774 (1959).
19. T. D. James, K. R. A. S. Sandanayake, R. Iguchi, and S. Shinkai, *J. Am. Chem. Soc.* **117**, 8982-8987 (1995).
20. H. Fang, G. Kaur, and B. Wang, *J. Fluorescence* **14**, 481-489 (2004).
21. T. D. James, H. Shinmori, and S. Shinkai, *Chem. Commun.*, 71-72 (1997).
22. S. Arimori, G. A. Consiglio, M. D. Phillips, and T. D. James, *Tetrahedron Lett.* **44**, 4789-4792 (2003).
23. S. Arimori, M. L. Bell, C. S. Oh, K. A. Frimat, and T. D. James, *J. Chem. Soc., Perkin Trans. 1*, 803-808 (2002).
24. V. V. Karnati, X. Gao, S. Gao, W. Yang, W. Ni, S. Sankar, and B. Wang, *Bioorg. Med. Chem. Lett.* **12**, 3373-3377 (2002).
25. S. Arimori, M. L. Bell, C. S. Oh, and T. D. James, *Org. Lett.* **4**, 4249-4251 (2002).
26. M. D. Phillips and T. D. James, *J. Fluorescence* **14**, 549-559 (2004).
27. R. Badugu, J. R. Lakowicz, and C. D. Geddes, *Bioorg. Med. Chem.* **13**, 113-119 (2005).
28. N. DiCesare and J. R. Lakowicz, *J. Phys. Chem. A* **105**, 6834-6840 (2001).
29. N. DiCesare and J. R. Lakowicz, *Chem. Commun.* 2022-2023 (2001).
30. N. DiCesare and J. R. Lakowicz, *Tetrahedron Lett.* **43**, 2615-1618 (2002).
31. S. Arimori, L. I. Bosch, C. J. Ward, and T. D. James, *Tetrahedron Lett.* **42**, 4553-4555 (2001).
32. X. Gao, Y. Zhang, and B. Wang, *Org. Lett.* **5**, 4615-4618 (2003).
33. H. Eggert, J. Frederiksen, C. Morin, and J. C. Norrild, *J. Org. Chem.* **64**, 3846-3852 (1999).
34. H. Cao, D. I. Diaz, N. DiCesare, J. R. Lakowicz, and M. D. Heagy, *Org. Lett.* **4**, 1503-1505 (2002).
35. H. Cao, T. McGrill, and M. D. Heagy, *J. Org. Chem.* **69**, 2959-2966 (2004).
36. W. Yang, H. He, and D. G. Druekhammer, *Angew. Chem. Int. Ed.* **40**, 1714-1718 (2001).
37. G. Springsteen and B. Wang, *Chem. Commun.*, 1608-1609 (2001).
38. J. T. Suri, D. B. Cordes, F. E. Cappuccio, R. A. Wessling, and B. Singaram, *Langmuir* **19**, 5145-5152 (2003).

39. Y. Kanekiyo and H. Tao, *Chem. Lett.* **34**, 196-197 (2005).
40. A. Yamauchi, T. Hayashita, S. Nishizawa, M. Watanabe, and N. Teramae, *J. Am. Chem. Soc.* **121**, 2319-2320 (1999).
41. A. Yamauchi, T. Hayashita, A. Kato, S. Nishizawa, M. Watanabe, and N. Teramae, *Anal. Chem* **72**, 5841-5846 (2000).
42. A. Yamauchi, T. Hayashita, A. Kato, and N. Teramae, *Bull. Chem Soc. Jpn.* **75**, 1527-1532 (2002).
43. A.-J. Tong, A. Yamauchi, T. Hayashita, Z.-Y. Zhang, B. D. Smith, and N. Teramae, *Anal. Chem.* **73**, 1530-1536 (2001).
44. T. Hayashita, A. Yamauchi, A.-J. Tong, J. C. Lee, B. D. Smith, and N. Teramae, *J. Inclusion Phenom. Mol. Recognit. Chem.* **50**, 87-94 (2004).
45. G. J. Kavarnos, *Fundamentals of Photoinduced Electron Transfer* (Wiley-VCH, New York, 1993).
46. H. Murakami, T. Nagasaki, I. Hamachi, and S. Shinkai, *J. Chem. Soc., Perkin Trans. 2*, 975 (1994).
47. M. Snare, P. J. Thistlethwaite, K. P. Ghiggino, *J. Am. Chem. Soc.* **105**, 3328-3332 (1983).
48. T. Hayashita, N. Teramae, T. Kuboyama, S. Nakamura, H. Yamamoto, and H. Nakamura, *J. Inclusion Phenom. Mol. Recognit. Chem.* **32**, 251-265 (1998).
49. B. D. Smith, S. J. Gardiner, T. A. Munro, M. F. Paugam, and J. A. Riggs, *J. Inclusion Phenom. Mol. Recognit. Chem.* **32**, 121-131 (1998).
50. K. A. Conners, *Binding Constants* (Wiley, New York, 1987), pp. 147-187.
51. T. D. James, K. R. A. S. Sandanayake, and S. Shinkai, *Nature* **374**, 345-347 (1995).
52. I. Suzuki, K. Obata, J. Anzai, H. Ikeda, and A. Ueno, *J. Chem. Soc., Perkin Trans. 2*, 1705-1710 (2000).

PLASMONIC GLUCOSE SENSING

Kadir Aslan[1], Joseph R. Lakowicz[2], and Chris D. Geddes*[1,2]

10.1. INTRODUCTION

Diabetes results in long-term health disorders including cardiovascular disease and blindness. One of the major challenges in the management of diabetes is the monitoring of glucose concentrations. Yet after several decades of intense research,[1-5] still no method is available for the continuous non-invasive monitoring of blood glucose, never mind a generic technology which could be applied across-the-board for glucose sensing in other physiological fluids. In fact the invasive nature of glucose monitoring in blood, primarily undertaken by "finger pricking", has further fueled the search for non-invasive technologies, [6-12] which can potentially monitor physiological glucose in fluids such as urine and tears based on the boronic acid / glucose [6-11] and glucose-binding protein / glucose[12] interactions.

Noble metal nanostructures, i.e., silver or gold, etc., have been studied extensively and are emerging as important colorimetric reporters due to their high extinction coefficients, which are typically several orders of magnitude larger than those of organic dyes[13-22]. Gold and silver nanoparticles display strong plasmon absorption bands depending on their size and shape, [16-17] and they are also shown to quench[21] and/or enhance fluorescence emission of fluorophore[22] within close proximity. Aggregation of these nanoparticles results in further color changes of their solutions due to mutually induced dipoles that depend on interparticle distance and aggregate size.[18-22] The fact that the plasmon resonance is a sensitive function of nanostructure geometry, coupled with synthetic advances that allow for the controlled and systematic variations in nanostructure geometry, is leading to the expanse of this sensing field called "plasmonics". Gold nanoparticle aggregation induced by analytes has been demonstrated for DNA, [14-16] proteins [17-18, 20] and antibodies.[21] In this chapter, the use of gold nanoparticles in glucose sensing is summarized.

[1]-Institute of Fluorescence, Laboratory for Advanced Medical Plasmonics (LAMP), University of Maryland Biotechnology Institute, [2]-Center for Fluorescence Spectroscopy, University of Maryland at Baltimore, 725 W. Lombard St., Baltimore, MD 21201 USA, * Corresponding author, geddes@cfs.umbi.umd.edu

10.2. OPTICAL PROPERTIES OF GOLD NANOPARTICLES

It is essential to present the optical properties of gold nanoparticles before summarizing their use in glucose sensing. As briefly mentioned in introduction, solutions of noble metal nanoparticles display intense colors due to the absorption and scattering of light. These properties are due to collective electron oscillations in the metallic particles, induced by the incident light field, giving rise to the so-called plasmon resonance (SPR) absorption.[23-24] The surface plasmon absorption depends on several factors, size, shape and composition. For sake of brevity, only the optical properties of spherical metal nanoparticles are presented here. Detailed information on the preparation and optical properties of anisotropic nanoparticles could be found in reference.[25]

Figure 10.1 shows the absorption spectra of gold nanoparticles with various diameters. Typically, the SPR peak for the 20 nm gold nanoparticles occurs around 520 nm and is red-shifted for the larger sizes of gold nanoparticles.

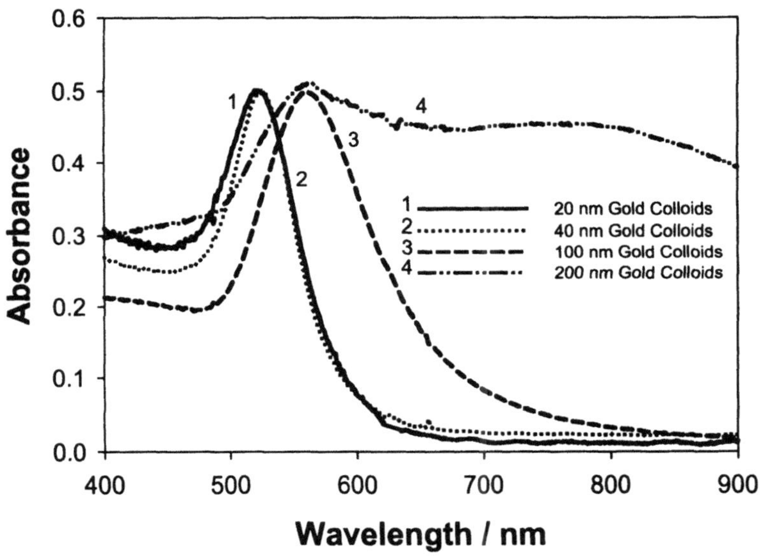

Figure 10.1. Absorption spectrum of gold nanoparticles in deionized water with various sizes.

Most applications of metallic nanoparticles to date have been concerned with measurements of the plasmon absorption, as seen by direct absorption or even by visual color, and the subsequent changes induced by either aggregation or flocculation of the nanoparticles in solution. Both experimental and theoretical studies have shown that the resonant wavelength of two close proximity and coupled nanoparticles is significantly red-shifted from that of the monomers, where the shift decays approximately exponentially with increasing particle spacing, decreasing to almost zero when the distance between the nanoparticles exceeds ≈ 2.5 times the particle size.

However one property of metallic nanoparticles which has received very little attention is their ability to efficiently scatter light.[23, 24] It has been reported that light scattered from individual colloids can be equivalent to the intensity of 10^5 fluorescein molecules,[23, 24] which has the potential to offer new approaches to bioaffinity based colloid sensing. While the scattered light from colloids does not have the information content of fluorescence, and therefore at first glance is unlikely to provide an opportunity for measurements which are not sensitive to total intensity such as anisotropy or wavelength–ratiometric measurements, it was found that changes in the plasmon absorption can be monitored by changes in the intensity of scattered light and its dependence on wavelength.[26]

Since the particle size, the degree of aggregation, and the dielectric constant of the media determine optical properties of metal nanoparticles, the applications of these particles exploiting the strong dependence of optical properties with aggregation and surrounding dielectric media have been implemented through proper surface modification of the nanoparticles or by embedding them in different dielectric media. This is also the essence of plasmonic glucose sensing.

The following section summarizes the methods used for the preparation of surface-functionalized gold that have been used for plasmonic glucose sensing.

10.3. PREPARATION OF LIGAND FUNCTIONALIZED GOLD NANOPARTICLES FOR GLUCOSE SENSING AND OTHER APPLICATIONS

Metallic nanoparticles that can be used for plasmonic glucose sensing are usually prepared by introducing a functional ligand (i.e., a monosaccharide derivative: glucose, mannose, lactose, or polysaccharide: dextran) to the surface of the nanoparticle. These functional groups interact reversibly with glucose-binding proteins, such as Concanavilin A (Con A), and results in significant changes in the optical properties of the nanoparticles that are directly related to the glucose concentration.

There are two generic strategies for the preparation of ligand functionalized metal nanoparticles: i) reduction of gold nanoparticles from their respective salts in the presence of the desired ligand and ii) attachment of the desired ligand to the gold nanoparticles after the synthesis of the nanoparticles. The examples for these techniques are as follows:

Otsuka *et al.*[27] described the preparation of gold nanoparticles (1-10 nm size range) with a narrow size distribution by in situ reduction of $HAuCl_4$ in the presence of heterobifunctional poly(ethylene glycol) (PEG) derivatives containing both mercapto and acetal groups (α-acetal-ω-mercapto-PEG). The α-acetal-PEG layers formed on gold nanoparticles provide for significant stability to the nanoparticles in aqueous solutions with elevated ionic strength and also in serum-containing medium. The PEG acetal terminal group was converted to aldehyde by gentle acid treatment, followed by the reaction with *p*-aminophenyl-β-D- lactopyranoside (Lac) in the presence of $(CH_3)_2NHBH_3$.

Lac-conjugated gold nanoparticles exhibited selective aggregation when exposed to *Recinus communis* agglutinin (RCA_{120}), a bivalent lectin specifically recognizing the β-D-galactose residue, inducing significant changes in the absorption spectrum with concomitant visible color change from pinkish-red to purple. Aggregation of the Lac-functionalized gold nanoparticles by the RCA_{120} lectin was reversible, recovering the original dispersed phase and color by addition of excess galactose. Further, the degree of aggregation was proportional to lectin concentration, allowing the system to be utilized to quantitate lectin concentration with nearly the same sensitivity as ELISA. This simple, yet highly effective, derivatization of gold nanoparticles with heterobifunctional PEG provides a convenient method to construct various colloidal sensor systems currently applied in bioassays and biorecognition.

In a similar fashion, mannose-modified gold nanoparticles were prepared by Lin *et al.*[28] In this regard, reduction of $HAuCl_4$ in the presence of a thiomannosyl dimer (synthesized by the authors) and $NaBH_4$ resulted in mannose-encapsulated gold nanoparticles (m-AuNP) with 6 ± 1 nm diameter. The authors[28] have studied the ability of the m-AuNP to bind mannose-specific adhesion FimH of type 1 pili *Escherichia coli,* and found that m-AuNP specifically bound to pili of the ORN178 strain of *E. coli* at room temperature.

In a recent paper, Noltin *et al.*[29] reported the synthesis, characterization, and biological evaluation of gold glyconanoparticles containing 100% galactosyl- and 100% glucosyl-β-*C*-glycosides linked to thiolated ethylene glycol spacers. The synthesis of gold glyconanoparticles was performed by the reduction of $HAuCl_4$ in the presence of above-mentioned linkers and $NaBH_4$ and the diameter of the resulting nanoparticles ranged between 1.6 and 2.1 nm. A biotin-NeutrAvidin adhesion assay was used to evaluate the relative ability of carbohydrate disulfides (galactosyl- and glucosyl-β-*C*-glycosides) in free form and immobilized on to gold nanoparticles to displace rgp120 (a HIV-associated glycoprotein) from plate-bound cellular receptors GalCer. The authors[29] data indicated that divalent disulfides were less than 12% as active as biotinylated GalCer, a water-soluble surrogate of GalCer. However, when these same carbohydrates were presented in a polyvalent display on gold, they were greater than 300 times more active than the disulfides and at least 20 times more active than biotinylated GalCer.

In 2001, de la Fuente *et al.*[30] described the synthesis of water-soluble gold nanoparticles functionalized with a monolayer of alkane thiol derivatized neoglycoconjugates of two oligosaccharides: lactose trisaccharide and trisaccharide Le^x antigen. The synthesis of gold glyconanoparticles was performed by the reduction of $HAuCl_4$ in the presence of above-mentioned linkers and $NaBH_4$ and the average diameter of the resulting nanoparticles were 2 nm. The authors showed how these tailored globular carbohydrate models can be used mimic glycosphingolipid clusters in plasma membrane, to investigate a novel mechanism of cell adhesion through carbohydrate-carbohydrate interactions in solution.[30] In a follow-up paper, [31] the same authors showed how these interactions can be used to guide the assembly of gold nanoparticles.

Only a few examples for the attachment of saccharide derivatives to the gold nanoparticles after the synthesis of the nanoparticles for plasmonic glucose sensing can be found in the literature. The lack of progress in this method is due to the difficulties involved in the surface modification process, such as the loss of stability of gold nanoparticles during the ligand attachment. The following paragraphs summarize the few successful attempts.

In 2003, Hone *et al.*[32] described the self-assembly of a mannose derivative onto preformed, citrate capped, water soluble gold nanoparticles. Through the use of a short (C2) hydrocarbon tether between the gold surface and the mannose recognition center, a selective, quantitative, and, importantly, rapid colorimetric detection method has been developed for the carbohydrate binding protein concanavalin A.

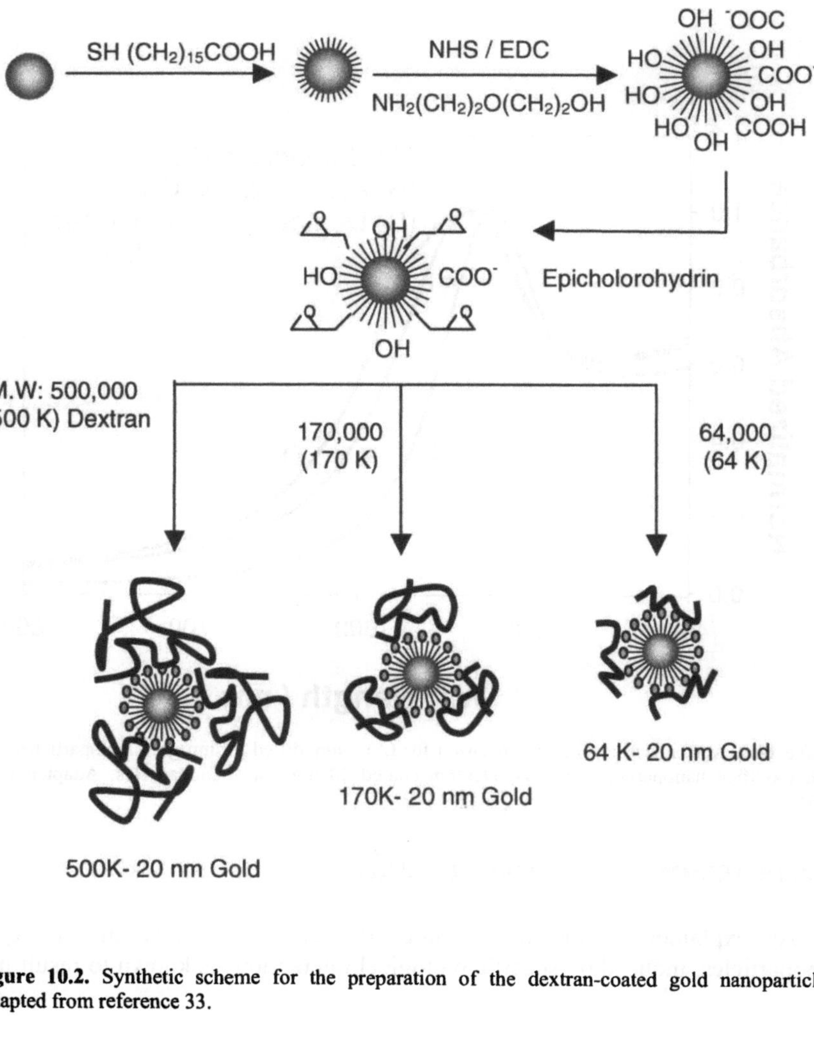

Figure 10.2. Synthetic scheme for the preparation of the dextran-coated gold nanoparticles. Adapted from reference 33.

In 2004, Aslan *et al.* described the immobilization of a polysaccharide, dextran, onto gold nanoparticles that were subsequently used for glucose sensing.[33, 34] The preparation scheme is presented in Figure 10.2. In summary, firstly, the surface of the gold nanoparticles is modified with a long-chain carboxyl-terminated alkane thiol (16-mercaptohexadecanoic acid).[19] The carboxylic acid groups are activated by N-3-(Dimethylaminopropyl)-N'-ethyl-carbodiimide (EDC) and N-Hydroxy-2,5-pyrrolidinedione (NHS) to form active NHS-esters, and the 2-(2-aminoethoxy)ethanol reacted with NHS-esters resulting in an addition of a hydroxyl-terminated second layer to the surface of the gold nanoparticles. The hydroxyl groups are activated using epicholorohydrin to which dextran (with 3 different molecular weight) is coupled covalently.[33] The excess dextran was removed by centrifugation. Figure 10.3 shows both a shift and broadening of the gold plasmon band at ≈ 520 nm as the colloids are homogeneously coated with dextran.

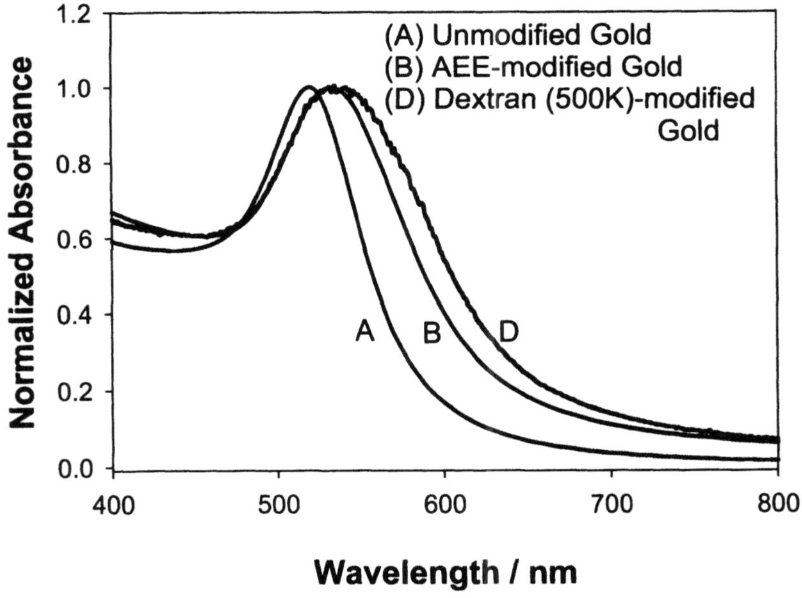

Figure 10.3. Normalized absorption spectrum for (A) Unmodified 20 nm gold nanoparticles, (B) AEE modified nanoparticles and (D) Dextran coated 20 nm gold nanoparticles. Adapted from reference 33.

10.4. PLASMONIC GLUCOSE SENSING

As explained in the Introduction, the close proximity of two gold nanoparticles, induced by specific biological interactions is known to result in a red-shifted resonance wavelength peak and increased scattered light intensity, due to near-field coupling. The only examples of plasmonic glucose sensing using these phenomena found in literature are by Aslan *et al.*[33-35] Their sensing strategy is based on the aggregation and disassociation of 20 nm gold particles, which result in the changes in plasmon absorption and light scattering of gold nanoparticles, induced by the presence of Con A and glucose, respectively.

Figure 10.4 shows the glucose sensing scheme reported by Aslan *et al.*[33-35] In this regard, high molecular weight dextran-coated nanoparticles are aggregated with Concanavalin A (Con. A), where the aggregation results in a significant shift and broadening of the gold plasmon absorption and in increased scattered light intensity. The addition of glucose competitively binds to Con A., reducing gold nanoparticle aggregation, and therefore the plasmon absorption/ scattered light when monitored at a near-red arbitrary wavelength.

The plasmonic type glucose nanosensors were optimized with regard to particle stability, pH effects, the dynamic range for glucose sensing and the observation wavelength, to be compatible with both clinical glucose requirements and measurements. In addition, by modifying the amount of dextran or Con A used in nanoparticle fabrication, the glucose response range was tuned to some extent, which means that a single sensing platform could potentially be used to monitor μ → mM glucose levels in many physiological fluids, such as in tears, blood and urine, where the glucose concentrations are significantly different. These results are summarized in this section.

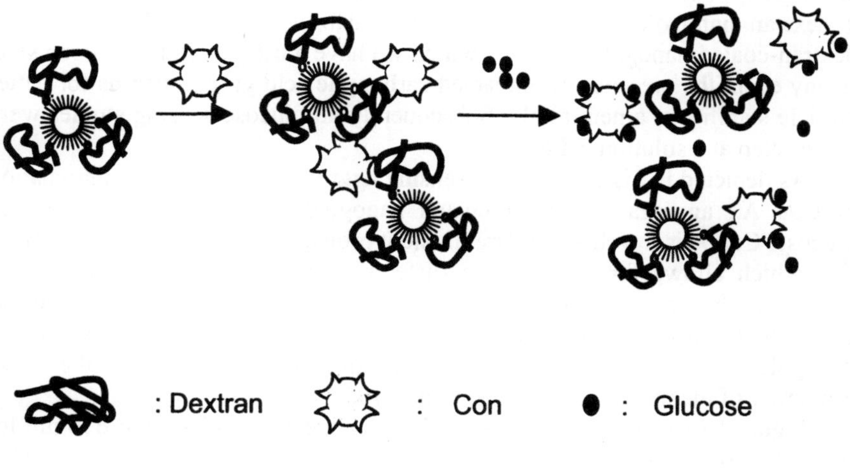

Figure 10.4. Glucose sensing scheme based on the dissociation of dextran-coated gold nanoparticles / Con A aggregates.

10.4.1. Fabrication of Glucose-Sensing Aggregates

For the building of plasmonic nanogold glucose sensors it is important to consider the pH effects on the stability of the sensing aggregates. Figure 10.5 shows the normalized absorption spectra of the dextran-coated nanoparticles as a function of pH. From Figure 10.5, it is seen that the dextran-coated nanogold plasmon absorption band is typically red-shifted and broader at lower pH values as compared to that in physiological pH for all three model systems. As has been previously reported,[26,36] the aggregation of nanogold results in a broadening of the absorption spectrum at wavelengths longer than 600 nm and a shift in the SPR peak. It was found that the aggregation of the dextran-coated nanoparticles in solution depends on several parameters, such as pH and ionic strength, which is thought due to the balance of forces, attractive van der Waals forces and repulsive electrostatic forces, between the nanoparticles. The flocculation parameter, [17, 19, 20] is the integrated absorption between 600-800 nm and provides further evidence for the extent of aggregation.

Figure 10.6 shows the flocculation, where we can see a notable difference in nanoparticle stability at ≈ neutral pH. For physiological glucose sensing, this is ideal, especially given that physiologies do not experience and notable changes in pH. The reduced flocculation parameter at pH 7 (improved particle stability) is thought to be a function of residual carboxylic acid groups that may be still present on the nanogold surface after reaction with AEE, Figure 10.2, step 2, and we have thus depicted this by the presence of carboxylic acid groups throughout Figure 10.2. We have guesstimated the conversion of the carboxylic acids groups shown in Figure 10.2, step 2, to be ≈ 80 %.[37] Interestingly, the 500 k dextran nanogold shows better particle stability than the 170 and 64 k dextran-coated nanogold particles, which we have attributed to the dextran size. In any event the presence of unreacted carboxylic acid groups affords for better particle stability at a neutral pH. Subsequently all glucose sensing studies were undertaken at a solution pH of 7.

As depicted in Figure 10.4, the sensing aggregate works by the dissociation of Con A., aggregated dextran-coated nanogold particles upon addition of glucose. Con A is a well-known multivalent protein (four binding sites at pH 7), [38-39] which allows for at least two different dextran-coated nanoparticles to cross-link, due to the affinity between dextran and Con A.[38-40] The addition of Con A both broadens and red-shifts the absorption spectra of the dextran-coated nanogold particles, a function of interparticle coupling due to the close proximity of the nanogold particles, Figure 10.7.

Figure 10.8 shows the time-dependent change of absorbance at 650 nm for the 500k dextran coated nanogold particles with different concentrations of Con A. The greatest changes in ΔA_{650} nm were observed for the 500 k dextran coated particles, with relatively smaller changes observed for the 64 k dextran coated nanogold particles (not shown).

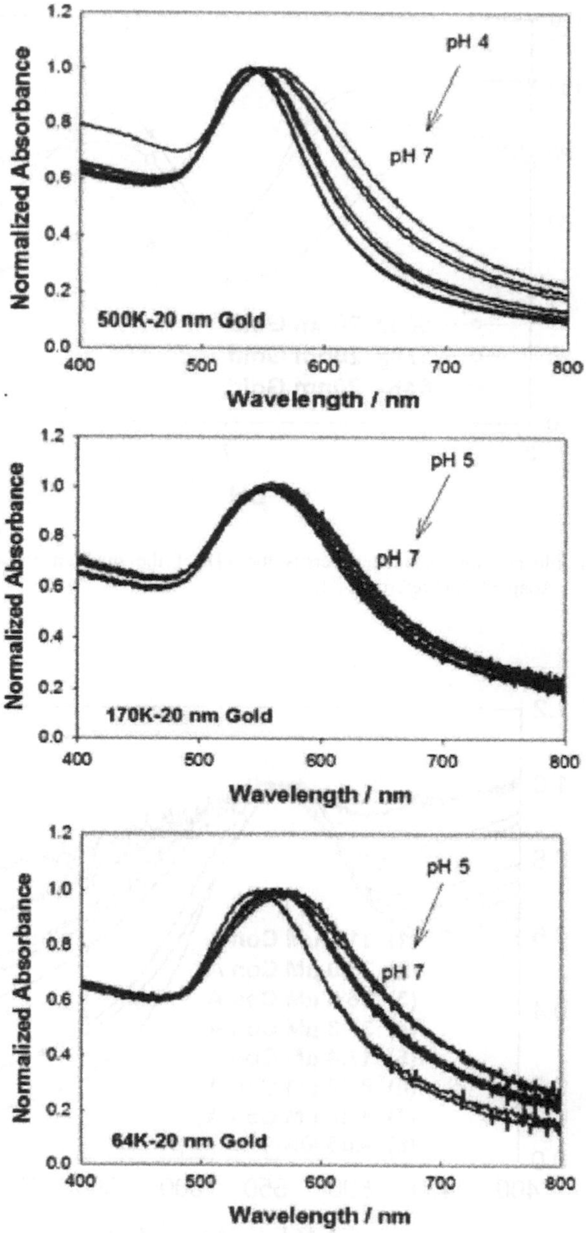

Figure 10.5. Normalized absorbance spectra of dextran-coated gold nanoparticles 500 K (Top), 170 K (Middle), and 64 K (Bottom) in different buffers with pH varying between 3 and 11. Adapted from reference 34.

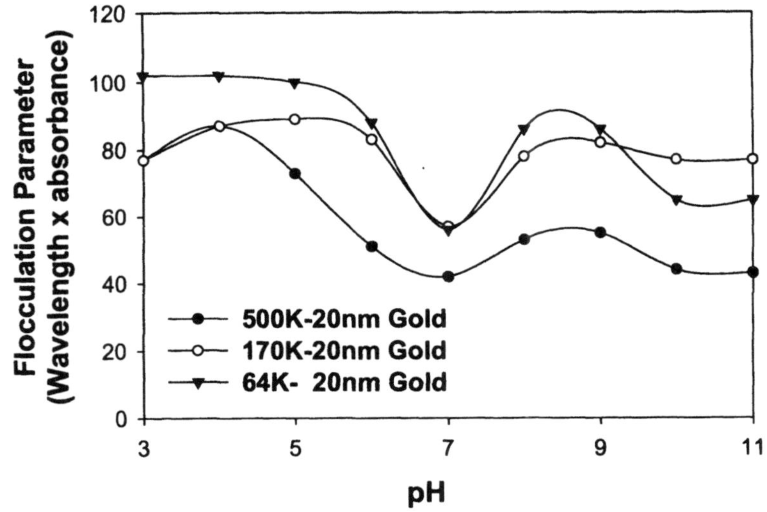

Figure 10.6. Flocculation parameter versus the pH of the medium for dextran-coated gold nanoparticles. . Adapted from reference 34.

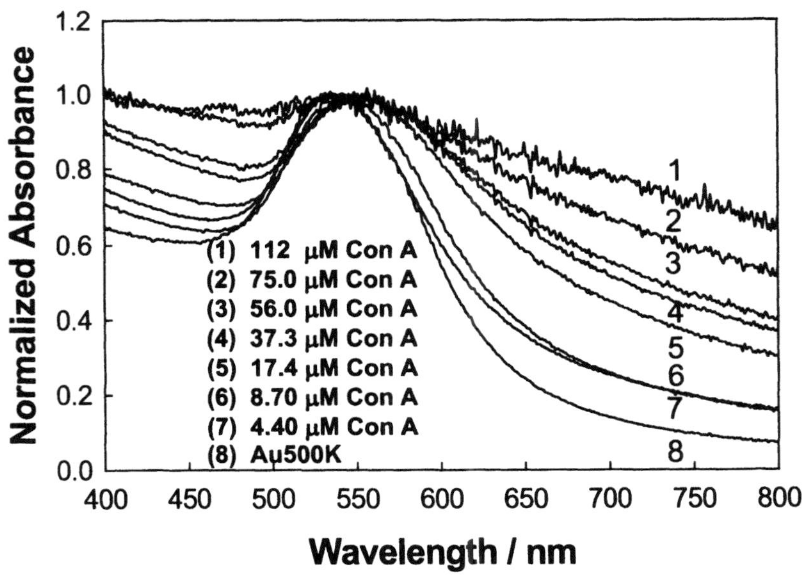

Figure 10.7. Normalized absorption spectra of 500 k dextran-coated 20 nm nanogold, crosslinked with different concentrations of Con A.. Adapted from reference 34.

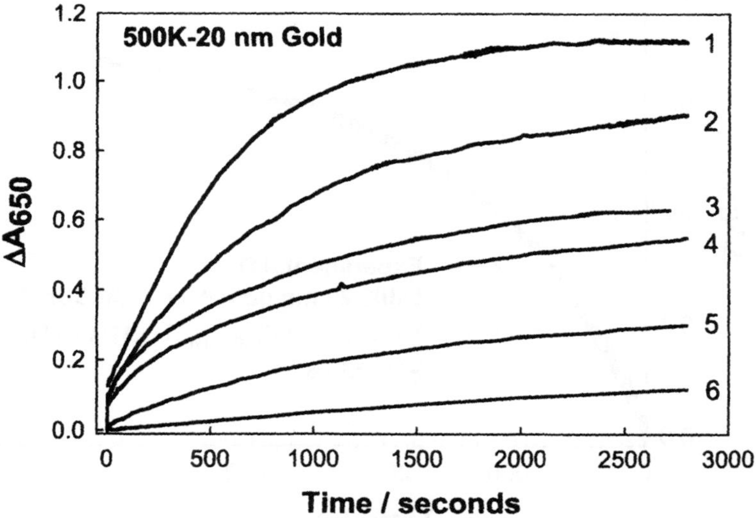

Figure 10.8. Time-dependent change in absorbance at 650 nm for 500 k dextran coated Gold nanoparticles) with different initial amounts of Con A: (1) 110 mM Con A, (2) 75.0 mM Con A, (3) 56.0 mM Con A, (4) 37.0 mM Con A, (5) 18.7 mM Con A, (6) 4.40 mM Con A. Adapted from reference 34.

The changes in absorption (ΔA_{650} = Measured abs – initial abs at time t = 0) could be modeled moderately well by a growth exponential function of the form:

$$\Delta A_{650} = \Delta A_{650(\text{final})}(1 - e^{-k_1 t}) \quad (1)$$

where ΔA_{650} is the Δ650 nm absorbance at time, t, $\Delta A_{650(\text{final})}$ is the Δ650 final plateau absorbance, and k_1 the rate constant for the rate of change of absorbance, due to dextran coated nanogold aggregation (units s^{-1}), Figure 10.9. For the three different dextran molecular weights considered for nanosensor fabrication, a greater rate of absorption change due to the aggregation of the nanoparticles as well as a greater final (plateau) ΔA_{650} nm value, i.e. $\Delta A_{650(\text{final})}$, Figure 10.10 and Table 10.1, for the 500 k dextran nanogold particles was seen. One possible explanation for this observation could be the availabIy of dextran on the surface of the nanogold for Con A binding, where the molecular weight of Con A is approximately 104,000 daltons (pH 7), and the extent of interaction between Con A and the dextran is limited by the size of the dextran. Hence as the size of the dextran is decreased, then the possibilities for crosslinking are reduced.

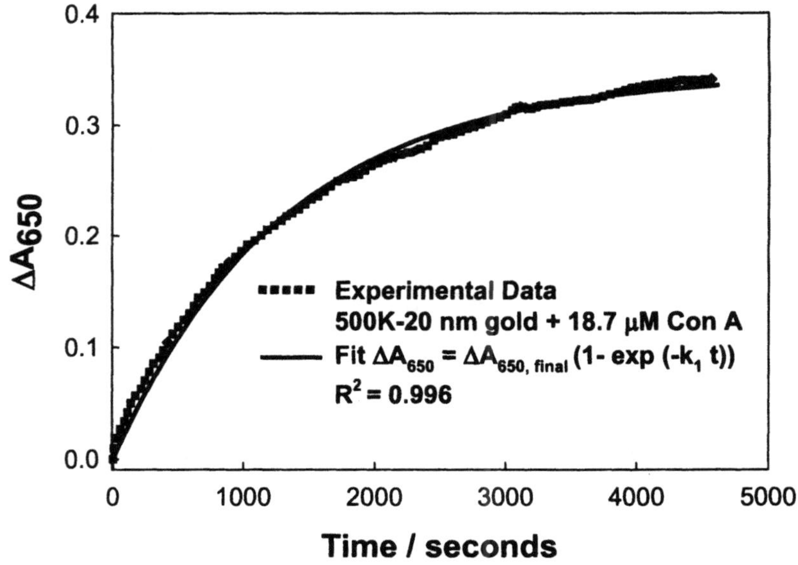

Figure 10.9. Change in absorbance at 650 nm for 500 k dextran coated gold nanoparticles, experimental data and the model fit. Adapted from reference 34.

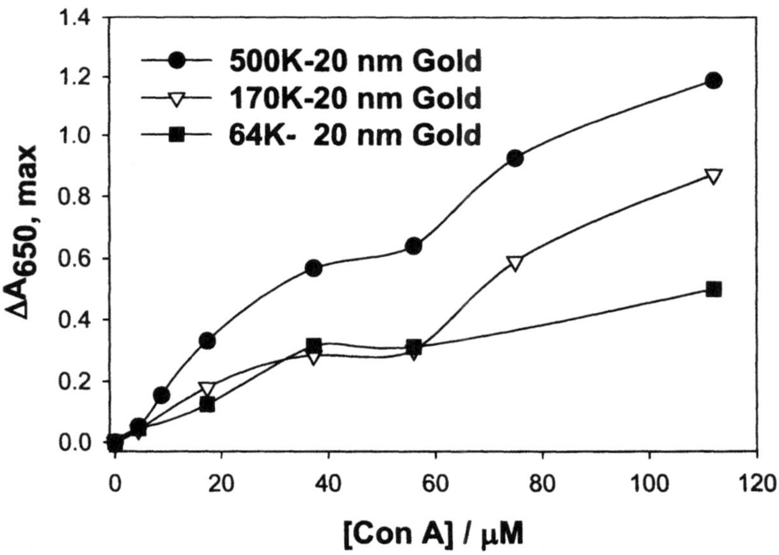

Figure 10.10. Maximum change in absorbance at 650 nm for dextran-coated gold nanoparticles versus the concentration of Con A used. Adapted from reference 34.

Table 10.1 – Kinetic data obtained from fitting to equation 1 with the data shown in Figure 10.8 and for 170 k and 64 k dextran coated gold nanoparticles aggregated with different amounts of Con A.

	500 K			170 K		
[Con A] μM	$\Delta A_{650,\,final}$	k_1 (s^{-1})	R^2	$\Delta A_{650,\,final}$	k_1 (s^{-1})	R^2
4.40	0.056	$3.87*10^{-4}$	0.988	0.037	$4.13*10^{-3}$	0.804
18.7	0.340	$7.65*10^{-4}$	0.996	0.167	$9.37*10^{-3}$	0.780
37.0	0.535	$14.0*10^{-4}$	0.956	0.311	$1.81*10^{-3}$	0.991
56.0	0.611	$16.0*10^{-4}$	0.944	0.271	$5.39*10^{-3}$	0.956
75.0	0.892	$18.1*10^{-4}$	0.912	0.551	$10.9*10^{-3}$	0.916
110	1.122	$19.4*10^{-4}$	0.996	0.912	$5.30*10^{-3}$	0.993

	64 K		
[Con A] μM	$\Delta A_{650,\,final}$	k_1 (s^{-1})	R^2
4.40	-	-	-
18.7	0.122	$0.29*10^{-3}$	0.500
37.0	0.311	$2.80*10^{-3}$	0.995
56.0	0.304	$0.69*10^{-3}$	0.560
75.0	-	-	-
110	-	-	-

10.4.2. Glucose Sensing based on the Dissociation of Dextran-coated Gold Nanoparticle / Con A Aggregates

There have been continued efforts to develop optical based methods for glucose detection.[1-4] Several solution-based fluorescence methods have been based on the glucose binding protein Con A, and a polysaccharide, typically dextran, which serves as a competitive ligand for glucose.[40] Typically for fluorescence based glucose sensors, the Con A is labeled with a fluorescent donor and the dextran with an acceptor, but the labels can readily be reversed.[40] The binding of Con A and dextran results in a both a decrease in fluorescence intensity and lifetime of the donor. However, the addition of glucose creates a competetion for the glucose binding sites on Con A, releasing Con A from the acceptor, resulting in an increase in the intensity, lifetime and phase angles.[40] In the system described in the previous section, the aggregation of the dextran coated nanogold with Con A results in both a red-shift and broadening of the gold plasmon absorption and in an increase in scattered light intensity. In an analogous manner to the FRET systems, based on Dextran and Con A as described above, the presence of glucose competes with dextran coated colloids for Con A binding sites, resulting in the dissociation of the Con A aggregated nanogold, Figure 10.4. This results in a decrease in the absorbance of the nanogold when monitored at an arbitrary near-red 650 nm wavelength as well the decrease in scattered light at 550 nm, Figures 10-12.

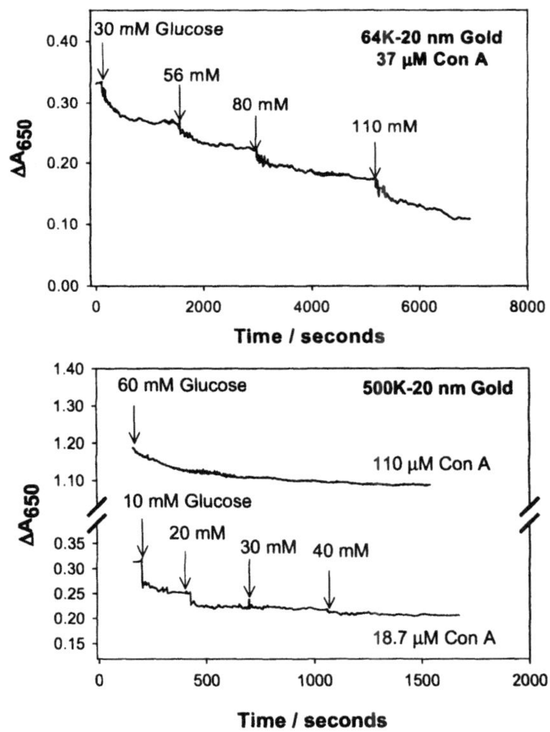

Figure 10.11. Time-dependent change in absorbance at 650 nm for (top) 64K dextran and (Bottom) 500 K dextran coated gold nanoparticles after the addition of glucose. Adapted from reference 34.

Upon addition of glucose to the Con A aggregated, Dextran-coated nanogold, the scattering spectra typically decrease in intensity and slightly blue shift, in accordance with the disassociation of the aggregates, Figure 10.12. Similar to the findings with the absorption based measurements,[34] it was found that the greatest scattering changes were obtained for a 170 K dextran sensing aggregate, which had been crosslinked with 20 μM Con A, Figure 10.12 – Bottom. These scattering spectra were taken after ≈ 45 mins, at the 90 % response times, i.e. the time after which the signal had changed by 90 % of its original value, as previously described by the authors.[34]

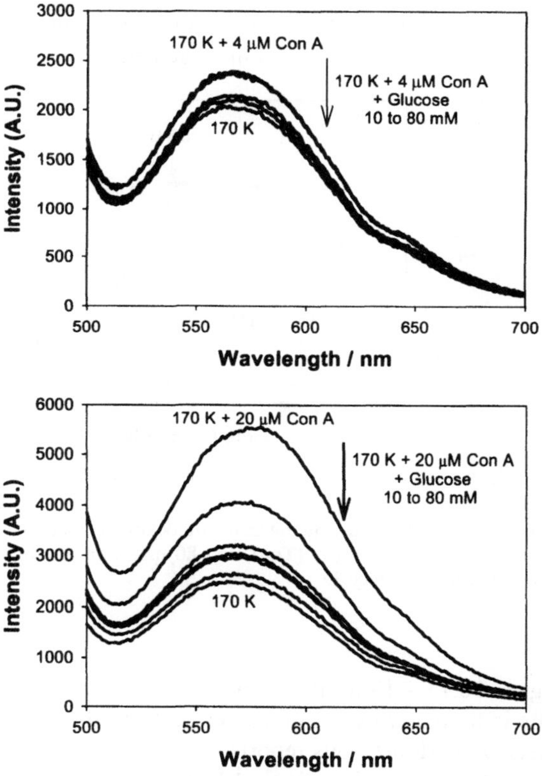

Figure 10.12. Wavelength-dependent scattering from dextran-coated gold nanoparticles after the addition of glucose with 4 mM Con A **(Top)**, and 20 mM Con A **(Bottom)** upon white LED illumination. Adapted from reference 35.

Figure 10.13 shows the time-dependent scattering at 560 nm for the sensing aggregate compositions shown in Figures 4 and 5. In Figure 10.13 – Top one can clearly see the slow aggregation of the system, indicated by the gradual change in 560 nm scatter, itself induced by 4 μM Con A. The addition of glucose decreases the I_{550} scatter as the aggregates disassociate. In comparison, much greater changes in scatter can be observed at 560 nm by the addition of 20 μM Con A, the addition of glucose after the 90 % response time, returning the scatter value to ≈ the initial unaggregated scatter value, i.e. before Con A

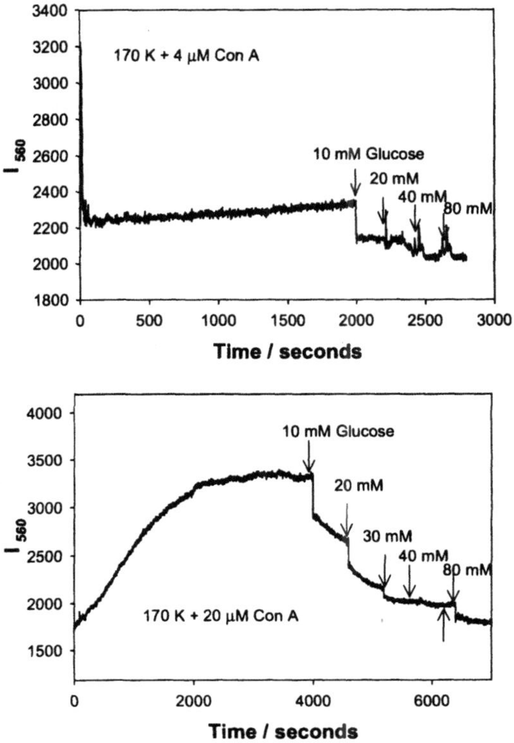

Figure 10.13. Time-dependent scattering intensity of dextran-coated gold nanoparticles in the presence of glucose and 40 mM Con A **(Top)**, and 80 mM Con A **(Bottom)** upon white LED illumination. Adapted from reference 35.

addition, Figure 10.13 – Bottom. As expected, increasing the concentration of Con A results in a greater extent of scattering from the larger aggregates. This result is consistent with previous reports which showed that the scattering cross section increased with colloid size.[23, 24] However, these reports referred to the size of the colloid monomers and not the aggregates as described here.

The changes for both I_{560} and I_{680} nm scatter vs. time, due to nanoparticle aggregation, for the 20, 40 and 80 μM Con A aggregates could be modeled moderately well by a growth exponential of the form:

$$I_{560} = I_{560(Final)}(1 - \exp(-k_1 t)) \quad (2)$$

where I_{560} is the uncorrected scattered intensity at 560 nm at time, t, $I_{560(final)}$ is the final plateau scattered intensity after 90 % Con A induced aggregation, and k_1, the rate constant for the rate of change of scattering, due to the dextran-coated nanogold aggregation (units s^{-1}), Figure 10.13 and Table 10.2. From

Table 10.2 one can see a greater rate of change of scatter for nanogold compositions containing more Con A, and indeed a greater final scattered intensity, i.e. $I_{560(final)}$. It is worth noting that the true form of the scattered light equation with time is likely to be:

$$I_{560} = I_0 + (I_{560(Final)} - I_0)(1 - \exp(-k_1 t)) \tag{3}$$

where I_0 is the initial scattered intensity of the sensing system and $\neq 0$. For the ease of data fitting, the data was normalized so that $I_0 = 0$.

The decrease in scattered light intensity upon cumulative glucose addition, shown in Figure 10.13, could be modeled well to a simple exponential, of the form:

$$I_{560} = I_{560(Final)}(\exp(-k_2 t) \tag{4}$$

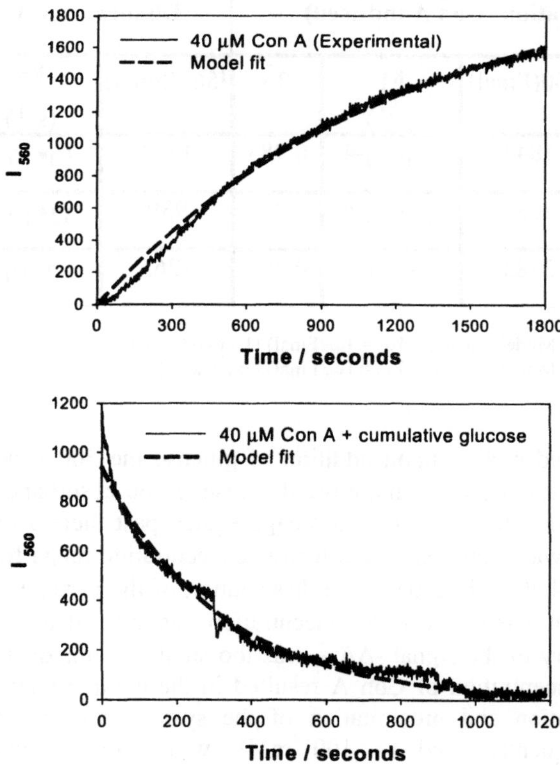

Figure 10.14. Fitting of equation 2 to the 560 nm time-dependent scattering for 40 µM Con A induced nanogold aggregation (**Top**), and equation 4 to the dissociation of the same sensing aggregate by glucose (**Bottom**). Adapted from reference 35.

where k_2 is the rate of change of scatter due to glucose addition, noting again the normalization to zero after complete dissociation by glucose, i.e. at $t = \infty$. In this regard, normalizing both curves for fitting, Figures 10.14 Top and Bottom is justified as the responses obtained in the scattering vs. time plots, Figure 10.13 reversibly change to the same initial scattered intensity value, the I_{560} values of Figure 10.14, being in essence ΔI_{560} scattering values. It should be noted that the $I_{560(final)}$ values for association (nanosensor Con A induced aggregation) and disassociation (by glucose) are not the same in Table 10.2. This was explained by the initial non-diffusion limited dissociation of the aggregates by glucose, Figure 10.13, just after the initial addition. Hence the fitting occurred for the data after the first glucose addition.

Table 10.2 – Parameters obtained from fitting the aggregation and disassociation curves to equations 2 and 4 respectively.

Aggregation (Con A induced)				Dissociation (Glucose)		
Con A	I_{560}(Final)	k_1 (s^{-1})	R^2	I_{560}(Final)	k_2 (s^{-1})	R^2
20 μM	1844	$6.6*10^{-4}$	0.983	1272	$10*10^{-4}$	0.982
40 μM	2088	$8.2*10^{-4}$	0.996	939	$32*10^{-4}$	0.989
80 μM	2383	$18*10^{-4}$	0.991	1265	$7.2*10^{-4}$	0.984

Aggregation Model Equation: $I_{560} = I_{560}(Final)\ [1 - \exp(-k_1\ t)]$
Dissociation Model Equation: $I_{560} = I_{560}(Final)\ [\exp(-k_2\ t)]$

As explained earlier, upon addition of glucose the 500 k dextran-coated nanogold particles show a cumulative decrease in the absorbance at 650 nm, where all systems studied had been preaggregated past there 95 % maximum ΔA_{650} value. It was found that the effective concentration range for Con A was using between 4.40 and 75 μM. The dissociation of the particles with glucose was insignificant when the Con A concentration was lower than ≈ 4.40 μM, and thus the recovery of the signal $-\Delta A_{650}$ was too small for practical applications. Too high a concentration of Con A resulted in the complete aggregation and almost flocculation and precipitation of the system, where the amount of glucose subsequently used (> 100 mM), was barely enough to cause competitive dissociation. Attractive glucose responses were however obtained when dextran-coated nanogold was aggregated with between 18 – 37 μM Con A, Figure 10.15. For the 18.7 μM Con A system, a most attractive glucose dynamic sensing range was observed (1 → 40 mM), noting that the blood glucose level is 3-8 mM for a healthy person, and increases to between 2 and 40 mM in diabetics.[22] For tear glucose, these values are typically 10 fold lower, ranging from 500 μM to several mM glucose, [13-15] well within the sensing capabilities of this system.

Similar responses to glucose were observed when 170 and 64 k dextran-coated nanogold were used, Figure 10.16. Again the change in absorption was negligible when both low and high concentrations of Con A were used. Interestingly the response to glucose saturated again at ≈ 40 mM glucose, providing for a useful physiological glucose sensing range. However, the response of the 64 k dextran-coated nanogold, aggregated with 37 μM Con A, reproducibly showed an increase in $-\Delta A_{650}$ as a function of glucose addition. This is a notable, almost linear, glucose response and is thought due to the fact that the initial extent of aggregation is smaller for these sensing aggregates, and thus the amount of glucose required for dissociation was less, i.e. the dissociation was easier. However, the 64 k dextran coated nanoparticles typically flocculated and precipitated after 6-8 hrs, where as the 500 k dextran coated nanoparticles were observed to be much more stable with time.

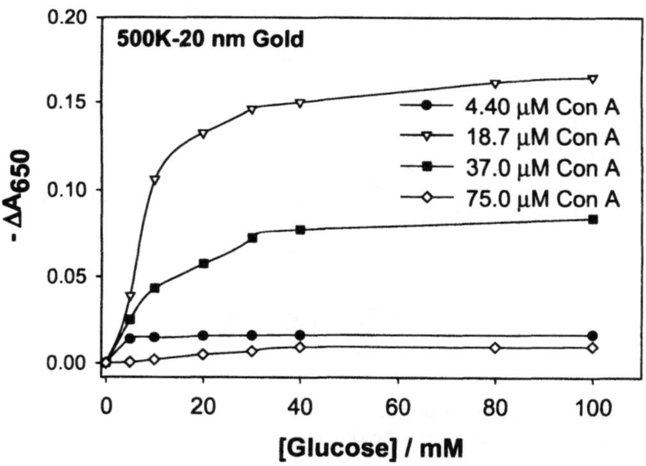

Figure 10.15. Cumulative change in absorbance 650 nm for 500 K dextran coated gold nanoparticles versus the concentration of glucose. Adapted from reference 34.

The wavelength ratiometric scattering responses towards glucose for the different sensing compositions of nanogold aggregates studied is shown in Figure 10.17. In addition, the ratiometric plots were normalized to enable the easy visualization of the glucose responses. Nanoaggregates crosslinked with 20 μM Con A showed the greatest glucose response, where the greatest and

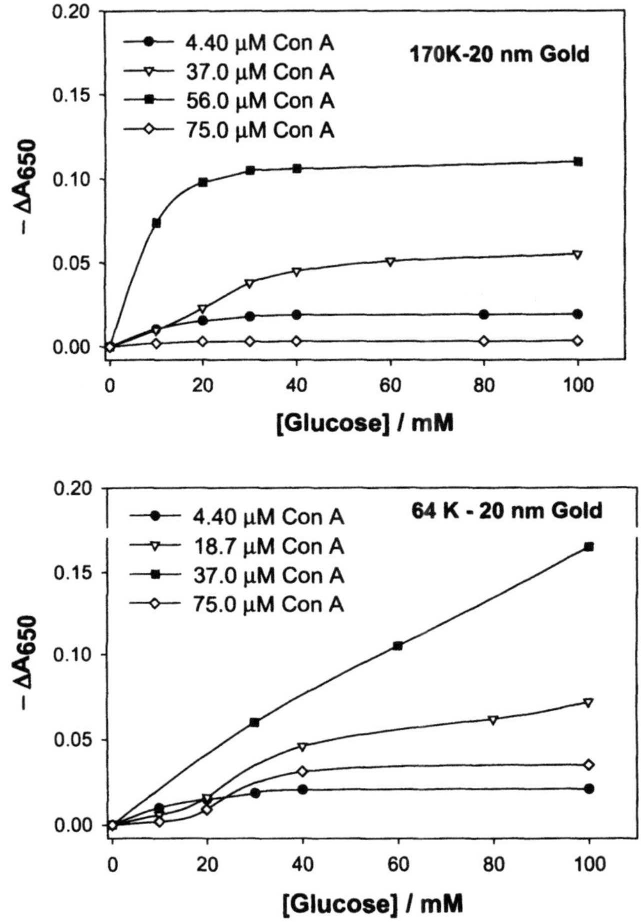

Figure 10.16. Cumulative change in absorbance 650 nm for (Top) 170 K dextran and (Bottom) 64 K dextran coated gold nanoparticles versus the concentration of glucose. Adapted from reference 34.

smallest amounts of Con A used, produced the smallest responses and therefore glucose dynamic sensing ranges. This finding was similar for the absorption based glucose measurements previously reported by the authors.[34] It was noted that this new approach requires much simpler instrumentation than previously reported, and the wavelength ratiometric approach is indeed independent of any excitation source fluctuations / drifts and the total nanogold sensing concentration used.

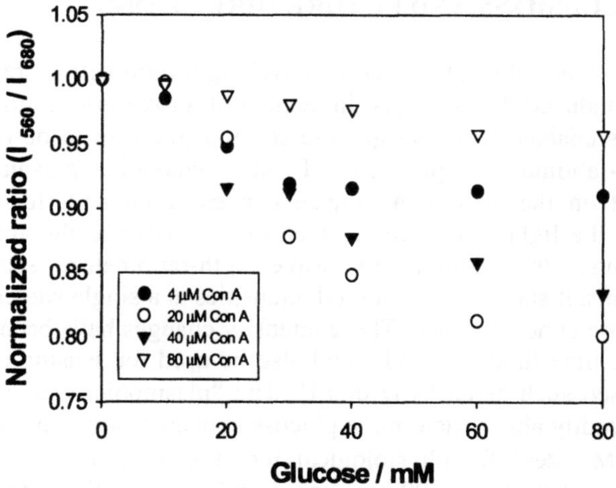

Figure 10.17 - Normalized ratiometric glucose response for various Con A, dextran-coated nanogold systems. Adapted from reference 35.

The role of reduced gold colloid size, i.e. sensing aggregates derived from only 10 nm nanogold was also studied, Figure 10.18. It was found that 500 k dextran coated 10 nm nanogold showed slightly greater ΔA_{650} values as compared to the 20 nm aggregates, where the reduced colloid size is thought to reduce the total amount of dextran in the sample, given the same number colloid density. While this result suggests the possibility of tunable glucose sensing ranges also by using different size, equally coated gold nanoparticles, our experience reveals a greater level of complexity when working with the smaller nanoparticles. One particular difficulty lies in their separation and recovery by centrifugation from the reactants.

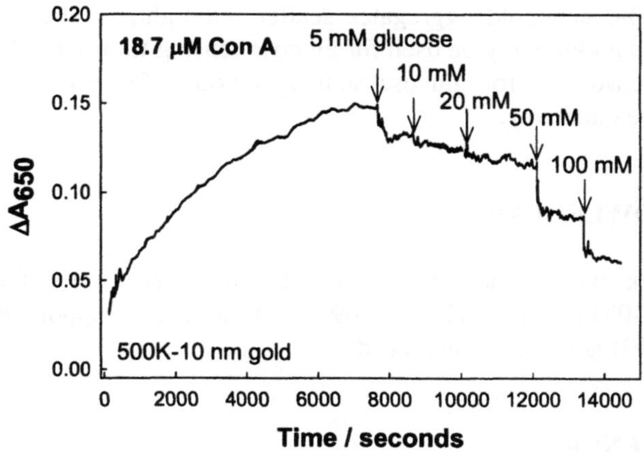

Figure 10.18 - Normalized Time-dependent scattering intensity of dextran-coated 10 nm gold nanoparticles in the presence of glucose and 18.7 µM Con A. Adapted from reference 33.

10.5. CONCLUSIONS AND FUTURE DIRECTIONS

Changes in absorption and wavelength-ratiometric resonance light scattering, induced by changes in nanogold proximity in the presence of glucose, has enabled a simple glucose sensing platform to be developed. The wavelength-ratiometric approach of the nanogold plasmon scatter is independent on the sensing aggregate concentration and fluctuations in the intensity of the light source and is therefore superior to the absorption based measurements. In addition to wavelength-ratiometric scattering based measurements, it shown that scattered intensities at a single wavelength can also report glucose concentrations. These intensity changes have been modeled with intensity vs. time functions and could also be used for sensing if a stable light source is used such as an LED or LD. The "plasmonic approach" to glucose sensing is readily able to determine glucose concentrations from several mM up to ≈ 60 mM, ideal for physiological blood glucose monitoring where red scattering wavelengths (> 600 nm) can be selected to eliminate the absorption of hemoglobin, water and melanin for transdermal-type glucose monitoring.

Over the past few years, the laboratories at University of Maryland and University of Maryland Biotechnology Institute have been engaged in developing new technologies to non-invasively and continually monitor physiological glucose, such as a range of glucose sensing contact lenses.[6-8] With these technologies there is a requirement to both simplify the technology, both from diabetic patient read-out and instrumentation perspectives. Incorporating the sensing aggregates within contact lenses could allow the simple colorimetric determination of tear glucose using ambient light excitation; a patient simply looks in a mirror to see the color change and compares with a precalibrated color strip. In addition, there has been research towards the development of glucose sensing tattoo's based on implantable hydrogel beads.[41] Given that gold and silver colloids are physiologically safe and already used in medicine,[42-43] and the fact the nanogold aggregates scatter red light, we believe that our approach could additionally be used for glucose sensing transdermaly, either for implantable sensors or for glucose sensing tattoo's. Further studies by our laboratories are underway.

10.6. ACKNOWLEDGMENTS

This work was supported by the NIH National Center for Research Resource, RR-08119 and NIH GM070929. Partial salary support to CDG and JRL from UMBI is also acknowledged.

10.7. REFERENCES

1. M. R. Robinson, R. P. Eaton, D. M. Haaland, G. W. Koepp, E. V. Thomas, B. R. Stallard, P. L. Robinson, Non-invasive glucose monitoring in diabetic patients: A preliminary evaluation, *Clin. Chem.* **38**, 1618- 1622 (1992).

2. H. M. Heise, R. Marbach, T. H. Koschinsky, F. A. Gries, Non-invasive blood glucose sensors based on near-infrared spectroscopy, *Ann. Occup. Hyg.* **18**, 439-447 (1994).
3. W. F. March, B. Rabinovitch, R. Adams, J. R. Wise, M. Melton, Ocular glucose sensor, Trans. Am. Soc. *Artif. Intern. Organs.* **28**, 232-235 (1982).
4. B. Rabinovitch, W. F. March, R. L. Adams, Non-invasive glucose monitoring of the aqueous humor of the eye, Part 1, Measurement of very small optical rotations, *Diabetes Care*, **5**, 254-258 (1982).
5. G.M. Schier, R.G. Moses, I.E.T. Gan, S.C. Blair, An evaluation and comparison of reflux ligand Glucometer II, two new portable reflectance meters for capillary blood glucose determination, *Diabetes Res. Clin. Pract.* **4,** 177-181 (1988).
6. R. Badugu, J. R. Lakowicz, C. D. Geddes, Towards the non-invasive continuous monitoring of physiological glucose using a novel monosaccharide-sensing contact lens, *Anal. Chem.* **76 (3),** 610-618 (2004).
7. R. Badugu, J. R. Lakowicz, C .D. Geddes, Opthalmic glucose sensing: A novel monosaccharide sensing disposable contact lens, *Analyst* **129(6),** 516-521 2004.
8. R. Badugu, J.R. Lakowicz, C.D. Geddes, A glucose sensing contact lens: A non-invasive technique for continuous physiological glucose monitoring, *Jn. Fluorescence* **13**, 371-374 (2003).
9. N. Dicesare, J. R. Lakowicz, Spectral properties of fluorophores combining the boronic acid group with electron donor group or withdrawing groups, Implication in the development if fluorescence probes for saccharides, *J. Phys. Chem. A.* **105.** 6834-6840 (2001).
10. N. Dicesare, J .R. Lakowicz, A new highly fluorescent probe for monosaccharides based on donor-acceptor diphenyloxazole, *Chem. Comm.* 2022-2023 (2001)
11. N. Dicesare, M. R. Pinto, K. S. Schanze, J. R. Lakowicz, Saccharide detection based on amplified fluorescence quenching of a water soluble poly(phenylene ethylene) by a boronic functionalized benzyl viologen derivative, *Langmuir* **18,** 7785-7787 (2002).
12. L .Talosa, I. Gryczynski, L. R. Eichorn, J. D. Dattelbaum, F. N. Castelleno, G. Rao, J. R. Lakowicz, Glucose sensors for low cost lifetime-based sensing using a genetically engineered protein, *Anal. Biochem.* **267,** 114-120 (1999).
13. A. Labande, D. Astrue, *Chem. Commun.*, 1007 (2000),
14. J. J. Storhoff, R. Elghanian, R. C. Mucic, C. A. Mirkin, R .L. Letsinger, One-pot colorimetric differentiation of polynucleotides with single base imperfections using gold nanoparticle probes, *J. Am. Chem. Soc.* **120**, 1959-1964 (1998).
15. R.A. Reynolds, C. A. Mirkin, R.L. Letsinger, Homogeneous, nanoparticle-based quantitative colorimetric detection of oligonucleotides, *J. Am. Chem. Soc.* **122**, 3795-3796 (2000).
16. R. Elghanian, J. J. Storhoff, R. C. Mucic, R. L. Letsinger, C. A. Mirkin, Selective colorimetric detection of polynucleotides based on the distance-dependent optical properties of gold nanoparticles, *Science*, **277**, 1078-1081 (1997).
17. M. Sastry, N. Lala, V. Patil, S. P. Chavan, A. G. Chittiboyina, Optical absorption study of the biotin- avidin interaction on colloidal silver and gold particles, *Langmuir* **14,** 4138-4142 (1998).
18. S. Cobbe, S. Connolly, D. Ryan, L. Nagle, R. Eritja, D. J. Fitzmaurice, DNA-controlled assembly of protein-modified gold nanocrystals, *J. Phys. Chem. B* **107**, 470-477 (2003).
19. K. Aslan, V. H. Pérez-Luna, Surface modification of colloidal gold by chemisorption of alkanethiols in the presence of a nonionic surfactant", *Langmuir* **18**, 6059-6065 (2002),.
20. K. Aslan, C. C. Luhrs, V.H. Pérez-Luna, Controlled aggregation of biotinylated gold nanoparticles by streptavidin, *J. Phys. Chem. B*, **108(40)**, 15631-15639 (2004)
21. K. Aslan, V. H. Pérez-Luna, Quenched emission of fluorescence by ligand functionalized gold nanoparticles, J. Fluorescence, **14(4)**, 401-405(2004).
22. K. Aslan, J. R. Lakowicz, H. Szmacinski , C. D. Geddes, Metal-enhanced fluorescence solution based sensing platform, J. of Fluorescence **14(6),** 677-679 (2004).
23. J. Yguerabide, E. E. Yguerabide, Light-scattering submicroscopic particles as highly fluorescent analogs and their use as tracer labels in clinical and biological applications. I. Theory, *Anal. Biochem.* **262,** 137-156 (1998).
24. J. Yguerabide, E.E. Yguerabide, Light-scattering submicroscopic particles as highly fluorescent analogs and their use as tracer labels in clinical and biological applications. II. Experimental Characterization, *Anal. Biochem.* **262**, 157-176 (1998).

25. V. H. Pérez-Luna, P. Betala, K. Aslan, Colloidal Gold, in *Encyclopaedia of Nanoscience and Nanotechnology*, Edited by H. S. Nalwa (American Scientific Publishers, California, 2004), pp. 27-49.
26. D. Roll, J. Malicka, I. Gryczynski, Z. Gryczynski, J. R. Lakowicz, Metallic colloid wavelength-ratiometric scattering sensors, *Anal. Chem.* **75**, 3108-3113 (2003).
27. H. Otsuka, Y. Akiyama, Y. Nagasaki, K. Kataoka, Quantitative and reversible lectin association of gold nanoparticles modified with α-lactosyl-ω-mercapto-poly(ethylene glycol), *J. Am. Chem. Soc.* **123**, 8226-8230, (2001)
28. C-C. Lin, Y-C. Yeh, C-Y. Yang, C-L. Chen, G-F. Chen, C-C. Chen, Y-C. Wu, Selective binding of mannose-encapsulated gold nanoparticles to Type 1 Pili in *Escherichia coli*, *J. Am. Chem. Soc.* **124**, 3508-3509, (2002).
29. B. Nolting, J-J. Yu, G-y, Liu, S-J. Cho, S. Kauzlarich, J. Gervay-Hague, Synthesis of gold glyconanoparticles and biological evaluation of recombinant Gp120 interactions, *Langmuir*, **19**, 6465-6473 (2003).
30. J. M. de la Fuente, A. G. Barrientos, T. C. Rojas, J. Rojo, J. Canada, A. Fernandez, S. Penades, Gold glyconanoparticles as water-soluble polyvalent models to study carbohydrate interactions, *Angew. Chem. Int. Ed.* **40**, 2258 (2001).
31. T. C. Rojas, J. M. de la Fuente, A. G. Barrientos, S. Penades, L. Ponsonnet, A. Fernandez, Gold glyconanoparticles as building blocks for nanomaterials design, *Advanced Materials*, **14(8)**, 585-588 (2002).
32. D. C. Hone, A. H. Holmes, D. A. Russell, Rapid, quantitative colorimetric detections of a lectin using mannose-stabilized gold nanoparticles, *Langmuir*, **19**, 7141-7144 (2003).
33. K. Aslan, J. R. Lakowicz, C. D. Geddes, Plasmonic tunable glucose sensing based on the dissociation of Con A-aggregated dextran-coated gold colloids, *Analytica Chimica Acta* **157(1-2)**, 141-146. (2004)
34. K. Aslan, J. R. Lakowicz, C. D. Geddes, Nanogold-plasmon-resonance-based glucose sensing, *Analytical Biochemistry*, **330(1)**, 145-155, (2004).
35. K. Aslan, J. R. Lakowicz, C. D. Geddes, Nanogold plasmon-resonance-based glucose sensing 2: wavelength-ratiometric resonance light scattering, *Anal. Chem.* **77(7)**, 2007-2014, (2005).
36. K.-H. Su, Q.-H. Wei, X. Zhang, J.J. Mock, D.R. Smith, S. Schultz, Interparticle coupling effects on plasmon resonances of nanogold particles, *Nano Letters*, **3**,1087-1090 (2003).
37. J. Lahiri, L. Isaacs, B. Grzyboowski, J.D. Carbeck, G.M. Whitesides, Biospecific binding carbonic anhydrase to mixed SAMs presenting benzenesulfonamide ligands: A model system for studying lateral steric effects, *Langmuir* , **15** 7186-7198 (1999).
38. R. Ballerstadt, J. S. Schultz, A fluorescence affinity hollow fiber sensor for continuous transdermal glucose monitoring, *Anal. Chem.* **72,** 4185-4192 (2000).
39. A. Yoshizumi, N. Kanayama, Y. Maehara, M. Ide, H. Kitano, Self-assembled monolayer of sugar-carrying polymer chain: sugar balls from 2-methacryloyloxyethyl D-glucopyranoside, *Langmuir* **15,** 482-488. (1999).
40. J. R. Lakowicz, Principles of Fluorescence Spectroscopy, 2nd edition Kluwer/Academic Plenum Publishers, New York, 1997.
41. R. J. Russell, M. V. Pishko, C. C. Gefrides, M. J. McShane, G. L. Cote, A Fluorescence-Based Glucose Biosensor Using Concanavalin A and Dextran Encapsulated in a Poly(ethylene glycol) Hydrogel, *Anal. Chem.*, **71**, 3126 (1999).
42. T. Lancaster, L. F. Stead, *Cochrane Database System Rev.*, *2*. (2000)
43. N. Hymowitz, H. Eckholdt, *Preventive Med.*, **25(5)**, 537 (1996).

OPTICALLY-BASED AFFINITY BIOSENSORS FOR GLUCOSE

Jerome S. Schultz*

11.1. INTRODUCTION

The first biosensor was developed by Leland Clark who enhanced his oxygen electrode technology (Clark and Lyons, 1962) by interposing another membrane bound region between the detector and the sample that contained an enzyme (glucose oxidase) to produce the first "biosensor" for glucose. This paradigm has been followed for most of the biosensor developments to date. Leland Clark has been recently recognized for his breakthrough technology by receiving the Russ Prize from the National Academy of Engineering in 2005. A brief review of the early research on biosensors was provided by Schultz (1991).

During this same period the first major use of bioreceptors for analytical purposes was the development of immunoassays by Berson and Yalow (1959). A critical attribute of the use of proteins as analytical reagents is the potential specificity of protein activity (whether as a binding agent, e.g. an antibody or as a catalyst, e.g. an enzyme). This unique property of these proteins allows their use as analytical reagents in complex mixtures. Thus the key feature of immunoassays is that specific analytes can be measured in very complex milieus such as blood.

Also she clearly pointed out that the method was not limited to antibodies (Nobel Address 1977).

"The RIA principle is not limited to immune systems but can be extended to other systems in which in place of the specific antibody there is a specific reactor or binding substance. This might be a specific binding protein in plasma, a specific enzyme or a tissue receptor site".

Yalow recognized that antibody-antigen reactions are reversible as illustrated in Figure 11.

Jerome S. Schultz, Distinguished Professor, University of California Riverside, CA 92521
jerome.schultz@ucr.edu

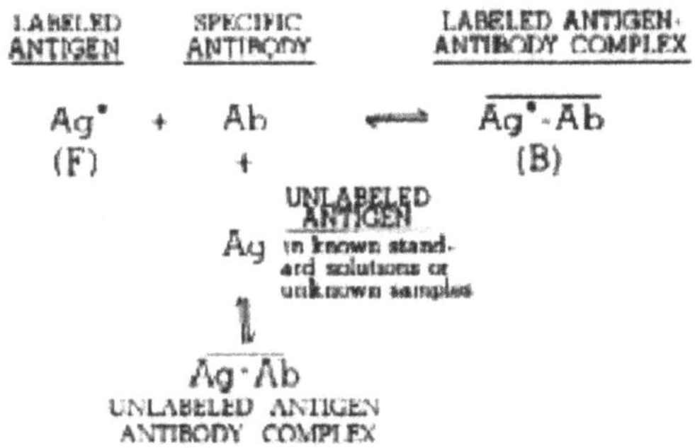

Figure 11.1. Competing reactions that form the basis of radioimmunoassay (RIA) (Yalow, 1977)

Most of the clinical immunoassay procedures that have been devised over the past 50 years were of the form of discrete tests, i.e. a single sample was exposed to the test system to obtain a result. After the result is obtained the materials are discarded. As there was increased interest in increasing the sensitivity of immunoassays, antibodies with very high affinities for the analyte of interest were sought, so that the binding between antibody an analyte approached irreversibility, i.e. a quick washing step did not dislodge the analyte from the antibody. The common perception that evolved, of antibodies having high affinities, reduced the appreciation that they could be effectively used as the key component of a sensor system that requires reversibility of the binding reaction.

Janata (1975) was one of the first sensor researchers to realize that antibody-analyte reactions could be used as the basis for sensor systems. He devised an "immunoelectrode" system utilizing Concanavalin A (Con A) as a surrogate for an antibody and mannan as a surrogate for an antigen. Con A was immobilized to an electrode and the differential potential versus a control protein electrode measured as a function of mannan concentration. In this device, the primary principle is the change in the electrical properties of the bioreceptor upon analyte binding. The property measured was the change in potential of the bioreceptor electrode. Although Janata clearly recognized value of antibodies as sensor elements he did not report on the reversibility of this device.

Another approach for the use of antibodies as sensor elements was described by Aizawa, et al (1976). Following the pattern of Clark's glucose electrode, these investigators placed an antibody on the surface of an oxygen electrode. In the sample containing the antigen to be measured they placed an antigen-analog that was labeled with the enzyme catalase. In this arrangement both the antigen and antigen-catalase competed for the surface immobilized antibody. The extent of antigen-catalase binding was measured by the addition of hydrogen peroxide that caused oxygen release at the surface of the oxygen electrode.

Although the above examples illustrated the potential use of antibodies for the development of detector systems, neither was a true sensor since the systems as configured could not be placed in sample environments where the analyte concentration was constantly changing.

However, these two early studies illustrate two general approaches for the use of bioreceptors for sensor elements in affinity sensors. Janata's immunoelectrode is an example of what I have termed a "direct" method (Schultz, 1986). In these situations one measures some attribute of the direct interaction of the analyte with the bioreceptor; in this case it was the potential of the adsorbed protein on an electrode. Other methods, to be discussed below, measure changes in optical properties of the bioreceptor upon analyte binding.

Aizawa's device falls into the category of an "indirect" or "competitive" method that more closely resembles the Yalow and Berson's immunoassay methodology. Here the enzyme labeled antigen competes with the "analyte" antigen for the binding sites of the selective antibody. To some extent this approach is more general in that the signal is generated by the modification of the competitive analog-analyte and there are many options to devise techniques to obtain high levels of sensitivity. In this report, we will focus on optical and particularly fluorescence methods.

11.2. DEVELOPMENT OF OPTICALLY BASED BIOSENSORS

In the early 1970s while a faculty member at the University of Michigan I was approached by several colleagues in the Medical School to join their project to develop and artificial pancreas. At that time insulin infusion pumps were being developed and it was clear that the key missing component for an artificial pancreas was a reliable implantable glucose sensor. Although aware of Clark's glucose electrode, I realized that there were some technical issues with this approach - namely that since the system consumed glucose, any deposition on the sensor surface would cause variations in the calibration of the sensor. So I looked for an approach that was not hampered by diffusional aberrations. Immunoassay methods seemed to be a robust technology since extent of binding is an equilibrium thermodynamic phenomena rather than a kinetic phenomena, as described by Yalow in her Nobel Lecture.

Yet as mentioned above, all immunoassays were single tests - not continuous monitoring devices. At that time I was conducting research on the behavior of dialysis/porous membranes as models for understanding the mechanism for reduced diffusion rates in tissues (Beck and Schultz, 1970) and realized that encapsulating the assay reagents within a dialysis chamber could be an approach for adapting many immunoassay techniques to function as continuous monitors. Sensors require at least two components: a recognition system and a transducer system. Previous to our work, most biosensors utilized electrochemical detection methods (as illustrated by the work of Janata and Aizawa above). I thought using optical measurement techniques (like fluorescence) might be more suitable for biosensors because of the many optical effects that could be harnessed for the transducer feature (fluorescence, optical polarization, colorimetry, etc). Fortunately during a visit to Goldstein's laboratory at NIH I was introduced to the use of optical fibers for sensors (Peterson, et al, 1980) and decided to use this approach for the construction of a glucose sensor.

The first documentation of our approach was in a patent application filed in 1978 (Schultz, 1982). The general methodology was named "Affinity Sensors" and first illustrated in a publication by Schultz and Sims (1979), Figure 11.2. Again, Con A was selected as a specific bioreceptor for glucose and the following scheme was proposed.

To quote from this paper:

> "The sensor element communicates with external solutions by means of a dialysis membrane. The membrane porosity is selected to allow free diffusion of low-molecular-weight species and to prevent the leakage of the high-molecular-

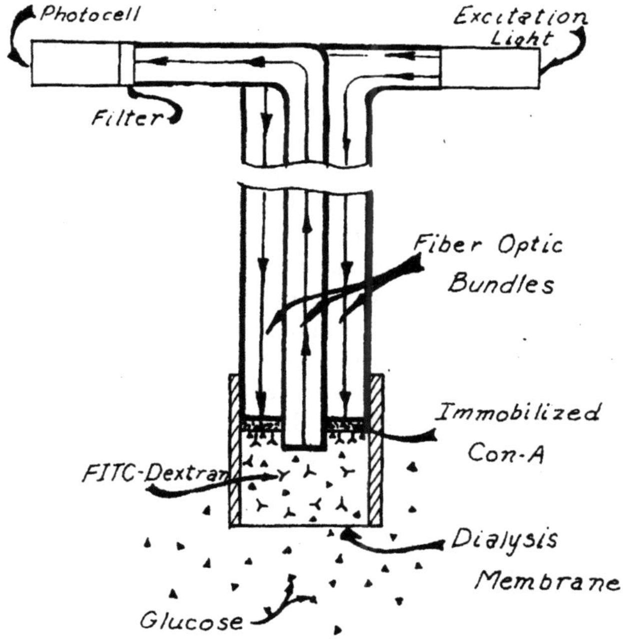

Figure 11.2. Schematic diagram of a fiber-optic affinity sensor for the measurement of glucose (not to scale). Increasing the concentration of glucose in the external solution increases the concentration of glucose in the sensor element. Fluorescent dextran is displaced from the immobilized Con-A and a greater emission intensity is measured. (Schultz and Sims, 1979)

> weight competitive ligand, in this case glucose and FITC-dextran, respectively.
>
> The total amount of Con-A and FITC-dextran in the element is constant, and the internal concentration of glucose will vary according to its concentration in the external solution. Thus the following reversible reactors take place within the sensor element:
>
> $$\text{Glucose+} \quad \text{Con-A} \rightleftharpoons \text{Con-A glucose} \qquad (1)$$
>
> $$\text{FITC-Dextran + Con-A} \rightleftharpoons \text{Con-A-FITC-dextran} \qquad (2)$$
>
> When the sensor element is exposed to higher concentrations of glucose reaction (1) is displaced to the right and reaction (2) is displaced to the left, resulting in an increase in free FITC-dextran. This increase provides a larger amount of fluorescence in view of the emission optical fibers and thus an increase in the output of the photodetectors. Conversely, a reduction in external glucose concentration results in a decrease in signal."

Preliminary experiments to demonstrate the efficacy of this approach were carried out by using immobilized Con-A on sepharose beads as illustrated in Figure 11.3.

A fixed amount of Con-A and FITC-dextran were mixed with different concentrations of glucose and the concentration of FITC-dextran in the supernatant measured by fluorescence methods. The middle curve illustrates the type of sensor behavior expected for these systems, namely at high analyte concentrations the bioreceptor becomes saturated, and thus the sensor system becomes insensitive to changes in analyte concentration. The upper curve in the figure demonstrates another important property of these bioreceptor based systems, namely in the presence of competing analytes (ligands) with higher affinity for the bioreactor, the responsiveness of the device towards the analyte of interest may be reduced. In this illustrative example the affinity of methyl-mano-pyranoside is 40 times greater than glucose and completely abolished the response to glucose. The lower curve represents the theoretical response if the system obeyed the simple stoichiometric equations shown above. But as will be discussed further below, since both Con-A and dextran are polyvalent, the actual binding behavior can differ significantly from a simple monvalent binding curve.

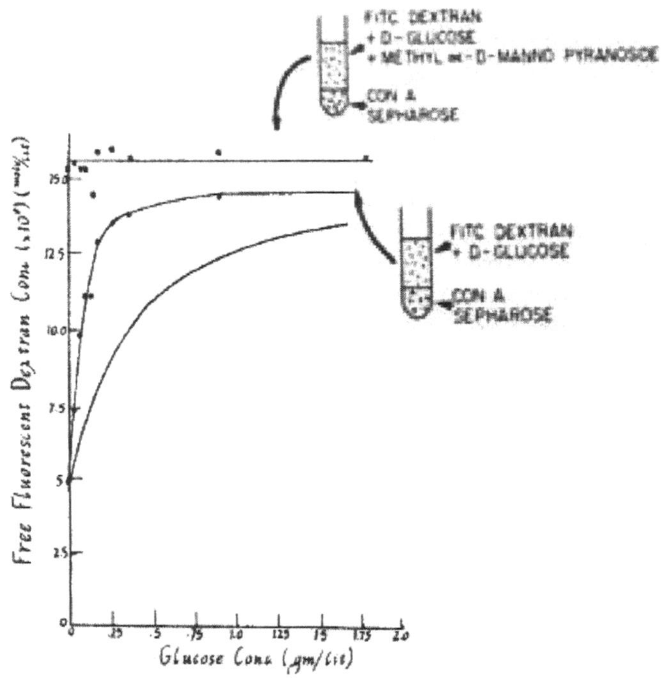

Figure 11.3. Displacement of FITC-dextran from Con-A bound to sepharose by increasing concentration of glucose. Middle curve - Free FITC-dextran concentration after glucose addition. Lower curve - Calculated response based on simple competition between glucose and dextran for Con-A sites. Upper curve - Free FITC-dextran concentration after methyl-mannoside addition. (Schultz and Sims, 1979).

Within the category of optical competitive affinity biosensors there are two general approaches for measuring the extent of binding of the analog-analyte (labeled component) to the bioreceptor. In the configuration shown in Figure 11.2, a form of spatial separation of the labeled analog-analyte from the bioreceptor is used to estimate the extent of binding. In this example Con-A is placed out of the field of view of the optical fiber detector. Many other configurations can achieve similar results, many of which are described in Schultz (1982).

To summarize, strategies for developing bioreceptor affinity biosensors can be categorized as either direct or indirect (i.e. using a competitive analog-analyte). In those techniques using a direct approach the sensor element can be inserted directly into the sample environment. In contrast, in those techniques using a competitive analog-analyte the sensor device must be isolated from the sample and communicate with it through a selective barrier, like a membrane, to prevent loss of the competitive ligand (usually high molecular weight). Thus these types of sensors are best for analytes that are not measurably restricted by the dialysis membrane, i.e. molecular with less than 5 kD.

11.3. COMPETITIVE OPTICAL AFFINITY GLUCOSE BIOSENSORS

The first demonstration of an optically based glucose biosensor was described by Schultz, et al (1982), and Mansouri and Schultz (1984). In this first implementation of our biosensor concept we elected to utilize spatial separation of the bound and free FITC-dextran (the analog-analyte) for monitoring the extent of binding. We took advantage of the effective focusing of illumination that occurs in the distal region of an optical fiber, as shown schematically by the yellow zone in Figure 11.4.

Actually light emanating from the distal end of an optical fiber diverges according to the numerical aperture of the fiber, but there is an effective focusing of the light energy along the central axis because of two effects: a) the diverging cone of light from an optical fiber has its highest intensity at the central axis and decreases towards the periphery, and b) the fraction of light emitted from fluorescent particles within this cone that can reenter the fiber decreases substantially off-center and at further distances from

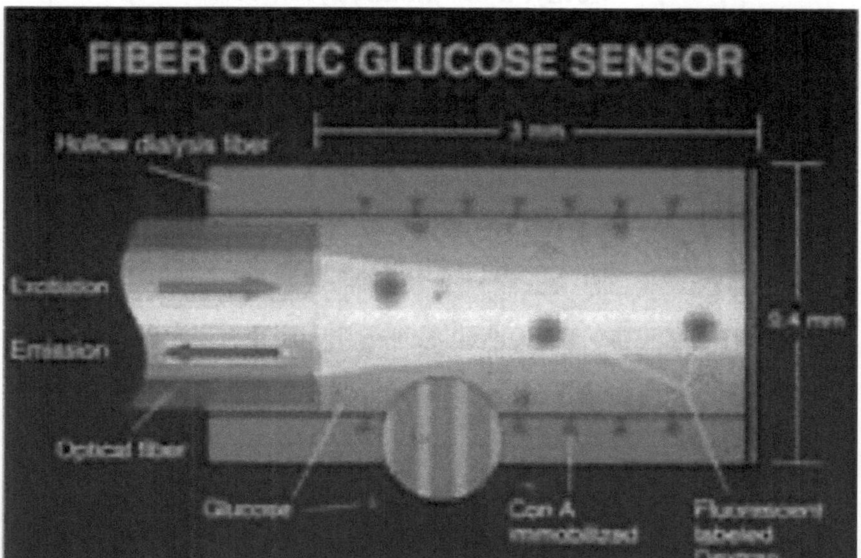

Figure 11.4. Schematic diagram of a fiber optic glucose sensor based on competitive binding of glucose and FITC-dextran for Con-A sites that are immobilized on the interior surface of a hollow dialysis fiber. (see the color insert after p. 429.)

the fiber face (see Komives and Schultz, 1992). The magnitude of emitted fluorescence that is detectable at the front face of an optical fiber levels off at about 3 mm of illumination depth. Typical calibration curves for this sensor configuration for different sugars are shown in Figure 11.5, and follow a similar saturation pattern as found in the preliminary experiments of Schultz and Sims (1979)(Figure 11.2).

As mentioned previously, a variety of sugars can bind to Con-A, and if a mixture of sugars were present in the sample, then the output of the sensor would be difficult to interpret since the one-to-one relationship between displaced FITC-dextran and glucose binding would no longer be valid. Fortunately, glucose is the only predominant sugar in blood and tissue; thus Con-A is satisfactory a bioreceptor for this particular application.

An in-vivo evaluation of our fiber optic glucose biosensor was conducted using the experimental arrangement shown in Figure 11.6. The sensor was placed in a vein of the animal, and the animal was injected with several boluses of a glucose solution.

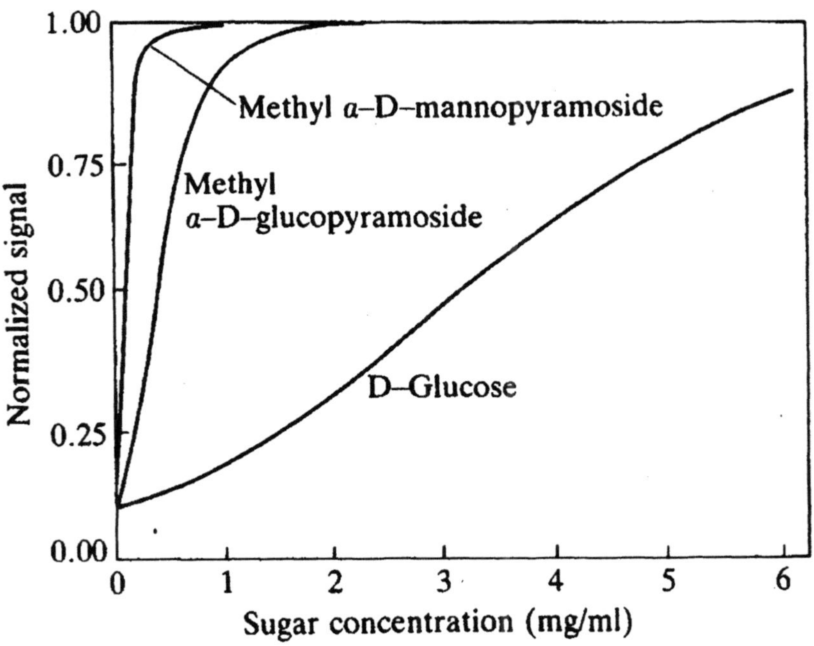

Figure 11.5. Typical response curves for the fiber-optic affinity sensor. The sensor was placed in different sugar solutions with different sugar concentration (Mansouri, 1983).

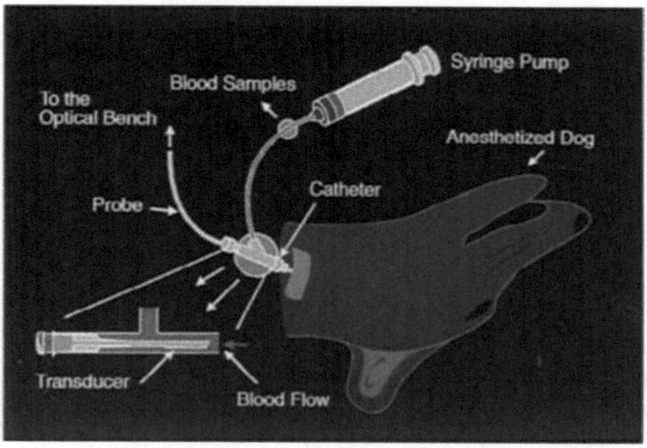

Figure 11.6. Experimental setup for measuring blood glucose levels in an animal. The sensor is placed in a "t" and venous blood continually withdrawn with a catheter. The animal is injected with a bolus of a glucose solution, and then samples removed periodically for glucose analysis. (see the color insert after p. 429.)

Blood samples were withdrawn periodically for conventional glucose analysis. As can be seen in Figure 11.7, the sensor response followed the changes in blood glucose with a slight lag of about 5 minutes.

An alternate method for separately viewing the free analog-analyte was described and patented by Komives and Schultz (1992).

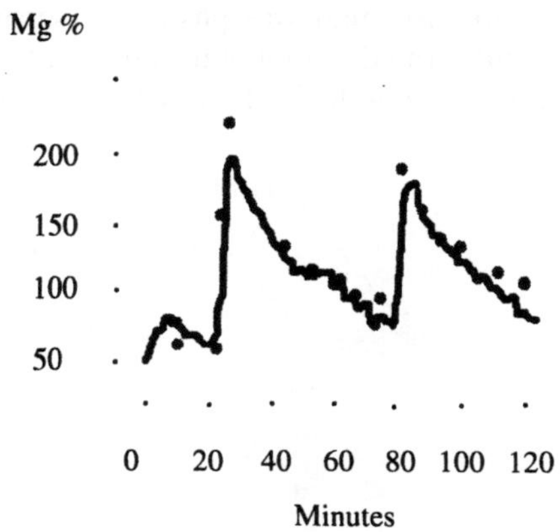

Figure 11.7. Comparison of glucose analyses with the fiber-optic sensor and laboratory clinical analyses (Mansouri and Schultz, 1984). Solid line: glucose biosensor, dots: withdrawn samples and analyzed by a clinical laboratory.

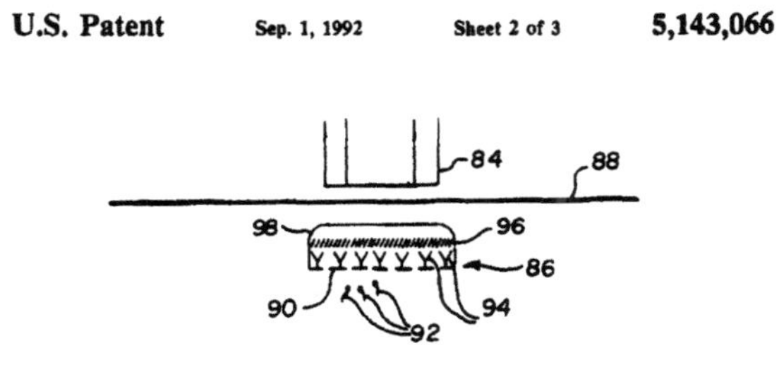

FIG. 3

Figure 11.8. Two-compartment sensor capsule. Lower section has the immobilized bioreceptor (94), and communicates with external analyte (92) through dialysis membrane (90). Lower section is separated from upper viewing chamber by an opaque porous membrane (96). The chamber is illuminated by a separated optical system (84) that also measures emitted light from the analog-analyte (Komives and Schultz, 1992).

As shown in Figure 11.8, the sensor chamber is composed of two compartments, separated by a porous opaque membrane. In this configuration the bioreceptor is placed in the lower chamber and most of the analog-analyte will be bound here in the absence of the analyte in the external fluid. The amount of fluorescence that is obtained from the upper illuminated chamber will be minimal under these circumstances. A similar method was described by Schultz (2001).

When the sensor chamber is exposed to analyte, some will diffuse through the dialysis membrane on the lower surface, analog-analyte will be displaced, diffuse to the upper chamber, and produce a fluorescent signal.

Experiments on a new concept - that of a freely floating biosensor "capsule" that could be placed under the skin - was initiated in my laboratory. This is the first time that a biosensor unit was physically separated from the detection unit and thus could provide minimum invasiveness for in-vivo monitoring. My concept for this type of device is shown in Figure 11.9.

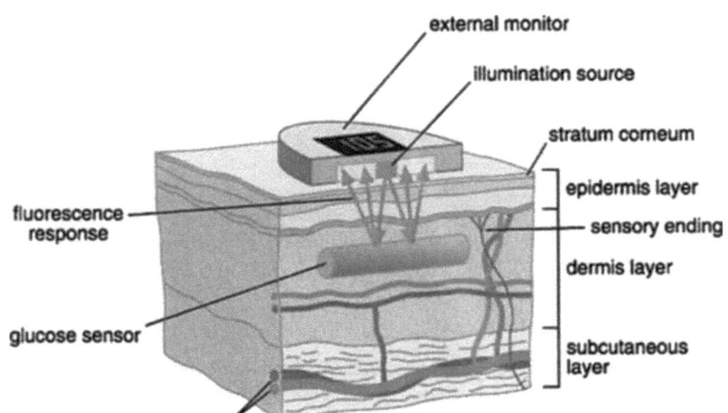

Figure 11.9. Concept for a subcutaneous glucose sensor based on an optical readout principle. (see the color insert after p. 429.)

The idea for a biosensor capsule derives from the earlier work of Lubbers who introduced "optode" technology for measuring the oxygen concentration in tissues (reviewed in Lubbers 1985). These optodes were polymer particles containing a fluorescent dye that exhibited fluorescence quenching in the presence of oxygen. Lubbers injected these particles into an animal and in surface tissues measured their fluorescence with a microscope/spectrofluorimeter apparatus.

In addition to the scheme of spatial separation for measuring the extent of binding, we developed alternative optical approaches to determine the fraction of bioreceptor sites occupied by the analog-analyte. The method introduced by Meadows and Schultz (1988, 1993) was based on fluorescence energy transfer (FRET) phenomena that we adapted from the previous work by Ullman, et al (1976) for immunoassays. One example of implementing this approach is the prototype glucose biosensor developed by Meadows and Schultz (1988) and illustrated in Figure 11.10.

The components of Rhodamine-labeled Con-A and FITC-dextran are placed in a porous dialysis hollow fiber, and sealed in with an optical fiber in one end. In the absence of glucose these two components bind to each other,

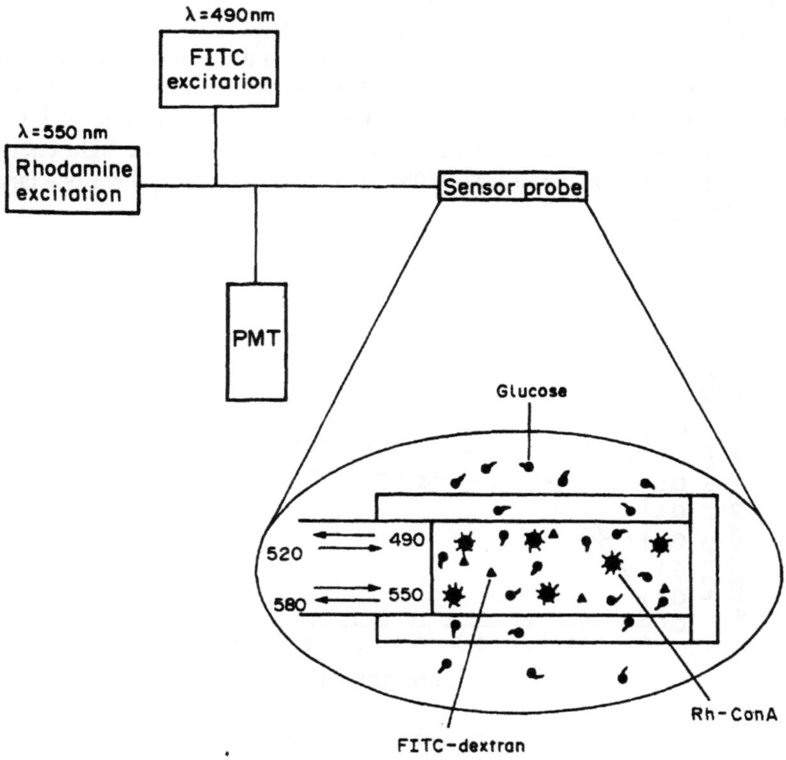

Figure 11.10. Schematic of homogeneous fiber optic affinity sensor, Meadows and Schultz (1998).

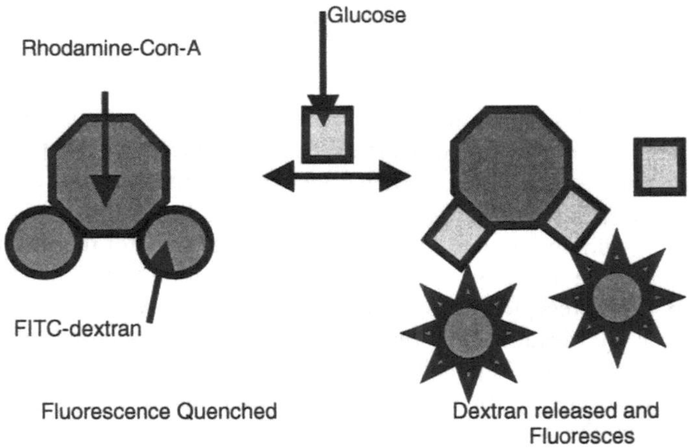

Figure 11.11. Fluorescence Energy Transfer method for a glucose biosensor. Symbols: octagon = Con-A; circle = FITC-dextran, square = glucose

and FRET takes place (Figure 11.11 Left). In the presence of glucose, some of the Con-A sites are occupied by glucose releasing FITC-dextran resulting in an increase in fluorescein fluorescence (Figure 11.11 – Right).

Here Con-A was labeled with rhodamine and dextran was labeled with FITC. One advantage of this method is that the system can be homogeneous, thus simplifying biosensor fabrication considerably. A disadvantage of FRET methods is that the reduction in signal intensity on binding of the light-energy accepting dye is often on the order of 30%-50%. In other words, there is a large background signal.

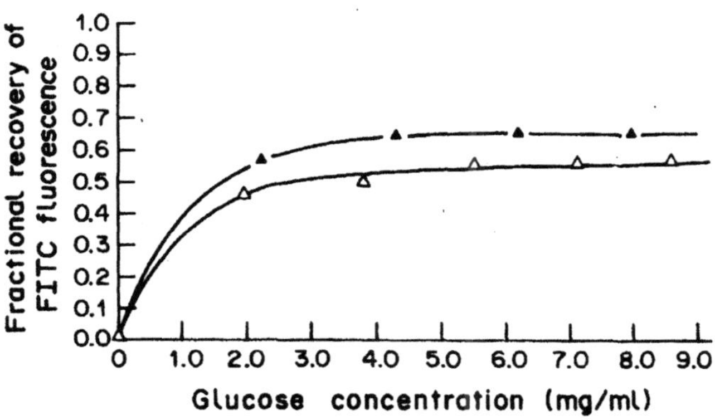

Figure 11.12. Typical calibration curves for the FRET based biosensor (Curves for different ratios of Rhodamine-Con-A to FITC-dextran, Meadows and Schultz (1993)

Since the same chemical reactions (Equations 1 and 2) take place in the Meadows-Schultz FRET system, the typical response to glucose concentration, Figure 11.12, is similar to those shown previously.

It is important to point out that the analysis of the response of competitive binding biosensors (indirect type) are more complex that the Langmuir isotherm binding behavior of the "direct" sensor types, e.g. those developed by the Lakowicz group (Gryczynski, 2003). Basically there are four degrees of freedom for optimizing the design of these sensor systems in so far as their response to analyte concentration, A. These factors are: (a) the binding constant for the analyte to the bioreceptor, K_a; (b) the binding constant for the analog-analyte to the bioreceptor, K_a^*; (c) the total concentration of the bioreceptor in the sensor chamber, R_t; and (d) the concentration of the analog-analyte in the sensor chamber, A^*_t.

The primary determining factor for the sensitivity and selectivity of these biosensors is the choice of the bioreceptor. Choosing a bioreceptor with the appropriate characteristics has the most impact on the performance of the biosensor. Key features are (a) an appropriate binding constant for the analyte in the expected concentration range for a particular application; (b) the selectivity of the bioreceptor for the desired analyte in the sample environment where the sensor will be used; and (c) long term stability of the bioreceptor.

Figure 11.5, above, illustrates the effect of binding constant on the sensitivity range of these biosensors to various analytes. In this example the analyte (sugar) was changed rather than the bioreceptor (Con-A). But the effect of changing binding constant is clearly seen. At the same concentrations of Con-A and FITC-dextran, the sensor is about 20 times more sensitive to methyl-glucopyranoside than to glucose.

A simplified model for competitive binding sensors was provided by Schultz (1987) based on the simple binding model shown in Equations 1 and 2. A graphical portrayal of this model (Figure 11.13) provides some intuitive appreciation for the behavior of these systems. One critical characteristic of sensor performance is the ratio of the background signal (in the absence of analyte) to the maximum signal obtained at high levels of analyte. Since Figure 11.13 is plotted in dimensionless units, this means that one desires low values of the ratio (A^*_t / R_t), or in functional terms, the concentration of available receptor sites should be much greater than the total concentration of the analog-analyte. One must balance this choice with the desired operational range for the sensor. For example the curve marked 0.01 has the desirable low background signal, but at the cost of lower sensitivity in the low concentration ranges of analyte. That is, for the range of analyte concentrations from 0 to 20 units, one prefers a sensor response that more closely resembles the curve marked 0.1 to the curve marked 0.01.

An example of the effect selecting analog-analytes with different affinities (case (b) mentioned above) is shown in Figure 11.14 (Mansouri, 1983). Here two glucose biopolymers (dextran and glycogen) were separately evaluated as the analog-analytes in glucose sensors similar in structure as

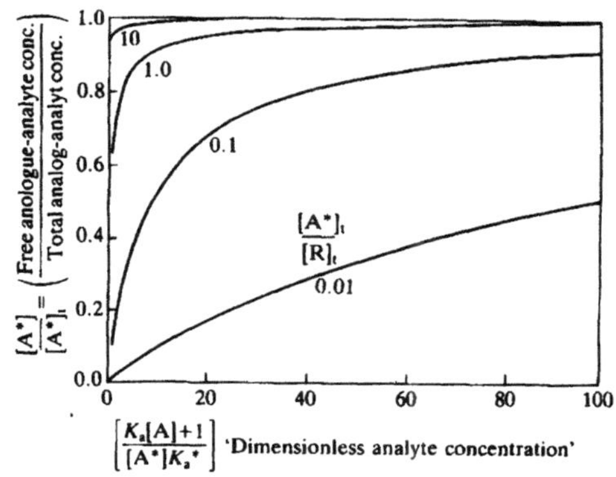

Figure 11.13. Dimensionless parametric plot for the response of competitive affinity biosensors to analyte concentration (Schultz, 1987).

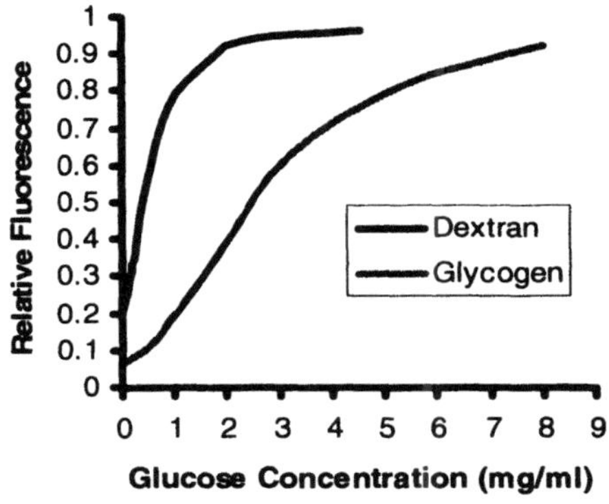

Figure 11.14. Effect of analog-analyte affinity on the response of the glucose biosensor (Mansouri, 1983).

shown in Figure 11.4. In this example glycogen shows a lower background signal, but dextran is more sensitive at lower glucose concentrations.

To further evaluate the effect of analog-analyte molecular weight and the degree and polyvalancy on the affinity of the analog-analyte to the bioreceptor, Liu and Schultz (1986) studied the binding behavior dinitrophenol (DNP) analogs with various DNP antibodies. In these experiments, the displacement of radioactively labeled DNP by different analytes was measured. Some of the results of this study are shown in Figure 11.15.

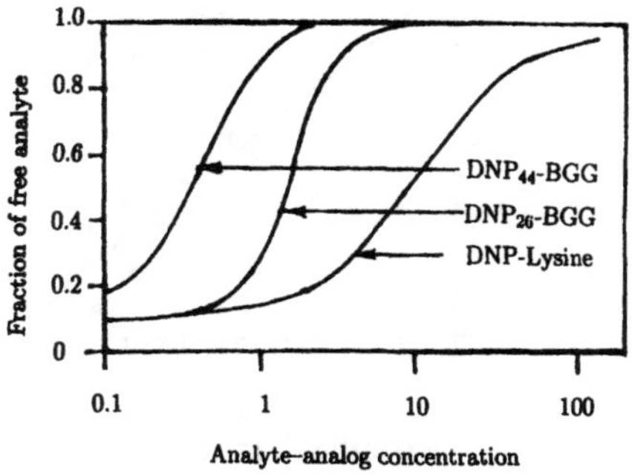

Figure 11.15. Examples of the effect of polyvalency on the affinity of an analog-analyte for an antibody, Liu and Schultz (1986).

Here it is seen that the effective affinity of the analog-analyte can be increase by up to 50 fold by increasing the valance of the analog-analyte from 1 to 44 DNP residues per molecule of bovine gamma globulin (BGG). Thus the range of the biosensor's sensitivity can be broadened by increasing the affinity of the analog-analyte for the bioreceptor. But, this extension of range of sensitivity may come at some decrease in response kinetics, as the dissociation of analog-analyte from the biorecptor is expected to be slower at higher levels of polyvalency.

The range in binding constants that may be available from different bioreceptors for the any given analyte is very extensive. For example a literature search of antibodies for dinitrophenol (DNP) revealed that antibodies have been found with binding constants that range between 10^4 to ^{11} M^{-1} (Schultz, 1987) or 10 million fold, Figure 11.16. Since the "mean" sensitivity of a biosensor is in the range $1/K_a$, one could possibly construct an antibody based biosensor that could measure DNP at concentrations of 10^{-11} M. The data in Figure 11.16 also shows that when one plots the equilibrium binding constant against the dissociation rate constant a surprisingly good inverse correlation is obtained. This correlation is evidence that the association rate for these binding reactions is fairly constant at 10^8 M^{-1} sec^{-1} and that the association process is primarily diffusion controlled.

Schultz (1987) showed that the fastest that a sensor can respond to a reduction in the analyte concentration is given approximately by the equation (where K_d is the dissociation rate constant):

$$T_{lag} = K_d^{-1}$$

Thus for an antibody with a binding constant of about 10^{11} M-1, the time for half removal of analyte is about 10,000 seconds or about 24 hours.

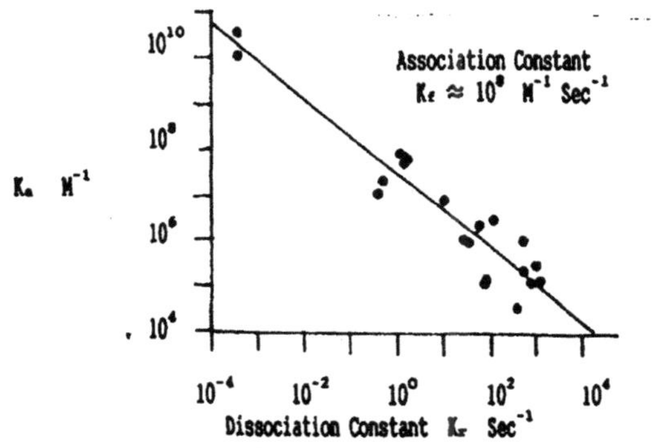

Figure 11.16. Direct correlation of the binding constant of DNP antibodies with the dissociation rate of DNP from these antibodies, Schultz (1987).

It should be pointed out that these results would only be valid for systems with simple binding behavior as depicted by equations 1 and 2. However, for the glucose sensing systems we developed (and followed by others), both the bioreceptor, Con-A, and the analog-analyte (dextran) have multiple binding sites per molecule. This changes the analysis dramatically, because multiple molecular complexes are possible giving rise to a somewhat intractable set of equations to solve. The analysis of multiple binding systems also arises in efforts to model the behavior of immune responses and was recently reviewed by Whitesides and colleagues where thermodynamic entropy concepts have been introduced to explain the behavior of these systems (Mammen, et al 1998). Nevertheless, a completely satisfactory predictive theory for polyvalency systems has yet to be formulated.

More recently combinatorial methods and site directed mutagenesis methods have been used to develop bioreceptors with a wide range of affinities for ligands of interest (Hellinga and Martin, 1998, Dwyer and Hellinga, 2004). Fehr et al (2003) have genetically engineered the maltose binding protein from E. coli, to produce bioreceptors for glucose with a wide range of affinity constants.

11.4. DYNAMICS OF AFFINITY BIOSENSORS

Factors that affect the response rate of these sensor systems are: a) diffusion rate of the analyte into or out of the sensor, b) the rate of reactions of the competing ligands for the receptor binding sites, and c) the diffusion rate of the analog-analyte (dextran) within the sensor chamber. Since the time constant for the reaction rate of glucose and Con-A is on the order of milliseconds, it was suggested that diffusional processes are responsible for the longer response times observed.

A thorough analysis of these factors (diffusion, reaction rates, and polyvalency) as related to sensor performances has not been achieved. Some attempts at a comprehensive analysis have been undertaken, for example Clark, et. al (1999) modeled the Mansouri-Schultz sensor assuming that the reactions were at equilibrium. However, one can get an approximate estimate of diffusional resistance of the dialysis membrane by this simple formula:

$$\text{Response time} = 0.1\ D / L^2$$

Where D = diffusion coefficient for analyte in the membrane (cm^2/sec), and L = membrane thickness (cm.)

11.5. RECENT DEVELOPMENTS IN AFFINITY BASED GLUCOSE BIOSENSOR SYSTEMS

The use of FRET methods for optical biosensors as introduced by Meadows and Schultz (1988) requires the labeling of the bioreceptor with either a dye or a fluorophore to interact with the fluorophore labeled analog-analyte. An alternative strategy devised by Ballerstadt and Schultz (1997, 1998) is to use two variants of the analog-analyte each labeled with a different fluorophore that comprise a FRET pair. Then the polyvalent biorecptor acts as a bridge to bring the two types of analog-analyte in close proximity to exhibit FRET behavior. This method bypasses the necessity to chemically modify the biorecptor, i.e. add a fluorophore. Many bioreceptors are already multivalent, e.g. antibodies, or can be made polyvalent by cross-linking them.

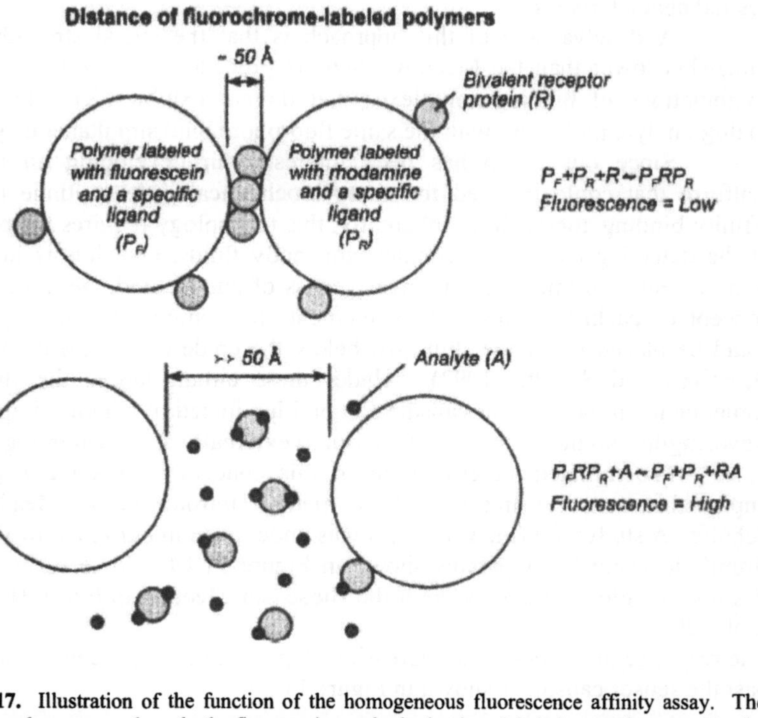

Figure 11.17. Illustration of the function of the homogeneous fluorescence affinity assay. The fluorescence decreases when both fluoroscein and rhodamine labeled complexes with the bioreceptor are formed. In the presence of the analyte, polymer complexes undergo disscociation and fluorescence increases (Ballerstadt and Schultz, 1997).

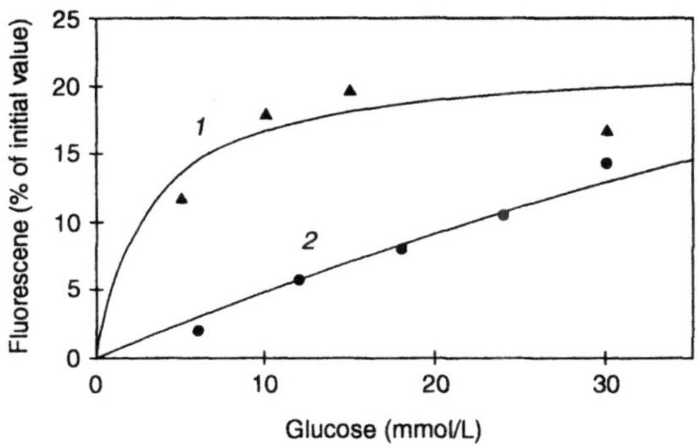

Figure 11.18. Plot of the glucose-induced fluorescence change vs. glucose concentration for two different fluorochrome-labeled dextran preparations. Curve - 1 FITC-dextran and RITC-dextran. Curve 2 – mannose grafted FITC and RITC-labeled dextran (Ballerstadt and Schultz, 1997).

The concept is illustrated in Figure 11.17, and typical calibration curves are shown in Figure 11.18. An additional advantage of this method is that the two analog-analytes can be selected with different binding affinities that then provide another degree of freedom to obtain a calibration curve with the desired characteristics.

A disadvantage of this approach is that the signal strength will be somewhat lower than the Meadows-Schultz procedure. Since there are other combinations of binding complexes that do not exhibit FRET (e.g. if two analog-analyte molecules with the same fluorphore bind simultaneously).

Since our group has been interested in developing an analytical platform that could be used for many biochemicals, we continue to pursue affinity-binding approaches. Inherently, this technology requires the placement of the detecting element in contact with body fluids, and thus is invasive to some extent. To minimize the invasiveness of our method, we introduced the concept of packaging the detector elements in a minute porous capsule that could be placed under the skin, just below the epidermis as mentioned above (Komives and Schultz, 1992). Under these circumstances the fluorescent components in the detector capsule can be illuminated by a light of appropriate wavelengths and the emitted light measured externally as shown in Figure 11.8.

We initiated an evaluation of this concept by making a prototype implantable sensor using the FRET method introduced by Meadows and Schultz. A study of this new concept was undertaken in our laboratory by Anne Brumfield using the apparatus shown in Figure 11.19. Some early results of this investigation were reported at the Biosensor Meeting in Berlin (Brumfield, et al, 1998).

The response of the device to increasing steps in glucose concentration pumped past the sensor capsule is shown in Figure 11.20.

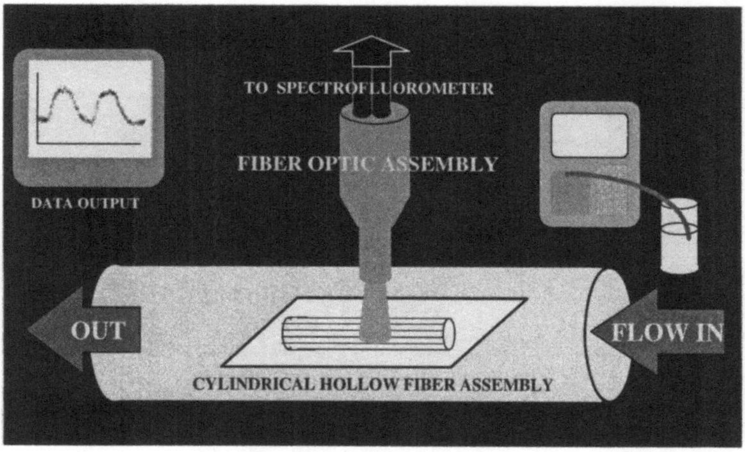

Figure 11.19. Apparatus for evaluation of sensor capsules designs for subcutaneous implantation.

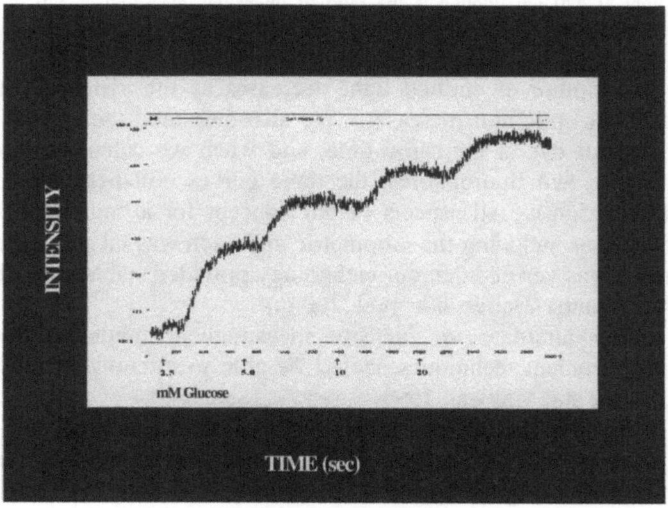

Figure 11.20. Response of glucose biosensor capsules to changes in glucose concentration in the external fluid.

In this type of configuration, where both the illumination of the capsule and emitted light from the capsule is scattered by the heterogeneous structure of skin, it requires a robust calibration technique to compensate for these and other variables from one sampling period to another.

Our method of providing compensation was to introduce an inert fluorescent dye with a different excitation spectrum into the capsule as an internal control. The data in Figure 11.20 shows that this method can be successful. Figure 11.20 (left) shows typical calibration curves for the sensor measured with the fiber optics placed at 2, 3, and 4 mm from the capsule. As

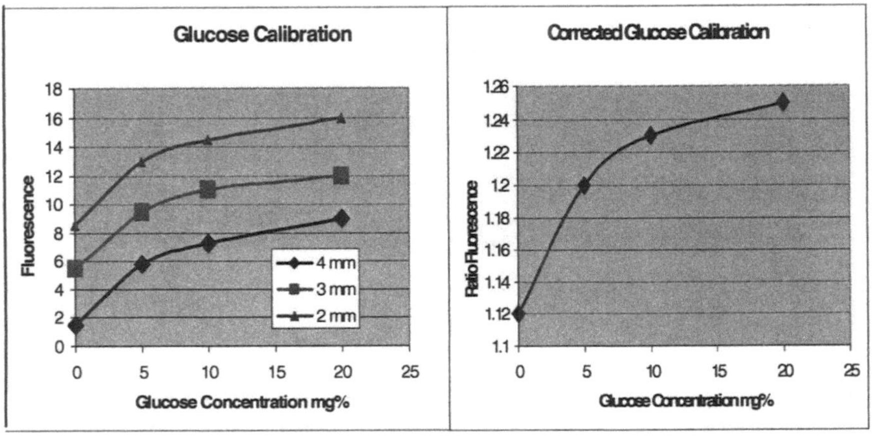

Figure 11.21. Method for compensating for variations in light intensity for subcutaneous sensors. Left panel – Fluorescence intensity measured at various distances between sensor capsule and light detector (emission at 520 nm). Right panel – Compensation obtained by using an inert fluorophore that emits light at 420 nm. Plotted is the ratio of intensities measured at 520 to that at 420 (Brumfield, et al, 1998) .

expected the capture of emitted light decreases as the distance between the sensor and fiber optic unit increases. We also measured the emission of light from our control dye at the same time, and when we calculated the ratio of emission of the two fluorophores, the three curves collapsed into one curve (Figure 11.20, right). All aspects of our concept for a "minimally-invasive" glucose biosensor including the ratiometric approach was taken by the group at BioTex and it was verified that our technology provided stable glucose readings for up to six months (Ballerstadt, et al, 2004a).

As an alternate to intensity measurement methods, fluorescence lifetime measurement techniques should be able to circumvent the scattering issue, Lakowicz and Maliwal, 1993.

With these promising results in hand, we continued our efforts to improve the response of the sensor beyond that achievable by FRET techniques. We returned to our original approach of spatially separating the bound and free form of one of the reactants. The strategy we took was similar in principle to the arrangement published by Schultz and Sims (1997). In our new method, Ballerstadt and Schultz (2000), the beads are infiltrated with dyes that absorb at the wavelengths of excitation light and also the emitted light from the fluorphores, Figure 11.22.

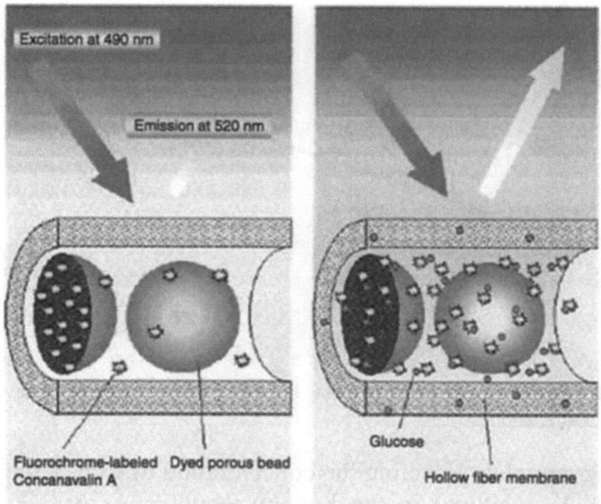

Figure 11.22. Microsphere glucose biosensor. In this configuration, the analog-analyte is a porous Sephadex bead (a polymer containing pendant glucose residues). The beads are colored with dyes that prevent the excitation light from penetrating into them and another dye that blocks emission from fluorochrome-labeled Con-A in the interior of the bead. (Ballerstadt and Schultz, 2000). (see the color insert after p. 429.)

In this example the biorecptor ConA is labeled with the fluorophore. In order to evaluate the performance of this method the beads, approximately 40 microns in diameter, are placed within a dialysis fiber to retain the Con-A. However, the system can be miniaturized even further by encapsulating individual beads within a porous membrane. Typically response of this device to changes in glucose concentration is on the order of 1 minute, Figure 11.23.

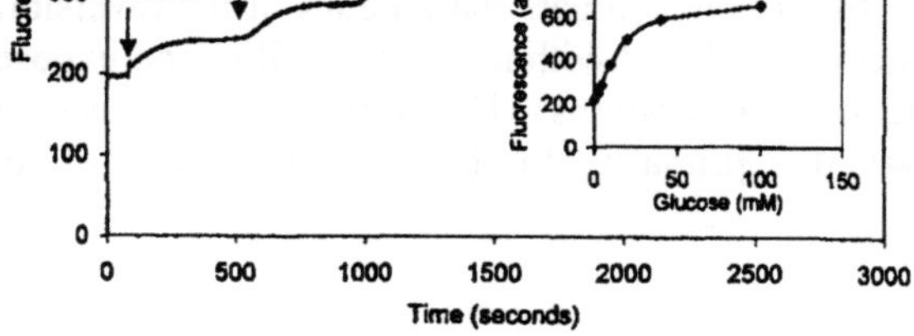

Figure 11.23. The time response of a hollow fiber sensor filled with colored Sephadex G200 beads. Arrows indicate a change in external glucose concentration. The inset shows a calibration curve. (Ballerstadt and Schultz, 2000).

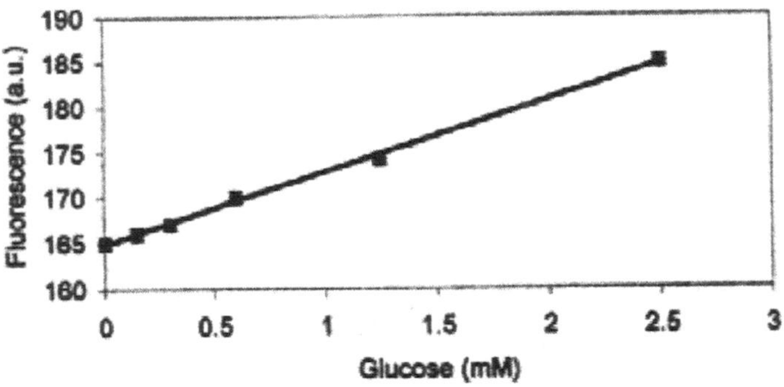

Figure 11.24. Response of the bead glucose biosensor to hypoglycemic levels of glucose. (Ballerstadt and Schultz, 2000).

By appropriately selecting the concentration of Con-A we were able to achieve excellent sensitivity in the hypoglycemic range of 1-3 mM, Figure 11.24.

11.6. INTEGRAL BIOSENSOR PROTEINS

Throughout our studies to develop various biosensor modalities, we were aware that one of the limitations of the competitive analog-analyte approach was the need to encapsulate the device within a porous chamber to prevent loss of the analog-analyte. This is particularly a problem if one desires to measure the glucose concentration within cells. To fulfill this need we initiated a program to develop a protein that had an integral reporter function for monitoring glucose binding, i.e. a "direct" approach. This approach utilizing FRET had been previously demonstrated for monitoring calcium intracellularly by Roger Tsien (Miyawaki and Tsien, 2000).

The structure of the protein that was engineered for this purpose is shown in Figure 11.25 (Ye and Schultz, 2003, Ye,et al 2004). There is an allosteric change in the protein structure when glucose binds, bringing the two green fluorescent proteins (GFP and YFP) in close proximity and allowing FRET to occur. Thus one mode of monitoring glucose concentration is to excite the GFP moiety with light at 395 nm and to monitor the emission from YFP at 527 nm. In order to evaluate the kinetics of the reversibility of glucose binding, the protein was placed within a dialysis fiber and exposed to solutions of different glucose concentrations. The response time was on the order of 5 minutes that was due primarily to the diffusion resistance of the dialysis membrane.

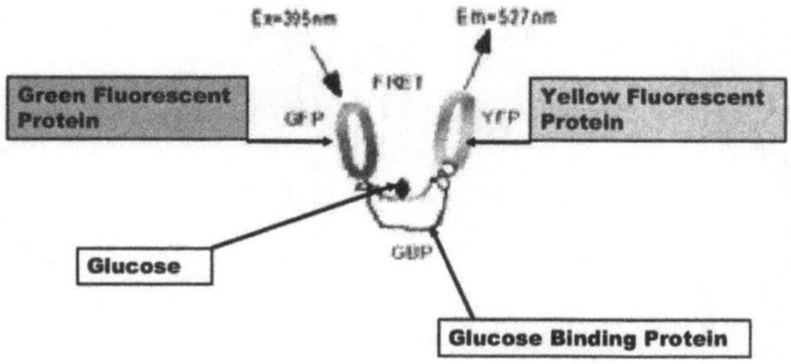

Figure 11.25. Integral biosensing protein for glucose constructed from an E. coli glucose binding protein.

This type of sensor protein opens many new opportunities for glucose sensing as shown in Figure 11.26. It can be expressed in cells to monitor intracellular variations in glucose concentration, it can be immobilized by various techniques and placed in various sample environments, or it can be placed within a porous membrane capsule to protect the protein from enzyme degradation while permitting exposure to analytes.

Fehr et al (2002) has shown, Figure 11.27, that this protein when expressed in yeast provides a measure on internal maltose concentration. Fehr et al (2003) also engineered a glucose sensitive protein and expressed it in mammalian cells.

11.7. ENHANCEMENTS OF OPTICAL GLUCOSE BIONSENSORS BASED ON AFFINITY PRINCIPLES

Over the last few decades an impressive array of techniques for producing biosensors based on our optical monitoring methods have been

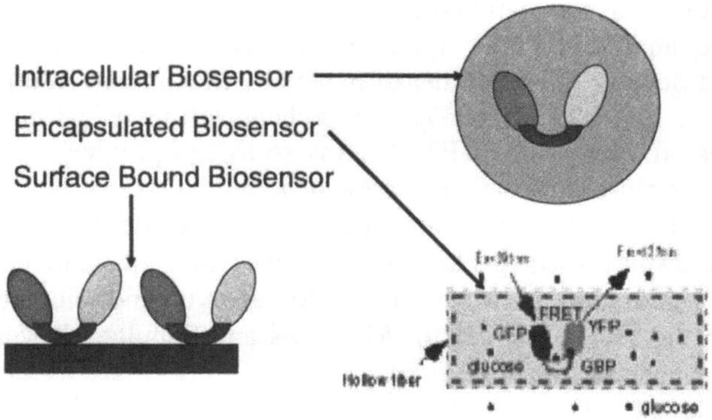

Figure 11.26. Various applications for integral biosensing proteins.

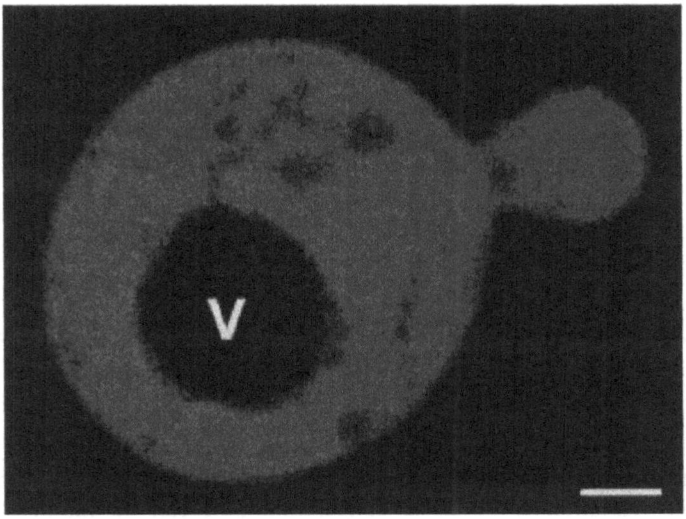

Figure 11.27. Confocal imaging of integral sensing protein in a yeast cell. No sugar is present in the vacuole (V). (Fehr, et al, 2002)

developed. Each of these techniques has its own favorable characteristics as well as some limitations.

Russell, et al (1999) showed that the FRET methodology of Meadows and Schultz also works in by immobilizing the reagents in hydrogels, although the response time is extended in this format. A patent on this method was issued to Cote et al (2002). In another alternative, Medintz et al (1999) developed a tethering technique for immobilizing the bioreceptor and analog-analyte for a FRET based sensor. Chinnayelka and McShane (2004b) have shown that the dimensions of the Meadows-Schultz FRET-type sensor can be reduced by encapsulating the reagents in microshells.

Lakowicz and Pickup (Lakowicz, 1993, Tolosa, 1997a, 1997b, McCartney 2001) have shown that phase-modulation fluorometry is a powerful method for quantitating FRET type sensor devices as those developed by Meadows and Schultz. In a number of papers by Pickup and Rolinski (Pickup, 2001; Rolinski, 1999, 2000, 2001) a comprehensive theoretical analysis of the molecular spatial aspects of FRET systems so as to optimize the response of these systems as biosensors has been presented.

Aslan and coworkers (2004a and 2004b) have demonstrated that gold particles that display surface plasmon resonance behavior can be utilized to monitor the association between the analog-analyte and bioreceptor as an alternative to the FRET methods of Meadows and Schultz. Reversibility has not yet been evaluated.

While at Motorola, Ballerstadt, et al (2004) implemented the methodology developed in my laboratory at Pittsburgh (Ballerstadt and Schultz, 2000) and showed the system can be used for several months.

Blagoi, et al (2005), have extended the FRET proximity methodology of Ballerstadt and Schultz (1997) for detection of carbohydrates and glycoproteins.

Burke, et al (2000) noted that one method for increasing the binding affinity of a bioreceptor such as Con-A for ligands to achieve higher sensitivity is to create multivalent ligands by a reversible cross-linking agent.

Chen and Hsu (1996) have extended the modeling of affinity biosensors previously presented by Schultz (1986). Kermis et al (2003) have investigated the use of HEMA membranes for affinity biosensor applications.

Chinnayelka and McShane (2004a,b) and Scognamiglio et al (2004) have shown that apo-enzymes can also be used as bioreceptors for affinity-based sensors.

Ge et al (2004) developed alternative fluorophores for the integral biosensor protein introduced by Fehr and Ye. While Hellinga (1999) has generated protein engineering techniques to produce internal FRET bioreceptors for biosensors.

Meledeo et al (2002) extended the studies of Mansouri (1983) on the effects of pH and temperature on the behavior of affinity biosensors.

11.8. CONCLUSION

Now the performance behavior of these various techniques is understood well enough that the appropriate choice for a specific application can be rationally determined.

11.9. REFERENCES

Aizawa, M., Morioka, A., Matsuoka, H., Suzuki, S., Nagamura, Y., Shinohara, R., Ishiguro, I., 1976, An enzyme immunosensor for IgG, *J. Solid-Phase Biochemistry*, **1**, 319-328.

Aslan, K., Lakowicz, J. R., Geddes, C. D., 2004a, TI Tunable plasmonic glucose sensing based on the dissociation of Con A-aggregated dextran-coated gold colloids, *Analytical Chimica Acta*. **517**:139-144.

Aslan, K., Lakowicz, J. R., Geddes, C. D., 2004b, TI Nanogold-plasmon-resonance-based glucose sensing, *Analytical Biochemistry*. **330**:145-155.

Ballerstadt, R., Polak, A, Beuhler, A, Frye, J., 2004b, TI In vitro long-term performance study of a near-infrared fluorescence affinity sensor for glucose monitoring, *Biosensors & Bioelectronics*. **19**:905-914.

Ballerstadt, R. and Schultz, J.S., 1997, Assay based on fluoresence quenching of ligands held in close proximity on a multivalent receptor. *Anal. Clinica Acta*., **345**:203-212.

Ballerstadt, R. and Schultz, J.S., 2000, A fluorescence affinity hollow fiber sensor for continuous transdermal flucose monitoring, *Anal. Chem.*, **72**: 4185-4192.

Ballerstadt, R. and Schultz, J.S., Homogeneous Affinity Assay for Quantitative Drug and Metabolite Determination. #5,814,449, Sept. 29, 1998

Ballerstadt, R. and Schultz, J.S., Method and Kit for Detecting an Analyte. #6,271,044, Aug. 7, 2001

Ballerstadt, R., Gowda, A., And Roger Mcnichols, R., 2004a, fluorescence resonance energy transfer-based near-infrared fluorescence sensor for glucose monitoring, *Diabetes Technology & Therapeutics*, **6**, 191-200.

Beck, R.E. and Schultz, J.S., 1970, Hindered diffusion in microporous membranes with known pore geometry, *Science* **170**:1302-1305.

Berson. S. A. and R. S. Yalow. 1959. "Quantitative Aspects of Reaction Between Insulin and Insulin-Binding Antibody." J, Clin. Invest. **38**, 1996-2016.

Blagoi, G., Rosenzweig, N., Rosenzweig, Z, 2005, TI Design, synthesis, and application of particle-based fluorescence resonance energy transfer sensors for carbohydrates and glycoproteins, *Analytical Chemistry*, **77**:393-399.

Brumfield, A.; Ballerstadt, R.; Schultz, J. S.; Schultz, J. S. 1998, *Fifth World Congress on Biosensors*; Berlin, Germany,; Elsevier: Amsterdam; p 48.

Burke, S. D., Zhao, Q., Schuster, M. C. and Kiessling, L. L., 2000, Synergistic formation of soluble lectin clusters by a templated multivalent saccharide ligand, *J. Am. Chem. Soc.*, **122**:4518-4519.

Chen, J. P. and Hsu, M. S., TI Mathematical analysis of sensors based on affinity interactions between competitive receptor-protein pairs, 1996, *Journal of Chemical Technology and Biotechnology*. **66**:389-397.

Chen, Y., Ji, T., and Rosenzweig, Z, 2003, Synthesis of Glyconanospheres Containing Luminescent CdSeZnS Quantum Dots, *Nano Letters*, 3: 581-584

Chinnayelka, S. and McShane, M. J., 2004a, TI Glucose-sensitive nanoassemblies comprising affinity-binding complexes trapped in fuzzy microshells, *Journal of Fluorescence*. **5**:585-595.

Chinnayelka, S. and McShane, M. J., 2004b, TI Resonance energy transfer nanobiosensors based on affinity binding between apo-enzyme and its substrate, *Biomacromolecules*. **5**:1657-1661.

Clark LC, and Lyons C., 1962. Electrode systems for continuous monitoring in cardiovas cular surgery, *Annals N.Y. Acad. Sci.* **102**, 29-45.

Clark, H. R., Barbari, T. A., and Rao, G, 1999, TI Modeling the response time of an in vivo glucose affinity sensor, *SO Biotechnology Progress*. **15**:259-266.

Cote, G. L, Pishko, M. V., Sirkar, K., Russell, R., and Anderson, R. R., 2002,Hydrogel particle compositions and methods for glucose detection. 56 pp. Application: US 99-354914 19990709. Priority: US 98-94980 19980731.

Dwyer, M.A., and Hellinga, H.W., 2004, Periplasmic binding proteins: a versatile superfamily for protein engineering, *Curr Opin Struct Biol.* Aug 14 (4), 495-504

Fehr, M., Frommer, W.B. and Lalonde, S., 2002, Visualization of maltose uptake in living yeast cells by fluorescent nanosensors, *Proc. Natl. Acad. Sci. USA*. **99**:9846-9851.

Fehr, M., Lalonde, S., Lager, I., Wolff, M.W. and Frommer W. B., 2003, In vivo imaging of the dynamics of glucose uptake in the cytosol of COS-7 cells by fluorescent nanosensors, *J. BiologicalChem.* **278**:19127-19133.

Ge, X. D., Tolosa, L., Rao, G., 2004, TI Dual-labeled glucose binding protein for ratiometric measurements of glucose, *Analytical Chemistry*. **76**:1403-1410.

Gryczynski Z, Gryczynski I, Lakowicz JR., 2003, Fluorescence-sensing methods, *Methods Enzymol.* **360**:44-75.

Hellinger, H. W., Biosensor, *International Patent number* WO 99/34212. Publication date July 8, 1999

Kermis, H. R., Rao, G, Barbari, T. A., 2003, TI Transport properties of pHEMA membranes for optical glucose affinity sensors, *Journal of Membrane Science*. **212**:75-86.

Komives, C. and Schultz, J.S, Optical Fiber Sensors for Continuous Monitoring of Biochemicals and Related Method.*U.S. Patent* # 5,143,066, Sept. 1, 1992. 12 pp.

Komives, C. and Schultz, J.S., 1992, Fiber-optic fluorimeter signal enhancement and application to biosensor design. *Talanta*, **39**: 429-441.

Lakowicz, J. R., Maliwal, B, 1993, Optical sensing of glucose using phase-modulation fluorometry, , *Analytical Chimica Acta.* **271**:155-164.

Liu, B. and Schultz, J.S., 1986, Equilbrium binding in immunosensors. *IEEE Trans. Biomed. Eng.* **53**, 133-138

Lubbers DW, 1995, Optical sensors for clinical monitoring, *Acta Anaesthesiol Scand Suppl*, **104**:37-54

Mammen, M., Choi, S., and Whitesides, G. M., 1998, Polyvalent interactions in biological systems: implications for design and use of multivalent ligands and inhibitor, *Angew. Chem. Int. Ed.* **37**:2754-2794.

Mansouri, S., 1983, Optical glucose sensor based on affinity binding. Ph.D. Thesis, The University of Michigan

Mansouri, S. and Schultz, J. S., 1984, A miniature optical glucose sensor based on affinity binding, *BIO/TECHNOLOGY*, **2**:385-390.

McCartney, L. J., Pickup, J. C., Rolinski, O. J., Birch, D. J. S., 2001, TI Near-infrared fluorescence lifetime assay for serum glucose based on allophycocyanin-labeled concanavalin, *Analytical Biochemistry*. **292**:216-221.

Meadows, D. and Schultz, J.S., 1988, Fiber optic biosensors based on fluorescence energy transfer. *Talanta* **35**: 145-150.

Meadows, D.L. and Schultz, J.S., 1993, Design, manufacture and characterization of an optical fiber glucose affinity sensor based on homogeneous fluorescence energy transfer assay system. *Analytica Chimica Acta*, **280**:21-30.

Meadows, D.L., and Schultz, J.S., 1991, A molecular model for singlet/singlet energy transfer of monovalent ligand/receptor interactions, *Biotechnology and Bioengineering*. **37**: 1066-1075.

Medintz , I. L., Anderson, G. P., Lassman, M.E., Goldman, E. R., Bettencourt, L. A. and Maurol, J. M., 1999, A general strategy for biosensor design and construction employing multifunctional surface-tethered components, *Anal. Chem.* **71**:3126-3132.

Meledeo, M. A., Ibey, B. L., O'Neal, D. P., Pishko, Michael, V., Cote, G. L., 2002, *Investigation of pH and temperature effects on FRET systems for glucose sensing, Proceedings of SPIE-The International Society for Optical Engineering, Optical Diagnostics and Sensing of Biological Fluids and Glucose and Cholesterol Monitoring II*, 4624:55-65.

Miyawaki A, Tsien RY, 2000, Monitoring protein conformations and interactions by fluorescence resonance energy transfer between mutants of green fluorescent protein. Methods Enzymol.327:472-500.

Peterson, J.I., Goldstein, S.R., Fitzgerald, R.V., and Ruckold, D., 1980, Fiberoptic pH probe for physiological use. *Anal. Chem.*, **52**, 864-869.

Pickup, J. C., 2001, Near-Infrared Fluorescence Lifetime Assay for SerumGlucose Based on Allophycocyanin-Labeled Concanavalin A *Analytical Biochemistry*. **292:**216–221.

Rolinski, O. J., Birch, D. J. S., McCartney, L. J., and Pickup, J. C., 2001, Fluorescence nanotomography using resonance energy transfer: demonstration with a protein–sugar complex, *Phys. Med. Biol.* **46**:221–226.

Rolinski, O. J., Birch, D. J. S., McCartney, L. J., and Pickup, J. C., 2000, A method of determining donor–acceptor distribution functions in Forster resonance energy transfer, *Chemical Physics Letters*. **324**:95–100.

Rolinski, O. J., Birch, D. J. S., McCartney, L., and Pickup J. C., 2001, Molecular distribution sensing in a fluorescence resonance energy transfer based affinity assay for glucose, *Spectrochimica Acta Part A* **57**:2245–2254.

Rolinski, O. J., Birch, D. J. S., McCartney, L. J., and Pickup, J. C., 1999, *Near-infrared assay for glucose determination. Proceedings of SPIE-The International Society for Optical Engineering, Advances in Fluorescence Sensing Technology IV*. **3602**:6-14.

Rolinski, O. J., Birch, D. J. S., McCartney, L. J., and Pickup, J. C., 2000, A time-resolved near-infrared fluorescence assay for glucose: opportunities for trans-dermal sensing, *J. Photochem. Photobiol.*, **54**:26-34.

Rosenzweig, Z., Rosenzweig, N., Blagoi, G., 2004, FRET-based luminescence sensors for carbohydrates and glycoproteins analysis. *Proceedings of SPIE-The International Society for Optical Engineering , Smart Medical and Biomedical Sensor Technology II*. **5588**:1-8.

Russell, R. J., Pishko, M. V., Gefrides, C. C., McShane, M. J., Cote, G. L., 1999, TI a fluorescence-based glucose biosensor using concanavalin A and dextran encapsulated in a poly(ethylene glycol) hydrogel, *Analytical Chemistry*. **71**:3126-3132.

Schultz, J.S., 1982, Optical sensor of plasma constituents. *U.S. Patent #4,344,438*. Aug. 17, 1982. 11 pages.

Schultz, J.S., 1986, Design of Fiber-Optic Biosensors Based on Bioreceptors, in: *Biosensors: Fundamentals and Applications*, A.P.F. Turner, I. Karube, and G.S. Wilson, eds., Oxford University Press, pp 638-654.

Schultz, J.S., 1987, Sensitivity and Dynamics of Bioreceptor-Based Biosensors, in: *Biochemical Engineering V., Annals N.Y. Acad. Sci.*, pp. 406.411.

Schultz, J. S., 1991, Biosensors, *Scientific American*. August 64-69.

Schultz, J. S., 1996, Biological and Chemical Components For Sensors, in: *Handbook of Chemical and Biological Sensors*, R.F. Taylor and J.S. Schultz, eds., Institute of Physics, Philadelphia, pp. 171.202.

Schultz, J.S. Sensors for continuous monitoring of biochemicals and related method, U.S. Patent #6256522 July 3, 2001

Schultz, J.S. and Ballerstadt, R.. Homogeneous Affinity Assay for Quantitative Drug and Metabolite Determination. U.S. Patent #5,814,449, Sept. 29, 1998

Schultz, J.S., and Mansouri, S., 1987, Optical Affinity Sensors, in: Methods in*: Enzymology, Vol 137 Immobilized Enzymes and Cells, part D.*, K. Mosbach, ed., Academic Press, pp. 349.365.

Schultz, J. S., Mansouri, S., and Goldstein, I. J., 1982, Affinity glucose sensor, *Diabetes Care,* **5**:245-253.

Schultz, J.S. and Sims, G., 1979, Affinity sensors for individual metabolites, *Biotech. and Bioeng., Symp.* **9**:65-71.

Scognamiglio, V., Staiano, M., Rossi, M., D'Auria, S., 2004, TI Protein-based biosensors for diabetic patients, *Journal of Fluorescence*. **14**:491-498.

Tolosa, L.,Malak, H.,Raob, G., and Lakowicz, J. R., 1997a, TI optical assay for glucose based on the luminescence decay time of the long wavelength dye Cy5 (TM), *Sensors and actuators B-Chemical*. **45**:93-99.

Tolosa, L., Szmacinski, H., Rao, G., Lakowicz, J. R., 1997b, TI lifetime-based sensing of glucose using energy transfer with a long lifetime donor, *Analytical Biochemistry*. **250**:102-108.

Ullman, E.F., Schwarzberg, M., and Rubenstein, K., 1976, Fluorescence excitation transfer immunoassay, a general method for determination of antigens, *J. Biol. Chem*., **251**, 4172.

Weber, A., and Schultz, J.S., 1992, Fiber-optic fluorimetry in biosensors: comparison between evanescent wave generation and distal-face generation of fluorescent light. *Biosensors and Bioelectronics*, **7**:193-197.

Ye, K. and Schultz, J.S., 2003, Genetic engineering of an allosteric-based glucose indicator protein for continuous glucose monitoring by fluorescence resonance energy transfer, *Anal. Chem.* **75**: 3451-3459.

RECENT CHEMILUMINESCENCE APPLICATIOINS FOR GLUCOSE SENSING

S. H. Lee[1] and M. M. Karim

12.1. INTRODUCTION

Carbohydrates constitute one of the three major chemical compound classes that are ingredients of nutrient products; they play a very important role in the life cycle and form the base of many foodstuffs (Evmiridis et al., 1999). The quality and the caloric value of many foodstuffs and diet products are partly dependent on the quantity and type of their carbohydrate content. In contrast, the presence of excess of sugar in the blood gives evidence of serious malfunction of the human organism (Evmiridis et al., 1999).

The determination of glucose content in human blood is an important analysis for the diagnosis and effective treatment of diabetes. Therefore, a large number of efforts have been concentrated to develop effective diagnostic tools for the benefit of diabetic patients in recent years (Lv et al., 2003). Conventional methods for the determination of carbohydrates are described in various textbooks (Horwitz, 1975; Heldrich, 1990; Watson, 1994; Egan, 1981; Hart, 1971) The gravimetric method using Fehling's reagent or the titrimetric using iodine solution are the most common in carbohydrate analysis (Evmiridis et al., 1999). However, they lack selectivity and sensitivity and in most cases are time consuming. CL analysis promises high sensitivity with simple instruments and does not need any light source, so CL have been one of attractive detection methods in recent years Otherwise, many flow chemical sensors and flow biological sensors which employed CL reagents in immobilized or solid-state format, have been developed. Those sensors had some advantages of simplicity of apparatus, little reagent consumption, and high sensitivities. In this chapter, the developments in glucose sensing by chemiluminogenic methods in the form of chemical sensors, application of capillary electrophoresis using CL detection

[1]S. H. Lee, and M. M. Karim, Department of Chemistry, Kyungpook National University, Taegu 702-701, Korea

technique and most recent technological approaches for determination of glucose are discussed.

12.2. CHEMILUMINESCENCE FOR THE DETERMINATION OF GLUCOSE

The chemiluminescence reaction of luminol with oxidizing agents was first reported in 1928 by Albrecht. Since that time the reaction has been used mostly to determine hydrogen peroxide, other oxidants and metal ions such as Cr(III), Cu(II), Fe(II), Fe(III), Ni(II), etc. Recently, several reviews have been reported on the application of chemiluminescence in chemical analysis (Robards and worsfold, 1992; Lewis et al., 1993; Rongen et al., 1994; Kricka, 1995; Bowie et al., 1996) In these reviews, the determination of reducing agents such as glucose using luminol is based on the measurement of the hydrogen peroxide produced by oxidizing enzymes when converting these reducing agents. Chemiluminescence is produced on oxidation of carbohydrates by acidic potassium permanganate in the presence of Mn^{II}. Glucose, galactose, fructose, arabinose ,xylose, lactose , sucrose and sugar all have been detected over the linear range 10^{-4} to 10^{-1} mol dm^{-3} (Agater et al., 1996).

Periodate oxidation is somewhat selective for carbohydrate compounds and can be applied for the determination of multicomponent samples (Malaprade, 1928; Pigman, 1956; Percival, 1962; Sklarz, 1967). In addition, the chemiluminescence (CL) emission detection method, apart from being selective, is very sensitive and extremely sensitive with existing photomultiplier technology (Robards and worsfold, 1992; Bowie et al., 1996; Townshend, 1995; Townshend and Wheatley, 1998; Babko et al., 1966; Chappelle and Picollo, 1975; Seitz et al., 1973; Kricka et al., 1983). The periodate oxidation method combining with CL detection has added advantages of relatively high sensitivity and increased selectivity. A kinetic method reported (Evmiridis et al., 1999) for the determination of glucose and fructose in mixtures is based on initial rate measurements and the overall rate constant of their oxidation by periodate using CL detection. The method is designed to determine hexose contents as low as 200 mg. However, the method can be tailored for even lower amounts (10 times less). When the total hexose concentration is known, the individual fructose and glucose content can be determined from the rate constant that best-fits the kinetic data; the relative standard deviation of three independent kinetic experiments with the same ratio of concentrations was found to be within 2% and the standard error of the rate constant from the fitting treatment was found to be less than 10% in all samples of different concentration ratios between glucose and fructose.

Hydrogen peroxide (H_2O_2) and potassium periodate (KIO_4) are two kinds of oxidants normally used to oxidize luminol (5-amino-2,3-dihydro-1,4-phthalazine dione) in a aqueous alkali solution to give a characteristic blue luminescence. Unfortunately, H_2O_2 suffers instability for operation in flow injection analysis (FIA) systems. KIO_4 is very stable, a low baseline with a narrow signal peak can be obtained. In addition, it does not generate bubbles in the tubes of the FIA system. From these reasons, luminol–KIO4 chemiluminescence (CL) system has attracted more attention recently and is of analytical interest for the determination of inorganic and organic species. A chemiluminescence (CL) system for the determination of hydrogen peroxide, glucose and ascorbic acid developed (Yanxiu et al., 1999) was based on hydrogen peroxide, which has a catalytic–cooxidative effect on the oxidation of luminol by KIO_4. The CL was linearly correlated with glucose concentration of 0.6–110 mg ml^{-1}. The relative standard deviation was 2.1% for 10 mg ml^{-1} (N =11). Detection limit of glucose was 0.08 mg ml^{-1}. The results obtained demonstrated the feasibility of using luminol–KIO_4–H_2O_2 CL system for the rapid, simple and sensitive detection of H_2O_2, glucose and AA with advantage of good reproducibility by a very small amount of sample. A unique feature of the method was that this CL system provided a low baseline.

12.3. CHEMILUMINOGENIC BIOSENSOR

More recently, the development of the immobilization techniques has provided the introduction of enzyme reactors, which can be positioned before the CL reaction takes place thus avoiding the lack of selectivity that may occur when a given CL reagent yields emission for a variety of compounds. In this alternative, the analyte is the substrate of the enzymatic reaction and one of the products will sensitively participate in the CL reaction. Substrates detected in this way include glucose, cholesterol, choline, uric acid, amino acids, aldehydes and lactate which generate H_2O_2 when flowing through a selective column reactor with immobilized oxidase enzymes in the presence of the necessary

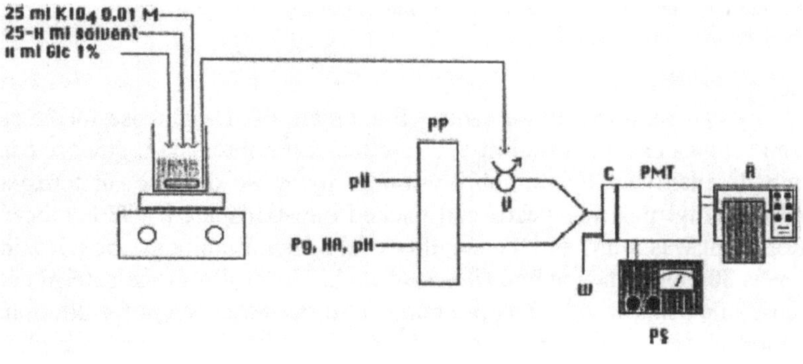

Figure 12.1. Apparatus layout. PP = peristaltic pump; V= six port valve; C= snailshell-like cell; W=waste; PMT= photomultiplier tube; PS = power supply; R= recorder; S1 = buffer carrier stream (1.4 ml min-1); S2 = reagent stream in buffer solution (1.4 ml min^{-1}).

oxidant, usually O_2, present in the samples. Luminol in the presence of a peroxidase catalyst is the best system so far for this post-column H_2O_2 determination. Figure 12.2 shows a typical manifold for glucose determination based on the CL emission produced when hydrogen peroxide formed by immobilized glucose oxidase reacts with luminol in the presence of potasium hexacyanoferrate (III) (Calokerinos et al., 1995).

Enzyme catalyzes the following reactions:

$$\beta\text{- D- glucose} + O_2 + H_2O \xrightarrow{\text{GOD}} \text{D- gluconic acid} + H_2O_2$$

$$\text{D- glucose} + O_2 \xrightarrow{\text{PyrOD}} \text{D- glucosone} + H_2O_2$$

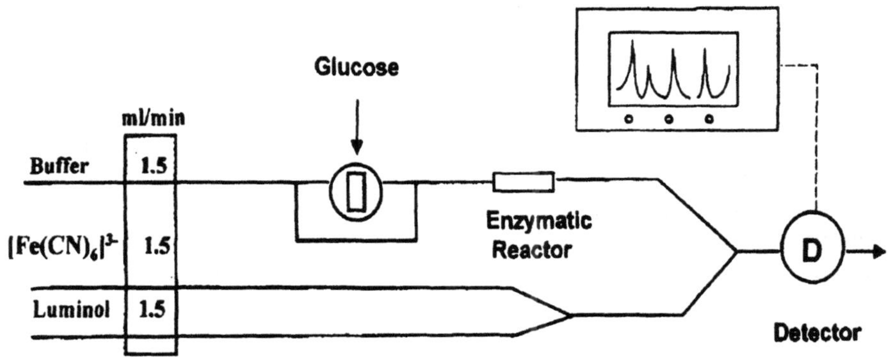

Figure 12.2. Schematic diagram for a FIA manifold for the analysis of glucose incorporating an enzymatic reactor with detection based on the luminol reaction.

PyrOD oxidizes the -α and – β anomers of D- glucose to the same extent and shows excellent stability. Using the luminol reaction, glucose can be determined sensitively by immobilization of pyranose oxidase on toresylate-hydrophilic vinyl polymer beads and packed into transparent PTFE tube. The detection limit was 3 nM and the stability of the chemiluminometric peroxidase sensor was 30 days. The method (Nobutoshi et al., 1997) allows the determination of glucose in plasma without any pretreatment procedure, except for dilution.

A optical sensor was prepared (Li et al., 2004) by entrapping glucose oxidase into TEOS – based sol- gel column. The silica sol- gel film was coated on alumina substrate. The optical sensor was based on the chemiluminescence intensity from the periodate and hydrogen peroxide in K_2CO_3 in medium and reported the performance as glucose sensing elements. The detection limit and the stability reported were 4.0 x 10^{-8} M and 3 months respectively. Figure 12.3 shows the flow CL sensing system of the above mentioned method.

A CL flow system combining was developed (Li et al., 2001)with microdialysis (MD) sampling presented to in vivo on-line monitoring of glucose in awake rabbit. The sol–gel method, which is very efficient for retaining the activity of HRP and is considered to be the best way of immobilizing GOD is introduced to co-immobilize HRP and GOD on the inside surface of the CL flow cell. The CL detection involved enzymatic oxidation of glucose to d-gluconic acid and H_2O_2, then H_2O_2 oxidizing luminol to produce CL in presence of HRP. The stability of the immobilized enzyme reactor was performed under in vitro conditions by repetitive injection of 0.4mM glucose standard solution for 4 h at a frequency of 10 h−1, using both a new reactor and one already used for 40 h. The leaking rate of HRP and GOD from the sol–gel membrane affects the stability of the enzyme reactor. The CL intensity of the newly prepared sol–gel membrane was compared with that of same membrane already used for 40 h, and no obvious decrease of the signal (about decreased by 10%) was observed because of large quantities of hydroxyl group in the sol–gel hybrid material capable of forming strong hydrogen bonds, which are favorable to prevent the enzyme from leaking out of the membrane.

Another flow-injection analysis detection method for glucose is based on oxidation of glucose by glucise dehydrogenase with concomitant conversion of NAD^+ to NADH followed by chemiluminescent detection of NADH (Martin and Nieman, 1993). In this method glucose dehydrogenase is immobilized via

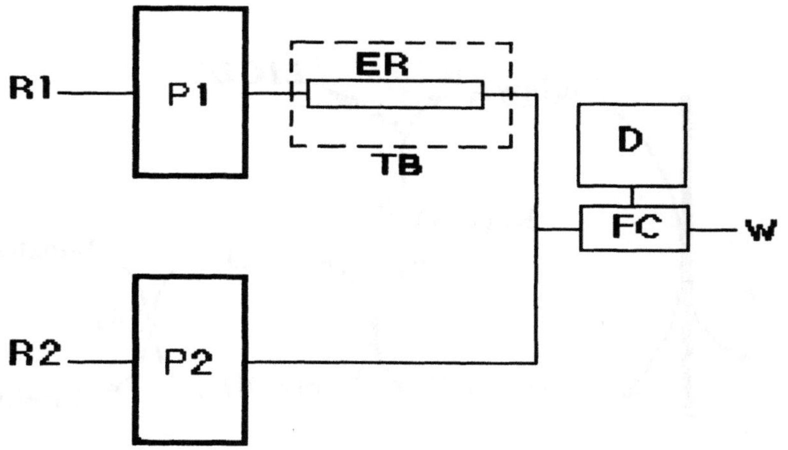

Figure 12.3. Schematic diagram of a flow sensor for chemiluminescent determination of glucose in aqueous medium: R1, water; R2, KIO_4 + buffer solution; P1 and P2, peristaltic pump; ER, enzyme reactor; TB, thermal box; FC, flow cell; D, detector; W, waste.

glutaraldehyde crosslinking to controlled pore glass to form an immobilized enzyme reactor. The chemiluminescent reagent, tris(2,2′-bipyridyl)ruthenium(II) [$Ru(bpy)^{2+}{}_3$] is immobilized in a Nafion film coated on a platinum electrode to form a regenerable chemiluminescent sensor. The immobilized $Ru(bpy)^{2+}{}_3$ is oxidized to $Ru(bpy)^{3+}{}_3$ which then reacts with NADH produced by the enzyme reactor to yield light and $Ru(bpy)^{2+}{}_3$. $Ru(bpy)^{2+}{}_3$ is thus recycled and made available again as shown in figure 12.4. Conditions for optimum enzyme reactor efficiency and chemiluminescent detection are determined and reported for pH (about 6.5), flow-rate (2 ml min^{-1}), and NAD^+ concentration (1–2.5 mM). At the optimum conditions a working curve is constructed where the upper limit for glucose detection is dependant on NAD^+ concentration and lower detection limit is 10 μM glucose. Signal reproducibility is 1–2% relative standard deviation. The method is very selective for glucose; some interference is seen from uric acid, ascorbic acid and catechol as well as species (such as oxalate and aliphatic amines) already known to chemiluminesce with the $Ru(bpy)^{2+}{}_3$ sensor.

Coimmobilization of glucose oxidase (GOD) and hematin (a catalyst) in carbon paste has several advantages. First, this method is very simple. Second, it is a bulk immobilization method, enzymes and other additives can be coimmobilized, and a new and fully active surface can be obtained by simple polishing or cutting in the case of fouling or deactivation. Third, many enzymes are very stable in carbon paste (Wang et al., 1997). Fourth, relatively strong intensity can be obtained by coimmobilizing enzyme and catalyst in CPE, since the catalyst, luminol, and hydrogen peroxide produced are mixed simultaneously and homogeneously on the surface of the carbon paste surface.

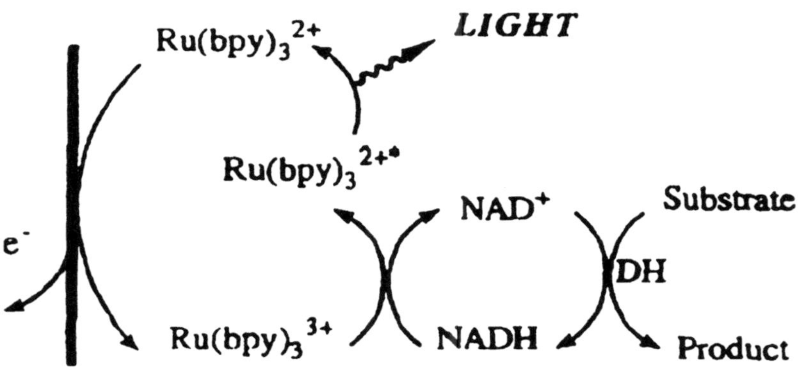

Figure 12.4. Reaction scheme for the coupling dehydrogenase (DH) enzyme selectivity to $Ru(bpy)^{3+}{}_3$ ECL.

Finally, it is a surface chemiluminescence analysis and thus ensures the conservation of reagent and a small volume of sample.

The method is based on the oxidation of glucose by glucose oxidase followed by chemiluminescent detection of hydrogen peroxide (Xu et al., 1999). Both glucose oxidase and hematin, a chemiluminescent reaction catalyst, were bulk-immobilized conveniently by direct mixing with carbon paste. Luminol in reagent solution passed through the flow cell and reacted with hydrogen peroxide produced by the enzyme reactor in the presence of the catalyst to yield light. An applied potential of 20.4 V avoided the electrode fouling effectively. Emitted light intensity vs glucose concentration was linear over the range of 1–100 mmol L^{-1} with a correlation coefficient of 0.992.

12.4. CAPILLARY ELECTROPHORESIS FOR GLUCOSE DETERMINATION

Capillary electrophoresis (CE) is one of most powerful techniques for carbohydrate analysis because it is highly efficient, high speedy and consumes nano liter sample . However, the poor chromophore properties of carbohydrate and the low detection length of the capillary present significant challenges to carbohydrate detection.

Incorporation of a UV-absorbing or fluorescing group into carbohydrate molecules by a labeling process is the most commonly used means to improve sensitivity for CE. However, conventional UV absorption detection produces poor concentration detection limits, which hindered the use of CE in the analysis of trace carbohydrates. Laser-induced fluorescence (LIF) detection has thus been developed and shown to be a powerful detection approach for trace components, but costly apparatus limited its wide utilization.

Chemiluminescence (CL) detection has been proven to be another sensitive detection method for CE. It features a simple, cheap optical system and an excellent sensitivity comparable to LIF. A method (Wang et al., 2003) based on pre-capillary derivatization with luminol (3-aminophthalhydrazide) was used for carbohydrate analysis using capillary electrophoresis with on-line chemiluminescence (CL) detection. The derivatives of seven monosaccharides were separated and detected by using 200 m*M* borate buffer containing 100 m*M* hydrogen peroxide at pH 10.0 as separation electrolyte and 25 m*M* hexacyanoferrate in 3 *M* sodium hydroxide solution as post-capillary chemiluminescence reagent with separation efficiencies ranging from 160,000 to 231,000 plates per metre. This type of separation is much better than those obtained in most post-capillary CL detections. The minimum amount of carbohydrate derivatized was 2 pmo(corresponding to the concentration of 2 m*M*). The method also provided a linear response for glucose in the concentration range of 0.1–250 m*M* with a mass detection limit of 420 amol or a concentration detection limit of 0.1 m*M*. Figure 12.5 shows the schematic diagram of a CE- CL system.

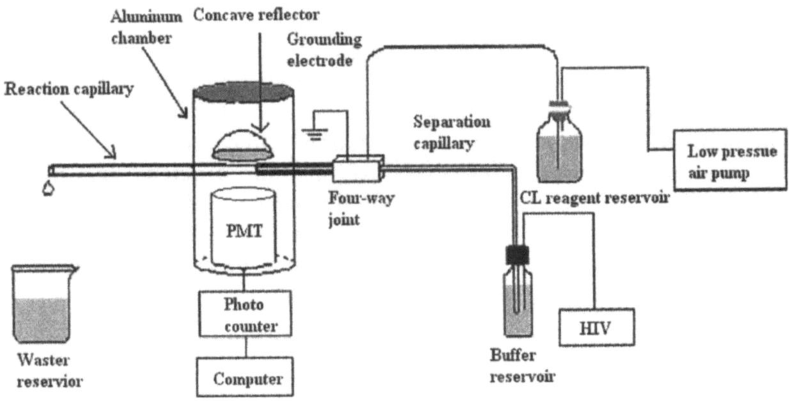

Figure 12.5. Schematic diagram of CE–CL

Recently, electrogenerated $Ru(bpy)_3^{3+}$-based ECL detection has been applied to CE and shown to be compatible with the nanoliter elution volumes characteristics of CE. Beside the advantage of using more times, ECL provides an extremely sensitive means of detection because of photons emitted from an ECL reaction at the surface of electrode.

The potential of CE coupled to $Ru(bpy)_3^{2+}$ ECL method was investigated to rapidly determine monosaccharides with low detection limits (Li, 2004). For the analysis of monosaccharides, the analytes were derivatized to tertiary amines prior to CE separation. Cyclic voltammogram of $Ru(bpy)_3^{2+}$ obtained using a Pt electrode as working electrode in presence of ruthenium complex in borate buffer solution for optimal electrolyzing potential of $Ru(bpy)_3^{2+}$ was $E_{pa} = 1.26$ *V* (*vs.* Ag/AgCl) (Fig. 6). In this system, $Ru(bpy)_3^{3+}$ was continuously *in situ* generated without the need of frequent replacing. The derivatization reaction for glucose takes about 30 *min* at room temperature when 2-diethylaminoethanethiol was used for derivatization reagent. The mole ratio of 10 is needed for 2-diethylaminoethanethiol and trifluoroacetic acid to obtain maximum ECL intensity. Under the optimum conditions (pH, 8.3; concentration of acetonitrile in 30 *mM* borate separation buffer solution, 20 %; concentration of $Ru(bpy)_3^{2+}$ in 200 *mM* borate buffer solution, 5.0 *mM*) a linear calibration range from 1.0×10^{-4} to 1.0×10^{-7} *M* (R=0.999) and detection limit of 6.0×10^{-8} *M* (S/N=3) were obtained for glucose. This system was successfully applied to the determination of monosaccharides in angelica. Figure 12.7 shows the schematic diagram of a CE- CL system.

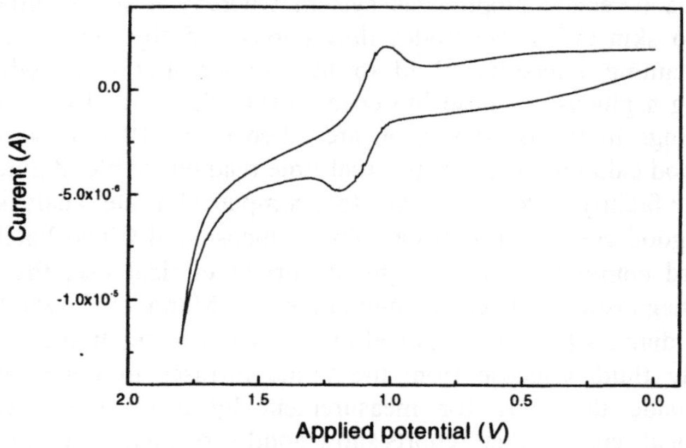

Figure 12.6. Cyclic voltammogram of 100 *mM* borate buffer solution containing 1.0 *mM* $Ru(bpy)_3^{2+}$; pH 8.0; scan rate 100 *mV/s*.

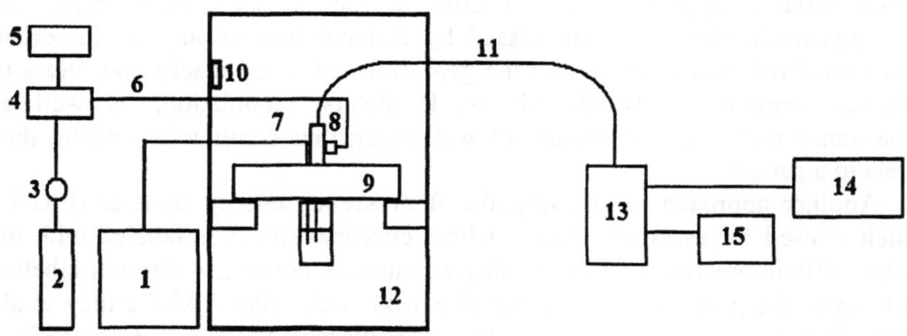

Figure 12.7. Block diagram of capillary electrophoresis. 1, high voltage supply; 2, gas tank; 3, gas regulator; 4, gas valve; 5, timer; 6, PE tube; 7, Pt electrode; 8, injection tee; 9, holder; 10, safety switch; 11, capillary; 12, blocking box; 13, ECL cell; 14, electrochemical analyzer; 15, interface.

12.5. GLUCOSE SENSING: RECENT APPROACHES FOR CLINICAL USE IN DIABETES CARE

Glucose monitoring is an essential component of modern diabetes management. Three in vivo glucose sensors are now available for clinical use: a subcutaneously implanted needle- type amperometric enzyme electrode coupled to a portable logger, the data from which can be down-loaded to a portable computer after up to 3 days sensing (Mastrototaro, 200). The sensor is based on the long-established technology of glucose oxidase immobilized at a positively charged base electrode, with electrochemical detection of hydrogen peroxide production; a reverse iontophoresis system whereby a small current passed between two skinsurface electrodes draws ions and (by electro-endosmosis) glucose-containing interstitial fluid to the surface and into hydrogel pads incorporating a glucose oxidase biosensor (Tamada et al., 1995; Garg et al., 1999) Readings in the latest version are taken every 10 min, with a single capillary blood calibration. There is a real-time read out of blood glucose levels as well as the facility to download data to a computer for later analysis. There is generally a good correlation between blood glucose and GlucoWatch readings and the third commercial technology in current clinical use, the GlucoDay (Menarini Diagnostics) is based on microdialysis (Maran et al., 2002) . Here, a fine, hollow dialysis fibre is implanted in the subcutaneous tissue and perfused with isotonic fluid. Glucose from the tissue diffuses into the fibre and is pumped outside the body for measurement by a glucose oxidase-based electrochemical sensor. Initial reports show good agreement between sensor and blood glucose readings, and good stability with a one-point calibration over one day. More trials are necessary to assess its clinical potential. Microdialysis is also being developed for clinical use by other companies such as Roche (Jungheim et al., 2001) . Though there have been clear difficulties with these first generation in vivo glucose sensors, subsequent improved versions will almost certainly appear. Several benefits include the detection of hypo- and hyperglycaemia that has been missed by conventional testing, an important educational role in demonstrating the glycaemic effect of snacks and meals of different composition which can teach dietary modulation, as well as reassurance and a greater feeling of well-being that continuous sensing data offers to a patient.

Another approach to studying the fluorescence energy transfer (FRET), which is used for assaying glucose where concanavalin A is labeled with the highly NIR-fluorescent protein allophycocyanin as donor and dextran labelled with malachite green as the acceptor (Rolinski et al., 2000; McCartney et al., 2001; Pickup et al., 2004). Competitive displacement of the dextran from binding to the lectin occurs when there is increasing glucose concentrations and leads to a reduction in FRET, measured as either intensity or lifetime (time-correlated single-photon counting).

Although Gluco- Watch approved by the FDA in 2001 is the first step towards the continuous and noninvasive monitoring of physiological glucose other emerging technologies include glucose-monitoring skin patches, implantable glucose sensor coupled insulin pumps, and laser blood drawing, which is deemed less painful than finger pricking with a lancet or needle.

A new technology which is noninvasive and continuous is the recent development of fluorescence-based glucose sensing contact lenses (Badugu et al., 2005). When worn by diabetics, who often require vision correction in any case, these lenses can potentially monitor tear glucose levels, which are known to track blood glucose levels. These contact lenses incorporate new monosaccharide fluorescent signaling boronic acid containing probes the unique chemical and photophysical features of these probes overcome lens environmental obstacles, such as pH and polarity. Recent findings show that this approach might indeed be suitable for the continuous non-invasive monitoring of tear glucose in the concentration range 250 mM to 5 mM, which is approximately tenfold lower than the concentration range found in diabetic blood.

12.6. ACKNOWLEDGEMENT

This work was supported by Korea Research Foundation Grant (KRF-2004-005-C00009).

12.7. REFERENCES

Agater, I. B., Roger, A. J., and Williams, K., 1996, Determination of carbohydrates by flow injection with direct chemiluminescence detection, *Anal. Commun.* **33(10)**:367 – 369

Bowie, A.R., Sanders, M.G., and Worsfold, P.J., 1996, Analytical applications of liquid phase chemiluminescence reactions, J. Biolumin. Chemilumin. 11(2):61-90.

Babko, A.K.,. Dubovenko, L.I., and Lukovskaya, N.M., 1966, *Chemiluminescent Analysis*, Tekhnika, Kiev.

Badugu, R., Lakowicz, J R., and Geddes, CD., 2005, A glucose-sensing contact lens: from bench top to patient- a review, *Current Opinion in Biotechnology* . **16**:1–8

Chappelle, E.W., and Picollo, G.L (Eds.)., 1975, *Analytical Applications of Bioluminescence and Chemiluminescence*, NASA Special Publication 388, NASA, Washington, DC.

Calokerinos, A.C., Deftereos, N.T., and Baeyens, W.R.G., 1995, Chemiluminescence in drug assay, *J.Pharm. Biomed. Anal.* **13**: 1063-1071.

Evmiridis, N. P.,. Thanasoulias, N. K., and. Vlessidis, A. G., 1999, Determination of glucose and fructose in mixtures by a kinetic method with chemiluminescence detection, *Analytica Chimica Acta.* **398**: 191–203.

Egan., H.,. Kirk, R.S., and. Sawyer, R., 1981, Pearson's Chemical Analysis of Foods, 8th edn., Churchill Livingstone, Edinburgh.

Garg, S.K., Potts, R.O., Ackerman, N.R., Fermi, S.J., Tamada, J.A., and Chase, H.P., 1999, Correlation of fingerstick blood glucose measurements with GlucoWatch Biographer glucose results in young subjects with type 1 diabetes, *Diab. Care.* **22**: 1708–1714.

Horwitz , W (Ed.)., 1975, Official Methods of Analysis, 12th edn., AOAC, Washington, DC.

Heldrich, K (Ed.), 1990, Official Methods of Analysis, 15th edn., AOAC, Washington, DC.

Hart, F.L., and Firscher, H.J., 1971, Modern Food Analysis, Springer, New York.

Jungheim, K., Wientjes, K.-J., Heinemann, L., Lodwig, V., Koschinsky, T., and Schoonen, A.J., 2001, Subcutaneous continuous glucose monitoring. Feasibility of a new microdialysis based glucose sensor system, Diab. Care. **24**: 1696–1697.

Kricka, LJ., 1995, Chemiluminescence and bioluminescence, *Anal Chem.* **67**:499R-500R.

Kricka, L.J., Thorpe, G.H.G., and Witehead, T.P (eds.)., 1983, Chemiluminescent and Bioluminescent Methods in Analitical Chemistry, *Analyst.* **108**:1274-1296.

Lv, Y., Zhang, Z., and Chen, F., 2003, Chemiluminescence microfluidic system sensor on a chip for determination of glucose in human serum with immobilized reagents, *Talanta* **59**: 571-576.

Lewis, S.W., Price, D., and. Worsfold, P.J., 1993, Flow Injection Assays with Chemiluminescence and Bioluminescence Detection , *J.Biolumin.Chemilumin.* **8**: 183-199.

Li, M.,. Lee, S. H.,. Bae, Z. U.,. Lee, K. P.,. Kim, D. H.,. Park, Y. C., and. Lee, M. S., 2004, Optical sensor for the determination of glucose based on KIO_4 chemiluminescence detection, *J. Fluores.* 14: 597-601.

Li, M 2004, Determination of monosaccharides, trimethylamine and carbamazepine by $Ru(bpy)_3^{2+}$-based chemiluminescence detection using capillary electrophoresis and flow Injection system, PhD Thesis, Department of Chemistry, Kyungpook National University, Korea.

Li, B.,. Zhang, Z., and. Jin, Y., 2001, Chemiluminescence flow sensor for in vivo on-line monitoring of glucose in awake rabbit by microdialysis sampling , *Anal. Chim. Acta.* **432**:95–100.

Malaprade, L., 1928, Malaprade reaction (Periodic Acid Oxidation), *Bull. Soc. Chim. France* [4] **43**:683.

Malaprade, L., 1928, Malaprade reaction (Periodic Acid Oxidation), *Compt. Rend.* **186**:32.

Martin, A. F., and Nieman, T. A., 1993, Glucose quantitation using an immobilized glucose dehydrogenase enzyme reactor and a tris(2,2′-bipyridyl) ruthenium(II) chemiluminescent sensor , *Anal. Chim. Acta* .**281**:475-481

Mastrototaro, J.J., 2000, The MiniMed Continuous Glucose Monitoring System. *Diab. Technol. Ther.* **2** (Suppl. 1): 13–18.

Maran., A., Crepaldi, C., Tiengo, A., et al., 2002, Continuous subcutaneous glucose monitoring in diabetic patients. A multicenter study, *Diab. Care.* **25**: 47–352.

McCartney, L.J., Pickup, J.C., Rolinski, O.J., Birch, DJS., 2001, Nearinfrared fluorescence lifetime assay for serum glucose based on allophycocyanin labelled concanavalin , A. *Anal. Biochem.* **292**: 216–221

Nobutoshi, K., Akiko, I., Satoru, F., Kazuya, S., and Motohisa, F., 1997, Highly sensitive flow-injection determination of glucose in plasma using an immobilized pyranose oxidase and a chemiluminometric peroxidase sensor, *Anal. Chim. Acta.* **354**:205-210.

Pigman, W. (Ed.)., 1956, The Carbohydrates, *Academic Press*, OH.

Percival, E.G.V., 1962, *Structural Carbohydrate Chemistry*, Garnett-Miller, London.

Pickup, J.C., Hussain, F., Evans, N. D., and Sachedina, N., 2004, In vivo glucose monitoring: the clinical reality and the promise- a review (in press), *Biosensors and Bioelectronics*

Robards, K., and Worsfold, P. J., 1992, Analytical applications of liquid-phase chemiluminescence, *Anal. Chim. Acta.* **266**: 147-173.

Rongen, H. A. H., Hoetelmans, R. M. W., Bult, A., and Van Bennekom, W. P., 1994, Chemiluminescence and immunoassays, *J. Pharm. Biomed. Anal.* **12**: 433-46.

Rolinski, O.J., Birch, DJS., McCartney., L.J., and Pickup, J.C., 2000, A time resolved near infrared fluorescence assay for glucose: opportunities for trans dermal sensing, *J. Photochem. Photobiol. B* **54**: 26–34

Sklarz, B., 1967, *Quart. Rev.* **21**: 3-28.

Seitz, W.R.,. Hercules, D.M in:. Cormier, M.J.,. Hercules, D.M., and Lee. J. (Eds.), 1973, *Chemiluminescence and Bioluminescence*, Plenum Press, New York.

Townshend, A (Ed.)., 1995, *The Encyclopedia of Analytical Science*, Academic Press, London.

Townshend, A., and Wheatley, R.A., 1998, Oxidative chemiluminescence of some nitrogen nucleophiles in the presence of formic acid as an ancillary reductant, *Analyst* (Cambridge, U. K.). 123(2): 267-272.

Tamada, J.A., Bohannon, NJV, and Potts, R.O., 1995, Measurement of glucose in diabetic subjects using noninvasive glucose extraction, *Nat. Med.***1**: 1198–1201.

Watson, C.A (Ed.)., 1994, Official and Standardized Methods of Analysis, 3rd ed., Royal Society of Chemistry, Cambridge.

Wang., J., Liu, J., and Cepra, G., 1997, Thermal stabilization of enzymes immobilized within carbon paste electrodes, *Anal. Chem.* **69:** 3124–3127.

Wang, X.,. Wang, Q., Chen, Y., and. Han, H., 2003, Determination of carbohydrates as their 3-aminophthalhydrazide derivatives by capillary zone electrophoresis with on-line chemiluminescence detection, *J. Chromatogr. A.* **992**:181–191.

Xu, G., Zhang, J., and Dong, S., 1999, Chemiluminescent determination of glucose with a modified carbon paste electrode, *Microchem. J.* **62:** 259–265.

Yanxiu, Z., Tsutomu, N., Feng, L., and Guoyi, Z., 1999, Evaluation of luminol–H_2O_2–KIO_4 chemiluminescence system and its application to hydrogen peroxide, glucose and ascorbic acid assays, *Talanta.* **48**:461–467.

THE GLUCOSE BINDING PROTEIN AS GLUCOSE SENSOR

Protein engineering for low-cost optical sensing of glucose

Leah Tolosa* and Govind Rao

13.1. INTRODUCTION

For the past several years now, our group has been developing a novel set of proteins for sensing applications. These are soluble proteins found in the periplasmic space of gram-negative bacteria such as *Escherichia coli*. Their main function is to shuttle small molecules from the outside world to membrane receptors, which then transports the molecules into the cytoplasm in an ATP-expending mechanism[1,2]. There is also evidence that these so-called binding proteins (BP's) are involved in chemotaxis[3,4]. Based on their functions, the binding proteins have evolved extremely high selectivity and sensitivity for their substrates[5]. As a general rule, the binding constants of the BP's for their substrates are in the micromolar range. BP's for glutamine, arabinose, histidine and many others, recognize *only* their respective substrates. The *E. coli* glucose binding protein (GBP), which is the subject of this chapter, mainly recognizes glucose but also binds galactose to a lesser degree. The sensitivity and selectivity of BP's make them ideal for biosensor development. But the main attractive feature of BP's is that they are not enzymes. Thus, they are considered as reagentless biosensors in that no additional ingredient is required for the sensing mechanism to proceed. For glucose sensing in particular, this is an improvement from glucose oxidase-based sensors where oxygen can be a limiting reagent. Additionally, glucose oxidase sensors produce hydrogen peroxide in the presence of glucose that in time degrades the enzyme itself. In contrast, GBP does not alter the chemistry of glucose. Rather, the process is a

* Center for Advanced Sensor Technology, Department of Chemical and Biochemical Engineering, University of Maryland Baltimore County, 1000 Hilltop Circle, Baltimore, MD 21250

simple equilibrium driven association between substrate and protein. For this reason GBP and other BP's retain their activity for longer periods even in solution.

This chapter will describe the sequence of steps in the development of a low-cost optical GBP-based biosensor for glucose. Additionally, practical applications for the glucose sensor will be presented.

13.2. PROTEIN ENGINEERING

The substrates for the BP's are very diverse: sugars (e.g., glucose, ribose, maltose), amino acids (e.g., glutamine, histidine), ions (e.g., phosphate, sulfate) and other small molecules (e.g., vitamins, dipeptides)[5]. There are very minor similarities in the amino acid sequences of the BP's. However, these similarities are not significant enough to explain their remarkably identical structures and mode of action[6,7]. In general, the BP structure is composed of two globular domains separated by a hinge region. Binding occurs by a change in the conformation of the protein from an "open" to a "closed" structure as the substrate is grasped in the binding site. To illustrate this change in conformation are the X-ray crystal structures of the glutamine binding protein as shown in Figure 13.1[8,9]. GBP cannot be shown in this way because the glucose-free GBP crystal structure was found to be in the "closed" conformation with water substituting for glucose in the binding site[10]. This was attributed to a lower energy of crystallization for this form than the "open" conformation. Nevertheless, spectroscopic data in solution clearly shows that the "open" and "closed" conformations apply to GBP[11-13].

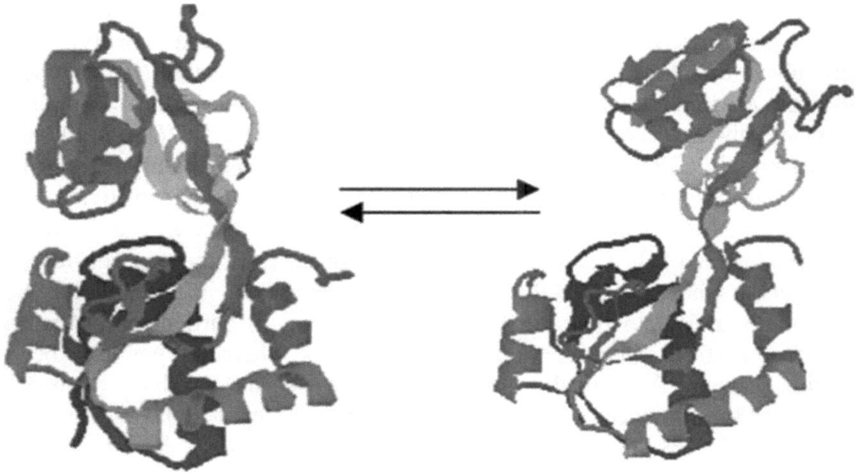

Figure 13.1. The structure of the glutamine binding protein showing the "closed" glutamine-bound form (left) and the "open" glutamine-free form (right). (see the color insert after p. 429.)

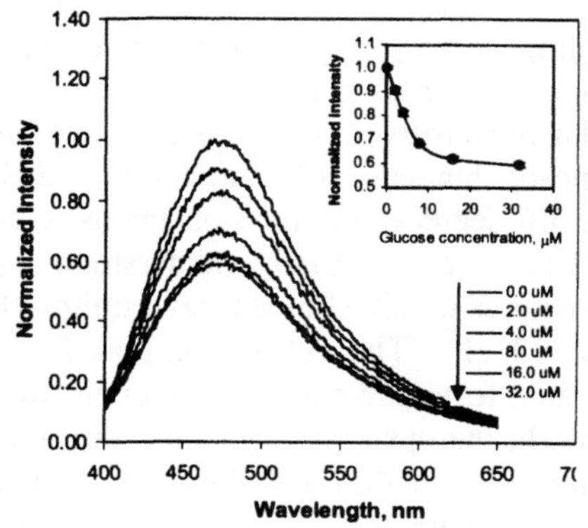

Figure 13.2. Emission spectra of ANS-labeled Q26C GBP in the presence of glucose.

These conformational changes associated with binding are put to good use by attaching polarity-sensitive fluorescent probes to specific sites on the protein. Some researchers have placed these probes close to the binding site[13, 14] to monitor binding. But this can result in a decrease in activity as the binding site can be blocked by the fluorophore. Others, including our group, have placed the fluorophores on the opposite side of the binding site in positions that are not directly involved in binding but are affected allosterically[11-13, 15,16]. The positions for the fluorophores can be selected by calculating the rearrangement of specific domains relative to the binding site[13]. But one can also easily overlay the open and closed conformations and pick out the amino acids that undergo the largest changes. With GBP, the lack of information on an "open" conformation can be resolved by substituting the structure of the arabinose binding protein[6], which is the closest to that of GBP. Cysteines are then introduced by site-directed mutagenesis to these positions followed by labeling with sulfhydryl-reactive probes. It is fortunate that due to the highly oxidative environment in the periplasm, the BP's have evolved with no cysteines in their amino acid composition.

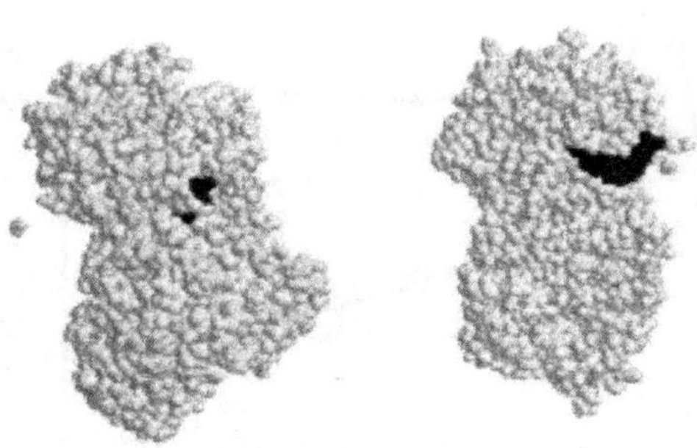

Figure 13.3. Space-fill models of ANS-GBP showing the degree of exposure of ANS to the buffer in the absence (left) and presence (right) of glucose.

13.3. POLARITY SENSITIVE PROBES

The choice of sites for labeling is very important in ensuring that signal transduction by GBP occurs upon addition of glucose. However, there is evidence that the interaction of the polarity-sensitive probe to the protein site is just as crucial. For example, the anilino-naphthalene sulfonate (ANS) label on the Q26C GBP undergoes an almost 50% decrease in emission intensity upon glucose saturation (Figure 13.2). This can be easily understood upon examination of the open and closed structures obtained by energy minimization calculations of the labeled protein in 13. 3[17]. The ANS is exposed to the aqueous environment in the glucose-bound form and explains the quenching of fluorescence in the presence of glucose. In the open conformation, the ANS is buried in the protein structure accommodated within a hydrophobic channel. Labeling the same position with the very similar 5-((2-aminoethyl) amino) naphthalene-1-sulfonic acid (EDANS) energy-minimized models reveal that the interactions of EDANS with the protein are mainly hydrophilic rather than hydrophobic in both open and closed conformations. The calculated dissociation constants, K_d, for the ANS- and EDANS-labeled Q26C GBP are 0.85 μM and 7.0 μM, respectively. There is an apparent decrease in sensitivity by an order of magnitude from ANS to EDANS. This implies that binding constants determined by the fluorescence changes of polarity probes are indirect measures of glucose binding. The same effects were observed in other mutants such as L255C, which has a K_d of 0.32 μM when labeled with acrylodan and 0.43 μM when labeled with 7-nitrobenz-oxa-1,3-diazole (NBD)[12]. It is likely that the label is affecting the protein structure to some extent. Labeling of the Q26C mutant with pyrene, for example, shows a more complicated binding isotherm that does not fit a straightforward single binding event (Figure 13.4). Nevertheless, with the right choice of probe, GBP can be an extremely useful glucose sensor.

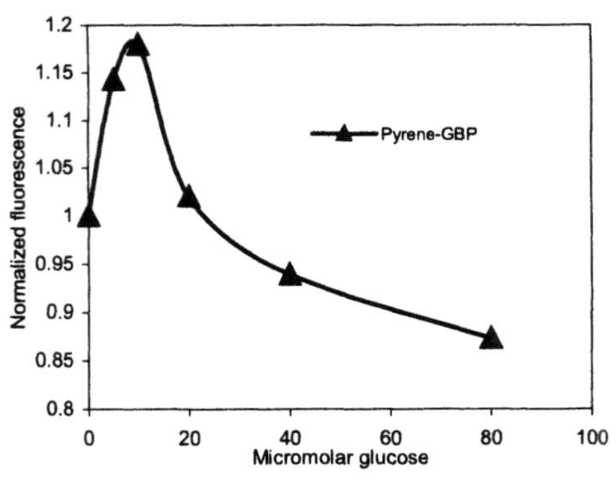

Figure 13.4. Response of Pyrene-GBP to glucose.

13.4. STRATEGIES FOR LOW-COST GLUCOSE SENSING

The changes in emission associated with glucose binding as shown in Figure 2[18], constitutes a serviceable intensity-based glucose sensor. However, when measuring fluorescence intensities at a single wavelength, it is common to encounter errors due to variations in the intensity of the light source, the length of the light path, sample positioning, concentration of the reagent, etc. One way to eliminate these systematic errors is to use the ratiometric approach,

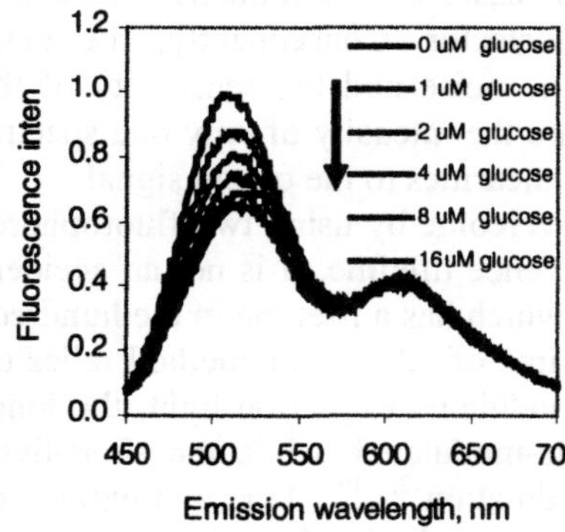

Figure 13.5. Emission spectra of the dual-labeled L255C GBP. Acrylodan emits at 510 nm, Ru emits at 610 nm.

that is, the measurement of the intensity ratio at two different wavelengths in the emission spectrum where the fluorescent probe exhibits different sensitivities to the analyte. Many pH sensors function in this way, but the design and synthesis of ratiometric probes for most analytes is a major challenge. Fortunately, when the sensing molecule is a large protein such as GBP, it is not too difficult to identify specific sites with different analyte sensitivities for fluorescent dye labeling. In the case of GBP, the isothiocyanate form of a ruthenium metal ligand complex was preferentially labeled to the N- terminal amino acid by controlling the pH. The emission spectra of L255C GBP labeled with acrylodan (Em = 410 nm) on the cysteine mutation and Ru (Em = 620 nm) on the N-terminal are shown in Figure 13.5[16].

This intensity-based ratiometric approach works very well for laboratory use. The caveat is that for field-testing, additional improvements are needed to make the measurements more robust and the instrumentation less expensive. By definition, the method requires measurements on two separate wavelengths. However, there are no photodetectors that can measure at a single wavelength, rather the light intensity is integrated over the spectral band that is passed by the monochromators (or band pass filter). The narrower the spectral band, the better is the sensitivity of the ratio determination. However, limiting the spectral band using band pass devices results in a decrease in the available intensity. As a result, high-sensitivity photodetectors are required (photo multipliers or avalanche photodiodes). These devices require high-voltage power supplies (up to 1000 V), which burdens the power budget and the

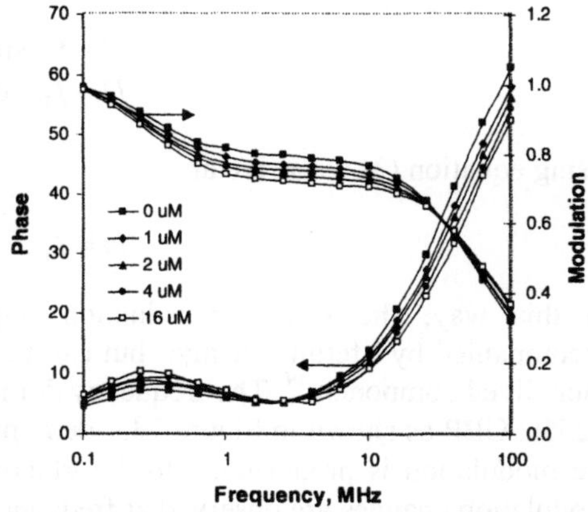

Figure 13.6. Frequency domain decay traces of the dual-labeled L255C GBP. Note the modulation changes between 1-10 MHz.

reliability of the device. Furthermore, the decreased emission intensity results in worse signal-to-noise ratio in the photodetector. Consequently, the plain spectral approach is a trade-off between the noise and the sensitivity of the sensor. Finally, it is very difficult to measure the intensity of only one spectral peak, as they typically overlap, adding non-linearities to the output signal.

All these problems can be easily overcome by using two fluorophores that have a large difference in their fluorescence lifetime. It is not an accident that the second probe chosen is ruthenium, which has a lifetime in the hundreds of nanoseconds, while acrylodan has a lifetime of ~2 ns. The method relies on the fact that at higher frequencies of the modulated excitation light, the long-lifetime fluorescence can be completely de-modulated, while the short-lived fluorescence still retains significant modulation[12, 19]. For a mixture of fluorophores, the phase and modulation can be calculated using the sine and cosine transforms of the intensity decays, N_ω and D_ω, respectively, at a given frequency ω

$$N_\omega = \sum f_i\, m_i \sin\phi_i \quad (1)$$
$$D_\omega = \sum f_i\, m_i \cos\phi_i \quad (2)$$

where f_i is the fractional steady-state intensity, ϕ_i is the phase, and m_i is the modulation. The modulation at frequency ω is given by

$$m = (N^2 + D^2)^{1/2} \quad (3)$$

In the case of Ru-GBP-Acr, the difference in lifetime between Ru and Acr is large. It is, therefore, reasonable to assume that a frequency ω can be identified where the modulation of Acr is close to 1.0 while the modulation of Ru is close to 0.0. At this frequency,

$$N = f_{Acr} \sin\phi_{Acr} \quad (4)$$
$$D = f_{Acr} \cos\phi_{Acr} \quad (5)$$

Using equation (3) we arrive at

$$m = f_{Acr} \quad (6)$$

In this way, the signal transduction required for sensing need not be accompanied by lifetime changes but can be limited to intensity changes of the short-lived component[19]. The frequency domain decay traces of the dual labeled L255C GBP as shown in Figure 13.6 demonstrate a range of frequencies where the modulation is proportional to the glucose concentration. Furthermore, the modulation changes are observed at frequencies lower than those required if the only emitting species is the short-lived component (e.g. 100 MHz for short lived dye alone vs. 1-10 MHz for short + long lived dyes). This allows for the use of simple and low-cost oscillators such as the 555-type. Additionally, this method involves the direct electronic modulation of an LED (light emitting diode) as the light source rather than the use of expensive lasers.

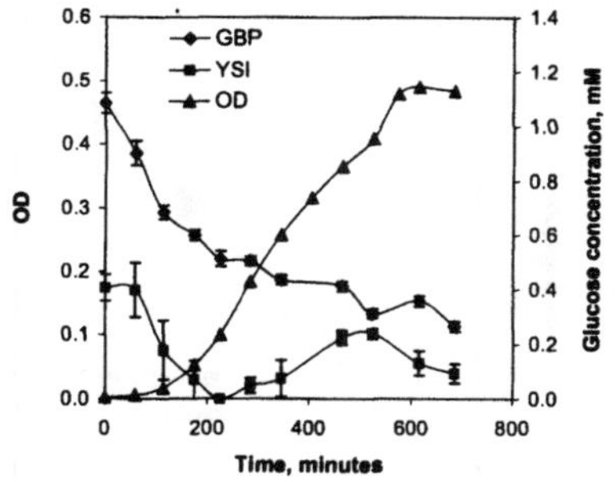

Figure 13.8. Glucose consumption in E. coli fermentation.

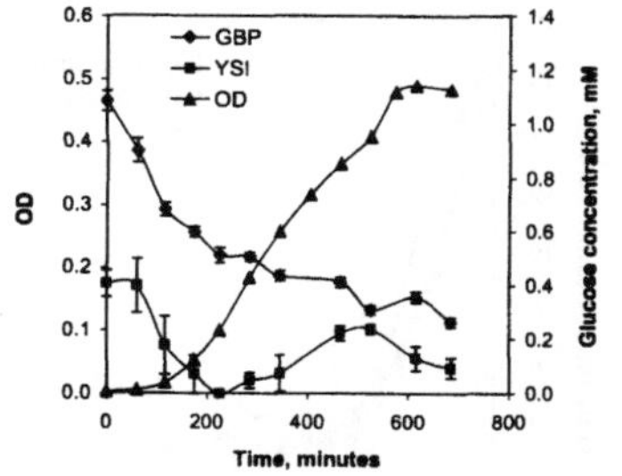

Figure 13.7. Glucose consumption in yeast fermentation.

13.5. APPLICATION OF THE GBP-BASED BIOSENSOR IN CELL CULTURE

The GBP sensor has been tested in monitoring glucose consumption in yeast fermentation and *E. coli* cell culture[18] (Figure 13.7). For comparison, the glucose concentrations were measured with the YSI 2700 Chemistry Analyzer (Yellow Springs Instrument Co., Inc.), which uses glucose oxidase-based electrodes. The YSI 2700 Chemistry Analyzer has a responsive range up to 140 mM and a lower detection limit (based on signal/noise=3) of 0.1 mM. GBP has a lower detection limit of 0.3-0.5 μM. Because of this extremely high sensitivity, the responsive range of GBP can be expanded to several orders of magnitude by simply diluting the sample. For example, in Figure 13.8 the samples analyzed using GBP were diluted 10,000 to 40,000 times while the samples analyzed with YSI were aspirated into the instrument as is. This dilution step does not only expand the responsive range, it minimizes/eliminates fluorescent interferences in the optically complex media. Fouling of the sensor surface can also be reduced significantly. It is not difficult to obtain reliable data even from the UV excitable ANS labeled GBP. The discrepancy between GBP and YSI data in Figure 13.7 can be explained by the sensitivity of the glucose oxidase electrode to other electroactive species such as urea and ascorbic acid (the preservative in Baker's yeast used here). GBP is not susceptible to these errors, but will detect galactose to a lesser extent.

For *E. coli* fermentation, GBP was able to follow the trend in glucose consumption as the cell density increased (Fig. 13.8). On the other hand, YSI gave erratic results because the glucose concentration in Luria-Bertani (LB) broth used for *E. coli* culture is too close to the lower detection limit of YSI, which was measured to be ~0.1 mM. Clearly, GBP is capable of monitoring glucose concentrations at the submillimolar levels that are beyond the capabilities of current industry standards. At the present time, monitoring of glucose levels in fermentations using LB medium is not routine because there is no method sensitive enough to do the measurements.

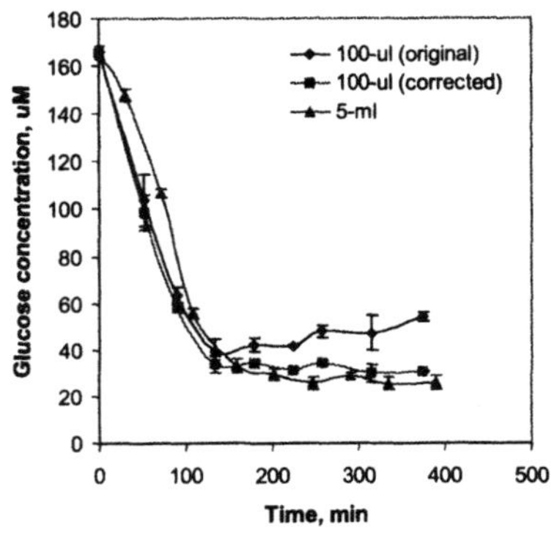

Figure 13.9. Glucose consumption in 100 μL fermentation volumes.

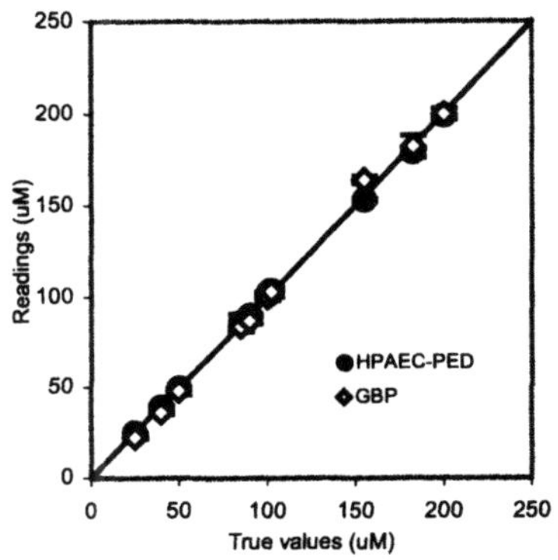

Figure 13.10. Comparison of HPAEC-PED and GBP methods of glucose sensing.

The high sensitivity of GBP offers another advantage in that very small sample volumes (<1 μl) are needed for analysis. Thus, there is a great possibility that GBP can be applied to samples derived by microsampling techniques such as iontophoresis, laser poration, ultrasound or microneedles. Another application is in microbioreactors where the fermentation volumes can be as small as 100 μl. In Figure 13.9 we show that this can be done with some correction for evaporation.

13.6. VALIDATION

As the use of binding proteins as biosensors begin to mature, there is a need to evaluate them against highly accepted methods that have been used by analytical chemists for many years. Glucose oxidase electrodes such as those manufactured by YSI are considered as such. However, the difference in sensitivities of GBP and glucose oxidase is three orders of magnitude. Thus, the limit of detection of YSI is about 0.1 mM while that of the glucose binding protein is about 0.4 μM. Clearly, there is a need to compare the assay to a method with comparable sensitivity. This we found in the High Performance Anion Exchange Chromatography with Pulsed Electrochemical Detection[20] (HPAC-PED), which has been in use for several decades and is a generally accepted technique. An additional advantage of this technique is that electrochemical conditions can be optimized to selectively detect glucose over other electroactive species. This complements the highly selective nature of the glucose binding protein. The correlation of the two methods can be seen in Figure 13.10[21]. Statistical analysis showed that there is no significant difference between the two techniques.

13.7. REFERENCES

1. C. van der Does and R. Tampe, How do ABC transporters drive transport?, *Biol. Chem.*, **385**(10), 972-933 (2004).
2. K. P. Locher, Structure and mechanism of ABC transporters, *Curr. Opin. Struct. Biol.*, **14**(4), 426-31 (2004).
3. L. Rodseth and F. A. Quiocho, Crystallization of the maltodextrin-binding protein for active transport and chemotaxis in several different liganded and mutant forms, *J. Mol. Biol.* **230**(2), 675-678 (1993).
4. S. L. Mowbray, Ribose and glucose-galactose receptors. Competitors in bacterial chemotaxis, *J. Mol. Biol.*, **227**(2), 418-440 (1992).
5. R. Tam and M. H. Saier, Jr., Structural, functional, and evolutionary relationships among extracellular solute-binding receptors of bacteria, *Micro. Rev.*, **57**(2), 320-346 (1993).
6. N. K. Vyas, M. N.,Vyas and F. A. Quiocho, Comparison of the periplasmic receptors for L-arabinose, D-glucose/galactose, and D-ribose. Structural and functional similarity, *J. Biol. Chem.*, **266**(8), 5226-5237 (1991).
7. F. A. Quiocho and P. S. Ledvina, Atomic structure and specificity of bacterial periplasmic receptors for active transport and chemotaxis: Variation of common themes, *Molec. Microbiol.*, **20**, 17-25 (1996).
8. C. D. Hsiao, Y. J. Sun, J. Rose, and B. C. Wang, The crystal structure of glutamine-binding protein from Escherichia coli, *J. Mol. Biol.*, **262**, 225-242 (1996).
9. Y. J. Sun, J. Rose, B. C. Wang and C. D. Hsiao, The crystal structure of glutamine-binding protein complexed with glutamine at 1.94 Δ resolution: Comparisons with other amino acid binding proteins, *J. Mol. Biol.*, **278**, 219-229 (1998).
10. M. M. Flocco and S. L. Mowbray, the 1.9 Δ x-ray structure of a closed unliganded form of the periplasmic glucose/galactose receptor from Salmonella typhimurium, *J. Biol. Chem.*, **269**(12), 8931-8936 (1994)
11. R. M. De Lorimier, J. J. Smith, M. A. Dwyer, L. L. Looger, K. M. Sali, C. D. Paavola, S. S. Rizk, S. Sadigov, D. W. Conrad, L. Loew and H. Hellinga, Construction of a fluorescent family, *Protein Science*, **11**, 2655-2675 (2002).
12. L. Tolosa, I. Gryczinski, L. Eichorn, J. Dattelbaum, F. N. Castellano, G. Rao and J. R. Lakowicz, Glucose sensor for low-cost lifetime-ased sensing using a genetically engineered E. coli binding protein, *Anal. Biochem.*, **267**, 114-120 (1999).
13. J. S. Marvin and H. W. Hellinga, Engineering biosensors by introducing fluorescent allosteric signal transducers: Construction of a novel glucose sensor, *J. Am. Soc.*, **120**(1), 7-11 (1998).
14. L. L. Salins, R. A. Ware, C. M. Ensor and S. Daunert, A novel reagentless sensing system for measuring glucose based on the galactose/glucose-binding protein, *Anal. Biochem.*, **294**(1), 19-26 (2001).
15. J. D. Dattelbaum and J. R. Lakowicz, Optical determination of glutamine using a genetically engineered protein, *Anal. Biochem.*, **291**, 89-95 (2001).
16. X. Ge, L. Tolosa, and G. Rao, Dual-labeled glucose binding protein for ratiometric measurements of glucose, *Anal. Chem.*, **76**(5), 1403-1410 (2004).
17. L. Tolosa, R. Harrison and G. Rao, unpublished data.
18. X. Ge. L. Tolosa, J. Simpson and G. Rao, Genetically engineered binding proteins as biosensors for fermentation and cell culture, *Biotech. Bioeng.*, **84**, 723-721 (2003).
19. J. R. Lakowicz, F. N. Castellano, J. D. Dattelbaum, L. Tolosa, G. Rao and I. Gryczynski, Low-frequency modulation sensors using nanosecond fluorophores, *Anal. Chem.*, **70**(24), 5115-5121 (1998).
20. W. R. LaCourse, *Pulsed Electrochemical Detection in High Performance Liquid Chromatography*, (John Wiley & Sons, New York, 1997).
21. X.Ge, S. J. Modi, W. R. LaCourse, G. Rao and L. Tolosa, Validation of an Assay Method for Micromolar Glucose Using an Engineered Glucose Binding Protein Labeled with a Fluorescent Probe, submitted (2005).

FLUORESCENT TICT SENSORS FOR SACCHARIDES

Laurence I. Bosch and Tony D. James†

14.1. INTRODUCTION

Boronic acid receptors with the capacity to bind selectively saccharides and signal this event by altering their optical signature have attracted considerable interest in recent years.[1-9] Boronic acids are known to bind saccharides *via* covalent interactions in aqueous media. The most common interaction is with *cis*- 1,2- or 1,3-diols of saccharides to form five- or six-membered rings respectively.[10]

14.2. RESULTS AND DISCUSSION

Compounds **1** a monoboronic fluorescent sensor, which shows large shifts in the emission wavelength on saccharide binding.[11, 12] When saccharides interact with sensors **1** in aqueous solution at pH 8.21 the emission maxima at 404 nm shifts to 362 nm (λ_{ex} 274 nm). (**Figure 14.1**).

The species responsible for the observed fluorescence properties are shown in **Scheme 14.1**. Species **A** contains a B-N bond and when excited at 274 nm emits at 404 nm. On addition of saccharide the B-N bond is broken to form boronate species **B** which emits at 362 nm when excited at 274 nm. (**Scheme 14.1**) The different fluorescent properties of species **A** and **B** can be ascribed to Locally Excited (LE) and Twisted Internal Charge Transfer (TICT) states of the aniline fluorophore.[13-16] Species **B** only shows the normal band associated with the LE state since the nitrogen lone pair is free to conjugate with the π-system while species **A** shows the anomalous band associated with the TICT state since the lone pair is coordinated with the boron and perpendicular to the π-system.

† Department of Chemistry, University of Bath, Bath BA2 7AY UK. (t.d.james@bath.ac.uk)

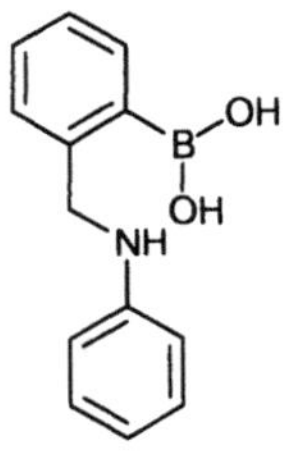

1

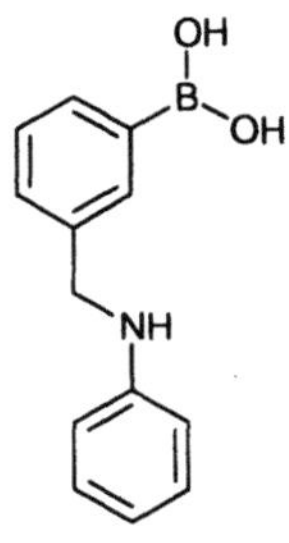

2

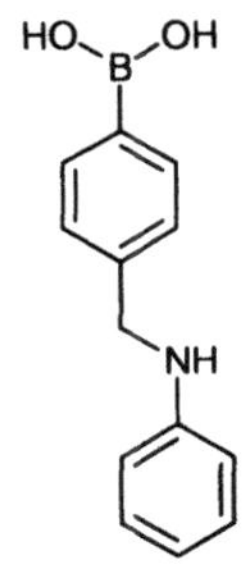

3

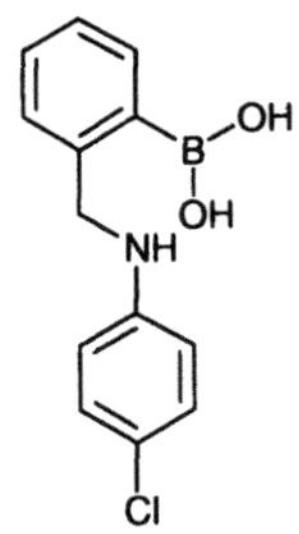

4

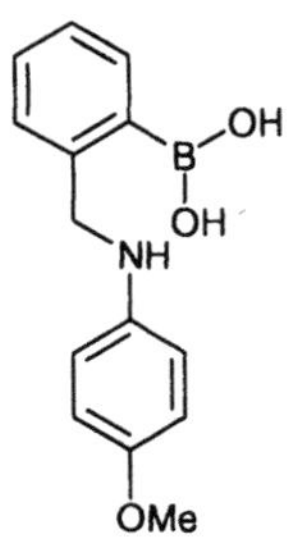

5

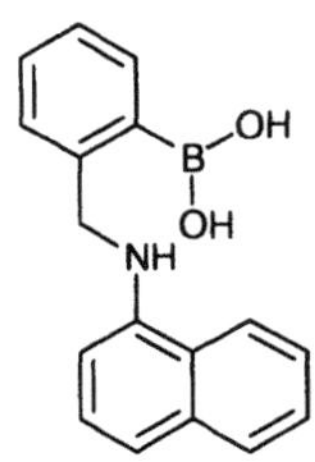

6

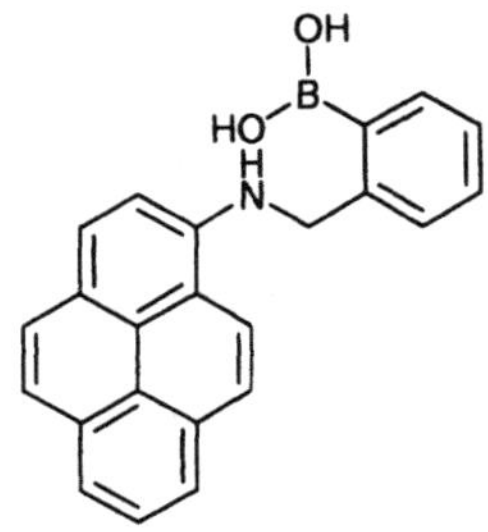

7

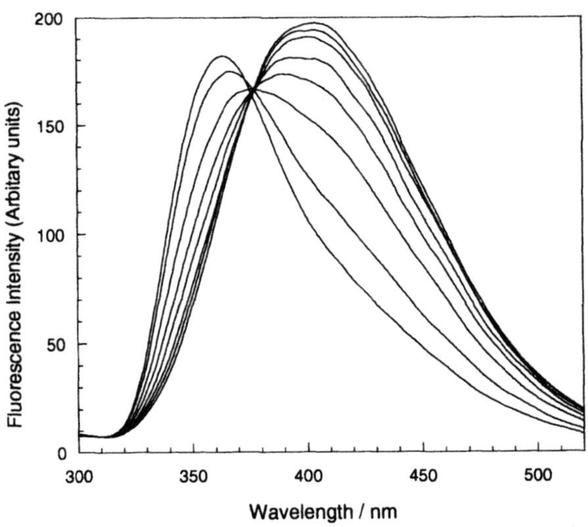

Figure 14.1 Fluorescence spectra change of **1** (2.0×10^{-5} mol dm^{-3}) with different concentration of D-fructose (0-0.1 mol dm^{-3}) in pH 8.21 buffer. λ_{ex} 274 nm.

From these investigations we realised that the species **B** emitting at 362 nm does not contain a B-N bond, but does show fluorescent enhancement. Therefore, we decided to investigate systems where formation of a B-N bond was not possible. (Compounds **2** and **3**) Compounds **1**, **2** and **3** were prepared according to **Scheme 2** from readily available starting materials, aniline and 2-, 3- or 4- formylbenzeneboronic acid. For compound **1** we were also able to obtain crystals suitable for X-ray analysis.[17] The structure is shown in **Figure 14.2**. It is clear from **Figure 14.2** that the boronic acid forms a cyclic trimeric anhydride, where only one of the units contains a B-N bond (1.747(2) Å). This value is similar to those found in similar cyclic boroxins[18] and within the 1.5 to 1.8 Å range expected for a strong B-N bond.[19] The structure supports **Scheme 14.1** in that it illustrates that a subtle balance exists for formation of a B-N bond.

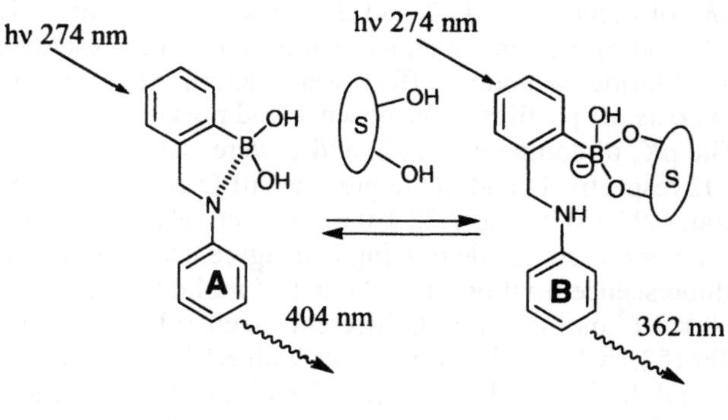

Scheme 1 Proposed fluorescent species

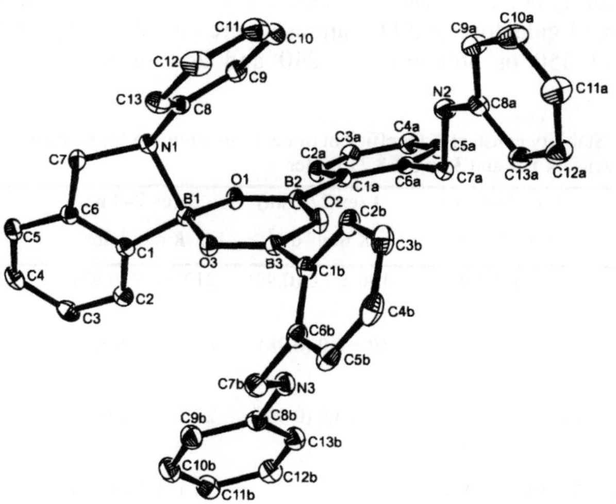

Figure 14.2 Crystal structure of **1a**. Ellipsoids are illustrated at the 30% probability level. Hydrogen atoms omitted for clarity.

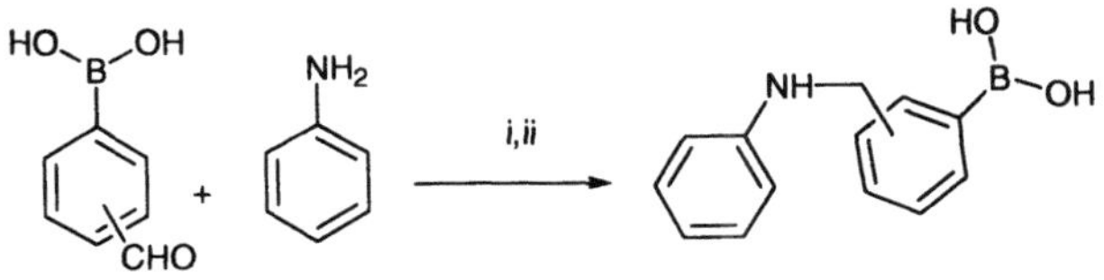

1-3

Scheme 14.2: Reagents and conditions: (i) EtOH/$PhCH_3$ reflux (**1**) or MeOH, rt (**2** and **3**); (ii) MeOH, $NaBH_4$ (5 eq), **1** 54% (two steps), **2** and **3** both 45% (two steps)

For compounds **2** and **3** when excited at 240 nm and 244 nm respectively an emission at 350 nm was observed. (Excitation at 274 nm resulted in no emission). Also, when compound **1** was excited at 244 nm, only an emission at 360 nm was observed. In all these cases only the LE state is formed.

The pK_a of compounds **1, 2** and **3** (1.0 x 10^{-5} mol dm^{-3}, 33.3wt% (**1**) and 52.1wt% (**2** and **3**) methanolic aqueous solution in the presence of 0.05 mol dm^{-3} sodium chloride osmotic buffer) were determined from the fluorescence intensity *versus* pH profiles in the absence and presence of D-fructose (0.05 mol dm^{-3}). The pK_a of compounds **1, 2,** and **3** were: 10.20 ± 0.01, 9.30 ± 0.03 and 9.58 ± 0.02 respectively and in the presence of D-fructose (0.05 mol dm^{-3}) were 7.96 ± 0.06, 6.95 ± 0.04 and 7.22 ± 0.02 respectively. The observed shift in pK_a to lower values on saccharide binding is in agreement with previous work.[2]

The fluorescence titration of **1** (2.0 x 10^{-5} mol dm^{-3}), **2** (1.0 x 10^{-5} mol dm^{-3}) and **3** (1.0 x 10^{-5} mol dm^{-3}) with different saccharides were carried out in a pH 8.21 buffer (52.1wt% methanol in water with KCl, 0.01000 mol dm^{-3}; KH_2PO_4, 0.002752 mol dm^{-3}; Na_2HPO_4, 0.002757 mol dm^{-3}).[20] The fluorescence spectra of **1**, **2** and **3** in the presence of D-fructose (0-0.1 mol dm^{-3}) are shown in **Figure 14.3**.

The stability constants (K) of fluorescence sensors **1**, **2** and **3** with D-fructose, D-glucose, D-galactose and D-mannose were calculated by fitting the emission intensity at 350 or 360 nm (λ_{ex} 240 and 244 nm) *versus* concentration of

Table 14.1: Stability constant K (coefficient of determination; r^2) for saccharide complexes of fluorescent sensors **1**, **2** and **3**, in pH 8.21 buffer.

Saccharides	**1** (ex 244 nm) K mol^{-1} dm^3	**1** (ex 274 nm) K mol^{-1} dm^3	**2** (ex 244 nm) K mol^{-1} dm^3	**3** (ex 240 nm) K mol^{-1} dm^3
D-Fructose	79 ± 1.7 (0.99)	106 ± 7.0 (0.99)	212 ± 6.9 (0.99)	129 ± 2.6 (0.99)
D-Glucose	6.4 ± 0.4 (0.99)	19 ± 6.0 (0.98)	8.7 ± 1 (0.99)	6.7 ± 0.5 (0.99)
D-Galactose	14 ± 0.6 (0.99)	27 ± 4.0 (0.99)	27 ± 1.3 (0.99)	18 ± 0.3 (0.99)
D-Mannose	7.8 ± 0.3 (0.99)	-*	14 ± 1.4 (0.99)	16 ± 0.8 (0.99)

*small changes in fluorescence

saccharides (**Figure 14.4**).[21] The stability constants for sensors **1**, **2** and **3** calculated from these titrations are given in **Table 14.1**.

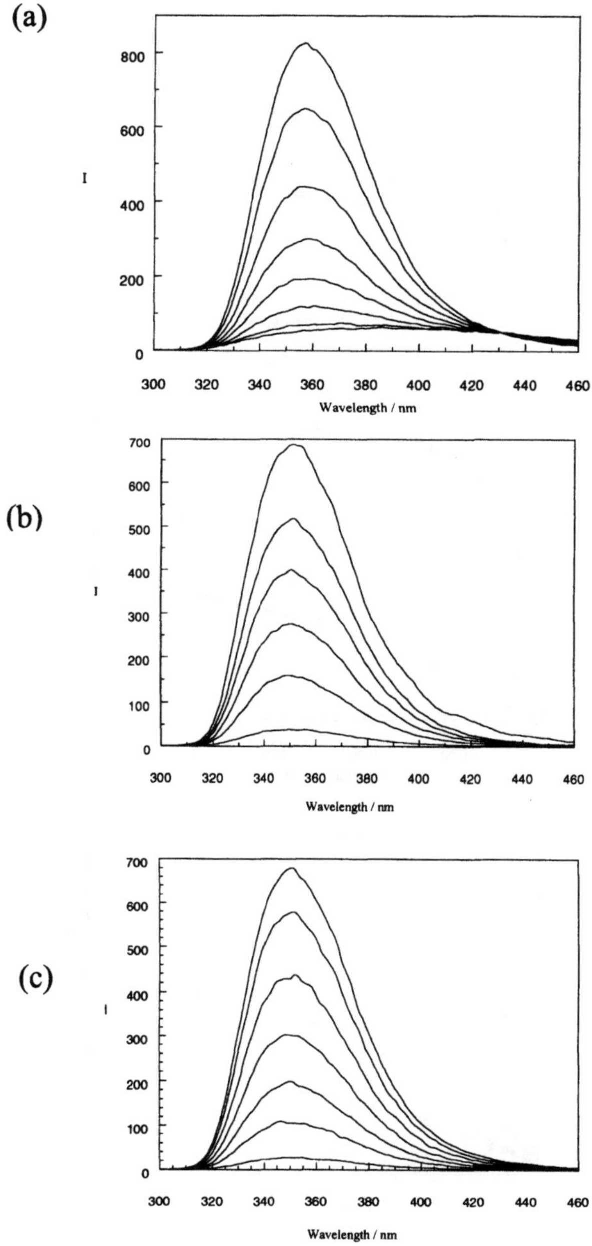

Figure 14.3 Fluorescence spectra change of **1** (a) (2.0×10^{-5} mol dm^{-3}), **2** (b) (1.0×10^{-5} mol dm^{-3}) and **3** (c) (1.0×10^{-5} mol dm^{-3}) with different concentration of D-fructose (0-0.1mol dm^{-3}) in 52.1wt% MeOH pH 8.21 phosphate buffer. λ_{ex} 244, 240 and 244 nm respectively.

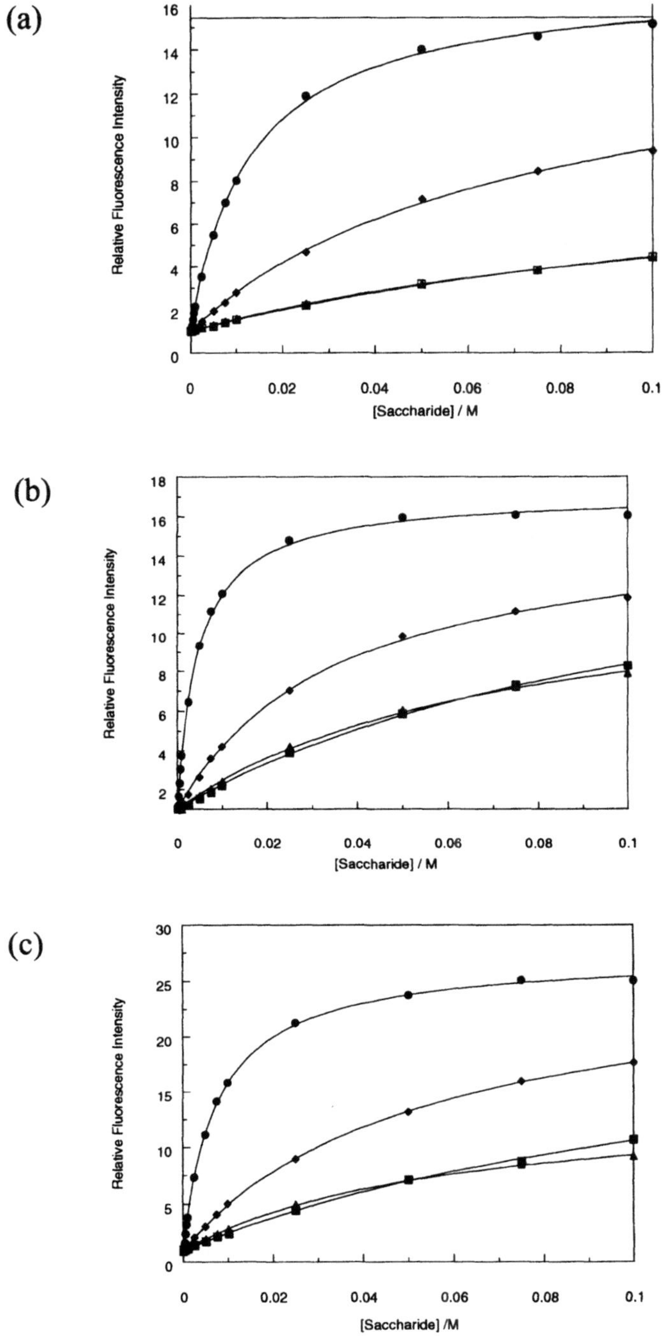

Figure 14.4: Relative fluorescence intensity *versus* saccharide concentration profile of **1** (a), **2** (b) and **3** (c) with (●) D-fructose, (■) D-glucose, (♦) D-galactose, (▲) D-mannose. The measurement conditions are the same as those in Figure 14.3. (a) λ_{ex} 244, λ_{em} 360 nm. (b) λ_{ex} 240, λ_{em} 350 nm. (c) λ_{ex} 244, λ_{em} 350 nm.

The fluorescence enhancements obtained for **1**, **2** and **3** on the addition of D-fructose are 15, 18 and 25 fold respectively. We believe that these large fluorescence enhancements can be attributed to fluorescence recovery of the aniline fluorophore. With these systems in the absence of saccharides the normal fluorescence of the LE state of the aniline donor is quenched by energy transfer to the phenylboronic acid acceptor. When saccharides are added, a negatively charged boronate anion is formed (*cf.* species **B** from **Scheme 14.2**), under these conditions energy transfer from the aniline donor is unfavorable and fluorescence recovery of the LE state of the aniline donor is observed.

This was confirmed by the quantum yield measurements of compounds **1**, **2** and **3**. The quantum yield Φ of aniline is 0.09[22] (in methanol), and the measured quantum yield Φ of **1** is 0.0082, **2** is 0.0087 and **3** is 0.0070 (in methanol).[23, 22]

Compounds **4** and **5** were also prepared (**Scheme 14.3**) in order to explore the effect of electron withdrawing and donating groups on the fluorescent properties of the system.

Sensor **4** contains an electron-withdrawing chlorine group on the aniline ring. The fluorescence of sensor **4** is quenched on addition of saccharides.

The pH-tritration was performed in an osmotic buffer (52.1% methanol aqueous solution with NaCl 0.05 mol dm^{-3}). p*K*a values are for sensors **4** are 10.47 ± 0.05 (0.99) and 7.82 ± 0.04 (0.99) in the absence and presence of D-fructose respectively.

The fluorescence titrations of **4** (2.0×10^{-5} mol dm^{-3}) were performed in pH 8.21 phosphate buffer at 279 nm excitation wavelength, with 4 different monosaccharides (D-fructose, D-glucose, D-galactose and D-mannose). The maximum fluorescence intensity was observed at 400 nm. The fluorescence is quenched on addition of D-fructose (6.7 fold). The bound species is non-fluorescent, so the quenched fluorescence observed is due to the decrease of the unbound species present in the media upon addition of saccharides. The stability constants (*K*) of fluorescence sensor **4** were calculated by fitting the emission at their maximum, λ_{em} 400 nm *versus* concentration of saccharide. The

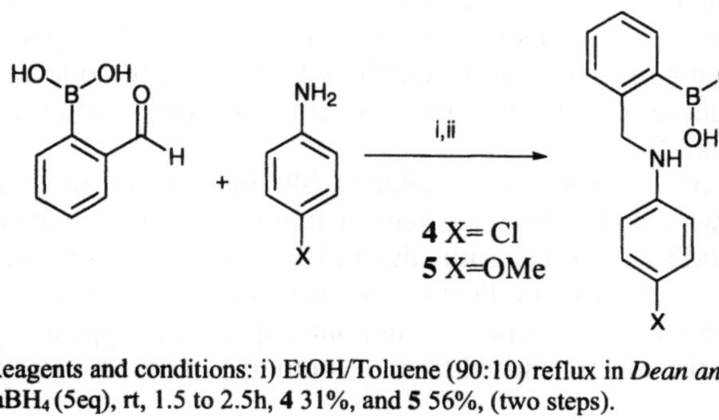

Scheme 14.3: Reagents and conditions: i) EtOH/Toluene (90:10) reflux in *Dean and Stark trap*, overnight; ii) $NaBH_4$ (5eq), rt, 1.5 to 2.5h, **4** 31%, and **5** 56%, (two steps).

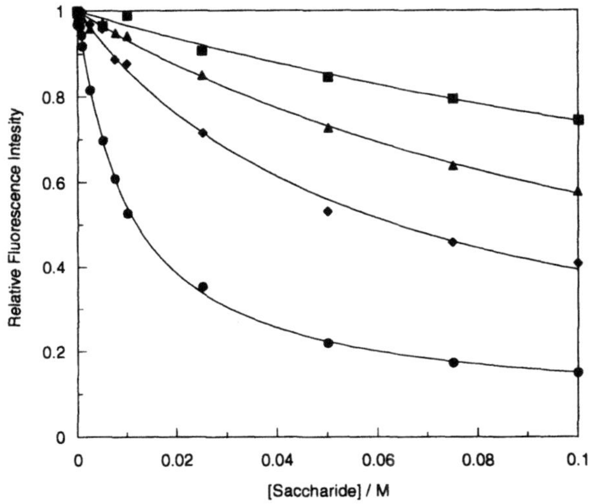

Figure 14.5: Relative fluorescence intensity *versus* saccharide concentration profile of **4** with (●) D-fructose, (■) D-glucose, (♦) D-galactose, (▲) D-mannose, pH 8.21 buffer, λ_{ex} 279 nm, λ_{em} 400 nm.

curves are shown in **Figure 14.5** and the calculated values are given in **Table 14.2.**

Sensor **5** has an electron-donating group, a methoxy group on the aniline ring. Sensor **5** displays an enhancement of fluorescence upon addition of saccharides.

The pH-tritration was performed in an osmotic buffer (52.1% methanol aqueous solution with NaCl 0.05 mol dm^{-3}) in the absence and in the presence of 0.005 mol dm^{-3} of D-fructose. For sensor **5**, the usual fitting curve, using Kaleidagraph failed, so another program, Scientist, has been used to fit the pK_a values. However this program cannot be used in the presence of D-fructose, due to large number of species involved. The two pK_a values in the absence of D-fructose for sensor **5** are 10.85 ± 0.06 (0.90), 4.99 ± 0.42 (0.90) respectively.

The fluorescence titrations of **5** were performed in pH 8.21 phosphate buffer at 240 and 286 nm excitation wavelength, with 4 different monosaccharides (D-fructose, D-glucose, D-galactose and D-mannose). The concentration used was 4.0×10^{-5} mol dm^{-3}.

The spectra of sensor **5**, when excited at 240 nm does not show any isostilbic point, and shows only the enhancement of fluorescence (4.7 fold) of the bound species upon addition of saccharide, due to LE state. However when sensor **5** is excited at 286 nm, the spectra displays two bands so two species are present, the unbound one and the bound one. The unbound species disappears upon addition of saccharides, the spectra shifted from 430 nm to 378 nm. An isostilbic point is observed at 409 nm upon addition of D-fructose. So sensor **5** behaves similar to sensor **1** with both TICT state and LE state depending excitation wavelength and condition used.

The stability constants (*K*) of fluorescence sensor **5** were calculated by fitting the intensity at their emission maximum *versus* concentration of saccharide. The curves are shown in **Figure 14.6** and the calculated values are given in **Table 14.2.**

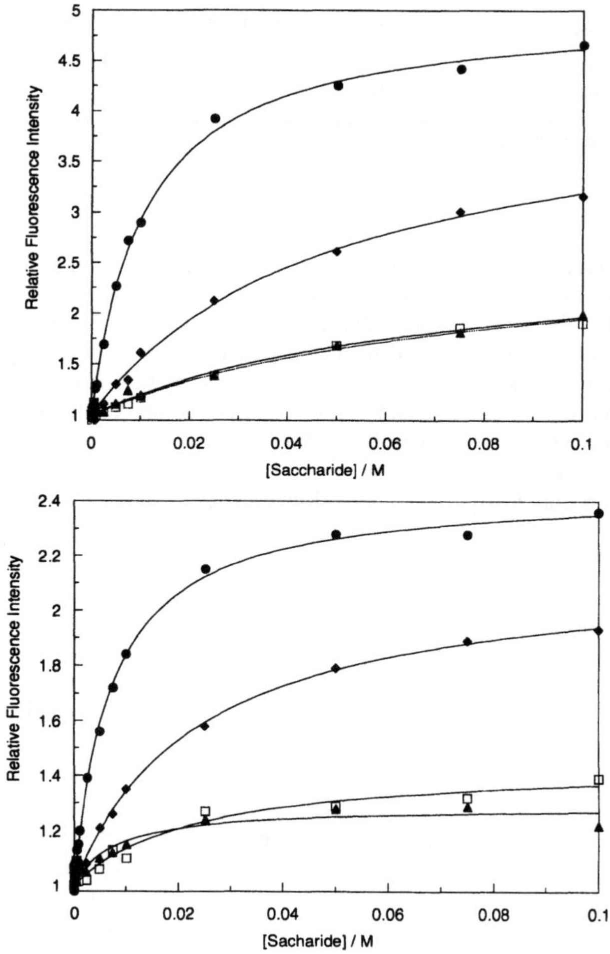

Figure 14.6: Relative fluorescence intensity *versus* saccharide concentration profile of **5** with (●) D-fructose, (□) D-glucose, (♦) D-galactose, (▲) D-mannose, pH 8.21 buffer, (a) λ_{ex} 240 nm, λ_{em} for each spectrum 370 nm. (b) λ_{ex} 286 nm, λ_{em} for each spectrum at their maximum 375-378 nm.

Table 14.2: Stability constant K (coefficient of determination; r) for saccharide complexes of fluorescent sensors **4** and **5**, in pH 8.21 buffer.

Saccharides	**4** (ex 279 nm) K mol^{-1} dm^{3}	**5** (ex 240 nm) K mol^{-1} dm^{3}	**5** (ex 286 nm) K mol^{-1} dm^{3}
D-Fructose	96 ± 3.8 (0.99)	93 ± 4.4 (0.99)	138 ± 6.8 (0.99)
D-Glucose	3.3 ± 3.0 (0.97)	12 ± 3.0 (0.98)	-*
D-Galactose	16 ± 2.6 (0.99)	20 ± 2.2 (0.99)	42 ± 5.4 (0.99)
D-Mannose	7.0 ± 1.4 (0.99)	13 ± 4.0 (0.97)	-*

*small changes in fluorescence

Similar results were found for sensor **1** and the sensors **4,** and **5**. Generally, sensor **1** has a better stability constant for all the 4 saccharides.

Sensor **4** is the only sensor where the fluorescence is quenched. The decrease of fluorescence observed is due to the decrease of the unbound species (TICT state), as the bound species (LE state) in this case is non-fluorescent. Also, the electron-donating group of sensor **5** does not give better results than sensor **1**.

To further explore the system, the aniline was replaced by a naphthalene **6** and a pyrene ring **7**. The sensors were successfully prepared as outlines in **Scheme 4.** Both sensors **6** and **7** display stronger fluorescence than sensor **1**.

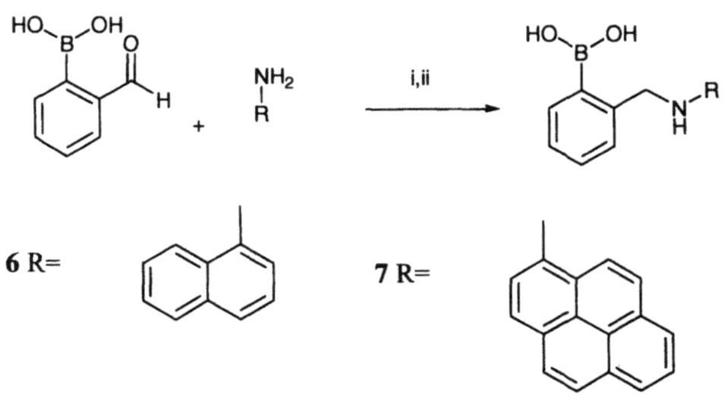

Scheme 14.4: Reagents and conditions: i) MeOH, rt, 1 h; ii) $NaBH_4$ (2eq), rt, 1h, **6** 68% (two steps). i) THF/MeOH (50:50) reflux in *Dean and Stark trap* under nitrogen atmosphere, overnight; ii) $NaBH_4$ (5eq), rt, 1h, **7** 47% (two steps).

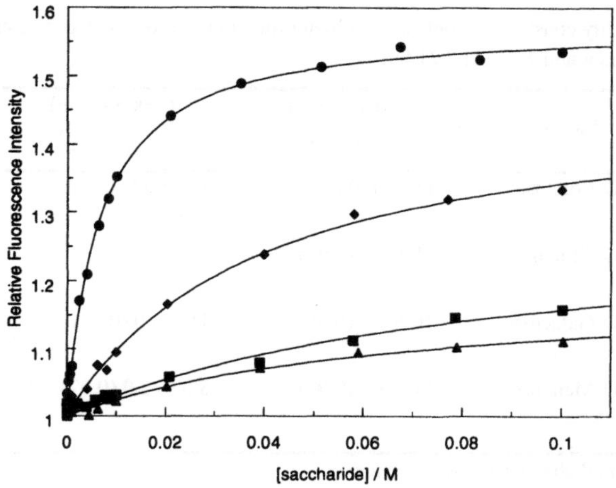

Figure 14.7: Relative fluorescence intensity *versus* saccharide concentration profile of **6** with (●) D-fructose, (■) D-glucose, (♦) D-galactose, (▲) D-mannose, pH 8.21 buffer, λ_{ex} 324 nm, λ_{em} 443 nm.

The fluorescence titration of **6** and **7** were performed in pH 8.21 buffer at 324 nm and 383 nm excitation wavelength respectively, with 4 different monosaccharides, (D-fructose, D-glucose, D-galactose and D-mannose) and sensors concentration (2.0×10^{-6} and 2.5×10^{-7} mol dm^{-3} for sensors **6** and **7** respectively). The spectra of sensor **6** shifted from 428 nm to 443 nm (15 nm) and increased upon addition of D-fructose. An isostilbic point is observed at 417 nm. The spectra of sensor **7** decreased and shifted from 441 nm to 454 nm (13 nm) upon addition of D-fructose. An isostilbic point is observed at 449 nm. The fluorescence enhancements for four sugars are shown in **Figure 14.7** and **Figure 14.8** for sensors **6** and **7** respectively. The stability constants are calculated from the previous curves and given in **Table 14.3**.

From **Table 14.3**, stability constants appear to have similar values for sensors **6** and **7**. If the stability constant values of sensors **6** and **7** are compared to sensor **1** (when excited at 274 nm), the values are in the same order and are close to each other. However, the spectra of sensor **1** shifted from 404 nm to 362 nm (42 nm) and increased upon addition of D-fructose. An isostilbic point is found at 377 nm. Sensors **6** and **7** produce a smaller shift but a higher emission

Table 14.3: Stability constant K (coefficient of determination; r^2) for saccharide complexes of fluorescent sensors **6** and **7**, in pH 8.21 buffer.

Saccharides	**6** (ex 324 nm) K mol^{-1} dm^3	**7** (ex 383 nm) K mol^{-1} dm^3
D-Fructose	154 ± 8.0 (0.99)	132 ± 15 (0.99)
D-Glucose	11 ± 3.0 (0.97)	-*
D-Galactose	26 ± 2.0 (0.99)	11 ± 3.0 (0.99)
D-Mannose	14 ± 3.0 (0.98)	8.0 ± 5.0 (0.97)

*small changes in fluorescence

wavelength, due to the two higher wavelength fluorophores used (naphthalene and pyrene). Sensor **1** produces the largest shift in wavelength (42 nm). In conclusion, changing the fluorophore from benzene to naphathalene and pyrene reduces sensitivity but extends the working wavelength of the sensors.

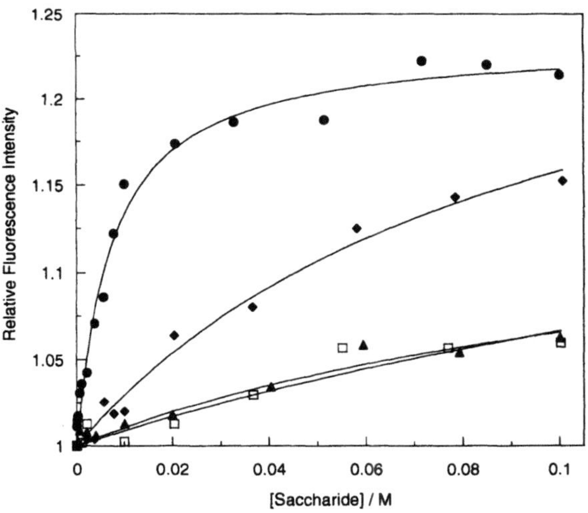

Figure 14.8: Relative fluorescence intensity *versus* saccharide concentration profile of **7** with (●) D-fructose, (□) D-glucose, (♦) D-galactose, (▲) D-mannose, pH 8.21 buffer, λ_{ex} 383 nm, λ_{em} 460 nm.

14.3. CONCLUSIONS

In conclusion, we have prepared several new systems which display large fluorescence enhancements on saccharide binding. (**1, 2, 3** and **5**) The fluorescence changes observed for the LE state at 360, 350, 350 and 370 nm for **1**, **2**, **3**and **5** on excitation at 244, 240, 244 or 240 nm has been ascribed to the fluorescence recovery of the aniline fluorophore. With sensor **1** on excitation at 274 nm a fluorescence increase at 360 nm due to the LE state of species **B** from **Scheme 1** and fluorescence decrease at 404 nm due to the TICT state of species **A** from **Scheme 14.1** is observed. (**Figure 14.1**) With sensor **4** on excitation at 279 nm only fluorescence quenching is observed at 400 nm due to the decrease of the TICT state species **A** from **Scheme 14.1**. With sensor **4** the LE state due to species **B** from **Scheme 14.1** is non-fluorescent. With sensors **6** and **7** the working wavelength has been extended to 443 and 460 nm on excitation at 324 and 388 nm respectively. We are confident that these discoveries will lead to the development of improved boronic acid based fluorescent saccharide sensors. Our ongoing research is directed towards exploiting these findings in other saccharide selective systems.

14.4. EXPERIMENTAL

14.4.1. General procedures

All the solvents and chemicals used as starting materials were reagent grade, unless stated otherwise and supplied by Frontier Scientific Ltd, Aldrich Chemical Co. Ltd., Lancaster Synthesis Ltd. and Fisher Scientific Ltd.

^{1}H and ^{13}C{^{1}H} NMR spectra were recorded on a Bruker AVANCE 300 (300.13 and 75.47 MHz respectively) and a Jeol-400MHz (399.78 and 100.52 MHz respectively) spectrometer. All the chemical shifts (δ) were reported in ppm using the deuterated solvent and or tetramethylsilane as the internal reference. All the following data use the abbreviations as follow: singlet (s), doublet (d), doublet of doublet (dd), apparent doublet (d app), multiplet (m), broad singlet (brs). *J*-values are given in Hertz.

The mentioned carbon masked by CD_3OD, is a carbon seen under CD_3OD but difficult to assign individually or shown by a 2D ^{1}H-^{13}C correlation.

The $_{13}$C{$_{1}$H} NMR spectra were subject to the PENDANT technique, resulting in primary and tertiary carbon atoms having a different phase to secondary and quaternary carbon atoms. The phase is presented in the following manner: (+) positive phase, (-) negative phase. Due to quadrupolar relaxation, the aryl carbon atoms attached directly to boron atoms of the boronic acids were not observed by ^{13}C-NMR.

Mass spectra including high-resolution spectra were recorded on a Waters Micromass Autospec Spectrometer using FAB (Fast Atom Bombardment) at the University of Bath and at the EPRSC mass spectrometry centre, Swansea.

Elemental Analysis were performed on a Exeter Analytical CE 440.

Melting points were determined using a Gallenkamp melting point apparatus and are reported uncorrected.

Thin Layer Chromatography (TLC) was performed on precoated aluminium-backed silica plates supplied by Fluka Chemie. Visualisation was achieved by UV light (254 nm).

Organic phases were dried with anhydrous $MgSO_4$, filtered and condensed with a Büchi evaporator. Further evaporation was carried out on high-*vacuum* line where necessary.

All the fluorescence measurements were performed on a Perkin Elmer Luminescence Spectrometer LS 50B with a Starna Silica (quartz) cuvet with 10 mm path lengths and four faces polished. Data was collected *via* the Pekin Elmer FL WinLab 4.00 software package.

UV-Visible absorption measurements used Sigma Spectrophotometer Silica (Quartz) Cuvets with 10 mm path lengths and were recorded on a Perkin-Elmer Lambda 20 UV/VIS spectrometer. The data was collected *via* the Pekin Elmer UV WinLab 2.70 software package.

The association constant *K* and p*Ka* were analysed with Kaleidagraph 3.51 using nonlinear (Levenberg-Marquardt algorithm) curve fitting. The errors reported are the standard errors obtained from the best fit.

The pH measurements in the pH-titration experiments were taken on a Hanna Instruments HI 9321 Microprocessor pH meter which was calibrated using Fisher Chemicals "colour coded" buffer solutions (pH 4.0- phthalate , pH 7.0- phosphate and pH 10.0- borate).

The pH measurements for the preparation of the HEPES buffer were taken on a Mettler DL 21 Titrator with a Mettler Toledo DG111-SC glass electrode which was calibrated using two 52.1% methanol/water buffers [pH 2.07 (0.01000 m KCl, 0.01101 m HCl) and 9.43 (0.00992 m KCl, 0.004996 m Na borate)].[20]

14.4.2. Preparation of the (2-boronobenzyl)-aniline (1)

2-Formylphenylboronic acid (1.00 g, 6.67 mmol) was dissolved in a mixture of absolute EtOH/Toluene (90/10) (50 ml) then aniline (621 mg, 6.67 mmol) was added to the reaction. A *Deans and Stark trap* was fixed to the reaction vessel and filled with the same solvent mixture EtOH/Toluene (90/10) (25 ml) to permit the azeotropic removal of water. The reaction was stirred under reflux for 16 hours. After cooling to 0°C with an ice bath, sodium borohydride (1.26 g, 33.35 mmol) was added slowly to the mixture and stirred for 2.5 further hours at room temperature.

The solvent was removed under reduced pressure. Water (100 ml) was added to the residue and the resulting solution was extracted with chloroform (3 × 100 ml). The organic phases were combined, washed with brine (150 ml) and dried over $MgSO_4$. The solvent was then removed *in vacuo*. Precipitation from chloroform/*n*-hexane afforded sensor **1** (813 mg, 54%) as a cream powder.

mp: 125-129°C; Found: C, 74.70; H, 5.97; N, 6.63. $C_{13}H_{14}BNO_2$-H_2O requires C, 74.72; H, 5.80; N, 6.70%; δ_H (300 MHz; CD_3OD) 4.21 (2H, s, CH_2), 6.60-6.70 (3H, m, Ar*H*), 6.95-7.25 (6H, m, Ar*H*); δ_C (75 MHz; CD_3OD) 51.6,

117.4 (2C), 121.5, 126.9, 127.9, 129.8, 130.2 (2C), 132.7, 146.0, 149.1; *m/z* (FAB) 497.3 ([M+2(3-$HOCH_2C_6H_4NO_2$)-$2H_2O$]$^+$, 100%); $\Phi = 0.0082$.

14.4.3. Preparation of the (3-boronobenzyl)-aniline (2)

3-Formylphenylboronic acid (200 mg, 1.33 mmol) was dissolved in methanol (20 ml) then aniline (136 mg, 1.46 mmol) was added to the reaction. The reaction was stirred overnight at room temperature. The reduction of the imine product was then carried out by the slow addition of sodium borohydride (250 mg, 6.61 mmol) to the solution. The reaction was stirred for 1.5 further hours at room temperature.

The solvent was removed under reduced pressure. Water (20 ml) was added to the residue and the resulting solution was extracted with chloroform (3 × 40 ml). The organic phases were combined, washed with brine (60 ml) and dried over $MgSO_4$. The solvent was then removed *in vacuo*. Precipitation from chloroform/*n*-hexane afforded sensor **2** (137 mg, 45%) as a white powder.

mp: 166°C; Found: C, 73.90; H, 5.56; N, 6.72. $C_{13}H_{14}BNO_2$-H_2O+0.03 $CHCl_3$ requires C, 73.60; H, 5.70; N, 6.58%); (HRMS: Found [M+2(3-$HOCH_2C_6H_4NO_2$)-$2H_2O$]$^+$, 497.1776. $C_{27}H_{24}BN_3O_6$ requires 497.1758); δ_H (300 MHz; CD_3OD) 4.31 (2H, s, C*H*$_2$), 6.55-6.65 (3H, m, Ar*H*), 7.00-7.10 (2H, m, Ar*H*), 7.30 (1H, m, Ar*H*), 7.42 (1H, m, Ar*H*), 7.60 (1H, m, Ar*H*), 7.76 (1H, m, Ar*H*); δ_C (75 MHz; CD_3OD) approximately 49.0 carbon masked by CD_3OD, 114.5, 118.3, 129.1, 130.3, 130.8, 133.8, 134.4, 140.8, 150.5; *m/z* (FAB) 498.1 ([M+H+2(3-$HOCH_2C_6H_4NO_2$)-$2H_2O$]$^+$, 100%); $\Phi = 0.0087$.

14.4.4. Preparation of the (4-boronobenzyl)-aniline (3)

4-formylphenylboronic acid (200 mg, 1.33 mmol) was dissolved in methanol (20 ml) then aniline (137 mg, 1.46 mmol) was added to the reaction. The reaction was stirred overnight at room temperature. The reduction of the imine product was then carried out by the slow addition of sodium borohydride (250 mg, 6.61 mmol) to the solution. The reaction was stirred for 1 further hour at room temperature.

The solvent was removed under reduced pressure. Water (20 ml) was added to the residue and the resulting solution was extracted with chloroform (3 × 40 ml). The organic phases were combined, washed with brine (60 ml) and dried over $MgSO_4$. The solvent was then removed *in vacuo*. Precipitation from chloroform/*n*-hexane afforded sensor **3** (135 mg, 45%) as a white powder.

mp: 164°C; Found: C, 73.90; H, 5.88; N, 6.15. $C_{13}H_{14}BNO_2$-H_2O+0.03 $CHCl_3$ requires C, 73.60; H, 5.70; N, 6.58%); (HRMS: Found [M+2(3-$HOCH_2C_6H_4NO_2$)-$2H_2O$]$^+$, 497.1772. $C_{27}H_{24}BN_3O_6$ requires 497.1758); δ_H (300 MHz; CD_3OD) 4.30 (2H, s, C*H*$_2$-NH), 6.55-6.65 (3H, m, Ar*H*), 7.00-7.10 (2H, m, Ar*H*), 7.30-7.40 (2H, m, Ar*H*), 7.56 (1H, m, Ar*H*), 7.69 (1H, m, Ar*H*); δ_C (75 MHz; CD_3OD) approximately 49.0 carbon masked by CD_3OD, 114.5,

118.4, 127.9, 130.3, 135.5, 144.1, 150.5; *m/z* (FAB) 497.1 ([M+2(3-$HOCH_2C_6H_4NO_2$)-$2H_2O$]$^+$, 100%); $\Phi = 0.0070$.

14.4.5. Preparation of the (4-chloro-phenyl)-(2-boronobenzyl) amine (4)

The 4-chloroaniline (84 mg, 0.66 mmol) was dissolved in a mixture of absolute EtOH/Toluene (90/10) (25 ml) then 2-formylphenylboronic acid (100 mg, 0.66 mmol) was added to the solution. A *Deans and Stark trap* was fixed to the reaction vessel and filled with EtOH/Toluene (90/10) mixture (25 ml). The reaction was stirred under reflux overnight. After cooling, sodium borohydride (125 mg, 3.30 mmol) was added slowly to the mixture and stirred for 1.5 further hours at room temperature.

The solvent was removed under reduced pressure. Water (30 ml) was added to the residue and the resulting solution was extracted with chloroform (3 × 30 ml). The organic phases were combined, washed with brine (45 ml) and dried over $MgSO_4$. The solvent was then removed *in vacuo*. Precipitation from chloroform/*n*-hexane afforded sensor **4** (54 mg, 31%) as a white powder.

mp: 148-150°C; (HRMS: Found [M+2(3-$HOCH_2C_6H_4NO_2$)-$2H_2O$]$^+$, 531.1349. $C_{27}H_{23}BClN_3O_6$ requires 531.1368); δ_H (300 MHz; CD_3OD) 4.19 (2H, s, CH_2-NH), 6.54 (2H, AA'BB'system, J_{AB} *8.8 Hz* highly second order, Ar*H*), 6.96 (2H, AA'BB'system, $J_{A'B'}$ *8.8 Hz* highly second order, Ar*H*), 7.08-7.25 (4H, m, Ar*H*); δ_C (75 MHz; CD_3OD) 50.8, 117.4 (2C), 124.8, 127.9, 130.0 (4C), 132.9, 145.7, 148.7; *m/z* (FAB) 531.1 ([M+2(3-$HOCH_2C_6H_4NO_2$)-$2H_2O$]$^+$, 100%); $\Phi = 0.1921$.

14.4.6. Preparation of the (4-methoxy-phenyl)-(2-boronobenzyl) amine (5)

2-Formylphenylboronic acid (200 mg, 1.33 mmol) was dissolved in a mixture absolute EtOH/Toluene (90/10) (25 ml) then *p*-anisidine (164 mg, 1.33 mmol) was added to the solution. A *Deans and Stark trap* was fixed to the reaction vessel and filled with mixture of EtOH/Toluene (90/10) (25 ml). The reaction was heated under reflux overnight. After cooling, sodium borohydride (250 mg, 6.61 mmol) was added slowly to the mixture and stirred for 2 further hours at room temperature.

The solvent was removed under reduced pressure. Water (40 ml) was added to the residue and the resulting solution was extracted with chloroform (3 × 40 ml). The organic phases were combined, washed with brine (60 ml) and dried over $MgSO_4$. The solvent was then removed *in vacuo* and dried under high *vacuum*. The desired sensor **5** (192 mg, 56%) was obtained as a pale brown foam.

Found: C, 69.8; H, 5.76; N, 5.94. $C_{14}H_{16}BNO_3$-H_2O requires C, 70.33; H, 5.90; N, 5.86%; (HRMS: Found [M+2(3-$HOCH_2C_6H_4NO_2$)-$2H_2O$]$^+$, 527.1858. $C_{28}H_{26}BN_3O_7$ requires 527.1864); δ_H (300 MHz; CD_3OD) 3.63 (3H, s, OCH_3), 4.24 (2H, s, CH_2-NH), 6.72 (2H, AA'BB'system, J_{AB} *9.2 Hz* highly second order, Ar*H*), 6.85 (2H, AA'BB'system, $J_{A'B'}$ *9.2 Hz* highly second order, Ar*H*), 7.08-7.26 (4H, m, Ar*H*); δ_C (75 MHz; CD_3OD) 54.5, 56.4, 116.1 (2C), 121.4

(2C), 128.1 (2C), 129.8, 132.4, 140.0, 145.0, 157.8; m/z (FAB) 527.2 ([M+2(3-$HOCH_2C_6H_4NO_2$)-$2H_2O$]$^+$, 100%);
$\Phi = 0.0149$.

14.4.7. Preparation of the 2-(boronobenzyl)-naphthalen-1-yl-amine (6)

2-formylbenzeneboronic acid (0.20 g, 1.33 mmol) was dissolved in methanol (20 ml) and 1-aminonaphthalene (0.19 g, 1.33 mmol) was then added as a methanol (2 ml) solution. A drying tube was fitted and the reaction stirred at room temperature for 1 hour. Sodium borohydride (0.10 g, 2.66 mmol) was added slowly as a solid. After stirring for a further hour, the reaction mixture was poured on to ice-water (20 ml). A few drops of sodium hydroxide solution (1 M) were added to aid precipitation of the product. The precipitate was collected by suction filtration and dried in a dessicator to afford **6** (0.25 g, 68%) as an off-white solid: mp 74-77 °C (dec.); (HRMS: Found 547.1937, [M-$2H_2O$ + 2 NOBA]$^+$. $C_{31}H_{26}BN_3O_6$ requires 547.1915); δ_H (300 MHz; CD_3OD; Me_4Si) 4.52 (2 H, s, ArCH_2-NHAr), 6.50 (1 H, d, *J* 6.0, Ar-*H*), 7.10-7.49 (8 H, m, Ar-*H*), 7.67-7.78 (1 H, m, Ar-*H*), 8.04-8.14 (1 H, m, Ar-*H*); δ_C (75 MHz; 50% $CDCl_3$ -50% CD_3OD; Me_4Si) 50.4, 107.4, 119.5, 122.7, 125.9, 126.2, 127.0, 127.6, 127.8, 128.2, 129.6, 130.1, 136.1, 145.3; *m/z* (FAB) 547 ([M-$2H_2O$ + 2 NOBA]$^+$, 100%).

14.4.8. Preparation of the (2-boronobenzyl)-pyren-1-yl-amine (7)

1-Pyrenemethylamine hydrochloride salt was added to aqueous NaOH (1 M) solution. The mixture was stirred for 30 min and the precipitate was extracted with ether. The organic layers were combined, dried over $MgSO_4$, filtered and evaporated to afford the product used in the next step.

1-pyrenemethylamine (200 mg, 0.92 mmol) was dissolved in a mixture of THF/MeOH (50/50) (10 ml) and 2-formylphenylboronic acid (165 mg, 1.10 mmol) was added to the solution. The solution was stirred at room temperature for 24 hours under nitrogen atmosphere. Sodium borohydride (125 mg, 3.30 mmol) was added slowly to the mixture and stirred for 1 further hour at room temperature. Then the solvent was removed under reduced pressure.

The residue was dissolved in chloroform (10 ml), washed with water (10 ml) and dried over $MgSO_4$. The solvent was removed under reduced pressure. Precipitation from chloroform/*n*-hexane afforded sensor **7** (151 mg, 47%) as an olive green powder.

mp: 129-130°C dec; Found: C, 81.8; H, 5.16; N, 4.25. $C_{23}H_{18}BNO_2$-H_2O+0.1$CHCl_3$ requires C, 82.07; H, 4.81; N, 4.14%; (HRMS: Found [M+2(3-$HOCH_2C_6H_4NO_2$)-$2H_2O$]$^+$, 621.2098. $C_{37}H_{28}BN_3O_6$ requires 621.2071); δ_H (400 MHz; CD_3OD+$CDCl_3$; Me_4Si) 4.68 (2H, s, CH_2-NH_2), 7.24-7.48 (4H, m, Ar*H*), 7.71-8.17 (9H, m, Ar*H*); δ_C (100 MHz; CD_3OD+$CDCl_3$) 49.7, 120.3, 123.3, 123.5, 123.7, 125.5, 125.8, 125.9, 126.0, 126.7, 127.6, 131.8, 132.3; *m/z* (FAB) 621.2 ([M+2(3-$HOCH_2C_6H_4NO_2$)-$2H_2O$]$^+$, 100%).

14.5. ACKNOWLEDGMENTS

Financial support from the Royal Society and Beckman-Coulter are gratefully acknowledged. We would also like to thank the University of Bath for support.

14.6. REFERENCES

1. A. P. de Silva, H. Q. N. Gunaratne, T. Gunnlaugsson, A. J. M. Huxley, C. P. McCoy, J. T. Rademacher, T. E. Rice *Chem. Rev.* 97(5), 1515-1566 (1997).
2. T. D. James, S. Shinkai *Top. Curr. Chem.* 218, 159 (2002).
3. W. Yang, H. He, D. G. Drueckhammer *Angew. Chem., Int. Ed.* 40(9), 1714-1718 (2001).
4. Z. Zhong, E. V. Anslyn *J. Am. Chem. Soc.* 124(31), 9014-9015 (2002).
5. A.-J. Tong, A. Yamauchi, T. Hayashita, Z.-Y. Zhang, B. D. Smith, N. Teramae *Anal. Chem.* 73(7), 1530-1536 (2001).
6. V. V. Karnati, X. Gao, S. Gao, W. Yang, W. Ni, S. Sankar, B. Wang *Bioorg. Med. Chem. Lett.* 12(23), 3373-3377 (2002).
7. N. DiCesare, J. R. Lakowicz *Chem. Commun.* (19), 2022-2023 (2001).
8. H. Cao, D. I. Diaz, N. Di Cesare, J. R. Lakowicz, M. D. Heagy *Org. Lett.* 4, 1503-1505 (2002).
9. H. Eggert, J. Frederiksen, C. Morin, J. C. Norrild *J. Org. Chem.* 64(11), 3846-3852 (1999).
10. J. P. Lorand, J. O. Edwards *J. Org. Chem.* 24, 769-774 (1959).
11. S. Arimori, L. I. Bosch, C. J. Ward, T. D. James *Tetrahedron Lett.* 42(27), 4553 - 4555 (2001).
12. L. I. Bosch, M. F. Mahon, T. D. James *Tetrahedron Lett.* 45, 2859-2862 (2004).
13. Z. R. Grabowski, K. Rotkiewicz, A. Siemiarczuk, D. J. Cowley, W. Baumann *Nouv. J. Chim* 3(7), 443 (1979).
14. W. Rettig, E. A. Chandross *J. Am. Chem. Soc.* 107, 5617 (1985).
15. W. Rettig *Angew. Chem. Int. Ed. Engl* 25, 971 (1986).
16. C. J. Joedicke, H. P. Luethi *J. Am. Chem. Soc.* 125(1), 252-264 (2003).
17. Crystallographic data (excluding structure factors) for the structures in this paper have been deposited with the Cambridge Crystallographic Data Centre as supplementary publication numbers CCDC 228172. Copies of the data can be obtained, free of charge, on application to CCDC, 12 Union Road, Cambridge CB2 1EZ, UK [fax: 144-(0)1223-336033 or e-mail: deposit@ccdc.cam.ac.uk].
18. J. C. Norrild, I. Sotofte *J. Chem. Soc., Perkin Trans.* 2, 303-311 (2002).
19. H. Hopfl *J. Organomet. Chem* 581, 129-149 (1999).
20. D. D. Perrin, B. Dempsey (1974) Buffers for pH and Metal Ion Control. Chapman and Hall, London
21. The *K* were analysed in KaleidaGraph using nonlinear (Levenberg-Marquardt algorithm) curve fitting. The errors reported are the standard errors obtained from the best fit.
22. G. Perichet, R. Chapelon, B. Pouyet *J. Photochem* 13, 67 (1980).
23. S. G. Schulman (1977) Fluorescence and Phosphorescence Spectroscopy: Physiochemical Principles and Practice. Pergamon Press, Oxford
24. The areas under the fluorescence spectra were calculated using KaleidaGraph version 3.51 for PC, published by Synergy Software and developed by Abelbeck Software, 2457 Perikiomen Avenue, Reading, PA19606. The quantum yields of sensors **1**, **2** and **3** were then calculated by comparison with aniline as standard using the following equation:[23]

$$\Phi_s = \Phi_{an} \times (FA_s / FA_{an}) \times (A_{an} / A_s)$$

Φ is quantum yield; FA is the area under the curve of the fluorescence peak; A is the absorbance of each sensor at 244 nm (for **1** and **3**) and 240 nm for **2**. $_s$ and $_{an}$ are the abbreviations for sensors (**1**, **2**, **3**) and the aniline reference respectively. Φ =0.09 (in methanol) [22] is used as the reference quantum yield of aniline in methanol.

GLUCOSE SENSING AND GLUCOSE DETERMINATION USING FLUORESCENT PROBES AND MOLECULAR RECEPTORS

Axel Duerkop, Michael Schaeferling, Otto S. Wolfbeis*[*]

15.1. ABSTRACT

We summarize our previous and recent work on the use of fluorescent probes and of synthetic molecular receptors (including polymeric receptors and bead-immobilized receptors) for determination of glucose. The following approaches were made: (1) for enzyme-based quantitation of glucose in solution, the fluorescent probes were added to the sample solution along with glucose oxidase (GOx) and kinetics was followed over time. Specifically, the metal ligand complex Ru(sulfo-dpp)$_3$ served as a water-soluble probe for oxygen, while the europium-tetracycline complex served as a probe for hydrogen peroxide; (2) for enzyme-based (continuous) glucose sensing, GOx is immobilized (along with indicator probes) on certain hydrogels which results in sensor layers that give an optical response to glucose via the indicator dye employed; (3) in an entirely different approach, a ruthenium-ligand complex with a boronic acid side group was synthesized and studied for its response to polyols and certain monosaccharides; (4) in yet another approach, a polymeric saccharide-responsive material was obtained by copolymerization of aniline with aniline boronic acid to give a copolymer with a near-infrared (600 – 900 nm) optical response (in absorption); (5) the enzyme-based schemes subsequently were applied to lifetime-based imaging of the distribution of glucose and GOx.

---

[*] Axel Duerkop, Michael Schaeferling, Otto S. Wolfbeis, Institute of Analytical Chemistry, Chemo- and Biosensors, University of Regens-burg, D-93040 Regensburg, Germany.
E-mail: Otto.Wolfbeis@chemie.uni-regensburg.de

15.2. INTRODUCTION

Rarely has there been more literature on a single analyte than on glucose and its determination. Given the significance of glucose, this does not come as a surprise. It simply reflects the fact that glucose has to be determined in numerous samples including blood, other body fluids, in feed and food, juice and drinks, in dairy products and fruit. Moreover, glucose is the degradation product of numerous other species which may be converted into it so that by detecting the quantity of glucose formed the concentration of its precursor may be determined. Glucose usually is present in relatively high concentrations so that most (but not all) assays have been developed for a rather high concentration range. However, analytical methods for relatively small concentrations of glucose are needed as well, for example when analyzing body fluids that do contain low concentrations (such as tear and interstitial fluid), or because a dilution step is needed during sample preparation so that the initial concentration of glucose is significantly lowered.

Fluorescent detection of glucose can be based on several approaches, some of which are presented here, and numerous others in further sections of this volume. Fluorescence (with its many parameters including fluorescence intensity, decay time, polarization, and its numerous assay formats that range from resonance energy transfer, time-resolved energy transfer to fluorescence fluctuation spectroscopy) and its various implementations (such as fiber optic and bead assays) are likely to remain among the most important detection approaches used for HTS due to their high sensitivity and amenability to automation. In terms of sensitivity, fluorescence correlation spectroscopy (FCS) is the method of choice. It combines maximum sensitivity with high statistical confidence. Chemiluminescence and bioluminescence (including bioluminescence resonance energy transfer) are known to be sensitive and to be applicable to both small and large volumes.

Measurement of fluorescence intensity is widely used but has limitations since the fluorescence intensity of a given sample represents a non-referenced analytical signal that depends on several factors. This can be seen when looking at Parker's law which relates fluorescence intensity (F) with the intensity (I_0) of the exciting (laser) beam, the molar absorbance (ε) of the label, the penetration length of the exciting beam (l), the quantum yield of the label (QY), and instrumental geometries (by introducing a geometrical factor k):

$$F = I_0 \cdot \varepsilon \cdot l \cdot c \cdot QY \cdot k \qquad (1)$$

Equation (1) only holds for solutions whose absorbances do not exceed ~ 0.05 but this condition is often fulfilled in trace analysis or when using thin sensor films. As a result of the many variables, fluorescence intensity often depends on parameters other than concentration. If, for example, the light source fluctuates in intensity (I_0) or the geometry of the system is slightly changed (thereby affecting k), the fluorescence intensity will change even at constant concentration of a probe or a fluorescent analyte.[1]

So-called self-referenced methods provide a solution to this limitation. Among these, fluorescence polarization and fluorescence resonance energy transfer (FRET) are particularly useful. In the first, the analytical signal is referenced to the plane of the exciting beam of polarized light. In the second, the ratio of the fluorescence intensities of two fluorophores at a single excitation wavelength is related to analyte concentration. Fluorescence polarization assay (FPIA) requires labeling of a single glucose receptor only. It has its strength when detecting changes of molecular mass following a binding event. Glucose, however, is a rather small molecule and antibodies are hardly accessible. FRET, in contrast to polarization, requires both binding partners of an affinity assay to be labeled. Competitive glucose assays employing the FRET scheme have been reported in which a species competing with glucose for a specific binding site (such as concanavalin) have been labeled along with the receptor.[2, 3] The need for double labeling is a major drawback of FRET assays even though they are quite versatile in that they can be based on measurement of either fluorescence intensity or decay time.

A fairly new self-referenced scheme is referred to as dual luminophore referencing (DLR).[3] Here, the luminescence of a first luminophore is related to that of an added reference fluorophore having a much longer decay time, and changes in intensity are converted into a phase shift when using a sine-modulated excitation source. In this method the phase shift of the emission is related to the respective excitation signal. A schematic of this approach is shown in Fig. 15.1. It is mandatory in this method that the excitation and emission spectra of indicator dye and of reference dye have at least some overlap.

Fluorescence decay time is a most attractive parameter in being self-referenced. It is preferably determined in either the time domain or the frequency domain[1] and has been used – as will be shown – in optical sensors and in imaging schemes where the respective indicator probes change their

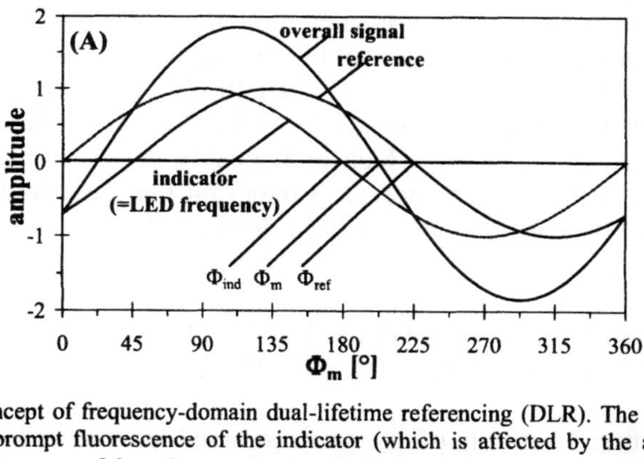

Figure 15.1. Concept of frequency-domain dual-lifetime referencing (DLR). The overall signal is the total of the prompt fluorescence of the indicator (which is affected by the analyte) and the delayed phosphorescence of the reference dye (usually in the form of beads). The phase angle (Φ_m) reflects the ratio of the amplitudes of the two components. Φ_{ref} is the phase angle of the of the reference dye; Φ_{ind} the phase of the of the indicator dye.

lifetime in response to pH or the concentration (or partial pressure) of oxygen. Indicator probes based on metal ligand complexes have found particular interest. Most ruthenium and iridium complexes, for example, have decay times in the μs to ms range and also display longwave emissions. Due to their long decay times, gated assays also have been reported, but these are intensity-based (rather than lifetime-based) schemes, and thus they are not self-referenced. The majority of self-referenced sensing schemes still are based on measurement of fluorescence intensity at two excitation or two emission wavelengths.

In summary, there are several methods for obtaining analytical signals in sensing glucose. Intensity is considered to be an error-prone parameter since it is affected by all too many variables. Decay time is clearly superior. Time-resolved assays can significantly reduce short-lived background fluorescence but require probes with long decay times. These usually are metal-ligand complexes derived from ruthenium or europium (see later). DLR is another and superb method for referencing signals but requires (a) the availability of reference beads of well defined and largely different decay time, and (b) the fraction of these beads/luminophores to remain strictly constant.

Glucose has been determined in the overwhelming majority of cases by making use of the enzyme glucose oxidase. However, dehydrogenases also have been used for optical sensing of glucose. When glucose dehydrogenase oxidizes glucose, the coenzyme NAD^+ is converted into NADH which has a fairly strong fluorescence at around 450 nm if excited at 365 nm. The increase in fluorescence over time is the analytical information that can serve to determine glucose.[4] This assay is irreversible unless NADH is recycled to NAD^+. Unfortunately, the background fluorescence of most biological matter if excited at 365 nm is fairly strong, as is its luminescence at 450 nm. A lifetime scheme for the NADH method has not been demonstrated so far.

Approaches to sense glucose via measurement of the intrinsic optical properties of glucose (such as optical rotation or NIR absorption) have had mixed success. Glucose, in being chiral, may be determined via its specific optical rotation but this scheme is hardly applicable to serum and other biofluids with their numerous chiral interferents. Glucose also has very characteristic absorption bands in the (near) infrared but this approach – even though pursued with large effort in the late 1990s in order to obtain a non-invasive glucose sensor – turned out to be compromised by "trivial" factors such as interferences by variable skin structure and a high individuality in the permeability of skin and tissue for NIR light. The approach based on the intrinsic absorption and fluorescence of the coenzyme FAD will be discussed later.

15.3. DETERMINATION OF GLUCOSE VIA GOx AND FLUORESCENT MOLECULAR PROBES

Oxidase-based sensing of glucose[5] is based on the following reaction catalyzed by the enzyme glucose oxidase (GOx):

$$\text{glucose} + O_2 + H_2O \xrightarrow{GOx} \text{gluconolactone} + H_2O_2 \quad (2)$$

$$\text{gluconolactone} + H_2O \longrightarrow \text{gluconate} + H^+ \quad (3)$$

When looking at the two reactions one can see that glucose can be sensed by one of the following schemes (provided the activity of GOx remains constant):

1. measurement of the consumption of oxygen
2. measurement of the production of hydrogen peroxide
3. measurement of the production of protons

Either approach has been investigated and successfully applied, some even in commercial products.

15.3.1. Determination of Glucose via Dissolved Glucose Oxidase and Changes in pH

Measurement of changes in pH is simply accomplished by adding a pH indicator to the sample solution and monitoring the decrease in pH due to the activity of GOx over time. The approach is limited because very often the initial pH and the buffer capacity of the sample (which affects the change in pH) are unknown. If however, the sample is diluted adequately (with a buffer of defined pH and ionic strength), the assay is fairly practicable and reproducible, albeit at the expense of the limits of detection.[6–8]

15.3.2. Determination of Glucose via Dissolved Glucose Oxidase and Measurement of the Consumption of Oxygen

Rather than measuring the formation of protons as described in 2.1., one may optically detect the consumption of oxygen in a glucose solution to which glucose oxidase (GOx) was added. Various probes have been reported whose fluorescence (phosphorescence) is quenched by molecular oxygen, many of them being excitable in the UV (for example pyrene) which however is disadvantageous since UV excitation causes substantial fluorescence background from most biological matter. In the worst case, the sample has an optical density so high that exciting UV light does not reach the fluorescent probe, or that emitted fluorescence is strongly reduced or even eliminated due to an inner filter effect. Probes with longwave absorption, in contrast, are well suited for this purpose as shown in work by Sasso et al.[9] and Vanderkooi et al..[10]

Most oxygen probes suffer, however, from poor solubility in the (usually aqueous) sample solution. We prefer to use ruthenium(II)tris-bathophenanthroline, referred to as Ru(sulfo-dpp)$_3$, as a water-soluble luminescent probe for oxygen. As can be seen from Fig. 15.2, it contains a classical ruthenium-tris-phenanthroline fluorophore that bears six sulfo groups

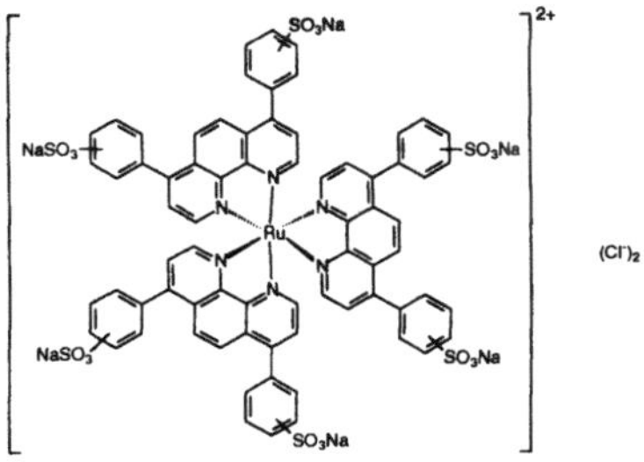

Figure 15.2. Chemical structure of ruthenium(II)-tris-bathophenanthroline-hexasulfonate dichloride hexasodium salt, Ru(sulfo-dpp)$_3$, a highly water-soluble, easily accessible and strongly luminescent probe for oxygen.

which render it water-soluble. Its luminescence (peaking at 607 nm) is efficiently quenched by oxygen.

If Ru(sulfo-dpp)$_3$ is added to a solution containing GOx, and glucose samples are also added in various concentrations, oxygen is consumed due to enzymatic action and this results in less quenching of luminescence. Fig. 15.3 shows the kinetics of the response to two concentrations of glucose. This is a very straightforward approach which – in our experience – works both in cuvettes and micro titer plates. One limitation consists in the diffusion of ambient oxygen into the wells of the microplates which becomes significant after a few tens of minutes. A second limitation (of oxygen-based detection schemes in general) is based on the varying background of oxygen in samples. This may be overcome by using a large excess of air-saturated buffer solution, or by compensating variable oxygen background by determining it independently (by the same protocol but without adding GOx).

15.3.3. Determination of Glucose via Formation of Hydrogen Peroxide

Determination of glucose via the hydrogen peroxide formed by GOx-assisted oxidation of glucose has the unique advantage of measuring an analyte (H_2O_2) against a "dark" background since H_2O_2 almost never forms an

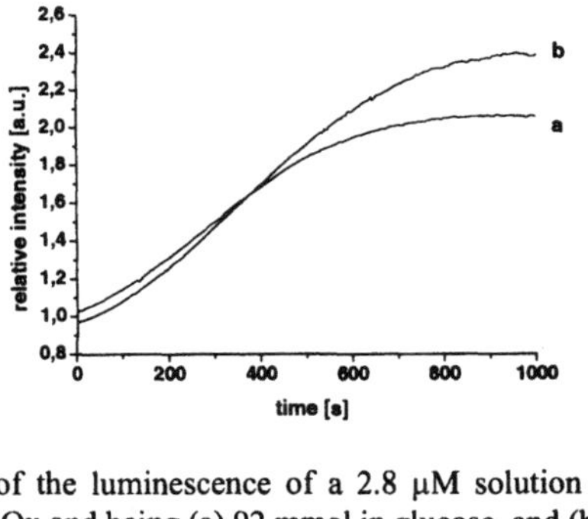

Figure 15.3. Time course of the luminescence of a 2.8 μM solution of Ru(sulfo-dpp)$_3$ also containing 72.6 μU/mL of GOx and being (a) 92 mmol in glucose, and (b) 138 mmol in glucose.

undesired background in samples including beverages and blood. However, in case of the latter, the absence of catalase (that will decompose H_2O_2) must be warranted. H_2O_2, in being a strong oxidant, is capable of oxidizing leuco dyes to their colored forms, and the formation of a colored species can be used as the analytical information in a kinetic assay.

In fact, chromogenic reactions form the basis for numerous (partially commercialized) glucose test strips. Chromogenic indicators for H_2O_2 include phenols and naphthols[11], phenylenediamines and benzidines,[12] antipyrine,[13, 14] hydrazonobenzo-thiazolones,[15, 16] bromopyrogallol,[17] and others.[18] These are sometimes referred to as Trinder reagents.[19–21] The same reagents also are widely used in photometric glucose assays, and of course may be used in all kinds of oxidase-based bioassays.

Fluorogenic indicators are less common in practice but more sensitive and selective. Reagents for detection of enzymatically formed hydrogen peroxide include 3,4-dihydroquinolones, 3,4-dihydrocoumarins, and o-hydroxyphenylfluorone,[22] variamine blue,[23] p-hydroxyphenylacetic acid[24], and its Tb^{3+}-EDTA complex,[25] homovanillic acid,[26, 27] tyramine and 3-(p-hydroxyphenyl) propionic acid,[28] aluminum aminophthalocyanines, and various porphyrins, resorufin,[29, 30], and 10-acetyl-3,7-dihydroxyphenoxazine (Amplex Red®).[31, 32]

The Amplex Red and Amplex Ultra Red reagents can be used in coupled enzymatic reactions to detect the activity of many different enzymes, or, when the substrate concentration is limited, to assay solutions for metabolically active constituents such as glucose. These reagents are fluorogenic substrates with low background color or fluorescence. A kit for determination of either glucose or glucose oxidase is commercially available. It can quantify glucose levels as low as 3 μM or 0.5μg/mL.[33]

A recently described assay[34] makes use of a new – and what appears to be first – reversible fluorescent probe for hydrogen peroxide, europium(III) tetracycline (Eu_3TC). Eu_3TC is weakly fluorescent and forms a 15 times more strongly fluorescent complex with hydrogen peroxide (referred to as Eu_3TC-HP). In addition to the change in intensity, the average decay time of Eu_3TC-HP is around 60 μs, while that of Eu_3TC is 30 μs only. Given this relatively long decay time, the fluorescence of Eu_3TC-HP can be detected in the time-resolved mode. The probe is added to the solution along with GOx, and the increase in fluorescence intensity (or the fraction of the slow-decaying component) is related to the concentration of glucose. Glucose can be determined at levels as low as2.2 μmol L^{-1}.

This scheme – though highly pH dependent – represents the first H_2O_2-based time-resolved fluorescence assay for glucose not requiring the presence of a peroxidase. The method also was adapted to higher glucose levels and applied to spiked serum samples (with glucose levels from 2.5 to 55.5 mmol L^{-1}). Features of the assay include easy accessibility of the probe, large Stokes' shift, a line-like fluorescence peaking at 616 nm (see Fig. 15.4), and a working pH of ~7. Figure 15.5 gives a typical calibration graph for glucose.

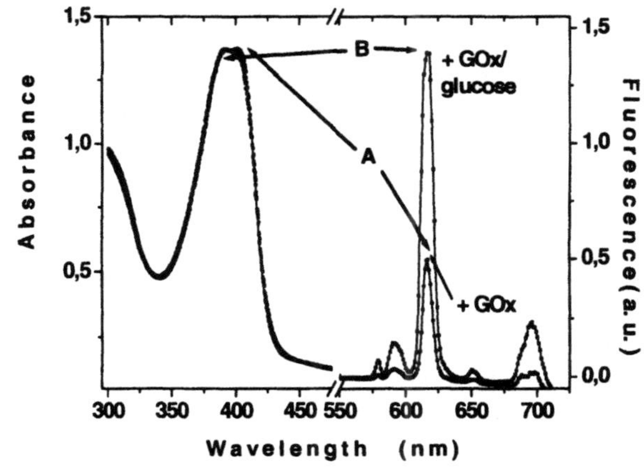

Figure 15.4. Absorption spectra (left side) and emission spectra (right side) of the system glucose oxidase/Eu_3TC before addition of glucose (A) and after its addition (B) which leads to formation of the Eu_3TC-HP complex. From ref. 35.

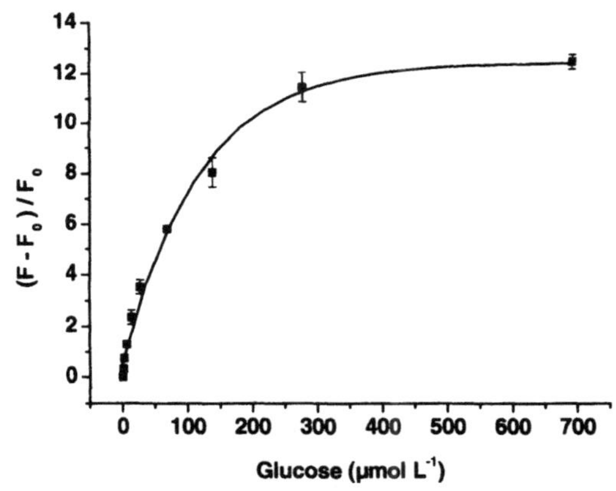

Figure 15.5. Calibration graph for the determination of glucose using dissolved GOx as the enzyme and dissolved Eu_3TC as the luminescent probe for enzymatically formed hydrogen peroxide. The assay was adjusted to match serum glucose levels but may be also be used for quantifying much smaller concentrations. From ref. (34).

15.4. CONTINOUS SENSING OF GLUCOSE USING FLUORESCENT SENSOR MEMBRANES

So far, methods have been described for determination of glucose in solution using indicators and enzymes in dissolved form. In order to save reagents, but in particular in order to enable continuous sensing of a sample passing by, the reagents may also be immobilized on an adequate polymer, and this results in glucose-sensitive sensor layers. The materials the sensors are made of may as well be deposited at the end of a fiber optic waveguide,[36] an integrated waveguide, or a capillary.[37]

While the same indicators and enzymes can be used that were found to be applicable to solution assays, the polymer must be selected carefully. In case of oxygen transduction, the choice of materials is particularly critical in that polymers for oxygen sensors typically are highly hydrophobic, whilst the polymers onto which GOx have been immobilized usually are hydrophilic.

Polymers most often applied in oxygen transducers for use in glucose biosensors include 1- and 2-component silicones, sol-gels, and ormosils such as co-condensates of phenyl-trimethoxysilane and tetramethoxysilane, ethyl cellulose, and polystyrene. These are predominantly hydrophobic and therefore good solvents for most oxygen probes. Typical hydrophilic (and sometimes cross linked) polymers include polyacrylamides, polyurethanes, poly(vinyl pyrrolidones), and certain copolymers. We (and others) find polyurethanes such as the D4 hydrogels and the polyacrylamide-polyacrylonitrile copolymers particularly useful.[38] Celluloses, dextranes, and synthetic poly-hydroxylated hydrogels have been applied as well. Domain polymers such as the one shown in Fig. 15.6 have found applications in many pH and ion sensors and in commercial instrumentation[39] since the material is biocompatible, has a very high water uptake (up to 89%), and is capable of retaining certain indicator dyes even without covalent immobilization.

15.4.1. Sensing Glucose via pH Sensor Layers Using Immobilized GOx

By immobilizing GOx onto sensors layers that respond to changes in pH, a glucose sensor with a pH transducer was obtained.[40] The layer was fixed at the end of an optical light guide to result in a fiber optic sensor capable of detecting 0.1 to 3 mM of glucose. Buffer capacity must be carefully controlled, however, since this heavily affects the total pH change observed. Table 15.1 gives an overview of pH sensors that have been used (or may be used) as transducers in enzyme-based glucose biosensors.

An alternative approach to sensing pH relies on the use of pH indicators dissolved in hydrophobic (rather than hydrophilic) matrices, for example plasticized PVC. In this case, a proton exchange-facilitating carrier such as tetraphenyl borate is added which assists in carrying the proton into (and out of)

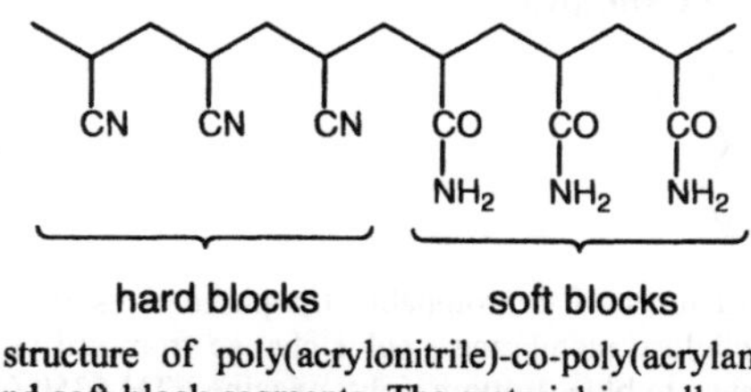

Figure 15.6. Chemical structure of poly(acrylonitrile)-co-poly(acrylamide) (= Hypan™), also showing the hard block and soft block segments. The material is well suited for sensing ionic and water-soluble organic species using appropriate indicator dyes.

Table 15.1. Composition and properties of typical pH sensitive and predominantly hydrophilic materials which along with immobilized glucose oxidase may be used for continuous sensing of glucose using via pH transduction

no.	pH indicator	polymer	excitation/emission wavelength	operational pH range	ref.
1	sodium fluorescein	sol gel	490/518 nm	5 – 9	41
2	chlorofluorescein octadecyl ester	polyurethane hydrogel (D4)	470/525 nm	5.2 – 8.5	42
3	HPTS	ion exchange resin beads in hydrogel	408,468/515 nm	6.2 – 8.8	43
4	HPTS	polyacrylamide film	460/520 nm	4 – 9	44
5	N9 (azo dye with reactive vinyl group)	cellulose acetate or hydroxy-polyurethanes	460 and 590 nm (2λ-reflectometry)	5 – 9	45,46
6	phenol red	polyacrylamide beads	540 and 630 nm (2λ-reflectometry)	6 – 8.5	47

the plasticized PVC, thus resulting in protonation/deprotonation of the dye contained in a hydrophobic membrane. A typical composition of such a membrane is 33 % PVC, 66 % of trioctyl phosphate (or long-chain and branched esters of certain organic acids), 0.5 – 0.9 % of potassium tetrakis(4-chlorophenyl) borate (these materials are commercially available), and 0.5 – 0.1 % of a protonable or deprotonable dye (see Fig. 15.7). PVC/plasticizer-soluble porphyrins also have been used for photometric and fluorescent sensing of pH.[48] The dynamic range of such sensors can be adjusted over a rather wide range by

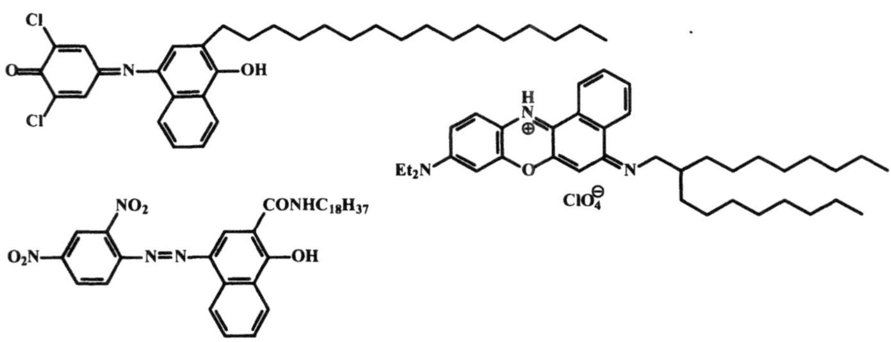

Figure 15.7. Chemical structures of deprotonable lipophilic dyes for use in pH sensors. Top: MEDPIN (used in ethyl cellulose membranes; color change from red to blue). Bottom left: KFU 111 (color change form orange to blue. Bottom right: oxazine ETH 5350 (color change from blue to red).

varying the ratio of PVC and plasticizer, and by the fraction of tetraphenyl borate added: Unlike the other dyes, the Nile Blue derivative fluoresces with a quantum yield of ~25 %.

Polyaniline (PANI) represents a rather unusual but easily manufactured material for use in reflectometric sensing of pH. Its best analytical wavelengths are between 600 and 1000 nm (!). PANI has been shown to be useful for sensing glucose via the changes in the pH caused by enzymatic oxidation and hydrolysis.[49]

15.4.2. Sensing Glucose via Oxygen Sensor Membranes

Transduction by sensing oxygen has the advantage of having available a highly reliable sensor type and being dependent on a single variable only (unlike pH transduction which also depends on buffer capacity and ionic strength). However, oxygen forms a strong and variable background in blood and – in the worst case – can become the rate-determining parameter in reaction (2).

While the feasibility of sensing glucose has been described already in 1983[50], these approaches suffered from incompatibility with low cost light sources such as light-emitting diodes (LEDs). Following the discovery of more longwave oxygen probes the way was paved for making the vision of reversible optical glucose sensing come true. The discovery of decacyclene as a sensitive probe for oxygen led to a glucose sensor with excitation/emission wavelengths of 430/500 nm.[51, 52] GOx was covalently immobilized on a nylon membrane or on carbon black, and the consumption of oxygen was measured by following, via fiber optic bundles, the changes in the fluorescence of decacyclene which was made silicone-soluble by butylation). The oxygen probe was dissolved in a very thin silicone membrane placed beneath the enzyme layer. The analytical ranges cover the clinical range (0.1 to 20 mM). A thin layer of carbon black acted as an optical isolation in order to prevent ambient light and sample fluorescence to interfere.

In 1986 the ruthenium-ligand complexes were discovered to be viable oxygen probes that are compatible with the blue LED as a light source.[53] This finding was immediately transferred to glucose sensing.[54] Specifically, ruthenium(II)-tris-(1,10-phenanthroline) was adsorbed onto silica gel, and the resulting material incorporated into a silicone matrix and covered with a layer of polymer-immobilized GOx. The resulting sensor responded to glucose in concentrations from 0.06 to 1 mM glucose. The range may be fine-tuned to clinical levels by proper choice of the polymers used. The sensor finds application in Osmetech's clinical glucose analyzer (www.osmetech.com), except that the ruthenium(II)-tris-(diphenyl-phenanthroline) complex (which has a better quantum yield) is used as a probe for oxygen.

Sol-gels represent an interesting matrix material for sensing purposes because of their specific network which is capable of retaining molecules having a diameter larger than 1 nm. The chemical and physical structure of

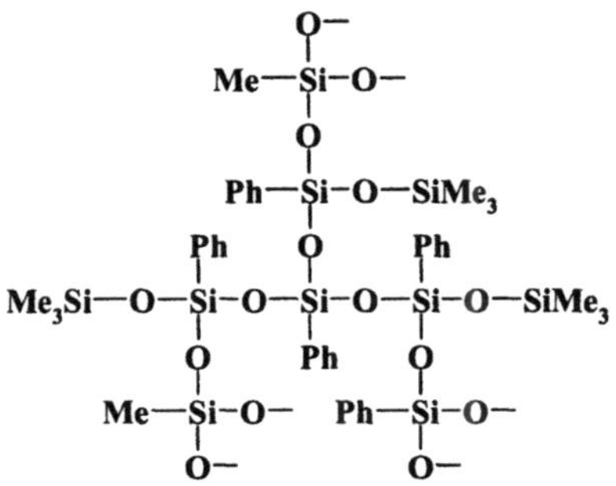

Figure 15.8. Chemical structure of a segment of a hydrophobic (chloroform-soluble) ormosil made by hydrolysis and polycondensation of phenyltrimethoxysilane with methyl-trimethoxysilane, followed by endcapping of terminal hydroxy groups with trimethylsilyl (-$SiMe_3$) groups. Indicator dyes can be entrapped into such microdomains. Obviously, there is ample space for chemical ingenuity in sol-gel based material sciences.

sol-gels varies substantially, depending on the reaction conditions applied during polycondensation of tetraalkoxysilanes of the type $Si(OR)_4$. Moreover, organically modified sol-gels may be produced that are derived from precursors of the general structure R'-$Si(OR)_3$, or R'R"-$Si(OR)_2$, where R' and R" denote alkyl or aryl substituents. Obviously, sol-gels can come in a large variety. Fig. 15.8 gives the structure of a typical ormosil that turned out to be useful in sensing oxygen[55] and glucose, respectively.

In a typical approach made to sense glucose, various types of thin-film glucose biosensors were prepared[56] by different methods for immobilizing the oxygen probe and the enzyme GOx. Specifically, three combinations of oxygen transducer and sol-gel immobilized GOx were found to be quite useful. In the first, GOx was sandwiched between a sol-gel layer doped with Ru(dpp) and a second sol-gel layer composed of pure sol-gel (the sandwich configuration). In the second, a sol-gel layer doped with Ru(dpp) was covered with sol-gel entrapped GOx (the two-layer configuration). In the third, both GOx and a sol-gel powder containing GOx were incorporated into a single sol-gel phase (the powder configuration). In all cases, it was found to be essential to add sorbitol which results in a more porous sol-gel in which diffusion is not impaired.

Oxygen, in being present in almost any practical sample, forms a background which also contributes to the quenching of the fluorescence of the probe even though it is not related to glucose concentration. If oxygen partial pressure varies (which is often the case in practice), this background must be taken into account, preferably by sensing oxygen via a second sensor (containing no GOx). Equations can be derived which describe the response of such sensors, how the effect of varying oxygen supply can be compensated for by making use of two sensors, one sensitive to oxygen only, the other to both oxygen and glucose, and how such sensors can be calibrated using 2 calibrators only.[56]

Table 15.2. Composition and properties of typical oxygen-sensitive fluorescent materials considered to be applicable to glucose sensing using GOx.

no.	oxygen indicator	polymer	excitation/emission wavelength	quenching efficiency[a]	ref.
1	Ru(phen)[b] on silica nanoparticles	suspended in silicone	455/615 nm	2.4	54
2	Ru(phen)	polyacrylamide	455/612	1.6	57
3	$Ru(dpp)(TMPS)_2$[c]	made from $Ph\text{-}Si(OEt)_3$ and $Si(OMe)_4$	455/620 nm	1.6	55, 58
4	$Ru(dpp)(TMPS)_2$ or $Ru(dpp)(LS)_2$[c]	(a) ethyl cellulose, (b) polystyrene (PS)	455/620 nm	1.8 – 2.0	59
5	Pt tetrakis(penta-fluoro-phenyl)-porphyrin	polystyrene	541/650 nm	1.9	60
6	PdOEP and PtOEP[d]	PS-co-fluoroacrylate	535/645 nm 546/664 nm	- -	61
7	metallo-ketoporphyrins	in polystyrene	630/680 nm	5 - 10	62
8	$Pb(sulfo\text{-}oxinate)_2$	on ion exchanger	385/625 nm	2.8	63

[a] expressed as the ratio of signals under nitrogen and air (I_0/I_{air}).
[b] Ru(phen) stands for ruthenium(II)-tris(1,10-phenanthroline)
[c] Ru(dpp) stands for ruthenium(II)-tris(diphenyl-phenanthroline)
d PdOEP and PtOEP stand for palladium-octaethylporphyrin and platinum octaethylporphyrin

The sensor layers described before have thicknesses ranging from 2 to 10 µm. Films may be much thinner (this resulting in much faster response times), however at the expense of signal intensity and signal-to-noise-ratio. An optical glucose biosensor was constructed on the basis of a Langmuir-Blodgett film[64] into which lipophilic fluorescent probes for oxygen or pH were incorporated. Co-immobilization of GOx results in an optical response to glucose since the surface pH is lowered in presence of glucose, and the oxygen partial pressure as well. LB layer sensors are easily prepared, have a well-defined thickness, and usually have quite short response times. The LB bilayers used in these experiments are, however, prone to mechanical deterioration and give poor signal-to-noise ratios. Solutions to these limitation will include the use of polymeric lipid membranes, of evanescent wave techniques for gathering more intense signals, and the coupling of LB techniques to waveguide technologies which enable multiple sensor-light interactions.

15.4.3. Sensing Glucose via Sensor Layers for Hydrogen Peroxide

Equation (2) indicates that glucose can be sensed via the hydrogen peroxide (HP) formed as a result of enzymatic action. HP has the advantage of not forming a background in the vast majority of samples. While HP is the species that causes the coloration of commercial but irreversible test strips to occur, optical and continuous (reversibly responding) sensors for HP are scarce. An indirect sensing scheme has been proposed that exploits the catalytic decomposition of HP to form water and oxygen[65] but has not been adapted to glucose sensing so far.

A reversible optical sensor membrane for hydrogen peroxide was reported only recently. It uses an immobilized fluorescent probe and has successfully been employed in a GOx-based glucose biosensor.[38] The HP probe europium tetracycline (Eu_3TC) was incorporated, along with GOx, into a polyacrylonitrile-co-polyacrylamide (*Hypan*) polymer matrix.[66] Upon optical excitation with 400-nm light, the Eu_3TC in the membrane displays fairly strong fluorescence peaking at 616 nm. Its intensity increases up to 3-fold once the sensor is exposed to solutions containing HP from GOx activity (Fig. 15.9). The effect is reversible and thus can be used for continuous sensing of glucose in 0.1 to 5 mM concentrations. The largest signal changes are observed at physiological pHs. Phosphate and citrate interfere, however, as do Cu(II) ions which quench the fluorescence of Eu_3TC. The scheme will be presented in more detail later in the *Imaging Section*.

15.5. SENSING GLUCOSE VIA METAL-LIGAND COMPLEXES WITH A BORONIC ACID AS THE RECOGNITION SITE

Boronic acids are capable of specifically recognizing simple polyols and – in particular – saccharides.[67, 68] We have exploited this finding by designing what turned out to be the first metal-organic probe for saccharides.[69] Specifically, a ruthenium metal-ligand (ML) complex was endowed with a boronic acid side group to give the saccharide probe whose chemical structure is given in Fig. 15.10. Its design was based on the need for (a) an LED-compatible and long-lived luminophore and (b) a binding site for glucose that affects luminescence. Boronic acids have been shown to bind to saccharides

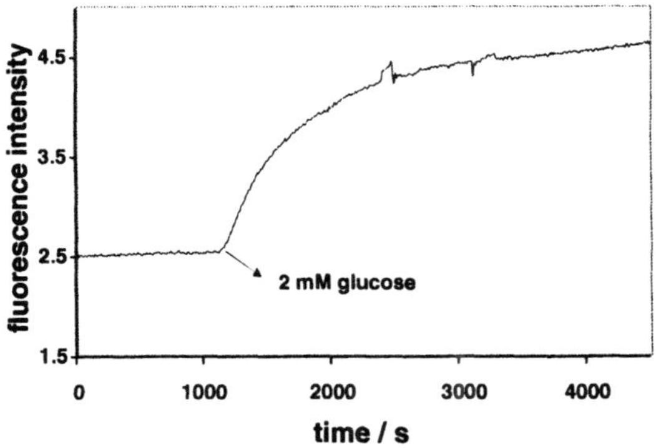

Figure 15.9. Time trace of the response of an enzymatic glucose biosensor based on an optical sensor for hydrogen peroxide. The tooth measured after about 2500 s is due to an air bubble. From ref. 66.

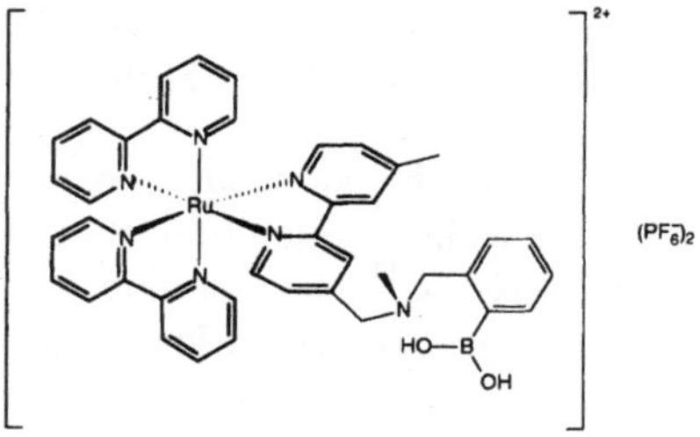

Figure 15.10. Chemical structure of a molecular receptor for glucose consisting of a boronic acid binding site and a ruthenium-ligand complex whose luminescence intensity and decay time is modulated by the binding process. From ref. 69.

with some selectivity. The molecular sensor is designed such that binding causes an excited state electron transfer from the nitrogen atom to the closely located boron atom. This process is called photo-induced electron transfer (PET). The nitrogen atom, in turn, modulates the luminescence of the ML probe by reducing its luminescence intensity and decay time. Like in other monoboronic acids, the response to fructose is stronger than to glucose in case of a single and sterically not optimized receptor. Diboronic acids are more specific and more responsive to glucose.

15.6. SENSING GLUCOSE VIA THIN FILMS OF ORGANIC CONDUCTING POLYMERS (OCPs)

Organic conducting polymers (OCPs) are obtained by chemical or electrochemical polymerization of organic monomers such as anilines, pyrroles or thiophenes. They occur in various forms depending on their molecular structure and display highly pH dependent electrical conductivity and absorption spectra. The spectral changes with pH occur in the near infrared range and may be used to monitor pH but also acidic gases such as HCl, SO_2, or ammonia.[70] OCPs usually are deposited on solid supports such as electrodes, organic semiconductors, indium-tin oxides (ITOs), or on optically inert plastic materials.

While most authors have used unsubstituted aniline to make PANI, it has been demonstrated that the range over which PANIs respond to pH can be varied to a wide extent by making use of substituted anilines.[71] Even better results were obtained[72] with a composite OCP built up from Prussian Blue and N-substituted polypyrroles. The resulting fully transparent, bright blue films are composed – in essence – of Prussian Blue incorporating a network of polypyrrole. It is sensitive to pH in the pH 5 – 9 range, with the largest spectral changes occurring at 720 nm. On covalent immobilization of certain enzymes, the respective biosensors are obtained. Methods of enzyme immobilization (using bifunctional crosslinking reagents) were optimized, and best results are obtained using the one-step carbodiimide method which results in highly active, stable and transparent biosensor films that allow optical and continuous determination of the respective substrates in the millimolar concentration range. Specifically, urease, acetylcholinesterase, trypsin and chymotrypsin were investigated.

OCPs not only respond to pH but also to oxidants such as hydrogen peroxide. We have coated the wells of microtiter plates with polyaniline layers and with layers of polyaniline/GOx.[49] The resulting wells can be used directly for enzymatic determination of glucose, either via pH or via hydrogen peroxide. Analyte detection is based on monitoring the absorption spectra of the polyaniline, which turn purple as a result of redox processes, or green on formation of acids by enzymatic reactions. Hydrogen peroxide (a species produced by all oxidases) and glucose (which yields protons on enzymatic oxidation) have been determined in the millimolar to micromolar concentration range.

The combination of the boronic acid receptor scheme with the technology of OCPs resulted in the design of a polyaniline with a near-infrared optical response to saccharides.[73] The material was obtained by copolymerization of aniline and 3-aminophenylboronic acid and was deposited in the wells of a micro titer plate. It represents a new type of sugar-binding polymer film whose absorption spectra between 500 and 800 nm undergo large changes on addition of saccharides including saccharose, fructose, glucose, sorbitol, mannitol, and glycerol at neutral pH. The spectral shifts depend on the concentration of the saccharides and – most notably – are fully reversible, thus allowing continuous sensing. Such films represent an interesting alternative to enzyme-based glucose sensors because of their ease of preparation, compatibility with LED and diode laser light sources, and their thermal and temporal stability. The unspecific response implies, on the other hand, that OCPs are not very selective, but there are numerous situations where interfering analytes are unlikely to be present.

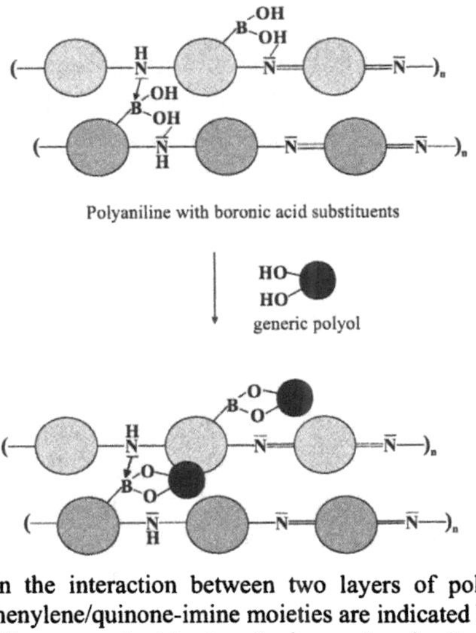

Figure 15.11. Changes in the interaction between two layers of poly(anilineboronic acid) on binding to a polyol. The phenylene/quinone-imine moieties are indicated by grey circles, the polyols by black circles. On binding a saccharide, two hydrogen bonds disappear, and the electronic interaction between electron-deficient boron atoms and electron-rich nitrogen atoms is changed. From ref. 73.

Alternatively, use may be made of so-called artificial neural networks that enable the unspecific response of a set of sensors to be analyzed such that one or more analytes may be quantitated rather precisely.

15.7. SENSING GLUCOSE VIA OXYGEN-SENSITIVE NANOPARTICLES

In one of the earliest papers on the use of nanoparticles in optical sensing, zeolite Y nanocrystals were used[74] as cavities for luminophores. The interior of the zeolite was first ion-exchanged with ruthenium(III)-chloride. Subsequently, the ligand 2,2'-bipyridyl (bipy) was allowed to diffuse into the cavities. Thus, the fluorescent oxygen probe $Ru(bipy)_3^{2+}$ was prepared inside the zeolite supercages from which it could not escape due to its size. It was found - with some surprise - that both the efficiency of quenching by oxygen and the long-term stability of the material were significantly improved. These crystalline nanoparticles, when incorporated into a polyurethane membrane that was soaked with a solution of GOx and subsequently cross linked with glutaraldehyde, report any changes in oxygen tension (as caused by enzymatic action) with very high sensitivity.

15.8. SENSING GLUCOSE VIA THE INITRINSIC FLUORESCENCE OF GLUCOSE OXIDASE

Glucose oxidase (GOx) has a fairly strong intrinsic green fluorescence which is characteristic of all enzymes having FAD or FMN as a coenzyme.[75] The fluorescences of FAD and its reduced form ($FADH_2$), respectively, have different quantum yields and decay times (double-exponential and in the 2 – 3 ns time domain). These findings have resulted in a reversible glucose biosensor that is based on the intrinsic fluorescence of GOx.[76, 77] The fluorescence of GOx (which can be excited at 450 nm and is measured at >500 nm) changes during interaction with glucose. The response is fully reversible because oxygen is accepted as a second substrate. The very specific feature of this kind of sensor relies on the fact that the recognition element is identical with the transducer element. The changes also occur in fluorescence polarization.[78]

In order to construct such a fiber optic glucose biosensor, solutions of GOx were entrapped at the end of a fiber using a semipermeable membrane which retains the enzyme, and fluorescence intensity was continuously recorded. Another specific feature of this sensor is the rather small glucose concentration range (typically 1.5 - 2 mM glucose) over which the full signal change does occur. Response times of 2 - 30 min and regeneration times of 1 - 10 min are observed. To achieve an extended analytical range (e.g. 2.5 - 10 mM) and shorter response times, kinetic measurements are suggested. Like sensors based on transduction of oxygen, this sensor also depends on oxygen supply since oxygen is the substrate for the back reaction (i.e. the oxidation of $FADH_2$).

15.9. IMAGING TECHNIQUES FOR GLUCOSE AND GLUCOSE OXIDASE

Classical sensing schemes were limited to a single site of sampling and/or sensing. This is more than adequate for many situations, in particular where legal requirements define the site of measurement. More recently, the availability of planar optical sensors (rather than optical fiber-based) sensors has paved the way for sensing parameters over certain areas, a technique referred to as (chemical) *sensing imaging*. Typical examples include imaging of oxygen in cancer research and therapy[79] and in microcirculation studies,[80] but also the testing of airplanes or racing cars in wind tunnels (since oxygen contributes to the total air pressure).

In sensing imaging, a sensing layer is placed in close contact to the sample to be imaged, or even sprayed onto the area of interest. Other important objects which can be imaged are microwell plates or sensor microarrays. All wells of a microplate or the sensor spots of a microarray format can be excited simultaneously and read out by a CCD-based detector in parallel, which turns imaging to an important method in high-throughput screening (HTS).[81]

In time-resolved fluorescent imaging, the area to be analyzed is illuminated (excited) with a pulsed light source (usually a set of LEDs), and fluorescence is collected by a CCD camera. One serious limitation in fluorescence imaging results from inhomogeneities of the light field and of the sensor layer (e.g. thickness and probe distribution). Therefore, methods based on the measurement of luminescence decay time like RLD (rapid lifetime determination)[82, 83] or PDR (phase delay ratioing)[84] are superior to intensity-based imaging in bioanalytical applications since they are independent of variation of indicator concentration in the sensing layer (as outlined in section 15.1). In order to facilitate the technical effort, probes with long decay times are preferred since decay times in the order of 1 -100 µs allow the modulation frequency of the LED and the synchronized CCD detector to be as low as 20 – 50 kHz.

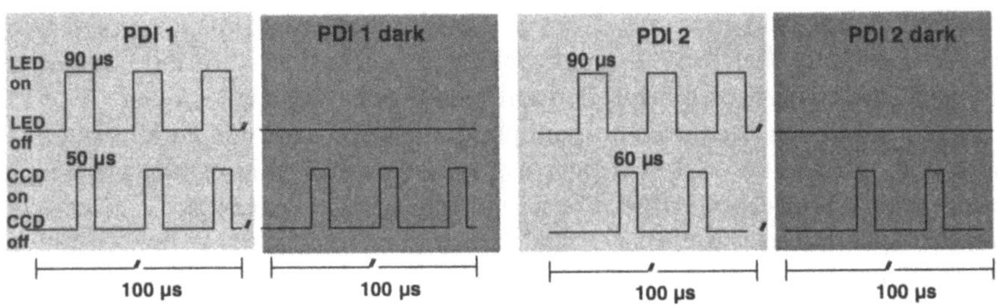

Figure 15.12. Schematic representation of the time gates applied in the phase delay rationing (PDR) imaging mode. Image PDI 1 is obtained during excitation of the probe [Eu_3TC] between 30 and 80 µs; image PDI 2 is taken in the decay phase in a time window between 130 and 190 µs.

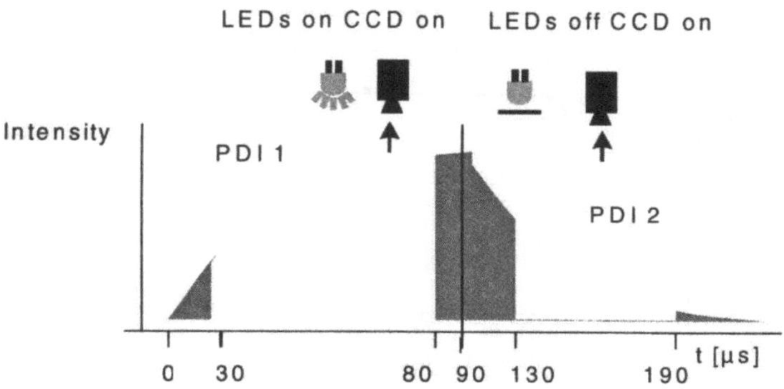

Figure 15.13. Time course of the data acquisition process. The images for the different gates (PDI 1, PDI 1 dark, PDI 2, PDI 2 dark) are acquired one after the other, added (stacked) and averaged separately. After subtraction of background ("dark pictures"), the ratio R_{PDR} is calculated.

15.9.1. Time-Resolved Imaging of Glucose via Enzymatically Produced H_2O_2

Time-resolved fluorescent sensing imaging of glucose was first demonstrated by means of a sensor membrane containing a hydrogen peroxide transducer.[85] Unlike in case of transduction via oxygen or pH, this method has the advantage being free of a background, since hydrogen peroxide is not present in appreciable amounts in clinical samples. However, the presence of catalase must be excluded. Enzymatically produced hydrogen peroxide (HP) is sensed via the fluorescent probe europium(III) tetracycline (in a molar ratio of 3:1) which is incorporated into a Hypan layer (see Fig. 15.6). For a fast parallel screening of many samples, the sensor layers were integrated into the wells of a 96-microtiterplate.[86] A typical calibration plot of a hydrogen peroxide sensor material obtained by time-resolved imaging is shown in Fig. 15.14. It shows a linear range from 0.5 to 10 ppm of H_2O_2.

Co-adsorption of GOx makes these sensor layers responsive to glucose via the HP produced by the GOx-assisted oxidation of glucose. Thus, enzymatic consumption of glucose could be visualized for the first time by means of time-resolved luminescence lifetime imaging. Unlike in previous non-imaging methods, the determination of HP does not require the addition of peroxidase and works at neutral pH. The lifetime-based images showed a dynamic detection range between concentrations from 5 to 1000 μmol L^{-1} of glucose. Lifetime-based imaging resulted in much better spot homogeneity and reproducibility than intensity-based methods, but at a comparable sensitivity and dynamic range.

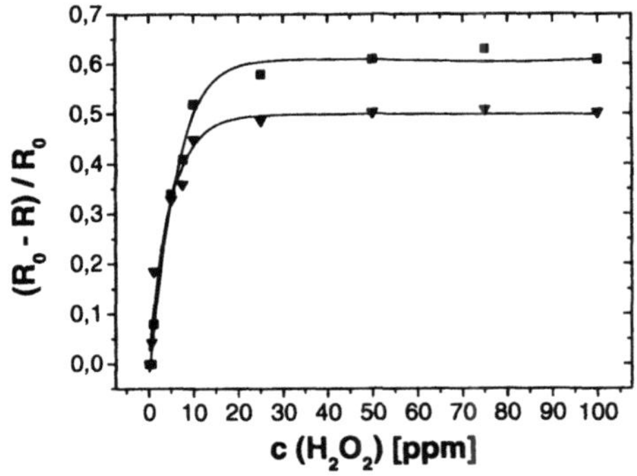

Figure 15.14. Referenced calibration plot $(R_0 - R)/R_0$ versus the concentration of hydrogen peroxide in MOPS buffer (pH 6.9). R_0 is the initial ratiometric fluorescence intensity of the sensor membrane, and R the intensity in presence of increasing concentrations of H_2O_2, obtained by PDR (▪) and RLD imaging (▾).

15.9.2. Time-Resolved Imaging of Glucose Using Oxygen Sensors

Since oxygen supply and changes in oxygen partial pressure are important analytical parameters for the monitoring of cell metabolism, many fluorescent probes for the determination of pO_2 have been described (also see Table 15.2) which are suited for lifetime imaging, including decacyclene[51] which can be excited at 440 nm, pyrene[87], ruthenium complexes[53, 88] and platinum or palladium (keto)porphyrins.[89, 90] Some of these probes may be administered to living cells for intracellular oxygen sensing[91] or can be used in the form of sensor layers, incorporated in different polymer matrices.[92] These transducer materials may also be placed at the bottom of microwell plates for the accomplishment of high-throughput screening assays.[93] Such microplates have been commercialized and can be applied in cellular drug screening, proliferation studies, and the screening of enzyme reactions (see: www.becton-dickinson.com; www.presens.de). The microwell plates can be screened with help of commonly used microplate fluorescence readers.

More recently, we have cast ~5 μm thick films of a poly(styrene-co-acrylonitrile; PSAN) doped with a fluorinated Pd-porphyrin of unusually long decay time (Pd-TPFPP) on the bottom of the single microwells. After drying and rinsing, the wells were filled with a solution of GOx (17.8 U mL^{-1}) and with glucose (dissolved in buffer of pH 6.8 in concentrations ranging from 0.1 to 10 mmol L^{-1}). Under the given experimental conditions, glucose could be imaged in the range from 0.5 to 5 mmol L^{-1} using the RLD method. Fig. 15.15 shows the corresponding ratiometric lifetime sensing image and the resulting calibration plot. The assay can be performed within 5 min. Sensitivity may be slightly increased by prolonging the reaction time which gives a maximum after 30 - 40 min.

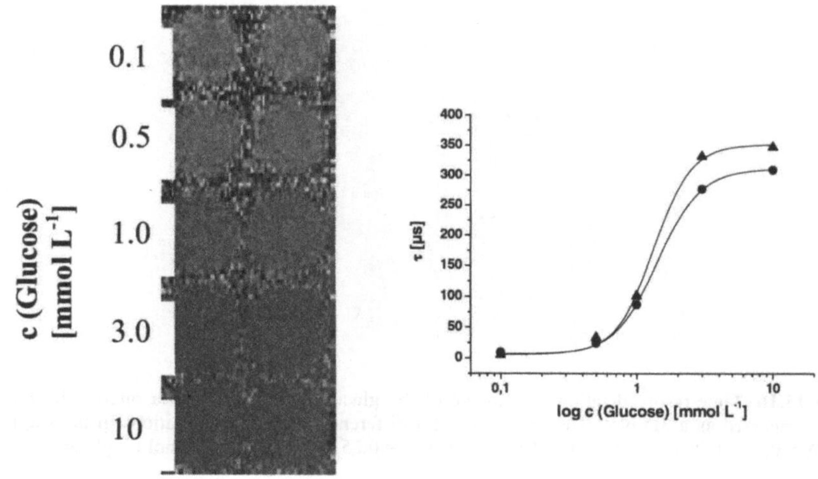

Figure 15.15. Left: Grey-scale fluorescence lifetime images of glucose in single wells of a microtiter plate coated with oxygen-sensitive films (probe Pd-TPFPP dissolved in PSAN) in presence of GOx (17.8 U mL^{-1}) at pH 6.0. Excitation wavelength: 405 nm, emission filter: Chroma 680. Excitation pulse: 2 ms; gate 1: 0-500 µs, gate 2: 600 - 1100 µs after excitation pulse. Right: Logarithmic calibration plot of fluorescence lifetime (τ) versus the concentration of glucose after a 5-min (●) and a 40-min (▲) incubation time.

15.9.3. Time-Resolved Imaging of Glucose Using Sensor Layers for Hydrogen Peroxide

Time-resolved fluorescent sensing imaging of glucose also was demonstrated for a glucose sensor membrane with a hydrogen peroxide transducer.[85] Enzymatically produced hydrogen peroxide (HP) is sensed via the fluorescent probe europium(III) tetracycline (molar ratio 3:1) that is incorporated into a hydrophilic polymer layer (see above). Co-adsorption of GOx makes these sensor layers responsive to glucose via the HP produced by the GOx-assisted oxidation of glucose. The hydrogel layers were integrated into a 96-microwell plate for a parallel and simultaneous detection of various samples. Glucose can be visualized by means of time-resolved lifetime imaging. Unlike in previous (however non-imaging) methods, the determination of HP does not require the addition of peroxidase and works at neutral pH. The lifetime-based images (see 15. 16) obtained are compared with conventional fluorescence intensity-based methods with respect to sensitivity and the dynamic range of the sensor layer.

15.9.4. Time-Resolved Imaging of the Activity of Glucose Oxidase

Fluorescent imaging of the activity of GOx (rather than of glucose) is a useful tool for GOx-based immunoassays with potential to high-throughput screening, to immobilization studies, and biosensor array technologies. GOx has been visualized by sensing/imaging by making use of a fluorescent

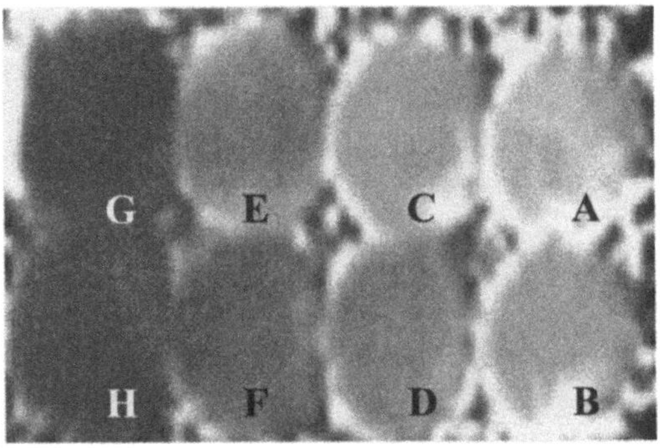

Figure 15.16. Time-resolved ratiometric image of the glucose sensor spots in a microwell plate format visualized as a 3D plot. The spots represent different glucose concentrations in the single wells. A = 0, B = 0.005, C = 0.01, D = 0.05, E = 0.1, F = 0.25, G = 0.5, H = 1 mmol L^{-1} glucose.

europium(III) tetracycline probe (Eu_3TC) for hydrogen peroxide[94] whose decay time in the μs range and whose large Stokes shift (210 nm) facilitate (a) intensity-based, (b) time-resolved and (c) decay time-based imaging of glucose oxidase. Since the probe Eu_3TC has an excitation maximum at around 400 nm, a set of violet LEDs (λ_{max} 405 nm) was used as the light source. A comparison of four methods for imaging revealed that rapid lifetime determination (RLD) imaging is most adequate in giving a linear range from 0.32 to 2.7 milli-units of GOx per mL of solution. The detection limit is 1.7 ng of GOx per mL which is similar to that of the time-resolved (gated) imaging using a microtiterplate reader.

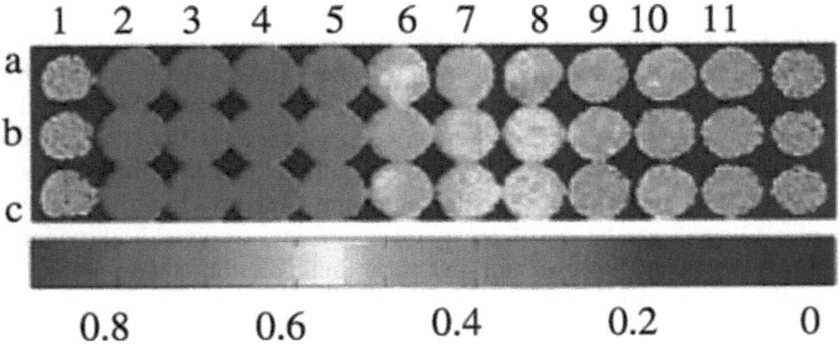

Figure 15.17. Rapid lifetime determination (RLD) imaging of the activity of glucose oxidase in the wells of a microtiterplate. The pictures reflect the normalized ratio of the two images as schematically represented in Figs. 15.12 and 15.13 (in pseudo colors). The cocktails in wells (from 1 to 12) had the following compositions: GOx activities 0 (blank), 135, 54.1, 27.1, 13.5, 5.4, 2.7, 1.35, 0.54, 0.27, 0.14 to 0.05 mUnit/mL, respectively; plus 100 μL of the EuTc stock solution, 15 μL of a 277.2 mmol L^{-1} glucose solution; total volume made up to 200 μL with MOPS buffer. From ref. 94. (see the color insert after p. 429.)

15.10. REFERENCES

1. B. Valeur, *Molecular Fluorescence: Principles and Applications*, (Wiley - VCH, Weinheim, 2002).
2. (a) J. S. Schultz, S. Mansouri and I. J. Goldstein, *Diabetes Care* **5**, 245 (1982); (b) R. Ballerstadt and J. S. Schultz, *Anal. Chem.* **72**, 4185 (2000).
3. G. Liebsch, I. Klimant, Ch. Krause and O. S. Wolfbeis, *Anal. Chem.* **73**, 4354 (2001).
4. R. Narayanaswamy, and F. Sevilla, *Anal. Lett.* **21**, 1165 (1988).
5. M. M. F. Choi, *Microchim. Acta* **148**, 107-132 (2004).
6. A. Kumagai, H. Suzuki, K. Ogawa and E. Kokufuta, *Chem. Sensors* **19**, 1 (2003).
7. C. C. Liu and M. R. Neuman, *Diabetes Care* **5**, 275 (1982).
8. I. Rocha and E. C. Ferreira, *Anal. Chim. Acta* **462**, 293 (2002).
9. M. G. Sasso, F. H. Quina, and E. J. H. Bechara, *Anal. Biochem.* **156**, 239 (1986).
10. J. M. Vanderkooi, G. Maniara, T. J. Green and D. F. Wilson, *J. Biol. Chem.* **262**, 5476 (1987).
11. W. Treiber, J. Siedel, and B. Vogt, *Ger. Offen.* DE 4309394 (1994).
12. M. V. Cattaneo and J. H. Luong, *Anal. Biochem.* **223**, 313 (1994).
13. F. Wang, Y. Wu, X. Wu, S. Sun, Y. Ci, *Fresenius' J. Anal. Chem.* **346**, 1011 (1993).
14. M. Mifune, K. Sugimoto, A. Iwado, H. Akizawa, N. Motohashi and Y. Saito, *Anal. Sci.* **19**, 569 (2003).
15. Y. Nakajima and Y. Goto, *Jpn. Kokai Tokkyo Koho* JP 121637 (2000).
16. D. J. Capaldi and K. E. Taylor, *Anal. Biochem.* **129**, 329 (1983).
17. Z.-X. Guo, H.-X. Shen and L. Li, *Microchim. Acta* **131**, 171 (1999).
18. W. Werner, H. G. Rey and H. Wielinger, *Fresenius' J. Anal. Chem.* **252**, 224 (1970).
19. D. Barham, P. Trinder, *Analyst* **97**, 142 (1972).
20. S. Charlton and E. Kurchacova, *US Pat.* 4,855,228 (1989).
21. M. R. Slaughter, P. J. O'Brien, *Clin. Biochem.* **33**, 525 (2000).
22. Z.-X. Guo, H.-X. Shen and L. Li, *Microchem. J.*, **61**, 134 (1999).
23. F. A. El-Essi, A. Z. A. Zuhri, S. I. Al-Khalil and M. S. Abdel-Latif, *Talanta* **44**, 2051 (1997).
24. G. G. Guilbault, S. S. Kuan and P. J. Brignac Jr., *Anal. Chim. Acta* **47**, 503 (1969).
25. J. Meyer and U. Karst, *Analyst* **125**, 1537 (2000).
26. G. G. Guilbault, D. N. Kramer and E. B. Hackley, *Anal. Chem.* **39**, 271 (1967).
27. G. G. Guilbault, P. Brignac and M. Zimmer, *Anal Chem.* **40**, 190 (1968).
28. G. G. Guilbault, P. Brignac and M. Juneau, *Anal Chem.* **40**, 1256 (1968).
29. H. Maeda, S. Matsu-Ura, M. Nishida, T. Senba, Y. Yamauchi and H. Ohmori. *Chem. Pharm. Bull.* **49**, 294 (2001).
30. S. Matsu-Ura, Y. Yamauchi, H. Ohmori and H. Maeda, *Bunseki Kagaku* **51**, 111 (2002).
31. M. Zhou, Z. Diwu, N. Panchuk-Voloshina and R. P. Haugland, *Anal. Biochem.* **253**, 162 (1997).
32. R. Haugland, *Handbook of Fluorescent Probes and Research Products* (9th ed.; Eugene, OR, 2002) pp.414, 443; http://www.probes.com/ media/pis/mp22189.pdf.
33. M. Zhou, C. Zhang, R. H. Upson and R. P. Haugland, *Anal. Biochem.* **260**, 257 (1998).
34. M. Wu, Z. Lin, A. Duerkop and O. S. Wolfbeis, *Anal. Bioanal. Chem.* **380**, 619 (2004).
35. O. S. Wolfbeis, A. Duerkop, M. Wu, Zh. Lin, *Angew. Chem. Intl. Ed.* **41** 4495 (2002).
36. O. S. Wolfbeis, *Fiber Optic Chemical Sensors and Biosensors* (CRC Press, Boca Raton, 1991, vol. 2).
37. B. H. Weigl and O. S. Wolfbeis, *Anal Chem.* **66**, 3323 (1994).
38. O. S. Wolfbeis, M. Schaeferling and A. Duerkop, *Microchim. Acta* **143** 221 (2003).
39. Ch. Huber, T. Werner, Ch. Krause and O. S. Wolfbeis, *Microchim. Acta* **142** 245 (2003).
40. W. Trettnak, M. J. P. Leiner and O. S. Wolfbeis, *Biosensors* **4**, 15 (1988).
41. M. Gouterman, J. Callis, L. Dalton, G. Khalil, Y. Mebarki, K. R Cooper and M. Grenier, *Meas. Sci. Technol.* **5**, 1986 (2004).
42. B. M. Weidgans, C. Krause, I. Klimant and O. S. Wolfbeis, *Analyst* **129**, 645 (2004).
43. H. R Kermis, Y. Kostov, P. Harms and G. Rao, *Biotechnol. Progress* **18**, 1047 (2002).
44. S.-H. Lee, K. G.. Chittibabu, J. Kim, J. Kumar, and S. K.Tripathy, *Polym. Mat. Sci. Eng.* **83**, 250 (2000).
45. G. J. Mohr and O. S. Wolfbeis, *Anal. Chim. Acta* **292** 41 (1994).
46. U. Kosch, I. Klimant, T. Werner, O. S. Wolfbeis, *Anal. Chem.* **70** 3892-3897 (1998) and refs. cited;

47. J. I. Peterson, S. R Goldstein, R. V.Fitzgerald and D. K. Buckhold, *Anal. Chem.* **52,** 864 (1980).
48. D. B. Papkovsky, G. V. Ponomarev, O. S. Wolfbeis, *J. Photochem. Photobiol. A* **104** 151 (1997).
49. S. A. Piletsky, T. L. Panasyuk, E. V. Piletskaya, T. A. Sergeeva, A. V. Elskaya, E. Pringsheim and O. S. Wolfbeis, *Fresenius' J. Anal. Chem.* **366,** 807 (2000).
50. D. W. Luebbers and N.Opitz, *Sens. Actuators* **4,** 641 (1983).
51. W. Trettnak, M. J. P. Leiner and O. S. Wolfbeis, *Analyst* **113,** 1519 (1988).
52 B. P. H. Schaffar and O. S. Wolfbeis, *Biosensors Bioelectron.* **5,** 137-148 (1990).
53. O. S. Wolfbeis, M. J. P. Leiner and H. E. Posch, *Microchim. Acta* **III,** 359 (1986).
54. M. C. Moreno-Bondi, O. S. Wolfbeis, M. J. P. Leiner and B. P. H. Schaffar, *Anal. Chem.* **62,** 2377 (1990).
55. I. Klimant, F. Ruckruh, G. Liebsch, A. Stangelmayer and O. S. Wolfbeis, *Mikrochim. Acta* **131,** 35 (1999).
56. O. S. Wolfbeis, I. Oehme, N. Papkovskaya and I. Klimant, *Biosensors Bioelectron.* **15,** 69 (2000).
57. Z. Rosenzweig, R. Kopelman, *Anal. Chem.* **68,** 1408 (1996).
58. C. McDonagh, B. D. MacCraith and A. K. McEvoy, *Anal. Chem.* **70,** 45 (1998)
59. A. Apostolidis, I. Klimant, D. Andrzejewski and O. Wolfbeis, *J. Combinat. Chem.* **6,** 325 (2004).
60. S. Lee and I. Okura, *Anal. Commun.* **34,** 185 (1997).
61. Y. Amao, T. Miyashita and I. Okura, *J. Porphyr. Phthalocyan.* **5(5),** 433 (2001).
62. D.V. Papkovsky, G.V. Ponomarev, W. Trettnak and P. O'Leary, *Anal. Chem.* **67,** 4112 (1995).
63. F. Alava-Moreno, M. Valencia-Gonzalez, A. Sanz-Medel and M. Diaz-Garcia, *Analyst* **122,** 807 (1997).
64. B. P. H. Schaffar and O. S. Wolfbeis, *Proc. SPIE* **990,** 122 (1989).
65. H. E. Posch and O. S. Wolfbeis, *Microchim. Acta* **I,** 41 (1989).
66. V. A. Stoy, *J. Biomaterials Appl.* **3,** 552, (1989)
67. G. Deng, T. D. James and S. Shinkai, *J. Am. Chem. Soc.* **116,** 4567 (1994).
68. J. Yoon, A. W. Czarnik, *J. Am. Chem. Soc.,* **114,** 5874 (1992).
69. O. S. Wolfbeis, I. Klimant, T. Werner, Ch. Huber, U. Kosch, Ch. Krause, G. Neurauter and A. Duerkop, *Sensors Actuat. B* **51,** 17 (1998).
70. Z. Ge, C. W. Brown, L. Sun and S. C. Yang, *Anal. Chem.* **65,** 2335 (1993).
71. E. Pringsheim, E. Terpetschnig and O. S. Wolfbeis, *Anal. Chim. Acta* **357,** 247 (1997).
72. R. Koncki and O. S. Wolfbeis, *Biosensors Bioelectron.* **14,** 87 (1999).
73. E. Pringsheim, E. Terpetschnig, S. A. Piletsky and O. S. Wolfbeis, *Adv. Mater.* **11,** 865 (1999).
74. B. Meier, T. Werner, I. Klimant and O. S. Wolfbeis, *Sens. Actuators* B **29,** 240 (1995).
75. O. S. Wolfbeis, *The Fluorescence of Organic Natural Products*, in: *Molecular Luminescence Spectroscopy: Methods & Applications,* Vol. 1, edited by S. G. Schulman (Wiley, New York, 1985), pp. 167-369.
76. W. Trettnak and O. S. Wolfbeis, *Anal. Chim. Acta* **221,** 195 (1989).
77. Sanz, V; Galban, J; de Marcos, S. and Castillo, J. R., *Talanta* **60,** 415 (2003).
78. Sanz-Vicente, I.; Castillo, J. R. and Galban, J., *Talanta* **65,** 946 (2005).
79. P. Babilas, V. Schacht, G. Liebsch, O. S. Wolfbeis, M. Landthaler, R. M. Szeimies and C. Abels, *Brit. J. Cancer* **88,** 1462 (2003).
80. P. Babilas, G. Liebsch, V. Schacht, I. Klimant, O. S. Wolfbeis, R. M. Szeimies and C. Abels, *Microcirculation, in press* (2005).
81. M. Schaeferling, *Luminescence lifetime-based imaging of sensor arrays for high-throughput screening applications*, in: *Springer Series on Chemical Sensors and Biosensors*, Vol. 3, edited by G. Orellana, M.C. Moreno-Bondi (Springer, Heidelberg, 2005).
82. R.J. Woods, S. Scypinski, L.J. Cline Love and H.A. Ashworth, *Anal. Chem.* **56,** 1395 (1984).
83. T. Q. Ni and L. A. Melton, *Appl. Spectroscopy,* **47,** 773 (1993).
84. P. Hartmann and W. Ziegler, *Anal. Chem.* **68,** 4512 (1996).
85. M. Schaeferling, M. Wu and O.S. Wolfbeis, *J. Fluorescence* **14,** 561 (2004).
86. M. Schaeferling, M. Wu, J. Enderlein, H. Bauer and O.S. Wolfbeis, *Appl. Spectroscopy* **57,** 1386 (2003).

87. J. R. Lakowicz, *Principles of Fluorescence Spectroscopy,* 2nd edn, (Kluwer Academic/Plenum Publishers, New York 1999).
88. E.R. Carraway, J.N. Demas, B.A. DeGraff and J.R. Bacon, *Anal. Chem.* **63**, 337 (1991).
89. D.B. Papkovsky, J. Olah, I.V. Troyanowsky, N.A. Sadowsky, V.D. Rumyantseva, A.F. Mironov, I. Yaropolov and A.P. Savitsky, *Biosens. Bioelectron.* **7**, 199 (1991).
90. P.J. Spellane, M. Gouterman, A. Antipas, S. Kim and Y.C. Liu, *Inorg. Chem.* **19**, 386 (1980).
91. W. Zhong, P. Urayama and M.-A. Mycek, *J Phys D: Appl. Phys.* **36**,1689 (2003).
92. Y. Amao, *Microchim. Acta* **143**, 1 (2003).
93. G. Liebsch, I. Klimant I, B. Frank, G. Holst and O.S. Wolfbeis, *Appl. Spectroscopy* **54**, 548 (2000).
94. M. Wu, Z. Lin, M. Schaeferling, A. Duerkop and O. S. Wolfbeis, *Anal. Biochem.*, **340**, 66 (2005).

BORONIC ACID-BASED FLUORESCENCE SENSORS FOR GLUCOSE MONITORING

Gurpreet Kaur[1], Na Lin[1], Hao Fang[2], and Binghe Wang*[1]

16.1. INTRODUCTION

As of today there is no cure of diabetes, and approximately 2.1% of world population is affected by this disease. [1] The inability for diabetic patients to appropriately control blood glucose level has long-term health consequences such as cardiovascular problems, renal complications, blindness, nerve damages, and foot and skin complications. These long-term consequences are the direct results of elevated glucose level for a prolonged period of time. Consequently, proper control of blood glucose concentration is the key to reducing complications and prolonging life. A critical step in controlling blood glucose level is appropriate monitoring. At present blood glucose monitoring involves finger pricking to draw blood sample. Another approach used today is the "Gluco Watch," which uses iontophoresis [2] as a way to extract biological fluid for analysis. This is minimally invasive, but has its disadvantages and is not a continuous monitoring approach.

An ideal glucose monitoring system should be (1) continuous, (2) non-invasive, (3) easy to use, and (4) affordable. There have been intensive efforts in developing such glucose monitoring methods, which will undoubtedly help patient compliance and optimal glucose concentration control. Current non-invasive methods under research and/or development include near-infrared (NIR) spectroscopy, [3, 4] contact lens type of sensors [5] and implanted sensor devices. [6] Except for the direct spectroscopic method, the others require the use of a sensor as the recognition and transducer unit. Currently commercially available glucose sensors use enzyme-based approaches. [7, 8] There is no doubt that the invention of such sensors has had a tremendous positive impact on diabetic patient care. However, the lack of reagent stability makes the enzyme

[1] Department of Chemistry and Center for Biotechnology and Drug Discovery, Georgia State University, 33 Gilmer St. S.E.Atlanta, Georgia, 30303. Phone: (404) 651-0289.
Fax: (404) 654-5827. E-mail: wang@gsu.edu
[2] School of Pharmacy, Shandong University, No. 44 Wenhuaxi Road, Jinan, China, 255012.

based-method undesirable for long-term implant and contact lens applications.[6,9] For such purpose, chemosensors have the advantage of enhanced stability, manufacturability, and high specificity. Particularly promising in this area is the use of fluorescent chemosensors.

Generally, selective sensors consist of three components: a) proper functional groups that afford strong intermolecular interactions (boronic acid), b) a 'reporter' event/moiety (stars) that allows the binding even to be recognized, and c) the appropriate three-dimensional scaffold that allow for necessary complementarily in terms of size, shape, and functional group orientation (bridge) as shown in Figure 16.1. In designing such chemosensors for glucose, boronic acids occupy a special place because of their strong functional group interaction with the diols that exist on glucose and other sugars. [10-14] In addition, boronic acids can be incorporated to change spectroscopic properties upon binding. This allows for the integration of the binding and signaling events into one unit. This chapter will start with an overview of the basic chemistry of boronic acid in its interactions with diols, and then discuss the different fluorescent types boronic reporter groups used for both signaling the binding and the three-dimensional scaffolds that are appropriate for the selective recognition of glucose in different forms. Specifically, the following sections will cover (1) factors that affect the boronic acid-diol binding, (2) some earlier boronic acid-based fluorescent reporter compounds that got this field started, (3) a new generation of water-soluble boronic acid fluorescent reporter compounds, (4) new boronic acids that are suitable for application in contact lens type of sensing devices, and (5) finally a *de novo* approach for the construction of the appropriate 3-dimentional scaffold for the selective recognition of glucose. This chapter will not attempt to be exhaustive as there are other reviews that cover in more depth of the various aspects of this area. [15-19] Instead, the focus will be on the basic idea and principles used in each approach. It should also be noted that boronic acid-based electrochemical sensors are not included in this chapter, although there have been some very important discoveries in that area. [20, 21]

16.2. FACTORS THAT AFFECT THE BINDING BETWEEN BORONIC ACIDS AND DIOLS

Boronic acids covalently reacts with *cis* 1,2- or 1,3-diols to form five or six-member cyclic esters in aqueous solution. This complexation is reversible,

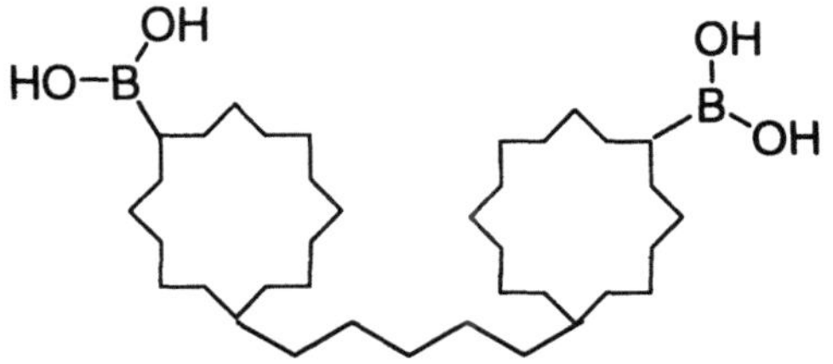

Figure 16.1. Three components of a "selective" sensor

which makes it an ideal interaction in sensor design. Boronic acids are Lewis acid and as such can react with water to go from the neutral trigonal form (**1**) to the anionic tetrahedral form (**2**) and release a proton at the same time (Scheme 16.1). Similarly the diol-boronic acid complex or the ester can be in either the trigonal (**3**) or tetrahedral (**4**) form. Because boronic acids and their esters can exist in two different ionization states, there are actually three different "binding constants" to consider. The first one relates to the conversion of the trigonal boronic acid (**1**) to the trigonal ester (**3**), termed K_{trig}. The second one refers to the conversion of tetrahedral boronate (**2**) to its ester counterpart (**4**), termed K_{tet}. However, neither of these two truly represents the overall binding constant between a diol and boronic acid for the purpose of sensor design. The third binding constant describes the overall binding strength regardless of the ionization state of the boron species, K_{eq} (Scheme 16.2). Therefore, it is very important that one knows exactly what type of binding constants a particular procedure gives before any comparison can be made with or between literature values. The binding constants mentioned in the subsequent discussions are overall binding constants (K_{eq}, Scheme 16.2) unless specified otherwise.[11]

It is known that phenylboronic acid (and other boronic acids) have different affinities for various diols depending on their structural features especially O-C-C-O dihedral angle. Smaller dihedral angles are often associated with higher binding constants. Another important factor to consider is the pKa of the boronic acids. Boronic acids with lower pKa's tend to have higher affinities for

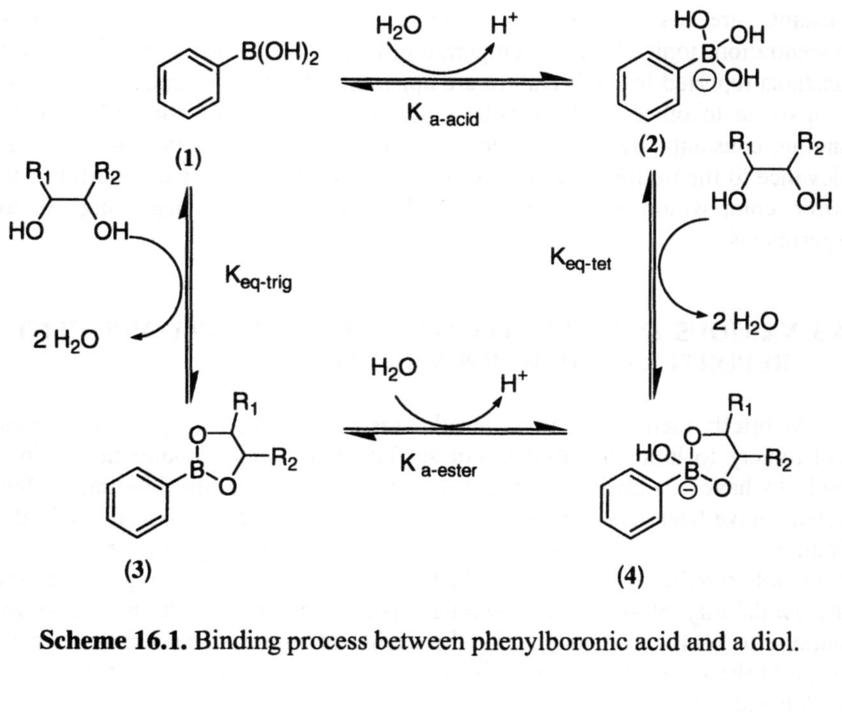

Scheme 16.1. Binding process between phenylboronic acid and a diol.

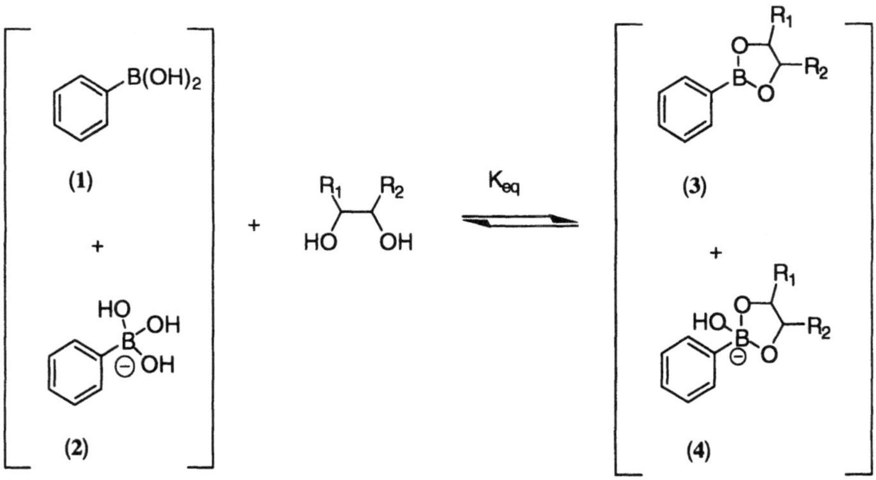

Scheme 16.2. Overall binding process between phenylboronic acid and a diol.

diols, [10, 12, 22, 23] although the optimal binding also depends on the pKa's of the boronic acid and diol, and the pH. [12]

Another thing that should also be noted is that boronic acid-diol binding constants are also influenced by factors such as solvent, buffer, buffer concentration, ionic strength, temperature, etc. [11, 12] Essentially all the binding constants reported in the literature are apparent binding constants, and are only comparable to others under similar conditions. The significant effect of the binding constants by these various factors also suggests that for the best relevance to the performance of the sensors under physiological conditions, the buffer composition should be carefully chosen at the beginning of the experiments.

16.3. VARIOUS TYPES OF BORONIC ACID-BASED FLUORESCENT REPORTER COMPOUNDS AVAILABLE

As briefly mentioned in the Introduction section, the ability for the boronic acid moiety to bind to a diol is not sufficient for sensor application. There needs to have a signaling event that is associated with the binding. Many systems have been designed over the years that allow for the boronic acid-diol binding event to regulate the spectroscopic properties of a chromophores/fluorophore, which in turn results in the signaling of binding. The availability of such spectroscopic reporter compounds has been and will continue to be the cornerstone for the design of future sugar sensors. In the design of these reporter compounds, several mechanisms have been used, which are detailed below.

16.3.1. Anthracene and pyrene based fluorescent reporter compounds

Excited state photoelectron transfer (PET) is a process in which an electron is transferred to the excited state chromophore, and therefore, results in the quenching of the excited state in a thermal process. Systems have been designed that modulated this process through boronic acid-diol binding to result in the signaling of the binding event.

Yoon and Czarnik first reported anthrylboronic acids-based fluorescence sensors for saccharides (**5**, **6**). These compounds showed significant fluorescence intensity changes upon binding with saccharides. The intensity change was lower for glucose than fructose. [14] Fluorescence intensity change upon binding could be due to the change in the hybridization state of the boronic ester, which has a lower pKa than the corresponding boronic acid. Specifically, the boronic acid (pKa about 8.8) should exist mostly in the neutral trigonal state at physiological pH. However, upon ester formation, the boron specie would exists in the anionic tetrahedral because of a decreased pKa (about 4.6 and 6.8 for the fructose and glucose esters, respectively. [11] Such a hybridization change resulted in what is called "chelation-enhanced quenching" and therefore lower fluorescence intensity for the complex. Such sensors represent very impressive initial success, although monoboronic acids of any kind would not be expected to have the selectivity needed for specific glucose sensing. [11, 24]

Based on anthracene system Shinkai and co-workers developed sensor (**7**) [25] (Figure 16.2). In this system, an amino group is positioned in a 1, 5-relationship with the boronic acid. It had previously been demonstrated by Wulff [26] that such an arrangement promotes dative B-N bond formation (Scheme 16.3). This bond formation helps to lower the apparent pKa of the boronic acid, which would in turn increase binding strength. [10, 11] Indeed, (**7**) showed significant changes in fluorescence intensity and much higher binding affinity to glucose and fructose compared with phenylboronic acid. [11, 25, 27] This change in fluorescence intensity was proposed to be due to the strengthening of the B-N bond upon ester (**8**) formation. [10, 13, 25] Specifically, the fluorescence of the anthracene system can be quenched by the nitrogen lone pair electrons through PET. However, binding to a diol is known to lower the pKa of the

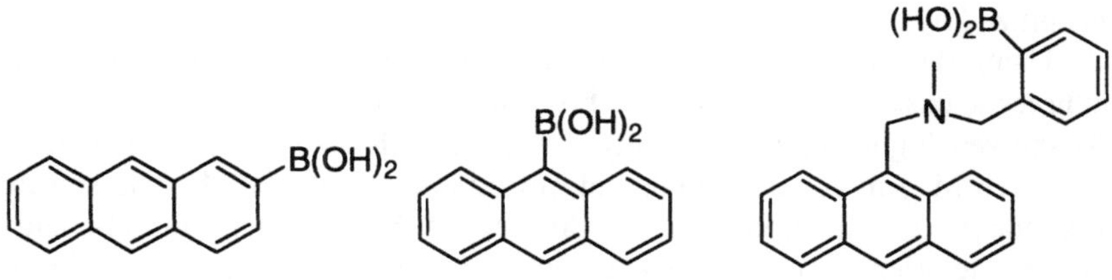

(5) (6) (7)

Figure 16.2. Anthracene-based boronic acid reporter compounds.

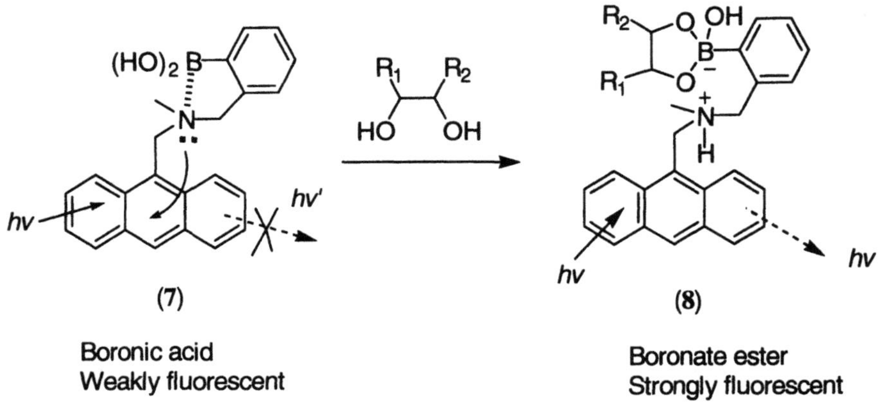

Scheme 16.3. Illustration of an anthracene-based photoinduced electron transfer system.

boron functionality,[10, 11] which was thought to consequently strengthen the B-N bond. This B-N bond strengthening was thought to "tie up" the nitrogen lone pair electrons and thus lead to a reduced PET fluorescence quenching. Recently, based on experimental results and computational work, our group has proposed a different mechanism that causes the change in fluorescence intensity upon ester formation.[28, 29] Rather than B-N bond formation, the hydrolysis mechanism is apparently in play (Scheme 16.3). However, the elucidation of a new mechanism does not change the fact that this system (**7**) is and has been a very useful reporter system for the development of fluorescent sensors for sugars.

Monoboronic acids prefer to bind fructose over glucose with about 50 fold selectivity at physiological pH.[11] In order to increase the binding affinity and selectivity to glucose, it is essential that a more selective recognition element should be built into the system. This can be achieved with the inclusion of an additional binding site, which can be either a) another boronic acid to be recognized by a second pair of diols or b) other recognition moieties.

Using the PET system as a starting point, Shinkai and coworkers designed anthracene based monoboronic and bisboronic acid compound (**9**).[13] Compound (**9**) showed higher binding affinity for glucose (Ka =3980 M^{-1}) than fructose (Ka= 316 M^{-1}) in aqueous methanol buffer solution at pH 7.8. The selectivity was due to the incorporation of the second boronic acid moiety and the appropriate orientation of the two boronic acid moieties relative to the diol pairs on glucose. The authors also deduced from NMR experiments that bisboronic acid (**9**) binds with glucose in the pyranose form in a 1:1 fashion. Upon further examination, Norrild and coworkers found that the solvents used have a tremendous effect on in which glucose binds with the diboronic acid sensor.[30,31] Glucose undergoes mutarotation in water faster than in methanol.[32] From extensive NMR studies Norrild concluded that bisboronic acid binds to glucose in the pyranose form in non-aqueous solution. However, in an aqueous environment it binds initially to glucose in its pyranose form and then slowly converts to its furanose form.

Shinkai and coworkers also developed other potential glucose sensors with bisboronic acids. They designed compound (**10**) which had two pyrene units separated by hexamethylene linker. The sensor was based on the idea that pyrene would normally give excimer formation due to stacking. However, upon formation of a rigid 1:1 complex between glucose and sensor, stacking becomes very difficult, which prevents excimer emission and only gives monomer emission. Indeed addition of glucose at different concentration gave high monomer emission than excimer emission with Keq of 1995 M^{-1}. Other saccharides were tested with this sensor; they all gave strong fluorescence with high stability constants. However, fructose and allose deviated from there isosbestic point due to competitive 1:2 complex formation between (**10**) and the sugar. This indicates that in binding with fructose and allose each boronic acid behaves like a monoboronic acid.[13] In this case, the spacing between the pyrene groups and the conformational constraints imposed by the linker are two key elements in determining the specificity for glucose. In another study, Appleton and Gibson further examined the influence of linker length in analogs of (**10**) and found that C_6 and C_7 linker displayed high specificity to D-glucose, while a longer linker (e.g. C_{12}) behaved like a monoboronic acid showing preference for fructose over glucose.[33]

Drawing from the lesson of compound (**10**), the James group developed pyrene based diboronic acid compound (**11**) and polymeric sensor (**12**) with only one pyrene unit as fluorophore linked via a hexamethylene linker to phenylboronic acid. Both compounds showed increase in fluorescence intensity upon addition of saccharides. Compound (**11**) showed a 22-fold selectivity for glucose over its pyrene based monoboronic acid whereas compound (**12**) was more selective for fructose but showed only 3 fold selectivity over pyrene monoboronic acid.[34]

Recently, the Tao group used pyrene stacking to their advantage in developing glucose sensors.[35] Their system involved utilization of a polycation group which induces a 1:2 complexation between glucose and two pyrene-based monoboronic acids. The complex formation in turn results pyrene stacking which causes excimer emission. The monomer, compound (**13**), showed no excimer emission in aqueous glucose solution. However, upon addition of polycation (**14**) the spectrum showed broad excimer emission in the range of 430-600 nm indicating that polycation caused 1:2 complexations between glucose and pyrene monoboronic acids. Fluorescence studies showed glucose had higher selectivity over other saccharides. Other saccharides, fructose, galactose, and ribose, showed low to no excimer emission due to their poor ability to a form 1:2 complex with these boronic acids. Further studies

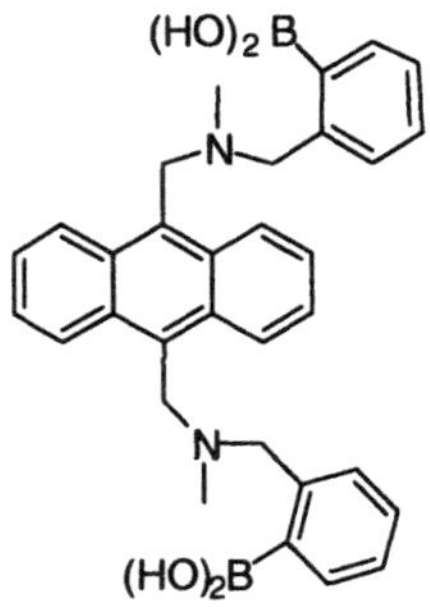

(9)

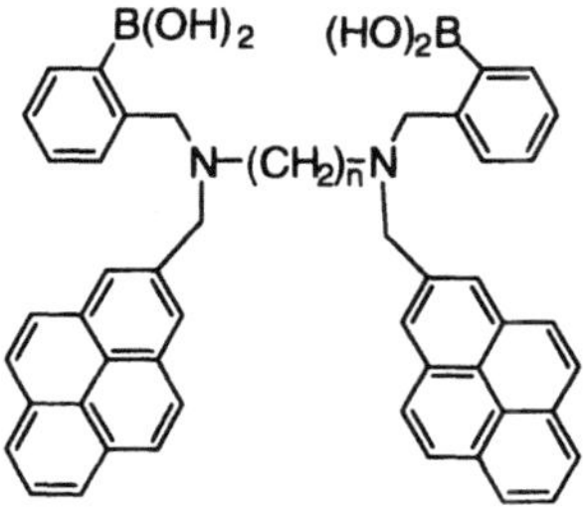

(10) n= 6, 7, 8, 12

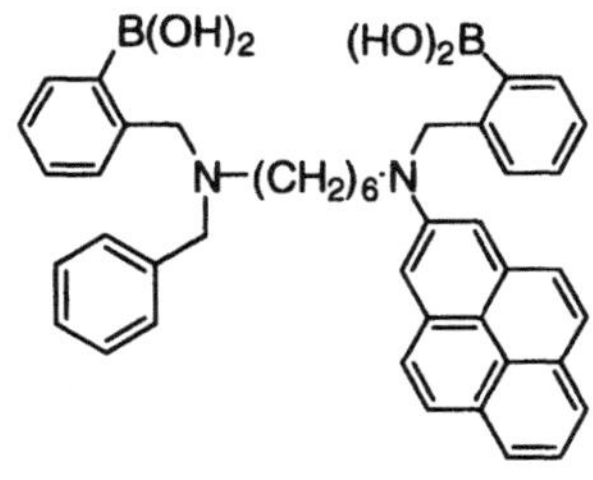

(11)

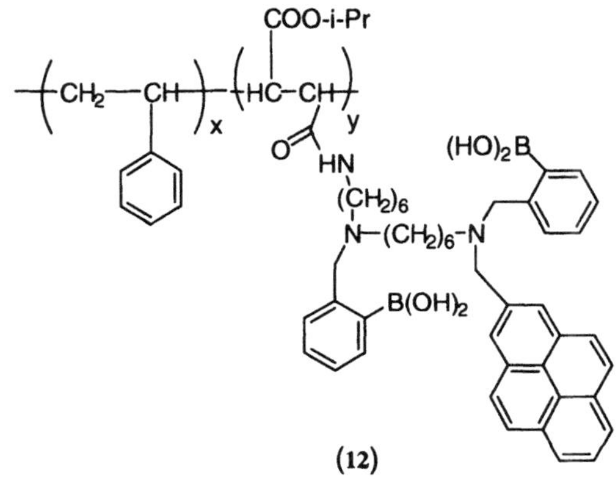

(12)

Figure 16.3. Anthracene and pyrene-based boronic acid sensors for glucose.

indicated that electrostatic interactions are present between **(13)**, which exists in its tetrahedral form, and polycation **(14)**.

James and coworkers developed a saccharide detection system using the concept of fluorescence resonance energy transfer (FRET). This detection system involves two different fluorophores: phenanthrene as the donor and

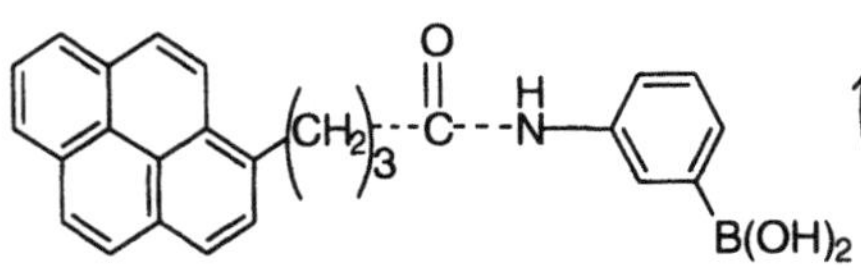

(13)

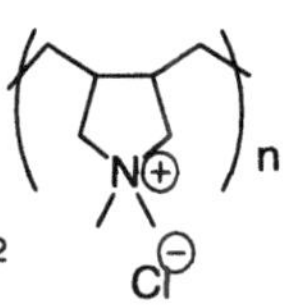

(14)

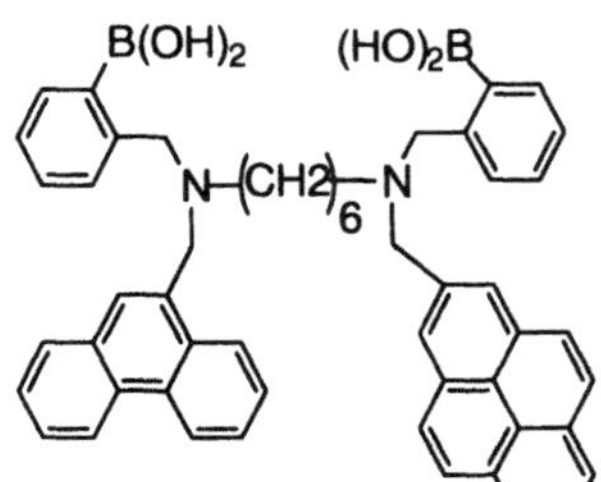

(15)

Figure 16.4. Pyrene-based glucose sensors.

pyrene as the acceptor which were separated by a linker (hexamethylene) with two phenyl boronic acids attached to each fluorophores (**15**). This sensor was designed in such a way that the emission wavelength (369 nm) of phenanthrene overlaps with the excitation wavelength (342 nm) of pyrene, which allows resonance energy transfer between these two fluorophores. The result of this transfer is that when excited at 299 nm (phenanthrene), the entire system would emit at 417 nm. Binding of a sugar would allow the two fluorophores to come closer and therefore increase the efficiency of the FRET process, which results in an increased fluorescence intensity at 417 nm. [36] The binding constant between (**15**) and glucose was determined as 142 M^{-1} compared to 30 M^{-1} and 44 M^{-1} for the individual phenanthrene and pyrene boronic acid, respectively.

Using Shinkai's anthracene fluorescent reporter system as the basic building block, our group synthesized different bisboronic acids (**16**) with different amide linkers. [24] Compound (**16a**) with two acetamides attached to the phenyl ring in an ortho relationship offers the appropriate diboronic acid orientation and distance for selective binding with glucose. It showed a high binding affinity (K_{eq} = 1472 M^{-1}) for glucose and a 43-fold selectivity over fructose. Neither (**16b**) nor (**16c**), although having the same number of carbons in the linker, showed the kind of selectivity and affinity for the glucose as (**16a**) did. This indicates that the rigidity of the linker plays a critical role in determining saccharide selectivity.

16.3.2. Internal charge transfer compounds

In a chromophore, excited state internal charge transfer (ICT) happens that can lead to thermal decay of the excited state. ICT systems contain an electron donor group and an electron acceptor group in the same chromophore/fluorophore. The ICT process can also be modulated to afford a signaling event in boronic acid-sugar binding. In such a case, the boronic acid

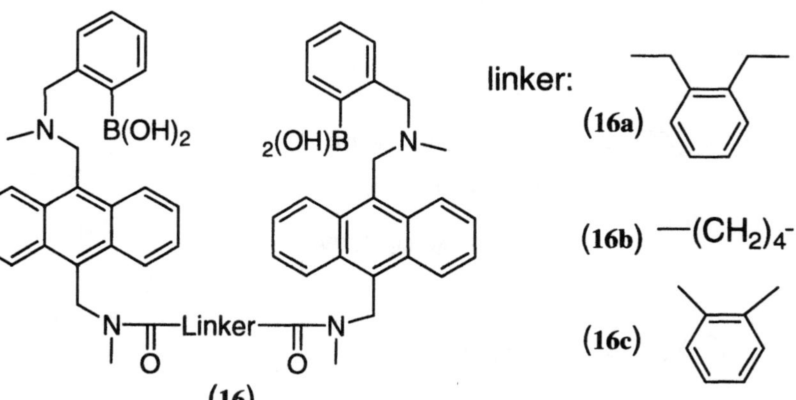

(**16**)

Figure 16.5. Anthracene-based bisboronic acid sensors for glucose.

moiety acts as an electron acceptor when it is in the neutral form. When the boronic acid group changes to its anionic form (tetrahedral form) upon binding with a sugar, it is no longer an electron acceptor. This leads to the spectral changes due to the perturbation of the charge transfer nature of the excited state. [37]

The Shinkai group developed the first generation of such ICT sensors based on a stilbene core. Inhibition of rotation of the ethylene linker of stilbene in its excited state is known to give fluorescence. Such compounds possess a large conjugated π system of electrons, which allows for the possibility of the 'terminal' substituent to affect the chromophoric properties of these systems. Four stillbene boronic acid analogs (**17a-d**) were synthesized and evaluated. Compounds (**17b**) and (**17d**) bear an electron-donating group and (**17c**) bears an electron-withdrawing cyano group. Changing the pH from low to high induced a blue shift in the emission spectrum of (**17b**) and (**17d**) and an increase in intensity by about one fold in the presence of sugar. This was thought to be due to the loss of the electron-withdrawing properties of the boron functionality when converted to the anionic form. Similarly, the polyene derivatives (**18**), [38] diphenyloxazole derivatives (**19**) [39, 40] and chalcone derivatives (**20**) [41] were prepared and tested for binding with sugars. The fluorescence intensity changed only by a maximum of five folds in these ICT systems. Understandably, these monoboronic acids showed preference for fructose over glucose as would have been expected for any monoboronic acids. [10, 11] However, these "reporter" systems could, in principle, be used for the construction of diboronic acid sensors that show selectivity for glucose.

16.3.3. Polymer based sensors

Above mentioned sensors have disadvantage of having lower excitation wavelength and not being completely water soluble. A system with longer wavelength is highly desirable. Polymer based fluorophores have been reported to have wavelengths in the range of 450nm to 630nm. [42, 43] Wolfbeis and group

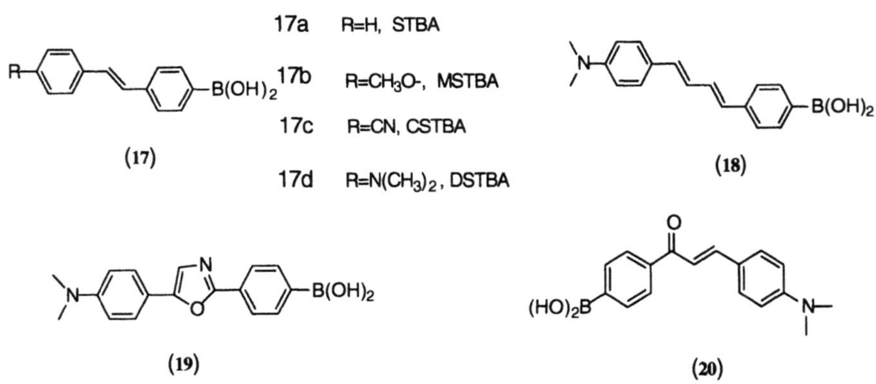

Figure 16.6. ICT sensors

devised there system based on organic conductive polymer (OCPs), polyaniline.[44] OCPs have long wavelength absorption and change in there color results from the strong interactions between the polymer chains. These interactions can be distorted with introduction of diol-binding receptor such as boronic acid. Wolfbeis group synthesized novel polymer (**P**) by copolymerizing aniline with 3-aminophenylboronic acid. The novel polymer absorption spectrum covered the region from visible to near-infrared and was pH dependent. Both polyaniline and (**P**) had apparent pKa around 7 and (**P**) showed absorption spectral changes with addition of saccharides whereas polyaniline did not show any spectral changes. 3-aninophenylboronic acid also did not show any absorption change in the presence of excess fructose or glucose. Saccharide sorbitol showed the major change in fluorescence followed by fructose, mannitol, glycerol, glucose, and saccharose but (**P**) showed low selectivity among saccharides.

16.3.4. Water soluble fluorescent boronic acid

Although many boronic acid-based fluorescent reporter compounds had been reported in the initial period of study, many of them lack the necessary water solubility for application under near physiological conditions. Some also lack the necessary photochemical, chemical, and metabolic stability. Therefore, in the last few years, there have been significant amounts of effort in developing water soluble boronic acid reporters that are chemically and photochemically stable. In this regard, several water soluble boronic acid reporters have been reported. The Heagy group developed the naphthalimide sensors (**21-26**). Naphthalic anhydride chromophore itself is highly fluorescent and photostable. With introduction of polar groups on the naphthalene ring, such compounds can be made water soluble at micro molar concentrations. These sensors showed decrease in fluorescence intensity upon addition of saccharides. The mechanism through which this fluorescent intensity changes is thought to be due to the modulation of the PET mechanism when the boron atom changes from the neutral sp^2 form to that of anionic sp^3 upon addition of saccharide at physiological pH. This saccharide complexation changes the orbital energy of the HOMO of the phenyl boronic acid π system by occupying the next highest molecular orbital. Sensor (**21**) showed the largest chelation-enhanced quenching (CHEQ) and selectivity toward fructose. Upon introduction of a nitro group to the naphthalimide (**21**), compound (**22**) showed selectivity in fluorescence change for glucose over fructose and galactose.

Table 16.1. Naphthalimide-based water-soluble boronic acid reporters

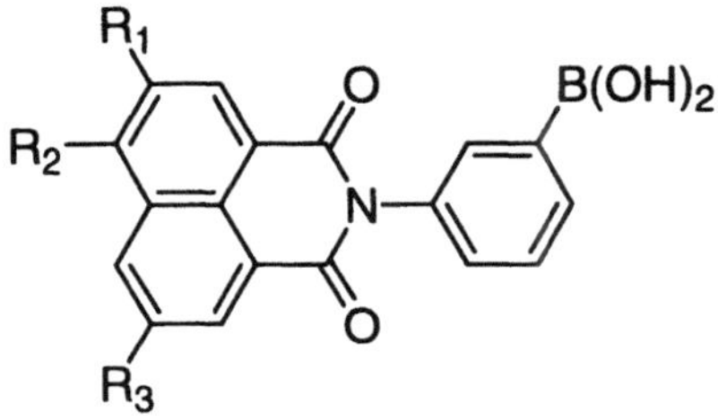

Entry	R_1	R_2	R_3
(21)	H	H	H
(22)	H	H	NO_2
(23)	H	H	NH_2
(24)	H	H	CH_3CONH
(25)	H	OCH_3	H
(26)	SO_3K	NH_2	SO_3K

Sensor (**23**) with an amino substituent showed enhanced fluorescence for galactose at low pH. Sensors (**24-26**) showed high quantum yield and excellent solubility in water, but did not show much change in fluorescence intensity with saccharides. Especially worth mentioning is sensor (**26**) that did not show effective PET quenching compared to other sensors. With the introduction of two sulfonic groups, the HOMO energy of the fluorophore was raised and in turn the hydroxyboronate anion could no longer effectively quench the fluorescence.[45, 46]

Norrild and coworkers also developed a water soluble bisboronic acid (**27**) with K_{eq} of 2512 M^{-1} for D-glucose. The sensor has charged cationic pyridine rings which made it not only water soluble but also helped to improve the binding with sugar at neutral aqueous solution.[47]

Our group has also developed several water soluble boronic acids with the quinoline and naphthalene chromophores. 8-Quinolineboronic acid (8-QBA, **28**) is a novel type of fluorescent sensor for carbohydrates which gives up to 40 fold fluorescence intensity increases at physiological pH upon saccharide addition, good water solubility, and chemical and photochemical stability. 8-QBA itself is non-fluorescent at pH above 5 and weakly fluorescent at lower pH. The mechanism through which fluorescence increases upon addition of saccharides is not known. However, ^{11}B-NMR indicates that the boron atom in both 8-QBA and its ester forms is in the tetrahedral form at physiological pH, so the change in fluorescence intensity is not due to change in hybridization state of the boron atom.[48]

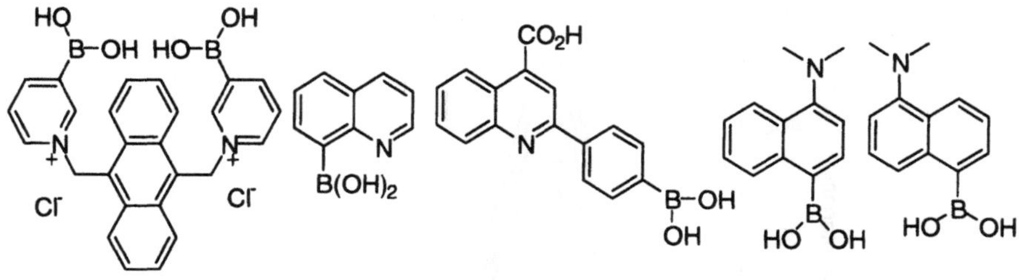

(27) (28) (29) (30) (31)

Figure 16.7. Water-soluble fluorescent boronic acid compounds

Along a similar line, another quinoline-based compound, **(29)**, was designed in which the boronic acid group was not directly attached to the quinoline ring. Compound **(29)** has a free carboxyl group which can be used as a "handle" for tethering other structural moieties. This sensor designed showed fluorescence intensity increases upon binding with sugars when the pH of the solution is 5 or higher. Compound **(29)** showed up to 25-fold fluorescence intensity changes with the addition of various sugars. [49] Since compound **(29)** has a carboxyl "handle" tethered to it, incorporation of other boronic acid moieties, which can be used to modify the selectivity, can be easily accomplished. Studies in our lab have already shown that converting the free acid to an amide does not change the fluorescence properties of **(29)**.

Two naphthalene-based new water-soluble fluorescent reporters discovered by our lab, **(30-31)**, also show large increase in fluorescence property changes upon addition of saccharides at physiological pH. Compound **(30)**, 4-dimethylaminonaphthaleneboronic acid (4-DMANBA), is thought to be an ICT sensor, where the amino group is the electron donor and the boron atom is the electron acceptor when in the trigonal form. The lone pair electrons on the donor amino group can form conjugation with the empty p orbital of the sp^2 boron atom directly attached to the naphthalene ring. [50] The ICT process is very sensitive to small perturbations in the electronic properties of the molecule by changing its spectroscopic properties. As mentioned earlier, upon addition of a diol to boronic acid, the ester formation lowers the pKa of the boron atom by about 2-3 pH units and converts the boron atom from its neutral sp^2 to the anionic sp^3 form. [11] The introduction of an anion to the system in turn leads to change in fluorescent properties. Binding studies shows that fructose and sorbitol induce up to 41-fold increases in fluorescence intensity whereas glucose results in a 17- fold increase at 1 M in 0.1 M aqueous phosphate buffer at pH 7.4.

Compound **(31)**, 5-dimethylaminonaphthaleneboronic acid (5-DMANBA), is a positional analog of 4-DMANBA **(30)**. However, this compound exhibit ratiometric fluorescent changes upon sugar addition. 5-DMANBA exhibits an emission maximum at 513 nm and absorption maximum at 300 nm in aqueous

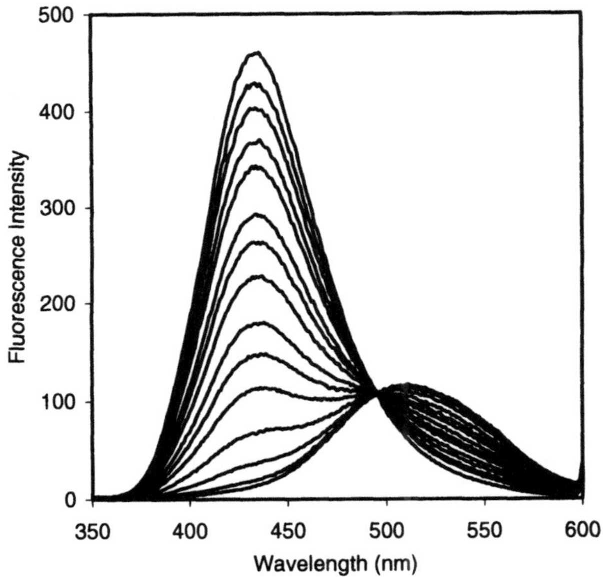

Figure 16.8. Ratiometric fluorescence change of 5-DMANBA (**31**) (1.0×10^{-5} M) with different concentrations of D-fructose (0-50 mM) in 0.1 M aqueous phosphate buffer at pH 7.4, λ_{ex} = 300 nm.

phosphate buffer at pH 7.4. Addition of a saccharide results in a significant decrease in the fluorescence intensity at 513 nm and large increase at 433 nm. Figure 16.7 shows ratiometric fluorescence changes upon addition of fructose to (**31**) at different concentration at pH 7.4.

Other saccharides induced similar ratiometric changes as fructose did. Sorbitol induced a 42-fold increase at 200 mM and glucose induced a 23-fold fluorescence intensity increase at 1 M concentration. [51]

16.3.5. Fluorescent Boronic Acids Suitable For Embedment In Contact Lens For Glucose Sensing

It has been known for a long time that tear glucose during hyperglycemia is elevated.[52] Therefore, there is the possibility of extrapolating blood glucose concentration by monitoring tear glucose levels. Such is the idea behind the possible development of non-invasive and continuous glucose sensing in a contact lens type of device. In this regard, the Geddes group reported their preliminary work on testing the feasibility of such an approach. [53] In this case, fluorescent sensors would be incorporated into a contact lens, and reading of the fluorescent signals from the contact lens would allow for non-invasive monitoring of glucose concentrations. They concluded that such approach is feasible if suitable fluorophores can be designed to response well in the low pH microenvironment of a contact lens. A series of boronic acids (**32**), (**33**) and (**34**) were synthesized for embedment into contact lens. The purpose for the use of a quinolinium moiety was to lower the pKa of the boronic acid, which as mentioned previously, is known to increase the boronic acid binding affinity with diols. Their initial results indicate that such an approach offers the possibility of continuous monitoring of glucose concentration in the range of 50-1000 μM, which is the normal physiological range in tear. [54, 55]

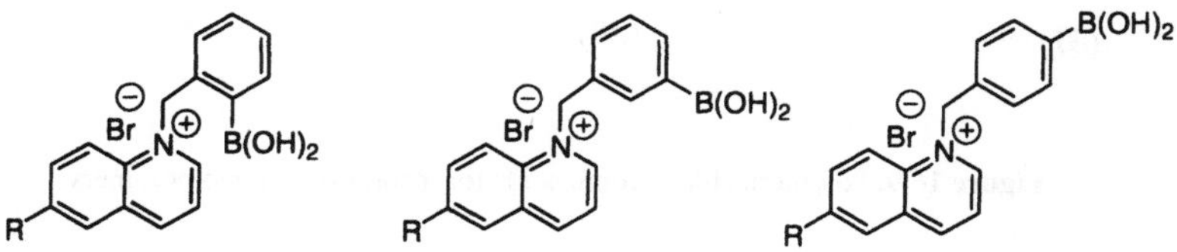

(**32**) (**33**) (**34**) R = OCH_3 or CH_3

A system which involves a bimolecular dye/quencher complex was developed by Singaram and coworkers. [55] This optical glucose sensor is a combination of pyranine (**35**) and a boronic acid substituted benzyl viologen (o-BBV) (**36**). When o-BBV binds to a sugar, the complex dissociates, results in an increase of fluorescence. The system utilizes a dye that is active in visible spectrum and is sensitive to glucose in the range of 0-1800 mg/dL at pH 7.4.

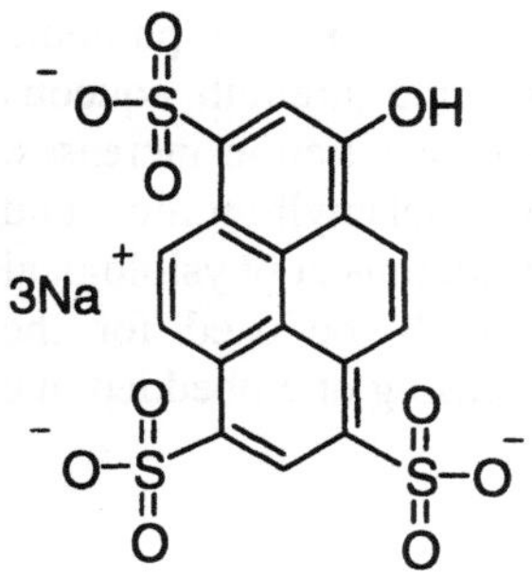

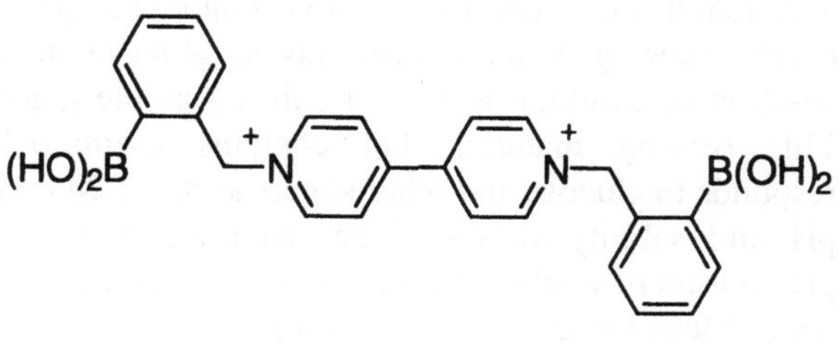

(**35**) (**36**)

Lowe and coworkers co-polymerized 4-vinylphenylboronic acid (**37**) with acrylamide and cross linker *N, N'*-methylenebisacrylamide to form a stable hydrogel film. This film was used to monitor glucose concentration at pH 9, 8 and 7.4. Addition of glucose to the sensor results in a red shift at pH 9, but no significant changes were observed at pH 8 and 7.4. [56] Based on this research two other polymer hydrogel systems were developed using 3-

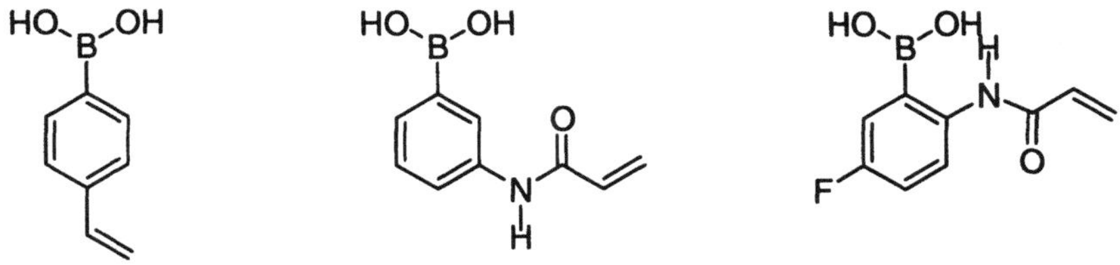

(37) (38) (39)

Figure 16.9. Polymerizable boronic acids for incorporation into polymers.

acrylamidophenylboronic acid (**38**) and 2-acryamido-5-fluorophenyl boronic acid (**39**), respectively. Both systems displayed an increased selectivity to glucose over lactate at physiological pH and could be deployed as holographic sensor for glucose detection. [57]

A new type of carbohydrate sensing material was reported by the Asher group, which consists of a crystalline colloidal array (CCA) incorporated into a polyacrylamide hydrogel (PCCA) with pendent boronic acid groups. [58, 59] They propose that upon addition of a sugar, a supermolecular bis-bidenate glucose-boronic acid complex was formed, which is stabilized by PEG or crown ether-capped sodium cations. This complexation correlated with the diffraction of wavelength and, therefore, visible color changes. Thus, this PCCA photonic crystal sensing material responds to glucose in low ionic strength aqueous solution by swelling and color shifting as the glucose concentration increases. This sensing material that contains 4-amino-3-fluorophenylboronic acid responds to glucose and other sugars at <50 □M concentrations at physiological pH and salinity values. Such materials could potentially be used for the preparation of implantable materials for blood glucose sensing.or embedded in a contact lens for tear glucose sensing.

16.4. APPROACHES AVAILABLE FOR THE CONSTRUCTION OF THE APPROPRIATE 3-DIMENSIONAL SCAFFOLD FOR THE SELECTIVE RECOGNITION OF THE TARGET SUGAR

With the availability of a large number of boronic acid fluorescent reporter compounds, there is still one important challenge that has not been discussed in detail, i.e., the methods that can be used to achieve the proper 3-dimensional structure, which affords the necessary selectivity. Generally speaking, there are three different approaches. The first is random screening of individual libraries of compounds generated through parallel synthesis. [60-62] The synthesis and purification of a large number boronic acid compounds is not a trivial issue, and, therefore, this approach may need the collective effort of many group. The second approach is the combinatorial chemistry.[63] One specific method that will be very useful is that of the one bead-one compound approach.[64] In this approach, a large number of compounds can be synthesized in a library, and the screening can be carried out by monitoring the fluorescent changes of the beads followed by sequencing with mass spectrometry. The third approach is that of *de novo* design based on computational chemistry.

Drueckhammer and coworkers reported the first example of a boronic acid-based fluorescence sensor for glucose in the glucopyranose form based entirely on *de novo* design.[65] This design approach used the CAVEAT program to develop a receptor with precisely positioned arylboronic acid to recognize 1,2 and 4,6 positions of glucopyranose. The search stated with the construction of complex **(40)**. Upon further quantum mechanical geometry optimization, the two methyl aryl bonds were considered as two vectors. Then the CAVEAT program was used to search for polycyclic structure with substituent bonds that would match the vector pairs identified in (**40**). The search, after taking into consideration of rigidity, fluorophoric property, ease of synthesis and water solubility, identified compound **(41a)** as a potential sensor for D-glucopyranose. Further geometry optimization with Hartree-Fork 6-31G* gave glucose complex **(41b)** as the lowest energy conformation which matched very well with the original compound **(41a**). Complex formation between (**41a**) and glucose in its pyranose form was confirmed by both ^{1}H-NMR and mass spectrometry studies. Binding of (**41a**) with all the saccharides showed a decrease in fluorescence intensity as concentration increased. The sensor showed high affinity and selectivity for glucose (Keq = 4000) over other saccharides (galactose, Keq = 100 and mannose Keq = 83.3). The high selectivity for D-glucose suggests that only glucose is able to form a bidentate complex involving both boronic acid moieties of the "receptor." This successful example indicates that feasibility of using computational chemistry-base rational *de novo* design approach for the construction of glucose and other sugar sensors.

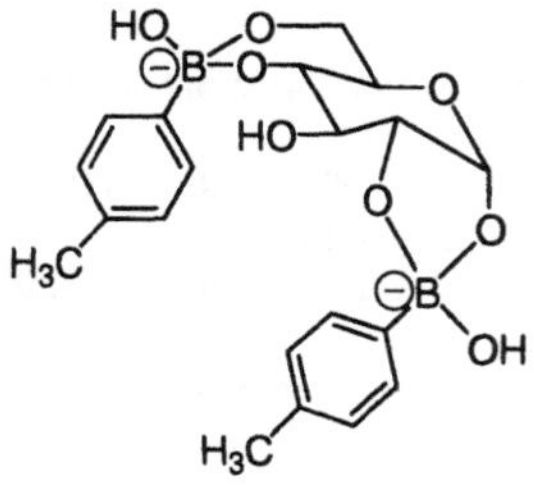

(40)

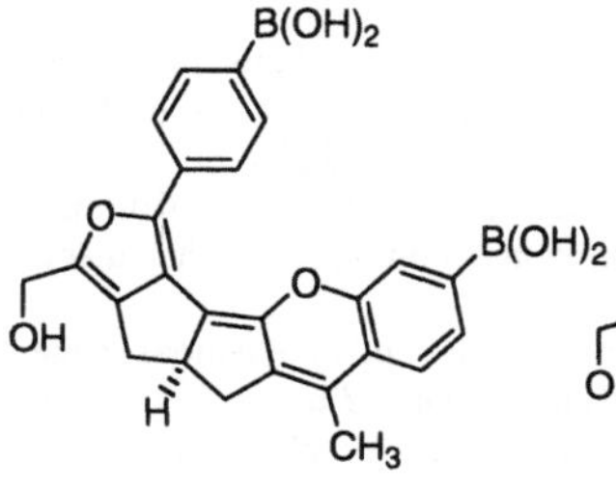

(41a)

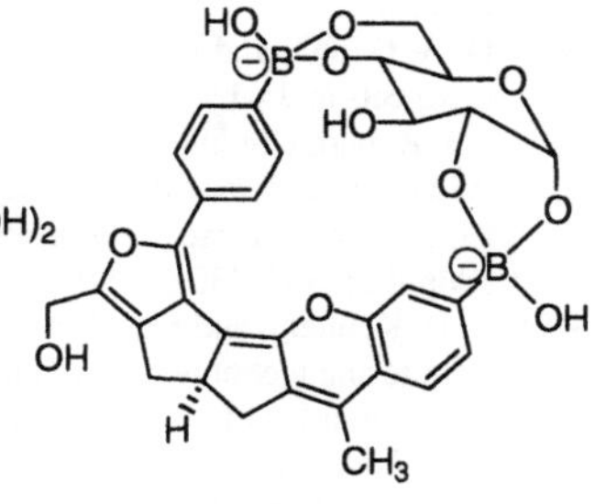

(41b)

16.5. CONCLUSION

During the last decade, much progress has been made in the synthesis of new boronic acid-based fluorescent sensors that could be used for the developing of non-invasive and continuous glucose monitoring devices. However, there are still many issues remaining, which include the short wavelengths (both emission and excitation) used for most of the available sensors, bio-fouling problems in implanted devices, the long-term chemical and photochemical stability of the available reporter compounds. It is our hope that this chapter will stimulate more research in this area, which will eventually lead to the construction of new glucose monitoring devices based on the selective recognition of sugars by fluorescent boronic acid compounds.

16.6. ACKNOWLEDGEMENT

Financial support from the National Institutes of Health (CA88343, CA113917, and NO1-CO-27184), the Georgia Cancer Coalition through a Distinguished Cancer Scientist Award, and the Georgia Research Alliance through an Eminent Scholar endowment and a Challenge grant is gratefully acknowledged. We also acknowledge the support of the Molecular Basis of Disease program at Georgia State University for a fellowship in support of GK.

16.7. REFERENCES

1. G. Hitman, *Type 2 Diabetes: Prediction and Prevention*. (Wiely;New York, 1999), 3-397.
2. R. T. Kurnik, B. Berner, J. Tamada, and R. O. Potts, Design and simulation of a reverse iontophoretic glucose monitoring device, *J.Eletochem.Soc,* **145**, 4119-4125 (1998).
3. H. M. Heise, R. Marbach, T. H. Koschinsky, and F. A. Gries, Non-invasive blood glucose sensors based on near-infrared spectroscopy, *Ann. Occup. Hyp.*, **18**, 439-447 (1994).
4. J. J. Burmeister, H. Chung, and M. a. Arnold, Phantoms for noninvasive blood glucose sensing with near infrared transmission spectroscopy, *Photochem. Photobiol,* **67**, (1998).
5. R. Badugu, J. R. Lakowicz, and C. D. Geddes, A glucose sesning contact lens: A non-invasive technique for continous physiological glucose monitoring, *J. Fluorescence,* **13**, 371-374 (2003).
6. D. A. Gough, and J. C. Armor, Development of the implantable glucose snesor: What are the prospects and why is it talking so long, *Diabetes,* **44**, 1005-1009 (1995).
7. J. J. Robert, Continuous monitoring of blood glucose, *Hormone research,* **57**, 81-84 (2002).
8. A. Maran, C. Crepaldi, A. Avogaro, S. Catuogno, A. Burlina, A. Poscia, and A. Tiengo, Continuous glucose monitoring in conditions other than diabetes, *Diabetes Metab Res Rev,* **20** (Suppl 2), S50-S55 (2004).
9. E. R. Kenneth, and K. J. Ernest, Issues and implications in the slection of blood glucose monitoring techniques, *Diabetes Technol. Ther.*, **1**, 3-11 (1999).
10. J. P. Lorand, and J. O. Edwards, Polyol Complexes and Structures of the Benzeneboronate Ion., *J. Org. Chem.*, 769-774 (1959).
11. G. Springsteen, and B. Wang, A Detailed Examination of Boronic Acid-Diol Complexation, *Tetrahedron,* **58**, 5291-5300 (2002).
12. J. Yan, G. Springsteen, S. Deeter, and B. Wang, The relationship among pKa, pH, and binding constants in the ineteractions between boronic acids and diols-it is not as simple as it appears, *Tetrahedron,* **60**, 11205-11209 (2004).

13. T. D. James, K. R. A. Sandanayake, R. Iguchi, and S. Shinkai, Novel Saccharide-Photoinduced Electron Transfer Sensors Based on the Interaction of Boronic Acid and Amine, *J. Am.Chem. Soc*, **117**, 8982-8987 (1995).
14. J. Yoon, and A. W. Czarnik, Fluorescent Chemosensors of Carbohydrates. A Means of Chemically Communicating the Binding of Polyols in Water Based on Chelation-Enhanced Quenching, *J. Am. Chem. Soc.*, **114**, 5874-5875 (1992).
15. W. Wang, X. Gao, and B. Wang, Boronic acid-based sensors, *Curr. Org. Chem.*, **6**, 1285-1317 (2002).
16. S. Shinkai, and M. Takeuchi, Molecular design of synthetic receptors with dynamic, (2004).
17. H. Cao, and M. D. Heagy, Fluorescent chemosensors for carbohydrates: a decade's worth of bright spies for saccharides in review, *Journal of Fluorescence*, **14**, 569-584 (2004).
18. S. Striegler, Selective carbohydrates recognition by synthetic receptors in aqueous solution, *Curr. Org. Chem.*, **7**, 81-102 (2003).
19. S. Wiskur, H. Haddou, J. Lavigne, and E. Anslyn, Teaching old indicators new tricks, *Acc. Chem. Res*, **34**, 963-972 (2001).
20. E. Shoji, and M. Freund, Potentiometric sensors based on the inductive effect on the pKa of poly(aniline): A nonenzymatic glucose sensor, *J. Am. Chem. Soc*, **123**, 3383-3384 (2001).
21. E. Shoji, and M. Freund, Potentiometric saccharide detection based on the pKa changes of poly(aniline boronic acid), *J. Am. Chem. Soc*, **124**, 12486-12493 (2002).
22. G. Springsteen, and B. Wang, Alizarin Red S. as a general optical reporter for studying the binding of boronic acids with carbohydrates., *Chem. Commun*, 1608-1609 (2001).
23. T. D. James, and S. Shinkai, Artificial Receptors as Chemosensors for Carbohydrates, *Topics in Current Chemistry*, **218**, 159-200 (2002).
24. V. Karnati, X. Gao, S. Gao, W. Yang, S. Sabapathy, W. Ni, and B. Wang, A Selective Fluorescent Sensor for Glucose, *Bioorg.Med. Chem. Lett.*, **12**, 3373- 3377 (2002).
25. T. D. James, K. R. A. S. Sandanayake, and S. Shinkai, Novel Photoinduced Electron-transfer Sensor for Saccharides Based on the Interaction of Boronic Acid and Amine, *Chem. Commun.*, 477-478 (1994).
26. G. Wulff, Selective Binding to Polymers via Covalent Bonds - The Construction of Chiral Cavities as Specific Receptor-Sites, *Pure. Appl. Chem*, **54**, 2093-2102 (1982).
27. W. Yang, J. Yan, H. Fang, and B. Wang, The first fluorescent sensor for D-glucarate based on the cooperative action of boronic acid and guanidinium groups., *Chem.Commun.*, 792-793 (2003).
28. S. Franzen, W. Ni, and B. Wang, Study of the Mechanism of Electron-Transfer Quenching by Boron-Nitrogen Adducts in Fluorescent Sensors, *J. Phys. Chem. B*, **107**, 12942-12948 (2003).
29. W. Ni, G. Kaur, G. Springsteen, B. Wang, and S. Franzen, Regulating the Fluorescence Intensity of an Anthracene Boronic Acid System: A B-N Bond or a Hydrolysis Mechanism?, *Bioorgan. Chem.*, **32**, 571-581 (2004).
30. J. C. Norrid, and H. Eggert, Evidence for Mono- and Bisdentate Boronate Complexes of Glucose in the Furanose Form. Application of $^{1}J_{C\text{-}C}$ Coupling Constants as a Structural Probe, *J. Am. Chem. Soc.*, **117**, 1479-1484 (1995).
31. M. Bielecki, H. Eggert, and J. C. Norrid, A fluorescent glucose sensor binding covalently to all five hydroxy groups of D-glucofurnaose. A reinvestigation., *J. Chem. Soc., Perkin Trans 2,*, 449-455 (1999).
32. F. P. Worley, and J. C. Andrews, Mutarotation. I. Velocity of Mutarotation of alpha Glucose in Methyl Alcohol and Water., *J. Phys. Chem.*, **31**, 742-746 (1927).
33. B. Appleton, and T. D. Gibson, Detection of toal sugar concentration using photinduced electron transfer materials: Development of operationally stable, reusable optical sensors, *Sens.Actuator.B Chem*, **65**, 302-304 (2000).
34. S. Arimori, L. I. Bosch, C. J. Ward, and T. D. James, Fluorescent Internal Charge Transfer (ICT) Saccharide Sensor, *Tetrahedron Lett.*, **42**, 4553-4555 (2001).
35. K. Yasumasa, and T. Hiroaki, Selective Glucose Sensing Utilizing Complexation with Fluorescent Boronic Acid on Polycation, *Chem Letters*, **34**, 196 (2005).
36. S. Arimori, M. L. Bell, C. S. Oh, K. A. Frimat, and T. D. James, Modular Fluorescence Sensors for Saccharides, *J. Chem. Soc., Perkin Trans. 1*, 803-808 (2002).
37. N. DiCesare, and J. R. Lackowicz, Charge transfer fluorenscent probes using boronic acid for monosaccharide signaling., *Journal of Biomedical Optics*, **7**, 538-545 (2002).

38. N. DiCesare, and J. R. Lackowicz, Spectral properties of Fluorophores Combining the Boronic Acid Group with Electron Donor or Withdrawing Groups. Implication in the Development of Fluorescence Probes for Saccharides, *J.Phys.Chem.A,* **105**, 6834-6840 (2002).
39. N. DiCesare, and J. R. Lakowicz, Wavelength-ratiometric probes for saccharides based on donor-acceptor diphenylpolyenes, *J. Photochem. Photobiol., A,* **143**, 39-47 (2001).
40. N. DiCesare, and J. R. Lakowicz, A new highly fluorescent probe formonosaccharides based on a donor-acceptor diphenyloxazole, *Chem. Commun.,* **19**, 2022-2023 , (2001).
41. N. DiCesare, and J. R. Lakowicz, Chalcone-analogue fluorescent probes for saccharides signaling using the boronic acid group, *Tetrahedron Lett,* **43**, 2615-2618 (2002).
42. A. Durkop, *Diploma Dissertation,* University of Regensburg, (1998).
43. G. Chen, Z. Guan, C. T. Chen, L. Fu, V. Sundareson, and F. H. Arnold, *Nature Biotechnol,* **15**, 354 (1997).
44. E. Pringhsheim, E. Terpetschnig, S. Piletsky, and O. Wolfbeis, A polyaniline with near-infrared optical response to saccharides, *Adv. Mater.,* **11**, 865-868 (1999).
45. H. Cao, D. I. Diaz, D. DiCesare, J. R. Lakowicz, and M. D. Heagy, Monoboronic Acid Sensor That Displays Anomalous Fluorescence Sensitivity to Glucose, *Org. Lett.,* **4**, 1503 -1505 (2002).
46. H. Cao, T. McGill, and M. D. Heagy, Substituent Effects on Monoboronic Acid Sensors for Saccharides Based on N-Phenyl- 1,8-napthalenedicarboximides, *J. Org. Chem.,* **69**, 2959-2966 (2004).
47. H. Eggert, J. Frederiksen, and J.C.Norrild, A New Glucose-selective Fluorescent Bisboronic Acid. First Report of Strong a-Furanose Complexation in Aqueous Solution at Physiological pH, *J. Org. Chem.,* **64**, 3846-3852 (1999).
48. W. Yang, G. Springsteen, J. Yan, S. Deeter, and B. Wang, A Novel Type of Fluorescent Boronic Acid that Shows Large Fluorescence Intensity Changes upon Binding with a Diol in Aqueous Solution at Physiological pH, *Bioorg. Med. Chem. Lett.,* **13**, 1019 - 1022 (2003).
49. W. Yang, L. Lin, and B. Wang, A New Type of Water-Soluble Fluorescent Boronic Acid Suitable for Construction of Polyboronic Acids for Carbohydrate Recognition, *Heterocycl. Commun.,* Manuscript accepted (2004).
50. X. Gao, Y. Zhang, and B. Wang, New Boronic Acid Fluorescent Reporter Compounds II. A Naphthalene-based Sensor Functional at Physiological pH, *Org. Lett.,* **5**, 4615-4618 (2003).
51. X. Gao, Y. Zhang, and B. Wang, Naphthalene-based waer-soluble fluorescent boronic acid isomers suitable for ratiometric and off-on sensing of saccharides at physiological pH, *New.J.Chem.,* **29**, 1-8 (2005).
52. D. Michail, P. Vancea, and N. Zolog, Sur l'elimination lacrymale du glucose chez les diabetiques, *C. R. Soc. Biol.,* **125**, 1095 (1937).
53. R. Badugu, J. R. Lakowicz, and C. D. Geddes, Noninvasive Continuous Monitoring of Physiological Glucose Using a Monosaccharide-Sensing Cintact Lens, *Anal. Chem.,* **76**, 610-618 (2004).
54. R. Badugu, J. R. Lakowicz, and C. D. Geddes, Boronic Acid Fluorescent Sensor for Monosaccharide Signaling Based on the 6-methoxyquinolinium Heterocyclic Nucleus: Progress Toward Noninvasive and Continuous Glucose Monitoring, *Bioorganic & Medicinal Chemistry,* 113-119 (2005).
55. R. Badugu, J. R. Lakowicz, and C. D. Geddes, Fluorescence Sensors for Monosacchrides Based on the 6-methylquinolinium Nucleus and Boronic Acid Moiety: Potential Application to Ophthalmic Diagnostic, (2005).
56. J. N. Camara, J. T. Suri, F. E. Cappuccio, R. A. Wessling, and B. Singaram, Boronic Acid Substituted Viologen Based Optival Sugar Sensors: Modulated Quenching with Viologen as a Method for Monosaccharide detection, *tetrahedron Letter,* **43**, 1139-1141 (2002).
57. S. kabilan, J. Blyth, M. C. Lee, A. J. Marshall, A. Hussain, X. P. Yang, and C. R. Lowe, Glucose-sensitive holographic sensors, *Jouranl of Molecular recognition,* **17**, 162-166 (2004).
58. S. Kabilan, A. J. Marshall, F. K. Sartain, M. C. Lee, A. Hussain, X. P. Yang, J. Blyth, N. Karangu, K. James, J. Zen, D. Smith, A. Domschke, and C. R. Lowe, Holographic glucose sensors, *Biosensors and Bioelectronics,* **20**, 1602-1610 (2005).

59. V. L. Alexeev, A. C. Sharma, A. V. Goponeko, S. Das, I. K. Lednev, C. S. wilcox, D. N. Finegold, and S. A. Asher, High Ionic Strength Glucose-Sensing Photonic Crystal, *Anal. Chem.*, **75**, 2316-2323 (2003).
60. B. A. Bunin, *The Combinatorial Index*. (Academic press;San Diego, 1998),
61. N. K. Terrett, *Combinatorial Chemistry*. (Oxford University Press;New York, 1998),
62. D. Hall, Combinatorial chemistry - a powerful approach to supermolecule discovery, *Canadian Chemical News*, **49**, 23-25 (1997).
63. H. M. Geysen, F. Schoenen, D. S. Wagner, and R. Wagner, Combinatotial Compound Libraries for Drug Discovery: An On-going Challenge, *Nat. Rev. Drug Discov.*, **2**, 222-230 (2003).
64. K. S. Lam, M. Lebl, and V. Krchnak, The "One-Bead-One-Compound" Combinatorial Library Method, *Chem. Rev.*, **97**, 411-448 (1997).
65. W. Yang, H. He, and D. G. Drueckhammer, Computer-Guided Design in Molecular Recognition : Design and Synthesis of a Glucopyranose Receptor., *Angew. Chem. Int. Ed.*, **40**, 174 (2001).

DEVELOPMENT OF SMART CONTACT LENSES FOR OPHTHALMIC GLUCOSE MONITORING

Ramachandram Badugu[1], Joseph R. Lakowicz[1,*], and Chris D. Geddes[1,2,*]

17.1. INTRODUCTION

Diabetes can lead to severe health complications over time, such as neuropathies, blindness, cancer, etc., [1,2] and it is one of the most concerned pathologies that cause patients death. The most effective way to manage diabetes is only possible by continuous, or at least, frequent blood glucose estimation, and subsequent drug and food administration. Consequently, the importance of diabetic control has resulted in worldwide efforts to develop non- or minimally-invasive methods for body glucose estimation [3-13]. Among these, the glucose monitoring using near infrared spectroscopy [3,4], optical rotation [5,6], colorimetric [7,8], and fluorescence detection [9-13] are notable. However, an enzyme based *"finger-pricking"* method is still the most commonly used technology in diabetic assessment. This method is relatively painful process and it does suffer from a few practical problems. The first one is inconvenience and the required compliance by patients, which is often difficult for both the young and old patients; while the second is that this is not a continuous monitoring method. Despite intensive efforts, no compatible device for the non-invasive and continuous monitoring of glucose estimation is available.

Body fluids such as tears track the blood glucose levels with time lag of about 30 mins. Although, the glucose concentration levels in the tears are not fully explored to date, it is about 10 times lower than in tears (of about 50-500 μM [14,15]) than that in blood and an elevation in the tear glucose concentration (up to 5 mM) during hyperglycemia has been demonstrated [16-18]. Giardini and Roberts have reported tear glucose concentrations to be ≈ 3

[1]-Center for Fluorescence Spectroscopy, Department of Biochemistry and Molecular Biology, [2]-Institute of Fluorescence, University of Maryland Biotechnology Institute, 725 W. Lombard St.Baltimore, MD, 21201. USA, * Corresponding authors, Lakowicz@cfs.umbi.umd.edu ; Geddes@umbi.umd.edu

mg/100 ml for subjects with a normal glucose metabolism [19]. Alternate to blood, if one has a capable device to monitor tear glucose, tears are easy and painless to use for glucose monitoring, and would offer a non-invasive and continuous monitoring technique. In this regard, we recently demonstrated an approach of using contact lens that is embedded with a monosaccharide sensing boronic acid probe, for the continuous and non-invasive notion of tear glucose monitoring [20-26]. The present approach is elegant and could be used efficiently for the monitoring of not only glucose but many other analytes of interest and could potentially become a platform for generic ophthalmic analyte monitoring. Herein this chapter we primarily focus our attention on the design and development of a glucose sensing contact lens. A schematic cartoon indicating the notion of the continuous glucose monitoring contact lens is shown in Figure 17.1. A glucose signaling boronic acid probe can be embedded in to the contact lens in various fashions. Here we have shown in the figure that one can modify the entire contact lens (which is most simplistic and easy for the experiments for the proof principle type, described in this chapter) with glucose sensing probe (Figure 17.1, top left) or probes embedded as sensor spots (Figure 17.1, bottom left). The later case provides an option of building a multi-analyte contact lens, i.e. each spot on the lens could be made to work specifically towards a different analyte of interest. Figure 17.1 also shows the possible diagram of the signal reading device. This can be a portable hand-held device having a suitable excitation source and detecting unit. However, our main interest is designing the glucose sensing contact lens, and the construction of a miniaturized electronic device is out of the scope of this review chapter.

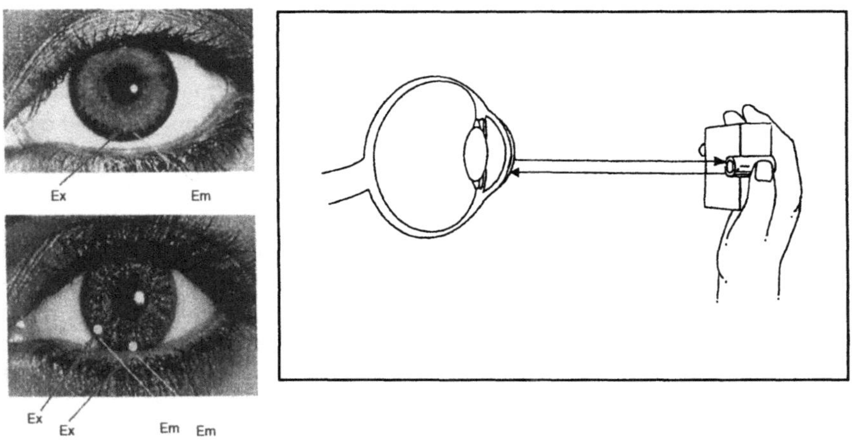

Figure 17.1. Potential methods for non-invasive continuous tear glucose monitoring. (Top left) BAF doped contact lens as described here, and (Bottom left) Sensor spots on the surface of the lens to additionally monitor other analytes in addition to glucose, such as drugs, biological markers, Ca^{2+}, K^+, Na^+, O_2 and Cl^-. Sensor regions may also allow for ratiometric, lifetime or polarization based fluorescence glucose sensing. (Right) Schematic representation of the possible tear glucose sensing device. Adopted from ref. 24.

17.1.1. Design Concept

As with any sensors, there are several issues that have to be addressed. The first is to identify suitable transduction elements, which in the presence of glucose, can report / produce suitable signals. The second is the design of the matrix to incorporate the transduction elements. For this, we have chosen an *off-the-shelf* disposable plastic contact lens, primarily because its physiological compatibility has already been assessed, and finally, the optimization of the sensor, with regard to sensitivity, response time, reversibility and shelf-life etc. The later two issues will be discussed throughout much of this chapter and indeed in past papers by the authors [20-26]. For the identification of suitable transduction elements, boronic acid has been known to have high affinity for diol-containing compounds such as carbohydrates [27-29], Scheme 1, where the strong complexation has been used for the construction of carbohydrate sensors [30-37], transporters [38] and chromatographic materials [39]. Boronic acid compounds have also been used for the synthesis of glucose sensors [40-45]. The reports from Shinkai [40,41], Norrild [31,32,], Lakowicz [42-45], groups on glucose sensing using boronic acid are most notable in this regard. As mentioned, we have been involved in the development of boronic acid based glucose sensors over last several years. However, very recently we have turned our attention toward continuous tear glucose monitoring using a modified contact lens. In this review chapter, we discuss how one can develop a suitable transduction element (which is a boronic acid fluorophore responsive to glucose within the contact lens) for continuous glucose monitoring, using a disposable, colorless and daily use contact lens and bornic acid based fluorophores.

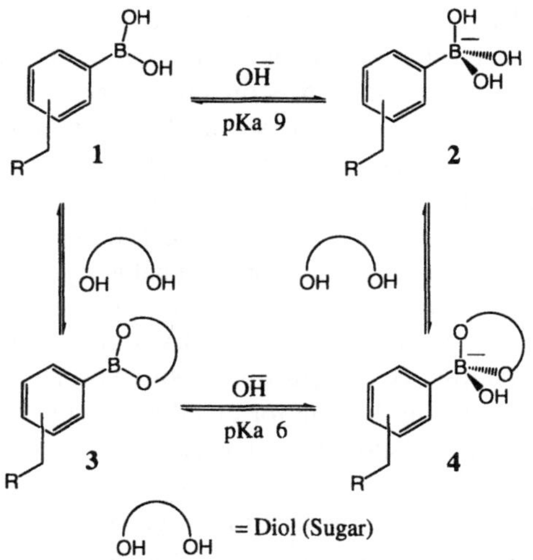

Scheme 17.1. Equilibrium between the various forms of the boronic acid in solution, and diol (sugar) interaction.

17.2. GLUCOSE SENSING USING BORONIC ACID PROBES IN SOLUTION

The potential applicability of the boronic acid probes to glucose sensing is evident from the vast amount of data published on this topic [20-45]. In this regard we are interested to provide an overview on the glucose sensing using boronic acid probes in solution, which may help the reader to understand the subsequent sections of this chapter. Boronic acid forms reversible covalent bonds with vicinal dihydroxy containing compounds such as saccharides, and during this, the electronic properties and geometry at the boron atom are altered (Scheme 1). Subsequently, the changes in electronic properties at the boron atom are being transferred to the appended fluorophore, leading to the spectral shift and/or intensity changes, depending on the transduction mechanism. The equilibrium involved with the boronic acid and diol interaction is illustrated in Scheme 1 and can be described as follows. Boronic acid is a weak Lewis acid composed of an electron deficient boron atom and two hydroxyl groups with trigonal planar geometry, **1** in Scheme 1, which can interact with strong bases like OH^- to from the anionic borate ester **2** (Scheme 1). Most boronic acids show typically high pK_a values of about 9 for this reversible reaction [40-45]. Boronic acid forms reversible covalent bonds with diols to form boronate diester **3**, Scheme 1 [29]. Also, the formed boronate diester interacts with OH^- ions to form **4**. However, in comparison to the boronic acid (**1**) the boronate diester (**3**) shows higher acidity due to a more electrophilic boron atom. Accordingly, boronate diester shows relatively reduced pK_a's (around 6). During the process of these reversible reactions, the boronic acid group undergoes electronic and structural changes as shown in the Scheme 1. In other words, boronic acid is initially an electron acceptor and after binding with the glucose it becomes an electron donor with in the same pH medium.

The affinity of the boronic acid is directly proportional to the number of available vicinal hydroxyl groups. Subsequently, the monophenylboronic acid shows higher affinity towards *D*-fructose over *D*-glucose because of stronger tridentate and weaker bidentate type binding with *D*-fructose and *D*-glucose respectively, with binding constants of ≈ 0.5 and 170 mM respectively [46]. The affinity of the boronic acid towards sugars is tunable by adjusting the geometry and substituents on the fluorophore moiety. A geometrically suitably placed diboronic acid containing fluorophore, ANDBA, shows higher affinity towards *D*-glucose over *D*-fructose. Hence glucose sensitive probes can be made with a variety of affinities, in the mM range for blood glucose [42-44], and in the μM range for tear glucose [47,48]. Also the pK_a of the boronic acid can be modulated according to the medium of interest. For example, as we can see in this section, the sugar response of most of the probes reported in the literature has been conducted in slightly elevated pH solutions such as pH 8. In contrast to this the new probes developed based on the quinolinium moiety show excellent sugar response in physiological pH solutions, also in slightly acidic media such as in a contact lens. This is due to the reduced pK_a of the new quinolinium probes. An increase and decrease in boronic acid pK_a can be observed with substituents with electron donating and accepting nature, respectively, on the phenyl ring. Here we have used strategic design logic by using an electron withdrawing quaternary nitrogen center within the interacting space of boronic acid to reduce the boronic acid pK_a. Also, in contrast to the other probes, because of the positively charged quinolinium moiety, these probes are readily water soluble and are therefore potentially useful for physiological applications. In essence we have several options to fine tune the glucose response of the probes within the medium of interest [47,48].

A wide range of transduction mechanisms have been employed in the design of boronic acid containing fluorophores including intramolecular charge transfer (ICT), photoinduced electron transfer mechanism (PET) and so on. The extent of fluorescence signal change during the sugar binding can be quantified to construct the calibration curves for accurate sugar detection. In the following section we have described the response of a few representative boronic acid probes towards monosaccharides in buffered solutions. The molecular structures of the probes are shown in Chart 1.

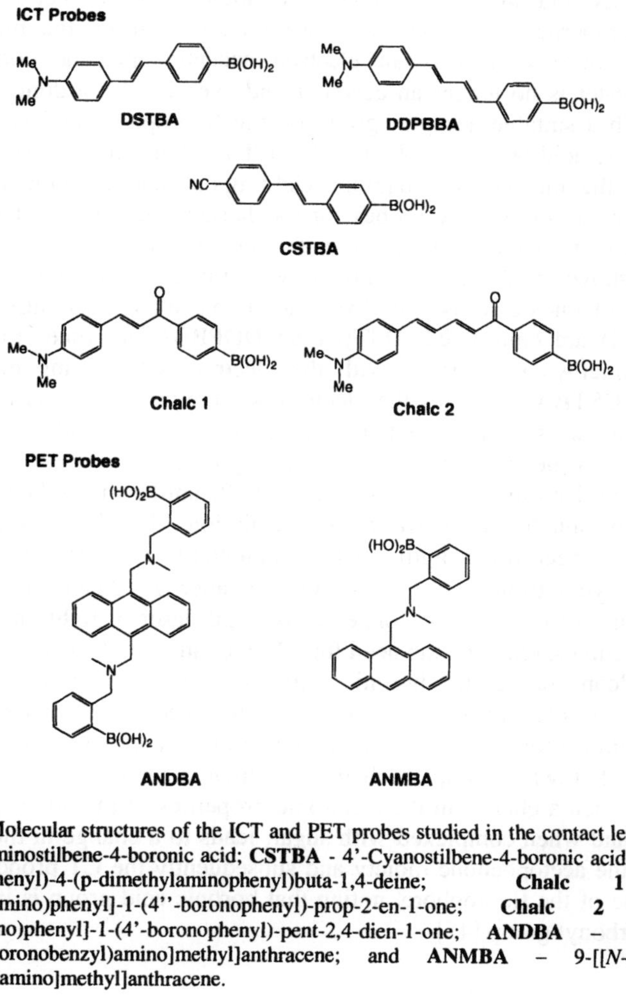

Chart 17.1. Molecular structures of the ICT and PET probes studied in the contact lens. **DSTBA** - 4'-Dimethylaminostilbene-4-boronic acid; **CSTBA** - 4'-Cyanostilbene-4-boronic acid; **DDPBBA** - 1-(*p*-Boronophenyl)-4-(p-dimethylaminophenyl)buta-1,4-deine; **Chalc 1** – 3-[4'(Dimethylamino)phenyl]-1-(4''-boronophenyl)-prop-2-en-1-one; **Chalc 2** – 5-[4'-(Dimethylamino)phenyl]-1-(4'-boronophenyl)-pent-2,4-dien-1-one; **ANDBA** – 9,10-bis-[[*N*-methyl-*N*-(*o*-boronobenzyl)amino]methyl]anthracene; and **ANMBA** – 9-[[*N*-methyl-*N*-(*o*-boronobenzyl)amino]methyl]anthracene.

17.2.1. Boronic Acid Probes with Intramolecular Charge Transfer Mechanism

Compounds with suitably placed electron donor and acceptor groups show intramolecular charge transfer (ICT) features in their absorption and emission spectra [49]. The extent of ICT efficiency in a system can be easily altered by changing the external parameters such as polarity of the medium; an increase in ICT efficiency is a common observation with increasing solvent polarity. An increase in ICT would result in a characteristic red shift in the spectrum. An opposite observation is noticed when the polarity of the medium is reduced. It is also commonly observed that most probes show reduced and increased fluorescence quantum efficiencies with an increase and decrease in ICT efficiency in a system, respectively.

In addition to the solvent polarity, modulations in the donating and accepting capabilities of the donor and acceptor units, respectively, result in spectral shifts and intensity changes. As mentioned in Section 17.2, the electronic properties of boronic acid were altered during the binding with saccharides; boronic acid is an electron acceptor and the corresponding boronate diester is no longer an acceptor and even acts as a donor when it is coupled with a suitable acceptor group on the fluorophore moiety [46]. As a result, boronic acid probes employing the ICT mechanism show spectral shifts along with the changes in quantum yield upon binding with saccharides, providing unique ratiometric probes for saccharide sensing [46]. Here we can demonstrate these features with a few examples. The molecular structures of the probes are shown in Chart 1. As shown, two stilbene derivatives (DSTBA and CSTBA) a polyene derivative DDPBBA and two chalcone derivatives (Chalc 1 and Chalc 2) are considered. DSTBA ad DDPBBA combines the electron-donating dimethylamino group with the electron withdrawing boronic acid group, and CSTBA combines the electron withdrawing cyano group with the boronic acid, in essence these two sets of probes show both decreased and increased ICT respectively, upon binding with glucose.

Chalcone derivatives, Chalc 1 and Chalc 2, unlike the stilbenes have the advantage of much longer wavelength emission [45]. This is particularly attractive as longer wavelength emission potentially reduces the detection of any lens or eye autofluorescence, as well as scatter (λ^{-4} dependence), and also allows the use of cheaper and longer wavelength laser or light emitting diode excitation sources, reducing the need for UV excitation in the eye.

In chalcone probes the boronic acid group does not produce resonance forms with the electron donating amino group. The CT occurs between the dimethylamino group (electron donating group) and the carbonyl group (electron withdrawing group) (Scheme 2). Upon sugar binding to the boronic acid group, then a change in the electronic properties of the boron group, both when free and when complexed with sugar, leads to a change in the electronic density of the acetophenone moiety and subsequently the CT properties of the excited state of the fluorophore, noting that boronic acid group is in resonance with the carbonyl group [45].

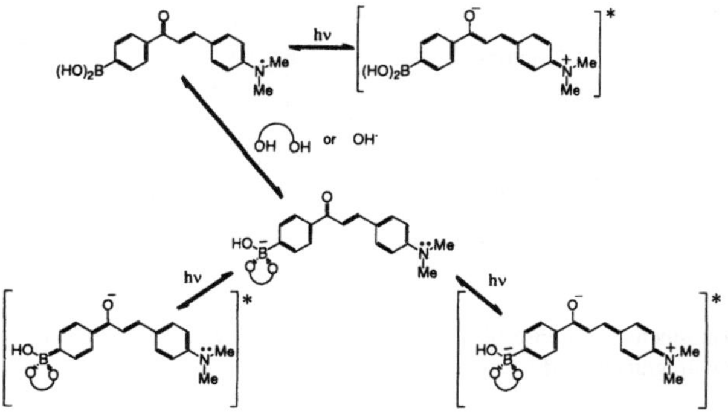

Scheme 17.2. Ground and excited state electronic distributions involved in the neutral and anionic forms of the boronic acid group of Chalc 1. For the case of interaction with OH^-, the diol should be replaced by two OH^- groups.

Figure 17.2 shows the effect of sugar on the emission properties of DSTBA in pH 8.0 buffer-methanol (2:1, v/v) solution. The emission spectrum shows a hypsochromic shift of about 30 nm and an increase in fluorescence intensity as the concentration of fructose increases (Figure 17.2 – left). These dramatic and useful changes can simply be explained by the loss of the electron withdrawing property of the boronic acid group following the formation of the anionic form as shown in Scheme 1. A similar response has been observed with the polyene derivative DDPBBA (Figure 17.2 – middle) [46].

The other stilbene derivative CSTBA possesses two electron withdrawing groups, boronic acid and cyano groups, in the absence of sugar. In the presence of sugar as shown in Figure 17.2 – right, we can observe a bathochromic shift of about 25 nm, and a decrease in the intensity at pH 8, which is opposite to that observed for DSTBA. This change has been attributed to an excited CT state present for the anionic form of CSTBA, where no CT state has been observed for the neutral form of the boronic acid group [46], suggesting that the anionic form of the boronic acid group can act as an electron donor group. Similar to that of DSTBA and DDPBBA, Chalc 1 and Chalc 2 show an excellent response to sugar and pH, resulting from the reduced ICT in the systems upon sugar binding [46].

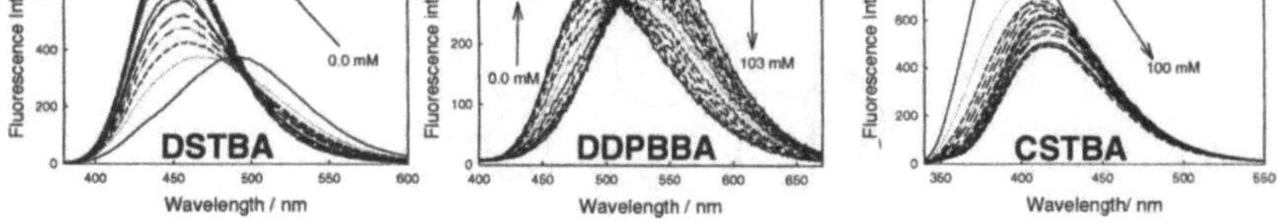

Figure 17.2. Emission spectra of DSTBA (left), DDPBBA (middle) and CSTBA (right) in pH 8.0 buffer/methanol (2:1) with increasing concentrations of fructose, λ_{ex} for DSTBA and DDPBBA is 340 and for CSTBA, 320 nm.

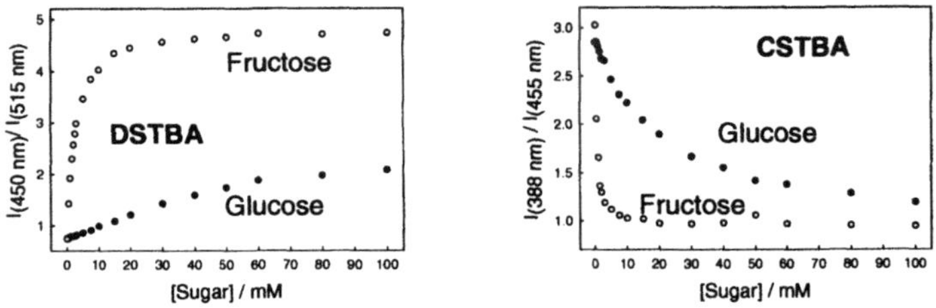

Figure 17.3. Ratiometric response plot for DSTBA (Left) and CSTBA (Right) in pH 8.0 buffer / methanol (2:1) with glucose and fructose.

A few representative titration curves for DSTBA and CSTBA are shown in Figure 17.3, the dissociation constants for DSTBA, DDPBBA, CSTBA, Chalk 1 and Chalc 2, are shown in Table 17.1. As previously mentioned, monoboronic acid derivatives show higher affinities for *D*-fructose, and the affinity decreases for *D*-glucose. As for the data provided in this section, one quickly realizes that the probes based on the ICT mechanism are potential candidates for physiological glucose monitoring.

Table 17.1. Dissociation constants (K_D, mM) of the boronic acid probes for glucose and fructose.

	DSTBA	DDPBBA	CSTBA	Chalc1	Chalc2	ANDBA	ANMBA
Glucose	98	17	18	34	30	0.51	21.3
Fructose	2.5	1.1	0.65	2.5	2.1	---	---

[a]The dissociation constant (K_D) values of DSTBA, DDPBBA and CSTBA were measured at pH 8.0 for that of Chalc 1 and Chalc 2 were obtained in pH 6.5 buffer.

17.2.2. Boronic Acid Probes having Photoinduced Intramolecular Electron Transfer Mechanism, (PET).

The anthracenediboronic acid derivative, ANDBA, is the one of the well-known boronic acid probes specific for glucose detection, developed by Shinkai *et al.* in 1995 [30]. The corresponding monophenylboronic acid probe ANMBA (Chart 1), is nonspecific for any saccharide, and shows high affinity towards fructose, as is the case with any monophenyl boronic acid based probe. The glucose specificity for ANDBA is due to the geometrically suitably appended two boronic acid moieties: the two boronic acid groups bind either side of the plane of the glucose ring as shown schematically in Figure 17.4 [30]. This novel approach of designing glucose specific probes has triggered vast amounts of interest in the research community over the last 10 years.

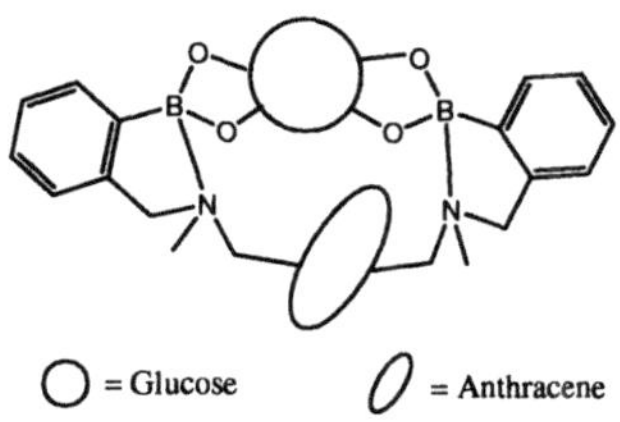

Figure 17.4. A Schematic representation of glucose binding with the diboronic acid derivative, ANDBA.

The fluorescence spectra of ANDBA and ANMBA in pH 8.0 buffer-methanol (2:1, v/v) with glucose are shown in Figure 17.5 [42]. The fluorescence intensity of these probes increases with increasing concentrations of the monosaccharides. This is because; originally in these systems a photoinduced electron transfer from the donor amine to the acceptor anthracene results in fluorescence quenching. Subsequent binding with monosaccharides, boronic acid shows increased acidity, which interacts more firmly with the nitrogen lone pair of electrons than before, thus reducing the PET interaction in the system. The suppressed PET interaction results in the emission enhancement of the probe. Subsequently, the probes show fluorescence enhancement in the presence of glucose. In contrast, the monophenylboronic acid derivative ANMBA shows a relatively weaker affinity towards glucose over the diboronic acid derivative ANDBA. The measured dissociation constants (K_D) are 21.3 and 0.51 mM for ANMBA and ANDBA, respectively, with glucose (Table 17.1) [42].

In addition to the fluorescence intensity enhancement, these probes show considerable increase in their fluorescence lifetimes with increasing sugar concentration [42]. Figure 17.6 shows frequency domain decay profiles of ANMBA and ANDBA in pH 8.0 buffer-methanol (2:1, v/v) with and without glucose. Similar to the wavelength ratiometric sensing of analytes, lifetime based sensing has intrinsic advantages for biomedical fluorescence sensing and is thus considered to provide for a more reliable analytical method [49]. As seen from Figure 17.6, these probes show increases in lifetime with increasing concentrations of glucose. A fluorescence lifetime change from 9.8 to 12.4 and 5.7 to 11.8 ns for ANMBA and ANDBA respectively has been observed. Although these probes are insoluble in water or indeed in physiological fluids, based on the results from steady-state and lifetime measurements, they serve to demonstrate that these probes, especially ANDBA, are potential PET based boronic acid probes for physiological glucose sensing applications [42].

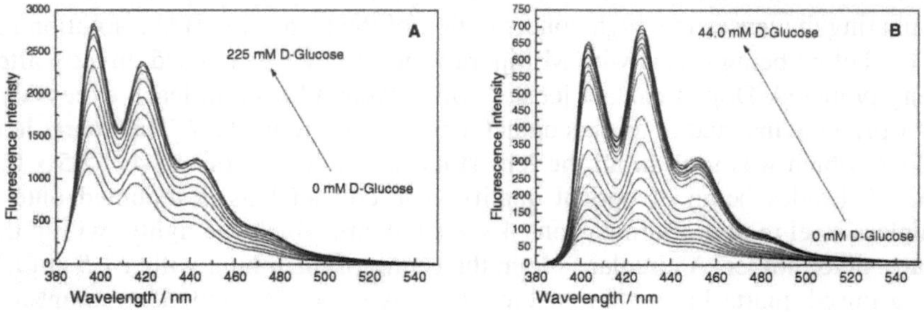

Figure 17.5. Fluorescence spectral changes of ANMBA (A) and ANDBA (B) in pH 8.0 buffer / methanol (2:1) with increasing concentrations of glucose. λ_{ex} = 365 and 380 nm for ANMBDA and ANDBA, respectively.

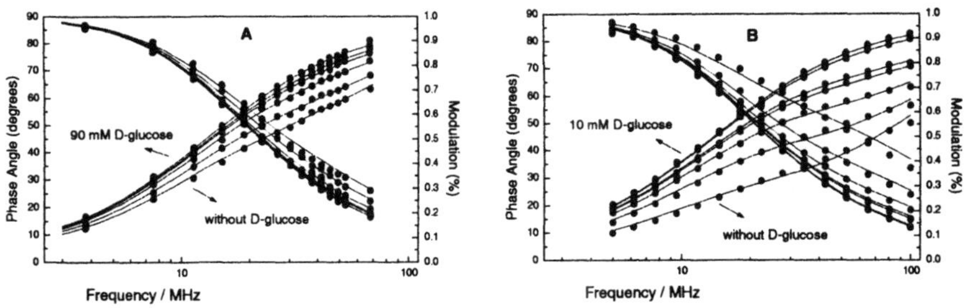

Figure 17.6. Effect of glucose on the frequency-domain decay profiles for ANMBA (A) and ANDBA (B) in pH 8.0 buffer / methanol (2:1) mixture. The lines represented the global fits with two lifetimes.

17.3. LENS FEASIBITY STUDY

As briefly mentioned in the introduction, the continuous monitoring of glucose may be possible using a contact lens embedded with a suitable glucose sensing probe, whose glucose sensing response is retained within the contact lens environment. In this regard we are interested to see the applicability of the probes discussed in Section 17.2 towards glucose monitoring within a disposable, off-the-shelf, contact lens.

17.3.1. Lens Doping and Contact Lens Holder

The details of the probe doping in to the contact lens has been described previously [20-26]. Briefly, the contact lenses were washed several times with Millipore water at 20°C to remove the salts and other preservatives from the contact lens. The contact lens is a polyvinyl alcohol type photo-cured polymer which swells slightly in water. Its hydrophilic character readily allows for the diffusion of the analytes in tears. The probe doping was conducted by incubating the lenses in a high concentration of the respective BAFs solution for 24 hrs before being rinsed with Millipore water. Lenses were used directly after being prepared. Doped contact lenses, which were allowed to leach excess dye for 1 hr, were inserted in to the contact lens holder, Figure 17.7. The quartz lens holder, which was used in all the lens studies, has dimensions of 4 x 2.5 x 0.8 cm, all 4 sides being of optical quality. The contact lens is mounted onto a stainless steel mount of dimensions 4 x 2 x 0.4 cm, which fits tightly within the quartz outer holder. A circular hole in the center of the mount with a 1.5 cm ID, has a raised quartz lip, which enables the lens to be mounted. The mount and holder readily allow for ≈ 1.5 cm^3 of solution to be in contact with the front and back sides of the lens for the sugar sensing experiments. Buffered solutions of sugars were then added to the lens. Fluorescence spectra were typically taken 15 mins after each sugar addition to allow the lens to reach equilibrium. Excitation and emission was performed using a Varian fluorometer, where the geometry shown in Figure 17.7 - right, was employed to reduce any scattering of the excitation light, the concave edge of the lens facing towards the excitation

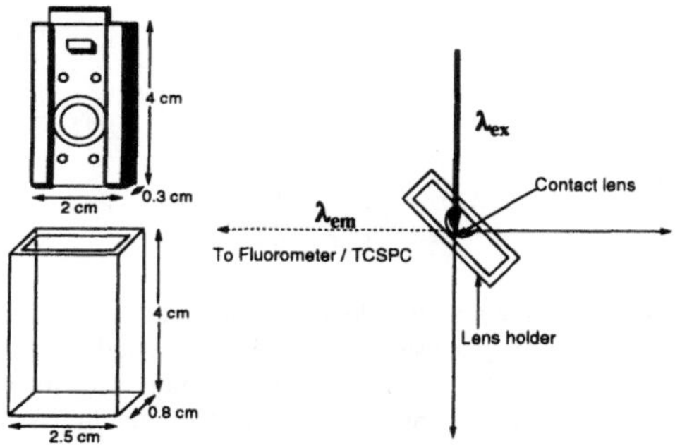

Figure 17.7. Contact lens mount and quartz holder (Left top and Bottom, respectively) and experimental geometry used for contact lens glucose sensing (Right). Adopted from ref. 24.

source. We additionally tested the lens excited from the convex edge, just as would be used in the eye, and encouragingly found identical results.

17.3.2. Response of ICT Probes within the Contact Lens

17.3.2.1. Stilbene Derivatives

Figure 17.8 shows the emission spectra and titration curves for a DSTBA doped contact lens towards both glucose and fructose. As expected the magnitude of the response towards fructose is greater, reflecting the higher affinity of monoboronic acids for fructose [46]. Comparing the response of DSTBA in both solution and lens (Figure 17.3 left and Figure 17.8 right), we can see that an opposite response is observed in the lens, where the emission spectra similarly shows a blue shift, accompanied by a decrease in intensity as the fructose concentration is increased. In addition, the sugar affinity is decreased slightly in the lens [20,26].

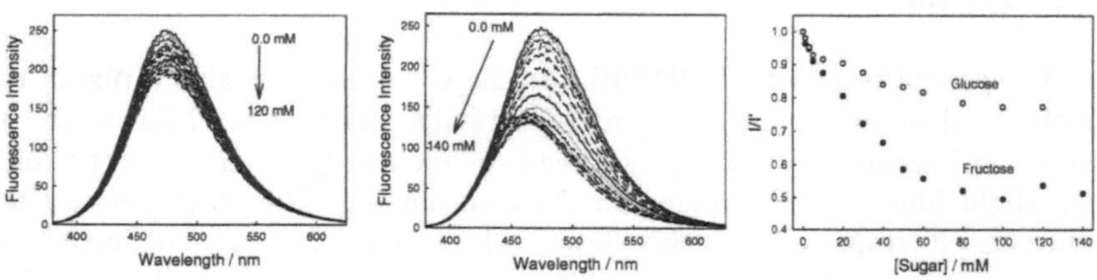

Figure 17.8. Emission spectra of the DSTBA doped contact lens, pH 8.0 buffer / methanol (2:1), with increasing concentrations of glucose (left) and corresponding spectra with increasing concentrations of fructose (middle). λ_{ex} = 340 nm. Intensity ratio plot for the DSTBA doped contact lens towards both glucose and fructose (right), where *I* and *I'* are the intensities in the presence and absence of sugar respectively at λ_{em} max.

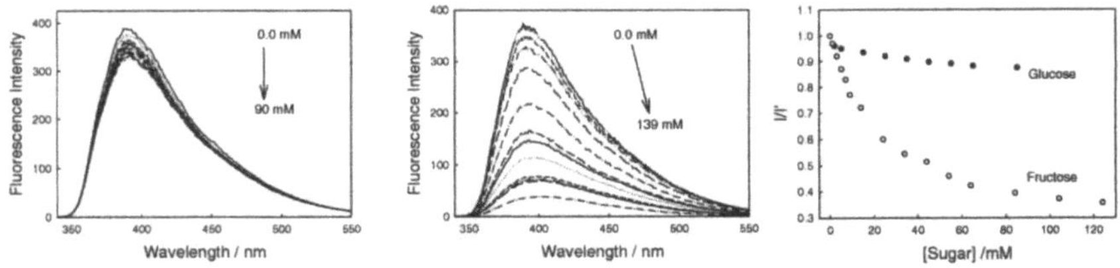

Figure 17.9. Emission spectra of the CSTBA doped contact lens, pH 8.0 buffer / methanol (2:1), with increasing concentrations of glucose (left) and corresponding spectra with increasing concentrations of fructose (middle). λ_{ex} = 320 nm. Intensity ratio plot for CSTBA doped contact lens towards both glucose and fructose (right), where I and I' are the intensities in the presence and absence of sugar respectively at λ_{em} max.

Similarly Figure 17.9 shows the response of CSTBA in the lens for both glucose and fructose. While a similar reduction in intensity is observed as compared to solution, no red shift in the emission is observed, indicative of a reduction in the electron donating capability of the anionic sugar bound form [20,26].

The lack of suitable spectral shifts in the presence of sugar eliminates, at this stage, the possibility of wavelength ratiometric sensing as shown for the solution based measurements in Figure 17.3. Subsequently, Figure 17.8-right and Figure 17.9-right compares the responses of the stilbene probes, DSTBA and CSTBA, based on a simple intensity ratio measurement. It is interesting to see the much greater response with fructose for CSTBA in the lens as compared DSTBA, where notable changes in intensity occur at < 20 mM [fructose]. However the glucose response of DSTBA in the contact lens appears more promising for [glucose] < 10 mM, where a 10 % fluorescence intensity change is observed for ≈ 10 mM glucose at pH 8.0. The greater response of CSTBA within the contact lens may arise from the relatively lower pK_a value for CSTBA over the DSTBA probe. The apparent pK_a values for CSTBA and DSTBA are 8.17 and 9.14, respectively [46].

17.3.2.2 Polyene Derivative

The spectral response of DDPBBA in the contact lens is also different to that observed in solution, c.f. Figure 17.2-middle and Figure 17.10, where a decrease in intensity is typically observed for increasing sugar concentration, and a slight blue shift is evident for fructose binding. This is in contrast to solution-based responses which show both a blue shifted and increased emission, Figure 17.2-middle. While the general spectral changes observed for both DSTBA (Figure 17.8) and DDPBBA (Figure 17.10), are similar, a greater dynamic response to sugar is observed for DSTBA as compared to DDPBBA, c.f. Figure 17.8 – right and Figure 17.10 – right. In addition, the response of DDPBBA towards both glucose and fructose are similar over the sugar concentration range studied, Figure 17.10 – right, as compared to the significantly different responses observed for both sugars for DSTBA and CSTBA, Figure 17.8- right and Figure 17.9-right [20].

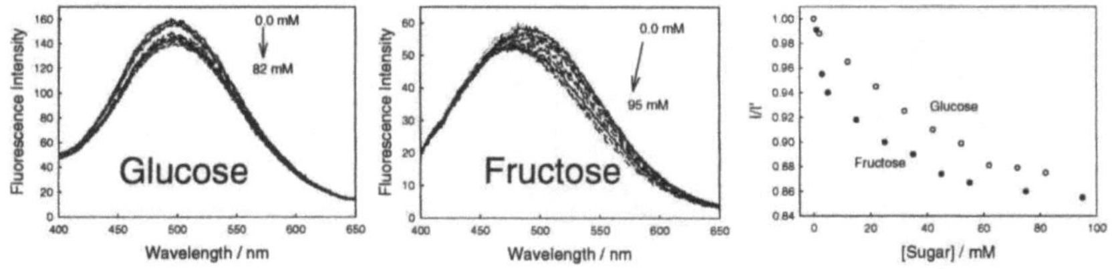

Figure 17.10. Emission spectra of the DDPBBA doped contact lens, pH 8.0 buffer / methanol (2:1), with increasing concentrations of glucose (left) and corresponding spectra with increasing concentrations of fructose (middle). λ_{ex} = 320 nm. Intensity ratio plot for DDPBBA doped contact lens towards both glucose and fructose (right), where *I* and *I'* are the intensities in the presence and absence of sugar respectively at λ_{em} max.

17.3.2.3 Chalcone Derivatives

The response of Chalc 1 and Chalc 2 doped contact lenses towards sugar are shown in Figures 17.11 and 17.12 respectively. Both chalcone doped lenses display similar responses to sugar, Figures 17.11 and 17.12 – right respectively, only their respective emission wavelengths differ. Chalc 1 shows an emission centered around 560 nm in the lens as compared to 580 nm in solution (not shown), while Chalc 2 shows an emission centered at 630 nm as compared to 665 nm in solution [20]. In contrast to the responses observed in solution, a reduction in fluorescence intensity is observed for both Chalc 1 and 2 doped contact lenses. Interestingly, the solution response for Chalc 2 towards 100 mM fructose at pH 8.0 produces a ≈ 3 fold increase in fluorescence emission, as compared to the ≈ 2.6 fold reduction for the same fructose concentration in the contact lens [20].

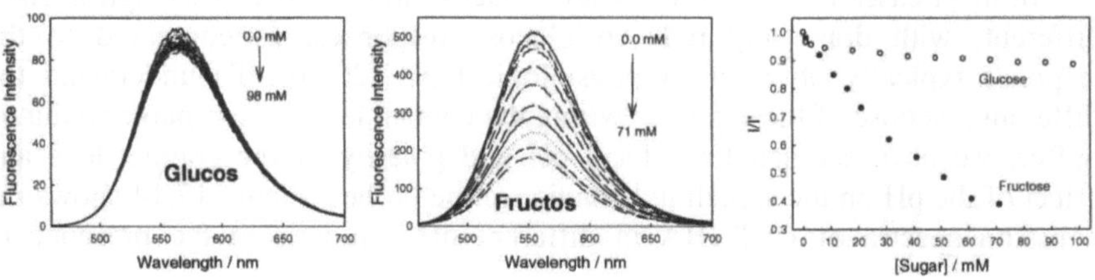

Figure 17.11. Emission spectra of Chalc 1 doped contact lens, pH 8.0 buffer / methanol (2:1), with increasing concentrations of glucose (Top left) and corresponding spectra with increasing concentrations of fructose (Top right). λ_{ex} = 430 nm. Intensity ratio plot for Chalc 1 doped contact lens towards both glucose and fructose (bottom), where *I* and *I'* are the intensities in the presence and absence of sugar respectively at λ_{em} max.

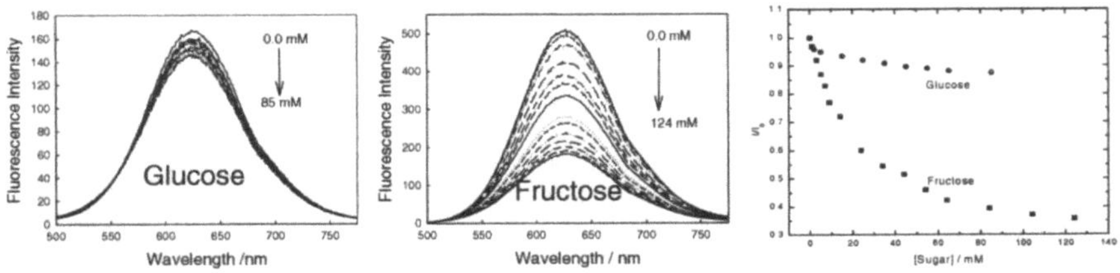

Figure 17.12. Emission spectra of Chalc 2 doped contact lens, pH 8.0 buffer / methanol (2:1), with increasing concentrations of glucose (Top left) and corresponding spectra with increasing concentrations of fructose (Top right). λ_{ex} = 460 nm. Intensity ratio plot for Chalc 2 doped contact lens towards both glucose and fructose (Bottom), where *I* and *I'* are the intensities in the presence and absence of sugar respectively at λ_{em} max.

17.3.3. Response of the PET Probes within the Contact Lens

The emission spectra of ANDBA in the lens with glucose are presented in Figure 17.13 [26]. The response of the probe towards glucose and fructose in the lens is compared in figure 17.13 -right. It is evident from the figure that ANDBA in acontact lens has lost its ability to respond to glucose, an insignificant spectral change is observed within the contact lens. Similar to ANDBA, the monophenylboronic acid derivative ANMBA shows non responsive spectral behavior towards glucose and fructose within the contact lens. Still it is of interest to note here that the diboronic acid derivative ANDBA still shows the glucose selectivity over the fructose, Figure 17.13-right.

17.3.4. Contact Lens Internal Properties and Effect on the Sensing Ability of the Probes

In most cases the response towards glucose in the lenses was significantly different, with drastically reduced glucose responses, as compared to the response typically observed at physiological pH [20,26]. To understand the different response of these probes within the contact lens, as compared to that in buffer, we have estimated the local pH and polarity of the contact lens and effect of the pH on the signaling behavior of the probes. Figure 17.14 shows the emission spectra of DDPBBA in different pH solutions in the presence of

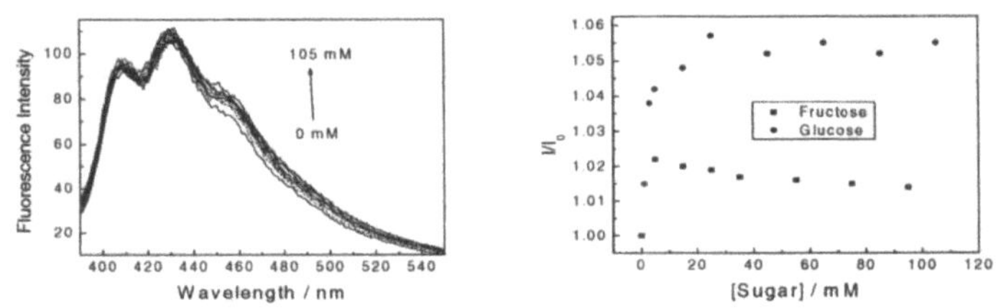

Figure 17.13. Emission spectra of ANDBA doped contact lens, pH 8.0 buffer / methanol (2:1), with increasing concentrations of glucose (Left). λ_{ex} = 365 nm. Intensity ratio plot for ANDBA doped contact lens towards both glucose and fructose (Right), where *I* and I_0 are the intensities in the presence and absence of sugar respectively at λ_{em} max.

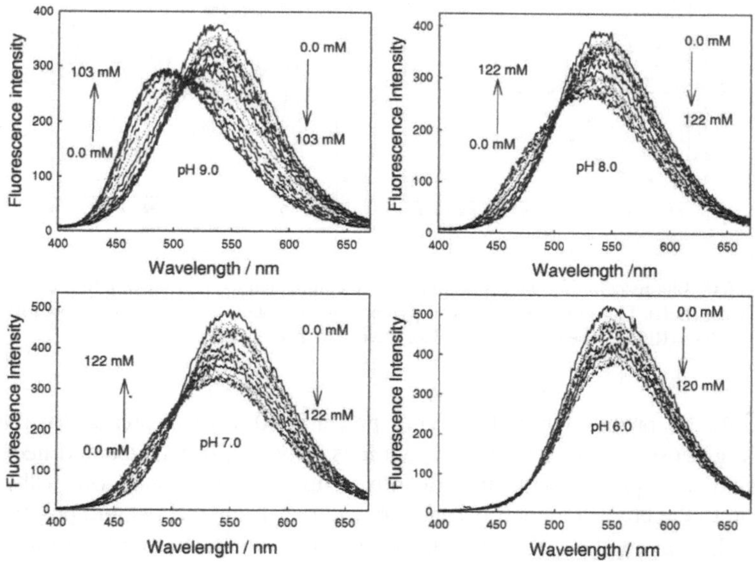

Figure 17.14. Emission spectra of DDPBBA in different pH media (buffer/methanol, 2:1) with increasing glucose concentrations. λ_{ex} = 340 nm.

increasing concentrations of glucose. As with all the other probes studied, we observed that the response towards glucose in the contact lens was similar to the response observed in ≈ pH 6.0 bulk solutions, c.f. Figure 17.10-left and 17.14 bottom right [20].

We subsequently doped the well-known pH sensitive probe fluorescein [49] within the lenses, and determined that the lenses had an unbufferable pH of ≈ 6.1, Figure 17.15-left. However, the influence of an acidic lens pH does not solely explain the spectral shifts observed in Figures 17.10 and 17.14, with DDPBBA additionally showing a 50 nm hyposchromic shift within the contact lens. An investigation of the polarity within the contact lens, using the pyrene I_1 and I_3 band ratio's (I_1/I_3) [20,50,51], Figure 17.15-right, revealed a lens polarity similar to that of methanol, which in hindsight was not too surprising, given the nature of the polymer, i.e. it is PVA based. Thus the difference in the polarity and pH may cause the difference in the response of the probes within the contact lens when compared to that in solution.

17.4. RATIONALE DESIGN OF NEW GLUCOSE SENSING PROBES

Feasibility studies of doped lenses using the above mentioned boronic acid probes produced poor glucose responses, rationaled as due to the mildly acidic pH and methanol like polarity within the contact lens, and the subsequent effect on the transduction mechanisms, ICT or PET [20]. This was also not surprising, given the fact that these probes were designed for sensing at a physiological pH

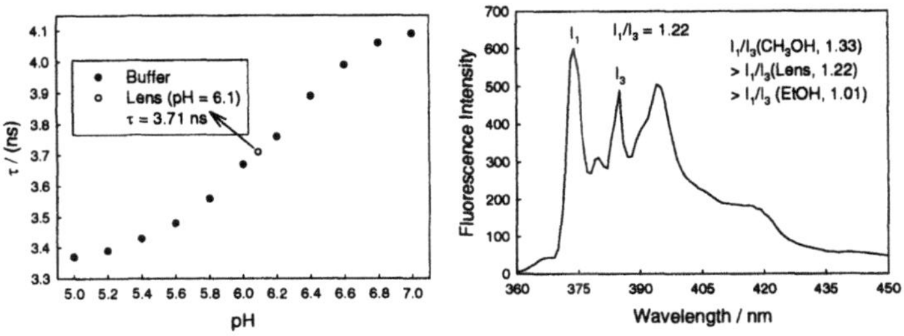

Figure 17.15. Fluorescein lifetime versus pH of the medium, and the lifetime of a fluorescein doped contact lens, (Left). Fluorescence spectra of pyrene doped contact lens to asses the polarity inside the contact lens (Right). The obtained I_1/I_3 data is close to that of methanol.

of ≈ 7.4, the probes typically having pK_a around 9 [46]. Hence to obtain a notable glucose response in the contact lens polymer, it was deemed necessary to design new probes with significantly reduced sugar-bound pK_a values. In addition to the environmental parameters and constraints of pH and polarity, the probes also have to be sensitive to the very low concentrations of tear glucose, ≈ 500 μM for a healthy person, increasing up to several mM for diabetics, recalling that the *blood glucose* levels for a healthy person are ≈ 10-fold higher [20-26].

As mentioned in section 17.2, the pK_a of phenyl boronic acid is known to be tunable with the appropriate substituents [52], for example, an electron withdrawing group reduces the pK_a of the sugar bound form, while an electron donating group increases it. This is clear from the apparent pK_a values of DSTBA and CSTBA (see section 17.3.2.1 and Table 17.2). We therefore considered the interaction between the electron accepting quaternary nitrogen of the 6-methyl- and methoxyquinolinium moieties and the boronic acid group, which reduces the pK_a of the probe [47,48]. In this regard we have synthesized 2 new classes of isomeric boronic acid containing probes, Chart 2, where the spacing between the interacting moieties, quaternary nitrogen of the 6-methyl- or methoxyquinolinium and boronic acid groups, enables both an understanding of the sensing mechanism to be realized, and the selection of the most suitable isomer based on its glucose binding affinity. In addition, control compounds (BMQ and BMOQ), which do not contain the boronic acid moiety, and are therefore insensitive towards sugar, were synthesized to understand the spectral properties and responses of the probes, Chart 2 [47,48].

17.5. GLUCOSE SENSING PROBES BASED ON THE QUINOLINIUM MOIETY

The synthesis and detailed spectral features of the new quinolinium based boronic acid containing probes (*o*-BMQBA–*N*-(2-boronobenzyl)-6-methylquinolinium bromide, *m*-BMQBA–*N*-(3-boronobenzyl)-6-methylquinolinium

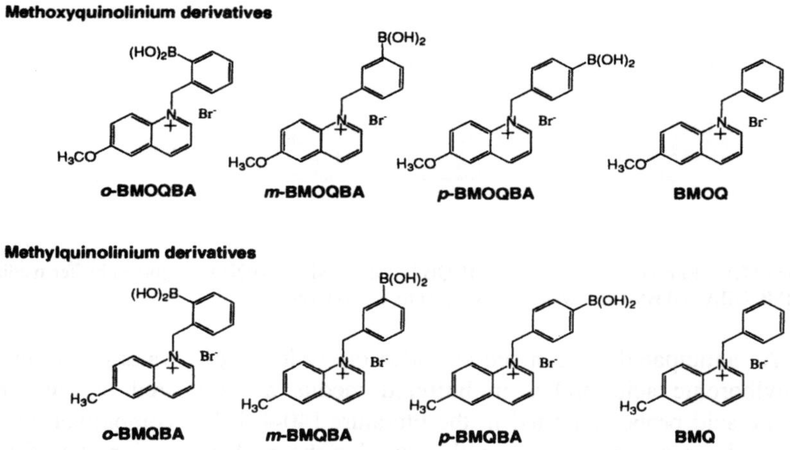

Chart 17.2. Molecular structure of the boronic acid probes based on quinolinium nucleus as the fluorophore. *o*-, *m*-, and *p*-BMOQBA – *N*-(2-, 3-, 4-boronobenzyl)-6-methoxylquinolinium bromide, and the control compound (BMOQ–*N*-benzyl-6-methoxyquinolinium bromide); *o*-, *m*-, and *p*-BMQBA – *N*-(2-, 3-, 4, -boronobenzyl)-6-methylquinolinium bromide, and the corresponding control compound (BMQ–*N*-benzyl-6-methylquinolinium bromide)

bromide, *p*-BMQBA–*N*-(4-boronobenzyl)-6-methylquinolinium bromide) and the control compound (BMQ–*N*-benzyl-6-methylquinolinium bromide), were described in recent reports [47,48]. These new probes consisting of a quaternized nitrogen center are readily water soluble, alleviating the need to use methanol to solubalize the probes as discussed in section 17.2. Also these probes have unique spectral properties, in particular very large Stoke-shift (about 100 nm), and are therefore most attractive to use in sensor applications.

17.5.1. Spectral Properties and Signaling Mechanism

Before we test the glucose sensing ability of the quinolinium probes we examined the pH response of these probes to estimate the boronic acid pK_a. Figure 17.17 shows the emission spectra of *o*-BMOQBA and *o*-BMQBA in buffer media whose pH is increased from pH 3 to 11. The emission spectra typically show a steady decrease in fluorescence intensity with increase in pH from 3 to 11. In contrast, the control compounds (BMOQ and BMQ) having no boronic acid group show no change in fluorescence intensity with varying the pH of the medium. Subsequently the apparent pK_a values of the boronic acid probes from the corresponding titration curves in the absence and in the presence of 100 mM glucose and fructose, obtained by plotting the normalized intensities at band maximum versus pH, are shown Table 17.2 [47,48].

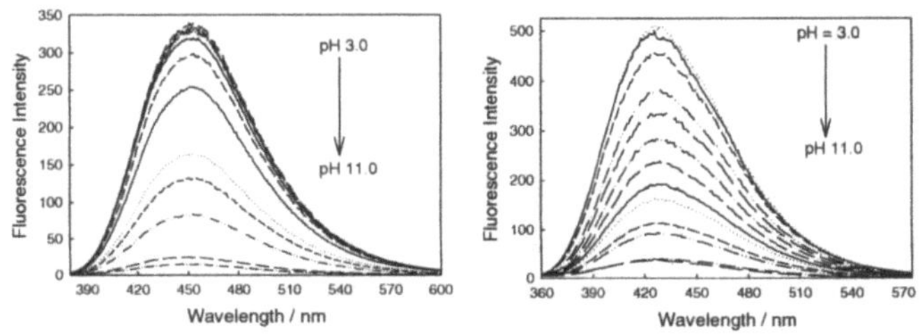

Figure 17.17. Emission spectra of *o*-BMOQBA (Left) and *o*-BMQBA (Right) in buffer media. λ_{ex} for BMOQBA and BMQBA was 345 and 320 nm, respectively.

As anticipated we can see considerably reduced pK_a values for the new phenylboronic acid probes in buffered media, as compared to the typical boronic acid probes reported in the literature [40-45]. For comparison the pK_a values of these probes along with those for the probes mentioned in Section 17.2 are depicted in Table 17.2. The interaction between the quaternary nitrogen center and the boronic acid group not only reduces the pK_a of the probes, but also serves to stabilize the boronatediester, formed upon sugar complexation. This in turn increase the affinity of the probes towards sugar. Hence the reduced sugar bound pK_a of these new probes, coupled with their increased glucose affinity, is most attractive for our glucose sensing contact lens application, noting our previous findings of a lens pH around 6.1 [20].

Table 17.2. Apparent pK_a values for boronic acid probes in buffer and the effect of 100 mM sugars.

Probe	In buffer	+100 mM Glucose	+100 mM Fructose
o-BMOQBA	7.90	6.62	4.80
m-BMOQBA	7.70	6.90	5.00
p-BMOQBA	7.90	6.90	5.45
o-BMQBA	6.70	6.10	5.00
m-BMQBA	7.75	6.85	5.05
p-BMQBA	7.80	6.95	5.45
DSTBA	9.14[a]	8.34	6.61
CSTBA	8.17	7.30	5.84
DDPBBA	8.90	6.97	6.20
Chalc 1	7.50	---	5.40
Chalc 2	7.50	---	5.20

[a]The pK_a values for the probes DSTBA, CSTBA, DDPBBA, Chalc 1 and Chalc 2 are from Ref. 46.

17.5.2. Signaling Mechanism in Quinolinium Probes

As mentioned in Section 17.2 and Scheme 1, the boronic acid group undergoes dramatic electronic and structural changes upon binding with a sugar. These dramatic changes in the boronic acid properties lead to the drastic changes in their absorption and fluorescence spectra. We can simply explain these effects on the spectral properties of the quinolinium probes in two ways. Firstly, phenyl group having a negatively charged boronate diester as in compound **4** of Scheme 1 is a relatively better donor over the corresponding phenylboronic acid group of **1** in Scheme 1, which substantially increases the photoinduced electron transfer (PET) efficiency in the system, leading to fluorescence quenching [53,54]. Secondly, it is well known that quinolinium derivatives exhibit higher fluorescence intensities over the parent quinolines. This is because, in the parent quinolines, the low-lying triplet n-π* state deactivates the excited fluorophores, considerably [55-57]. Upon quarternization, i.e. in quinolinium derivatives, such a deactivation pathway is removed. However partial neutralization of the quinolinium nitrogen by the electrostatic type interaction between the nitrogen and anionic boronate diester, formed upon sugar binding, (Figure 17.18) could result in fluorescence quenching. Based on the results obtained with several structurally similar systems, we have recently proposed an alternate mechanism called *"charge neutralization / stabilization"* mechanism that could also contribute some extent to the fluorescence quenching of these probes in the presence of glucose [47,48]. This mechanism can be explained schematically as shown in Figure 17.18.

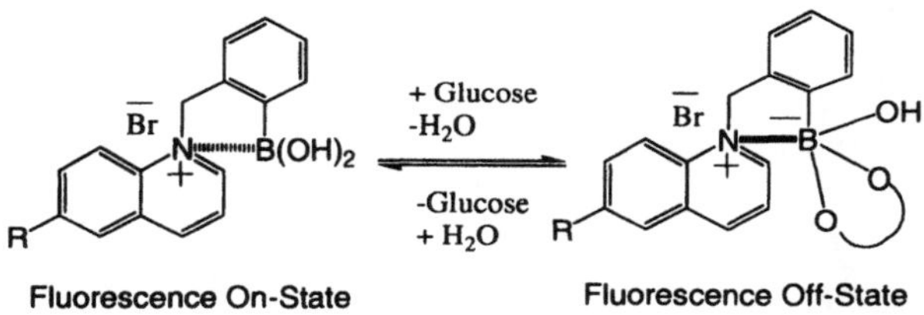

Figure 17.18. A schematic representation of the charge neutralization-stabilization mechanism with regard to glucose sensing. The bold-line between the N^+ and boron atom in the structure shown in the right side of the equation indicates the increased interaction between them, and is not intended to show covalent bond formation between the two atoms.

To understand the nature and extent of the quenching mechanisms, we developed isomeric boronic acid probes, where the spacing between the boron and nitrogen is varied. This is because among the both quenching mechanisms discussed above, the PET mechanism is independent on the space between the positively charged nitrogen and negatively charged boronate diester, while the charge neutralization/stabilization mechanism does depend on this space: charge stabilization will reduce with increasing space. However, based on the present results we are not to draw the distinctions between both mechanisms, and are working together making these probes useful for the tear glucose estimation [47,48].

17.5.3. Sugar Response of the Quinolinium Probes in Solution

The monosaccharide induced spectral changes of the probes are shown in Figure 17.19. In an analogous manner to that described above for increasing pH, we observed a systematic decrease in fluorescence intensity of the boronic acid probes with increasing glucose concentrations in pH 7.5 phosphate buffer [47,48]. The corresponding titration curves obtained by plotting I' divided by I, where I' and I are fluorescence intensities at 427 nm

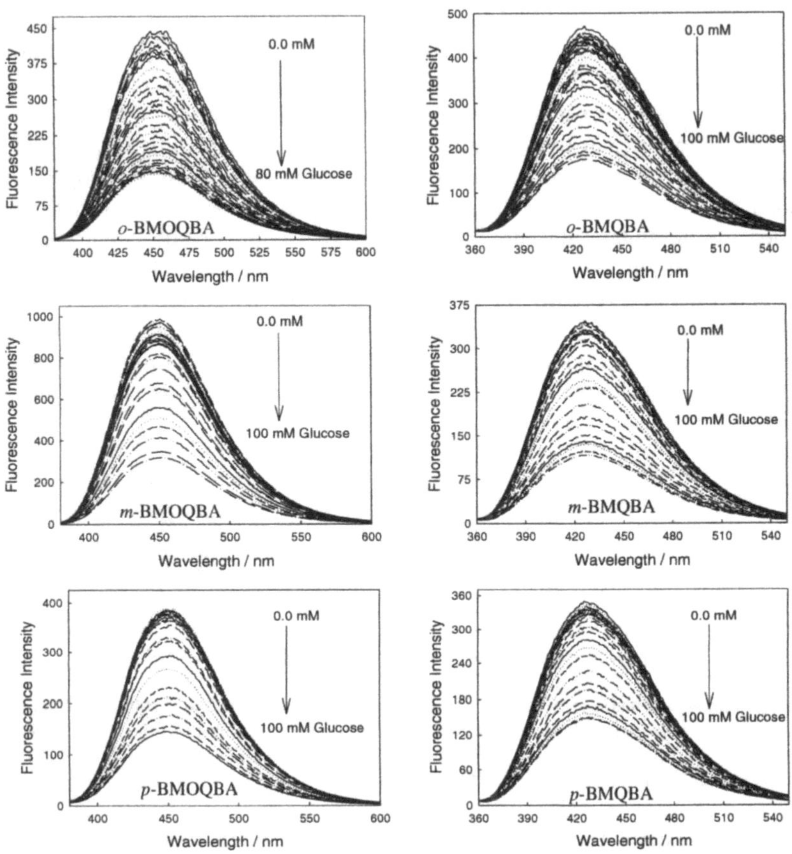

Figure 17.19. Emission spectra of BMOQBAs (Left column) and BMQBAs (Right column) in pH 7.5 phosphate buffer with increasing glucose concentration. The λ_{ex} for BMOQBAs was 345 nm and for BMQBAs 320 nm.

for BMQBA and 450 nm for BMOQBA, in the absence and presence of sugar respectively, versus glucose concentration, are also shown in Figure 17.20. The right-hand column of Figure 17.20 shows the response of the probes in the tear glucose concentration range.

For the BMOQBA probes we typically see a greater response for the *para*-isomer, with a 2.4-fold change in signal for 50 mM glucose. Interestingly, a ≈ 13 % change in signal is observed up to 2 mM glucose, noting that glucose levels in tears can change from ≈ 500 μM for a healthy person, and up to 5 mM for diabetics. For *ortho*-BMQBA, then a ≈16 % change in fluorescence signal can be observed over a similar glucose range [47,48].

The dissociation constants of the probes with both glucose and fructose in pH 7.5 phosphate buffer are presented in Table 17.3. As expected, a higher affinity for fructose is observed (lower K_D value), which is a general observation for monophenyl boronic acid derivatives [40-46], but it should be noted that the concentration of fructose in tears is substantially lower than glucose [20-26]. A comparison of the trends in glucose response observed in Figure 17.20, and the recovered K_D values in Table 17.3, show some differences, which we have attributed to the difficulties encountered during data fitting. While beyond the scope of this text, these fitting difficulties clearly reflect the need for a new kinetic sugar binding function with our new probes. Further studies are underway in this regard. Based on the data presented here, these probes having lower pK_a values, may be useful for the continuous monitoring of glucose using disposable contact lenses.

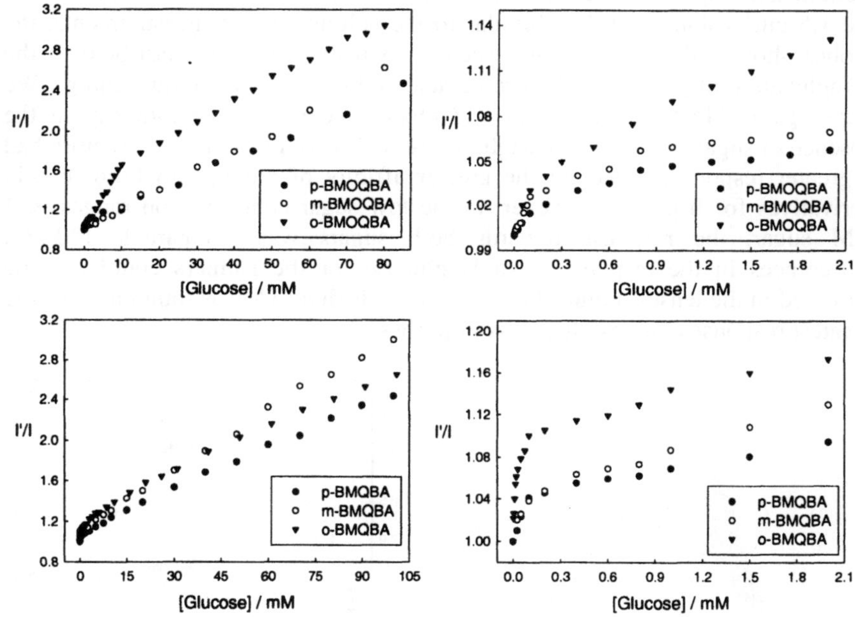

Figure 17.20. Respective intensity ratio for all three isomers of BMOQBAs in the absence, *I'*, and in the presence, *I*, of glucose, (Top left), and in the *tear glucose concentration* range (Top right). The corresponding plots for BMQBAs are also shown in the blood glucose concentration range and tear glucose concentration range, (Bottom right and Bottom left, respectively).

Table 17.3. Dissociation constants, K_D (mM), of the probes with glucose and fructose in buffer and in the contact lens.

	Glucose		Fructose	
Probe	*Buffer*	*Lens*	*Buffer*	*Lens*
***o*-BMOQBA**	49.5	322.6	0.65	84.7
***m*-BMOQBA**	1000	54.6	1.8	4.9
***p*-BMOQBA**	430.0	111.1	9.1	34.7
***o*-BMQBA**	100	17.9	4.7	34.8
***m*-BMQBA**	476	58.1	13.2	21.6
***p*-BMQBA**	370	128.2	13.8	12.9

17.5.4. Response of the New Glucose Signaling Probes in the Contact Lens

Doped contact lenses, which were previously washed and allowed to leach dye for 1 hour, were tested towards both glucose and fructose. Buffered solutions of sugars were added to the lens, pH 7.5 phosphate buffer, in an analogous manner to ocular conditions. Fluorescence spectra were typically taken 15 mins after each sugar addition to allow the lens to reach equilibrium. The 90 % response time, the time for the fluorescence signal to change by 90 % of the initial value, was ≈ 10 minutes.

Figure 17.21 shows the response of *o*-BMOQBA and *o*-BMQBA, Top and Bottom left respectively, for increasing concentrations of glucose injected into the 1.5 cm^3 volume [21-23]. Similar to the solution based measurements, the probes show a decrease in fluorescence intensity, which we attribute to the complexation of glucose with boronic acid and the subsequent quenching. We were again able to construct the *I'/I* plots, where *I'* is the intensity in the absence of sugar, Figure 17.21 Right. As was observed in solution, Fructose had a greater response, reflecting the greater affinity of monophenylboronic acid derivatives for fructose. However, in the low sugar concentration ranges, < 2 mM sugar, the response towards both sugars was comparable [20-26]. Differences in the response towards glucose for the isomers could also be observed in the lenses, Figure 17.22, where *m*-BMOQBA was found to have the greatest response amongst this class of probes.

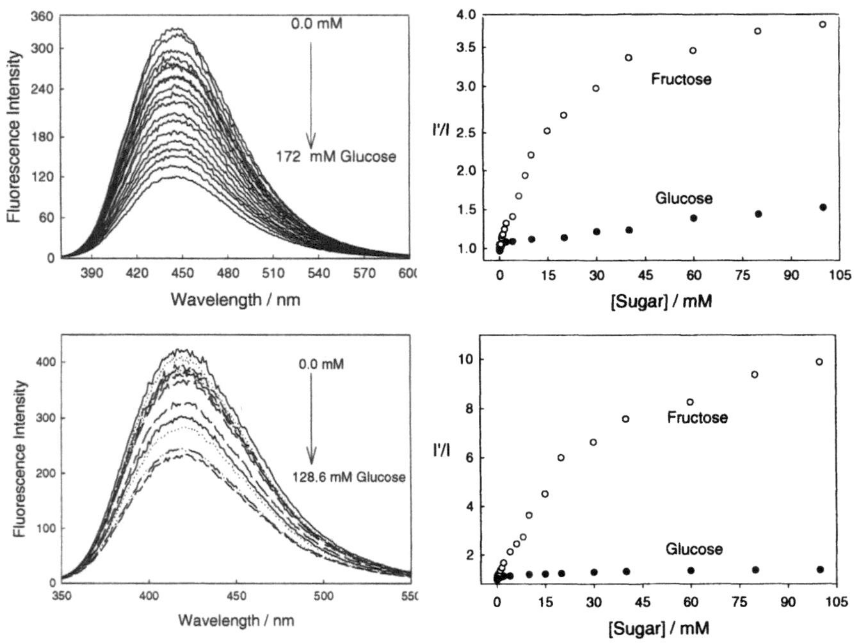

Figure 17.21. The emission spectra of an *o*-BMOQBA and *o*-BMQBA doped contact lens in the presence of increasing glucose concentrations (Top left and Bottom Left, respectively). λ_{ex} = 345 nm for *o*-BMOQBA and 320 nm for *o*-BMQBA. The corresponding emission intensity ratio at band maximum for *o*-BMOQBA and *o*-BMQBA doped contact lens in the absence, *I'*, and presence, *I*, of both Glucose and Fructose (Top right and Bottom right, respectively).

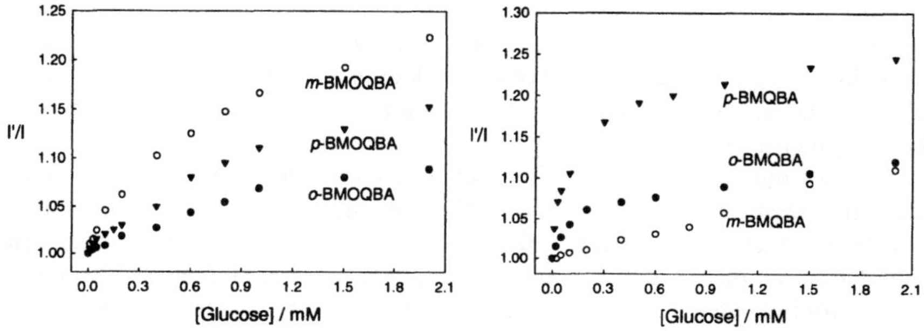

Figure 17.22. The response of BMOQBAs (Left) and BMQBAs (Right) in the contact lens and in the *tear glucose concentration range*.

From Figure 17.22 we can clearly see a greater response in the lens towards sugars than in our solution based studies at pH 7.5, with *p*-BMQBA showing a greater than 20 % fluorescence signal change with as little as 2 mM glucose. This wasn't unexpected, and is simply explained by the pK_a of the probes being < 7, the probes being compatible with the mildly acidic lens environment. The dissociation constants (K_D) obtained for all six probes with glucose and fructose within the contact lens are shown in Table 17.3. As seen from the table, the affinity of the probes towards sugar is relatively less in the contact lens, and that may be due to the nonbufferable pH as mentioned earlier in Section 17.3.

Figure 17.23 directly compares the response of the probes in both the contact lens and buffer, where we can see a comparable if not better response towards glucose in the lens in the low concentration range of sugar. However, in the high concentration range, the probe doped contact lens shows a weaker response towards sugar, and we are uncertain at this time but we speculate that the binding efficiency of the probes or the local environment such as pH and polarity of the lens still play a role on the binding interaction of the probes. Also, the probe location in the contact lens, glucose diffusion,

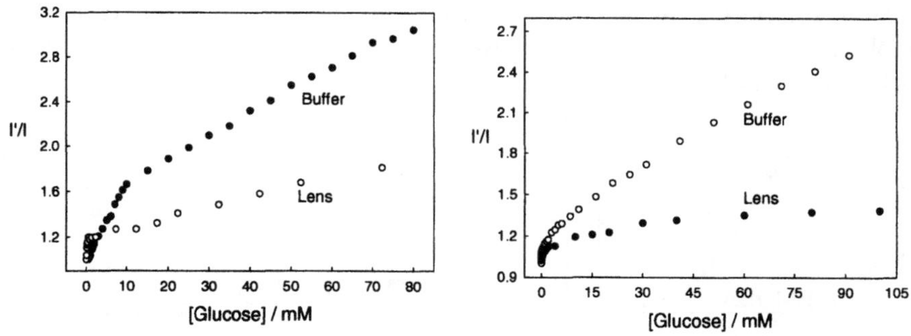

Figure 17.23. A comparison of the emission intensity ratio for the *o*-BMOQBA doped contact lens with that obtained in pH 7.5 phosphate buffer, in the absence, *I'*, and presence, *I*, of Glucose (Left) and the corresponding plot for *o*-BMQBA (Right).

and probe leaching from the lens may well complicate the sensing response of the probes from within the contact lens as evident from the *I' / I* Vs [sugar] plots. In this regard, the response of doped lenses also shows complex behavior towards mM fructose concentrations, with much simpler kinetics observed in the tear glucose concentration range (data not shown).

For all our glucose contact lens sensing studies we repeated these doped lens experiments several times, and in all cases the trends were reproducible. It is difficult to assess the effect of the PVA hydroxyl groups of the contact lens polymer on the response of boronic acid to sugar, but our studies with solutions of glycerol indicated that sugar had much higher binding affinities than glycerol hydroxyl groups. We therefore speculate that sugars will preferentially bind boronic acid groups in the PVA lens polymer. In any event the boronic acid probes function well (reversibly) towards sugars in the lens, in what is likely to be an environment saturated with PVA hydroxyl groups.

17.6. PROBE LEACHING, INTERFERENTS AND SHELF LIFE

In the present study, as the probes are post doped into the contact lens one might expect to have the probe leaching during the glucose estimation. To understand the leaching properties of the probes from the contact lens polymer we have undertaken the study using the lens holder shown in Figure 17.7, which contained ≈ 1.5 cm^3 buffer at 20^0 C. A Varian fluorometer measured the intensity change as a function of time to determine the percentage signal change, corresponding to dye leaching. It should be noted that with no sample present, no intensity fluctuations or drifts were observed, indicating stability of the fluorometer Xenon-arc source.

We were able to observe up to about 8 % change in fluorescence intensity, attributed due to dye leaching, for the BMOQBA class of probes, with very little change after about 25 minutes, Figure 17.24. In contrast, the BMQBA probes show a much greater extent of leaching over the same time period and under identical conditions. Given that the BMQBA probes typically showed a greater response towards glucose in the lens, then this suggests that the BMQBA probes may be more accessible in the lens, than the BMOQBA probes. In addition, similar results were obtained at 38^0C but with a different leaching rate [20-26].

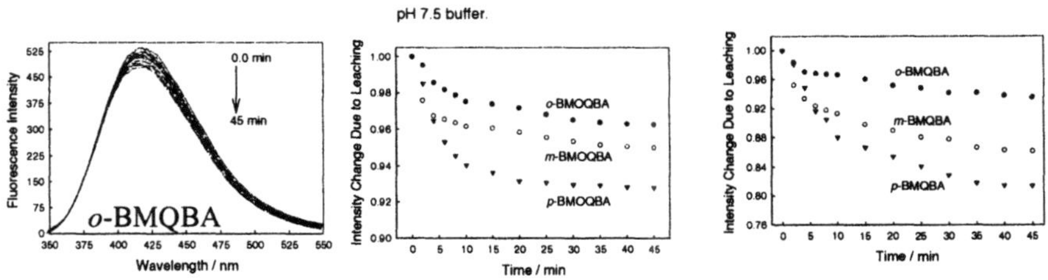

Figure 17.24. Emission spectra of *o*-BMQBA doped contact lens immersed in pH 7.5 buffer with time (Top left). λ_{ex} = 320 nm. Normalized intensity change at band maximum of BMOQBA with time due to leaching (Top right), and that for the BMQBAs (Bottom).

In all our lens response studies described here, lenses were pre-leached to a steady-state fluorescence intensity before use. After glucose measurements were undertaken the outer lens fluid volume surrounding the contact lens was found to be non-fluorescent indicating that dye had not leached from the lens during actual glucose sensing measurements. It should be noted that while chemistries are available to covalently label our probes within the contact lens polymer, which would eliminate any leaching, it is an important design concern for our approach that the lenses remain unmodified, so that their physiological characteristics and compatibilities remain unchanged. In fact our approach is targeted at reducing future lens redesign costs for industry, by using simple probe doping. However, we have recently developed new set of second generation probes in which all the prerequisite conditions, such as low pK_a, imposed by the contact lens were implemented [58]. Additionally these probes with an appending carbon-carbon double bond are useful as the contact lens immobilizable probes to eliminate the leaching. The molecular structure of the probes is shown in Chart 3. The solution based response of these probes is promising (data not shown) and further study of these probes within the contact lens is under consideration.

As with all sensors it is important to consider the effects of potential interferents and sensor shelf-life on the working response of the device. Throughout much of this paper we have shown the response of the probes towards fructose, primarily because of its well- known greater affinity for the boronic acid moiety [40-46]. However, the concentration of fructose in blood is ≈ 10 times lower than glucose [20], a relationship which is also thought to occur in tears [20]. Hence fructose is not thought to be a major interferent in tears, simply shown here to place the binding trends in context. However, tears are a complex mixture of proteins and other analtyes, such as sodium (120-170 mM), potassium (6-26 mM) and chloride (100 mM) [59]. We subsequently tested these new contact lens probes towards various aqueous halides, given that Na^+ and K^+ are unlikely to perturb our fluorophores, Table 17.4. As expected, the BMOQBA probes are modestly quenched by chloride, with steady-state Stern-Volmer constants in the range 170 - 182 M^{-1}, the BMQBA probes having significantly smaller quenching constants in the range 17–44 M^{-1}. This result is

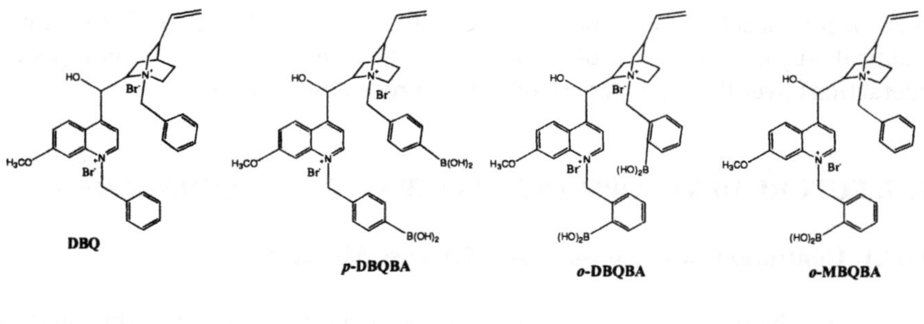

Chart 17.3. Molecular structure of the second generation probes based on naturally occurring quinine.

simply explained by the shorter lifetime of the BMQBA probes as compared to the BMOQBA probes, and the probability of an excited-state chloride ion encounter [55-57]. Encouragingly, the BMQBA probes typically showed a greater response towards glucose, as well as being the least perturbed by aqueous chloride. In any event, simple corrections in the fluorescence signal can readily account for chloride interference on the glucose response. This is shown in Figure 17.1. Sensors spots on the surface of the lens could contain a reference chloride compound or indeed another probe sensitive to both glucose and chloride. By employing the extended Stern-Volmer equation for multiple analytes [55], one can easily correct for background interferences.

Table 17.4. Stern-Volmer (K_{SV}, M^{-1}) constants of the probes with halides in water.

Probe	Cl^-	Br^-	I^-
o-BMOQBA	170	332[a]	471[a]
m-BMOQBA	182	413	540
p-BMOQBA	177	370	595
BMOQ	222	384	520
o-BMQBA	44.0	55.0	97.0
m-BMQBA	20.0	32.0	48.0
p-BMQBA	17.0	26.5	42.0
BMQ	35.0	55.0	71.0

[a]The concentrations of Br^- and I^- in tears is extremely low and is unlikely to be an interferent in our glucose measurements.

It is also informative to consider the pH of tears as a potential interferent, given the response of these probes to pH as shown in Figure 17.14. It is known that unstimulated tear pH levels can vary in the range 7.14 – 7.82 measured from healthy subjects at different times of the day, with a typical mean value around pH 7.45 [59]. However, a more acidic pH of less than 7.3 is found following prolonged lid closure, e.g. after sleep, which is thought to result from carbon dioxide produced by the cornea and trapped in the tear pool under the eye lids. While solutions of these new probes would be susceptible to these changes in pH, we have found that the doped lenses we studied were indeed unbufferable, hence external changes in pH are most unlikely to affect the glucose response

With regard to the glucose-sensing contact lens shelf-life, lenses that had been doped, leached and stored for several months both wet and dry, gave identical sugar sensing results, indicating no lens polymer – fluorophore interactions over this time period, or indeed probe degradation.

17.7. FUTURE DEVELOPMENTS BASED ON THIS TECHNOLOGY

17.7.1. Continuous and Non-invasive Glucose Monitoring

In this chapter, we have shown that fluorescent probes can be fabricated to be compatible with in the commercially available daily use disposable contact lenses, which have already been assessed and optimized with regard to vision correction and oxygen permeability. This has enabled the first prototype based on this new approach to be realized. With regard to glucose monitoring by this approach, we speculate on several future improvements to this technology:

- *Clear or Colored Contact Lenses*

  Many boronic containing fluorophores have a visible absorption [40-46], which apart from their lack of glucose sensitivity in the lens as discussed earlier [20], would introduce color into a doped lens. While colored lenses are attractive to a few people as sports or even fashion accessories, the majority of contact lenses worn today are clear; hence our colorless lenses described here are ideal in this regard. One disadvantage however with our lenses is the requirement for an excitation and detection device as shown in Figure 17.1. One improvement to our technology could be the use of colored contact lenses, which change color due to the concentration of tear and therefore blood glucose. This can be achieved by the ground-state binding of glucose to boronic acid and the subsequent changes in fluorophore absorption spectrum. A patient wearing the lenses could simply look in the mirror, or the color even determined by an on-looker, and compared to a precalibrated color strip, one could assess the extent of hyperglycemia. This technology would be most attractive to parents of young diabetic children or for care workers of the elderly. Work in currently underway in our laboratories in this regard.
- *Sensor Spots or Doped Lenses*

  As briefly mentioned earlier and shown in Figure 17.1, sensor spots on the surface of contact lenses could correct signal responses for interferents such as aqueous chloride. Indeed the spots could either be visible to the wearer (self-readout) or readable by an external monitoring device.
- *Detection Methods*

  While simple colorimetric methods are likely to be the most simple to introduce to the market place, other fluorescence sensing methodologies, such as polarization, lifetime and ratiometric sensing, offer many spectroscopic sensing advantages over the simple intensity measurements described for our lenses [49]. For example, fluorescence lifetime and ratiometric measurements are independent on total light intensity or indeed fluctuations in ambient room light.

17.7.2. Clinical Condition and Diagnosis from Tears

Given that tear glucose can be continuously and non-invasively monitored using our lens approach, then it may be possible for both clinical condition assessment and disease diagnosis using the contact lens sensing platform, and suitably designed fluorophores. In fact, as compared to saliva, tears represent a more stable body fluid of low protein concentration and with only modest variations in pH.

For example, for clinical condition assessment, Na^+, K^+, Ca^{2+}, Mg^{2+}, pH, Histamine, Urea, Lactate, Cholesterol and Glucose in tears are known to directly track or relate to the serum levels [59]. Indeed, one could potentially even track body core temperature using thermochromic type probes [60], embedded within contact lenses or even sensor spots.

For disease diagnosis, the possibility of diagnosing; Glaucoma, Sjogrens disease, Lysosomal storage diseases, corneal ulceration and bacterial infections could be realized by designing lenses to detect; Catecholamines, Lysozyme, Lysosomal enzymes, Collagenase and α-Antitrypsin respectively. It is possible that many other clinical conditions and diseases could also be either monitored or diagnosed via this approach, although relatively little tear biochemistry is known [59].

17.7.3. Drug Testing, Compliance and Screening

Saliva is and has been used for therapeutic drug monitoring, by predicting the free fraction of drugs in blood from that determined in saliva. Changes in the free drug levels can have important clinical consequences, since either toxic or sub-therapeutic levels may exist, even when the total drug concentration is in the normal range. Coupled with the fact that saliva concentrations varying greatly in both pH and composition, obviates the need for a novel clinical sensing platform for drug testing, compliance and screening. As mentioned earlier, tear fluid has a relatively lower protein concentration with only slight changes in pH, where the passage of drugs from plasma to tears takes place by diffusion of the non-protein bound fraction [59]. Subsequently, tears have already been used to assess the concentration of antibiotics such as Ampicillin and the Anticonvulsants, Phenobarbital and Carbamazepine [59]. However, these drug concentration measurements are inherently difficult due to the 5 -10 μL total tear volume to be sampled [59]. Indeed, while tear glucose levels have been known to be elevated during hyperglycemia for nearly 70 years, it is the difficulties associated with tear collection, which has limited the practical use of tears for diabetes mellitus assessment. However, similar to our glucose sensing contact lenses, it may be possible to develop lenses for drug testing, based on either colorimetric or other fluorescence spectroscopic based methodologies.

17.8. CONCLUDING REMARKS

We have developed a range of new glucose sensing contact lenses, by doping strategically designed fluorescent probes into commercially available contact lenses. The probes are completely compatible with the new lenses and can readily detect glucose changes up to several mM glucose, appropriate for the tear glucose concentration range for diabetics, i.e. $\approx$ 500 μM $\rightarrow$ 5 mM [20].

The lenses have a 90 % response time of about 10 minutes, allowing the continuous and noninvasive monitoring of ocular glucose. This is a significant improvement over enzymatic methods based on blood sampling by finger pricking, with many diabetics begrudgingly testing between 4 and 6 times daily.

With diabetes being widely recognized as one of the leading causes of death and disability in the western world, we believe our boronic acid doped contact lens approach and findings, are a notable step forward towards the continuous and non-invasive monitoring of physiological glucose.

17.9. ACKNOWLEDGEMENTS

The authors would like to thank the University of Maryland Biotechnology Institute and the NIH, National Center for Research Resources, RR-08119, for financial support.

17.10. REFERENCES

1. The diabetes control and complications trial research group, *Diabetes*, **48**, 271-286 (1997).
2. The diabetes control and complications trial research group, *N. Engl. J. Med.*, **329**, 977-986 (1993).
3. M. R. Robinson, R. P. Eaton, D. M. Haaland, G. W. Koepp, E. V. Thomas, B. R. Stallard and P. L. Robinson, "Non-invasive glucose monitoring in diabetic patients: A preliminary evaluation," *Clin. Chem.*, **38**, 1618-1622 (1992).
4. H. M. Heise, R. Marbach, T. H. Koschinsky, and F. A. Gries, "Non-invasive blood glucose sensors based on near-infrared spectroscopy," *Ann. Occup. Hyg.*, **18**, 439-447 (1994).
5. W. F. March, B. Rabinovitch, R. Adams, J. R. Wise and M. Melton, "Ocular Glucose sensor," *Trans. Am. Soc. Artif. Intern. Organs*, **28**, 232-235 (1982).
6. B. Rabinovitch, W. F. March and R. L. Adams, "Non-invasive glucose monitoring of the aqueous humor of the eye, Part 1, Measurement of very small optical rotations, *Diabetes Care*, **5**, 254-258 (1982).
7. G. M. Schier, R. G. Moses, I. E. T. Gan, and S. C. Blair, "An evaluation and comparison of reflolux Iiand Glucometer II, two new portable reflectance meters for capillary blood glucose determination," *Diabetes Res. Clin. Pract.*, **4**,177-181 (1988).
8. W. Clarke, D. J. Becker, D. Cox, J. V. Santiago, N. H. White, J. Betschart, K. Eckenrode, L. A. Levandoski, E. A. Prusinki, L. M. Simineiro, A. L. Snyder, A. M. Tideman and T. Yaegar, "Evaluation of a new system for self blood glucose monitoring," *Diabetes Res. Clin. Pract.*, **4**, 209-214 (1988).
9. W. Trettnak and O. S. Wolfbeis, Fully reversible fiber-optic glucose biosensor based on the intrinsic fluorescence of glucose-oxidase, *Anal. Chim. Acta*, **221**,195-203 (1989).
10. D. Meadows and J. S. Schultz, "Fiber optic biosensor based on fluorescence energy transfer," *Talanta*, **35**, 145-150 (1988).
11. L. Tolosa, H. Malak, G. Rao, and J. R. Lakowicz, "Optical assay for glucose based on the luminescence decay time of the long wavelength dye Cy5, *Sensors Actuators B.*, **45**, 93-99 (1997).
12. L. Tolosa, I. Gryczynski, L. R. Eichorn, J. D. Dattelbaum, F. N. Castellano, G. Rao and J. R. Lakowicz, Glucose sensors for low cost lifetime-based sensing using a genetically engineered protein, *Anal. Biochem.*, **267**,114-120 (1999).
13. S. D'Auria, N. Dicesare, Z. Gryczynski, I. Gryczynski, M. Rossi and J. R. Lakowicz, A thermophilic apoglucose dehydrogenase as a nonconsuming glucose sensor, *Biochem. Biophys Res. Commun.*, **274**, 727-731 (2000).
14. A. R. Gasser, L. E. Braverman, M. C. Fleming, R. A. Arky and B. R. Alter, *Am. J. Ophthalmology*, 1968, **65**(3), 414-420.
15. B. N. Das, S. Sengupta, B. K. Das and N. R. Goswami, *J. Indian Med. Assoc.*, 1995, **93**(4), 127-128.
16. D. Michail, N. Zolog, *Compte Rendu Soc. Biol.*, Paris, 1937, **126**, 1042.
17. D. Michail, P. Vancea, and N. Zolog, *C. R. Soc. Biol.*, Paris, 1937, **125**, 1095.
18. N. J. Van Haeringen, *Survey of Ophthalmology*, 1981, **29**(2), 84-96.
19. A. Giardiai, J. R. E. Roberts, *Br. J. Ophthalmol*, 1950, **34**, 737-743.
20. R. Badugu, J. R. Lakowicz, and C. D. Geddes (2004). The non-invasive continuous monitoring of physiological glucose using a novel monosaccharide-sensing contact lens, *Anal. Chem.*, 76, 610-618.

21. R. Badugu, J. R. Lakowicz, and C. D. Geddes (2003). A Glucose Sensing Contact Lens: A Non-Invasive Technique for Continuous Physiological Glucose Monitoring, *J. Fluorescence*, 13, 371-374.
22. C. D. Geddes R. Badugu, and J. R. Lakowicz, (2004). Contact lenses may provide window to blood glucose, *Biophotoincs international*, February (2), 50-53.
23. R. Badugu, J. R. Lakowicz, and C. D. Geddes (2004). Ophthalmic glucose sensing: A novel monosaccharide sensing disposable and colorless contact lens, *The Analyst*, 129, 516-521.
24. R. Badugu, J. R. Lakowicz, and C. D. Geddes (2005). Ophthalmic glucose monitoring using disposable contact lenses, *Reviews in Fluorescence 2005*, C. D. Geddes and J. R. Lakowicz, (Eds.), Kluwer Academic / Plenum Publishers, **2005**, Vol. 2, pp 363-398.
25. R. Badugu, J. R. Lakowicz, and C. D. Geddes (2004). Ophthalmic Glucose Monitoring using Disposable Contact Lenses – A Review, *J. Fluoresc.*, 14(5), (**2004**) 617-633.
26. R. Badugu, J. R. Lakowicz, and C. D. Geddes (2004). A Glucose sensing contact lens: A new approach to non-invasive continuous physiological glucose monitoring, In *Optical Fibers and Sensors for Medical Applications IV*; Israel Gannot, Ed., *Proceedings SPIE*, 5317, (**2004**), 234-245.
27. J. M. Sugihara and C. M. Bowman (1958). Cyclic Benzeneboronate Esters, *J. Am. Chem. Soc.*, 80(10), 2443-2446.
28. J. P. Lorand and J. O. Edwards (1959). Polyol Complexes and Structure of the Benzeneboronate Ion, *J. Org. Chem.*, 24(6), 769-774.
29. G. Springsteen and B. Wang (2002). A detailed examination of boronic acid–diol complexation *Tetrahedron*, 58(26), 5291-5300.
30 T. D. James, K. R. A. S. Sandanayake and S. Shinkai, (1995). Chiral discrimination of monosaccharides using a fluorescent molecular sensor, *Nature*, 374, 345.
31. J. C. Norrild and H. Eggert (1995). Evidence for monodentate and bidentate boronate complexes of glucose in the furanose form – application of (1)J(C-C)-coupling-constants as a structural probe, *J. Am. Chem. Soc.*, 117(5), 1479-1484.
32. H. Eggert, J. Frederiksen, C. Morin and J. C. Norrild (1999). A new glucose-selective fluorescent bisboronic acid. First report of strong alpha-furanose complexation in aqueous solution at physiological pH, *J. Org. Chem.*, 64(11), 3846-3852.
33. W. Yang, H. He, D. G. Drueckhammer (2001). Computer-guided design in molecular recognition: Design and synthesis of a glucopyranose receptor, *Angew. Chem. Int. Ed.*, 40(9), 1714-1718.
34. W. Wang, S. Gao, and B. Wang (1999). Building Fluorescent Sensors by Template Polymerization: The Preparation of a Fluorescent Sensor for D-Fructose, Org. Letts, 1(8) 1209-1212.
35. S. Gao, W. Wang and B. Wang (2001). Building Fluorescent Sensors for Carbohydrates Using Template-Directed Polymerizations, *Bioorg. Chem.*, 29, 308-320.
36. J. J. Lavigne, E. V. Anslyn (1999). Teaching Old Indicators New Tricks: A Colorimetric Chemosensing Ensemble for Tartrate/Malate in Beverages, *Angew. Chem. Int. Ed.*, 38(24), 3666-3669.
37. J. Yoon and A. W. Czarnik (1992). Fluorescent chemosensors of carbohydrates. A means of chemically communicating the binding of polyols in water based on chelation-enhanced quenching, *J. Am. Chem. Soc.*, 114, 5874-5875.
38. B. D. Smith, S. J. Gardiner, T. A. Munro, M. F. Paugam and J. A. Riggs (1998). Facilitated transport of carbohydrates, catecholamines, and amino acids through liquid and plasticized organic membranes, *J. Incl. Phenom. Mol. Recogn. Chem.* 32, 121-131.
39. S. Soundararajan, M. Badawi, C. M. Kohlrust, J. H. Hagerman (1989).Boronic acids for affinity – chromatography – spectral methods for determinations of ionization and diol-binding constants, *Anal. Biochem.*, 178, 125-134.
40. T. D. James, K. R. A. S., and S. Shinkai (1994). A glucose-selective molecular fluorescence sensor, *Angew, Chem. Int. Ed.*, 33(21), 2207-2209.
41. T. D. James, K. R. A. S. Sandanayake, R. Iguchi, and S. Shinkai (1995). Novel saccharide-photoinduced electron-transfer sensors based on the interaction of boronic acid and amine, *J. Am. Chem. Soc.*, 117(35), 8982-8987.
42. N. Dicesare and J. R. Lakowicz (2001). Evaluation of two synthetic glucose probes for fluorescence-lifetime based sensing, *Anal. Biochem.*, 294, 154-160.
43. N. Dicesare and J. R. Lakowicz (2001). Wavelength-ratiometric probes for saccharides based on donor-acceptor diphenylpolyenes, *J. Photochem. Photobiol. A: Chem.*, 143, 39-47.

44. N. Dicesare and J. R. Lakowicz (2001). New color chemosensors for monosaccharides based on Azo dyes, *Org. Lett.*, 3(24), 3891-3893.
45. N. Dicesare and J. R. Lakowicz (2002). Chalcone-analogue fluorescent probes for saccharides signaling using the boronic acid group, *Tet. Lett.*, 43, 2615-2618.
46. N. Dicesare and J. R. Lakowicz (2002). Charge transfer fluorescent probes using boronic acids for monosaccharide signaling, *J. Biomedical Optics*, 7(4), 538-545.
47. R. Badugu, J. R. Lakowicz, C. D. Geddes (2005). Fluorescence sensors for monosaccharides based on the 6-methylquinolinium nucleus and boronic acid moiety: application to ophthalmic diagnostics, *Talanta*, 65(3), **(2005)** 762-768.
48. R. Badugu, J. R. Lakowicz, C. D. Geddes (2005). Boronic acid fluorescent sensors for monosaccharide signaling based on the 6-methoxyquinolinium heterocyclic nucleus: Progress towards noninvasive and continuous glucose monitoring, *Bioorg. Med. Chem.* 13(1), **(2005)** 113-119.
49. J. R. Lakowicz, *Principles of Fluorescence Spectroscopy*, 2nd Edition, Kluwer/Academic Plenum Publishers, New York, 1997.
50. N. J. Turro, B. H. Baretz and P. I. Kuo (1984). Photoluminescence probes for the investigation of interactions between sodium dodecylsulfate and water-soluble polymers, *Macromolecules*, 17(7), 1321-1324.
51. K. Kalyanasundaram and J. K. Thomas (1977). Environmental effects on vibronic band intensities in pyrene monomer fluorescence and their application in studies of micellar systems, *J. Am. Chem. Soc.*, 99(7), 2039-2044.
52. N. Dicesare, J. R. Lakowicz (2001). Spectral properties of fluorophores combining the boronic acid group with electron donor or withdrawing groups. Implication in the development of fluorescence probes for saccharides, *J. Phys. Chem. A*, 105(28), 6834-6840.
53. M. A. Fox, M. Chanon, Eds. *Photoinduced Electron Transfer;* Elsevier: New York, 1998, Parts A-D.
54. G. J. Kavarnos, *Fundamentals of Photoinduced Electron Transfer;* VCH: New York, 1993.
55. C. D. Geddes (2001). Optical halide sensing using fluorescence quenching: theory, simulations and applications-a review, *Meas. Sci. and Tech.*, 12(9), R53.
56. O. S. Wolfbeis, E. Urbano (1982). *J. Heterocyclic Chem.*, 19, 841-843.
57. C. D. Geddes, K. Apperson, J. Karolin, D. J. S. Birch (2001). Chloride sensitive probes for biological applications, *Dyes & Pigments*, 48, 227-231.
58. R. Badugu, J. R. Lakowicz, C. D. Geddes (2005). Modified contact lens with quinine based boronic acid probes: progress towards the design of the non-leaching probe from the contact lens for tear glucose monitoring, *J. Fluoresc.* In Press.
59. N. J. Van Haeringen (1981). Climical Biochemistry in Tears, *Survey of Ophthalmology*, 26 (2), 84-96,
60. N. Chandrasekharan and L. Kelly, Progress towards fluorescent molecular thermometers, in *Reviews in Fluorescence 2003*, edited by C. D. Geddes and J. R. Lakowicz, Kluwer Academic Plenum Publishers, New York, 2004.

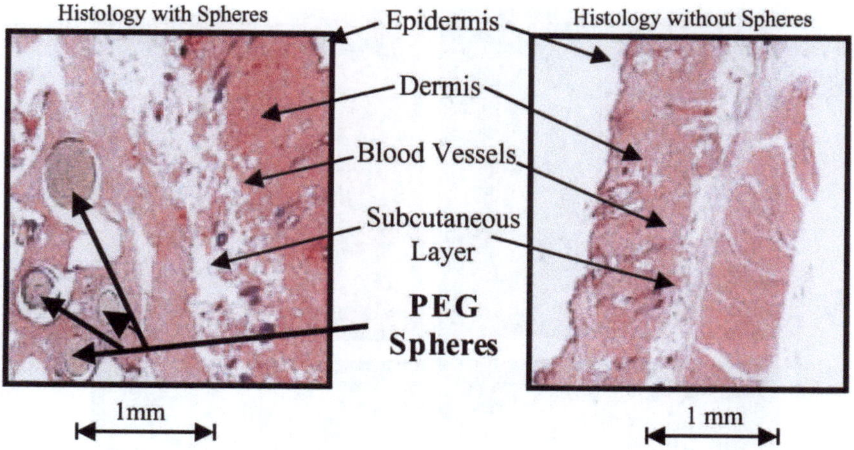

Figure 4.20. (Page 110, Ibey *et al.)* Histological sectioning of the implant site of the spheres and a control site allowed for immunological screening of the sensor site. No evidence of an immune response was seen at the implantation site despite overshooting the target implant depth.

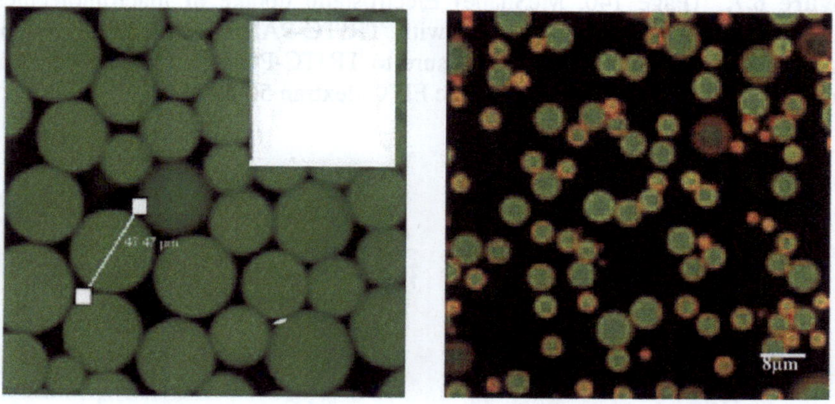

Figure 6.5. (Page 138, McShane) Top left: FITC-GOx loaded calcium-crosslinked microspheres formed by emulsion. Top right: Smaller alginate microspheres doped with FITC-GOx and coated with RITC-poly(allylamine) nanofilm.

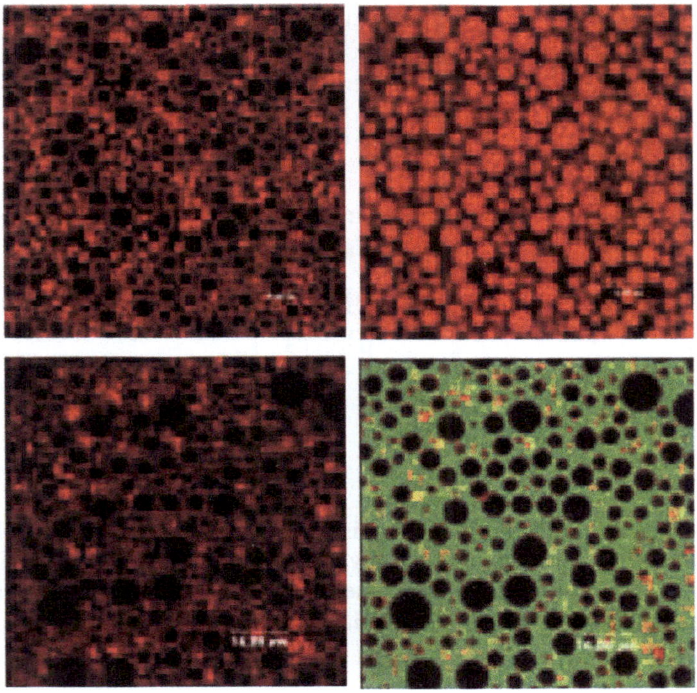

Figure 6.7. (Page 140, McShane) Electrostatic uptake of macromolecules. Left: alginate microspheres coated with TRITC-PAH/PSS nanofilms. Top right: same capsules following exposure to TRITC-POx. Bottom right: same capsules following exposure to anionic FITC-dextran 500kDa.

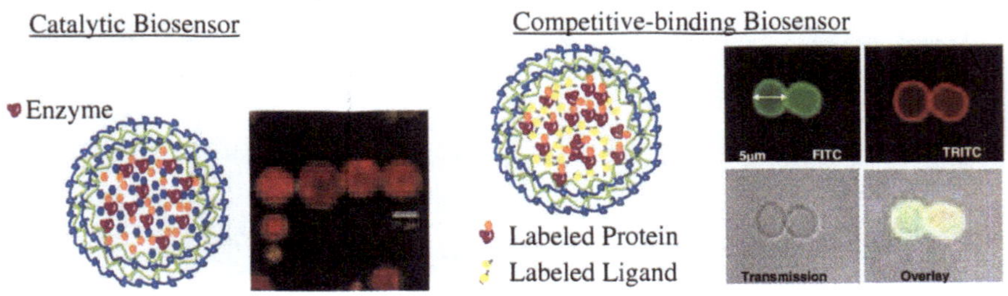

Figure 6.9. (Page 142, McShane) General description of microcapsule-based sensors. Top left: polyelectrolyte capsule, hollow or filled with polymer matrix; Bottom left: enzyme-based microcapsule sensor containing enzyme and dyes; Bottom right: microcapsule sensor employing competitive-binding FRET assay, comprising glucose-binding protein labeled with acceptor and glucose analog labeled with donor.

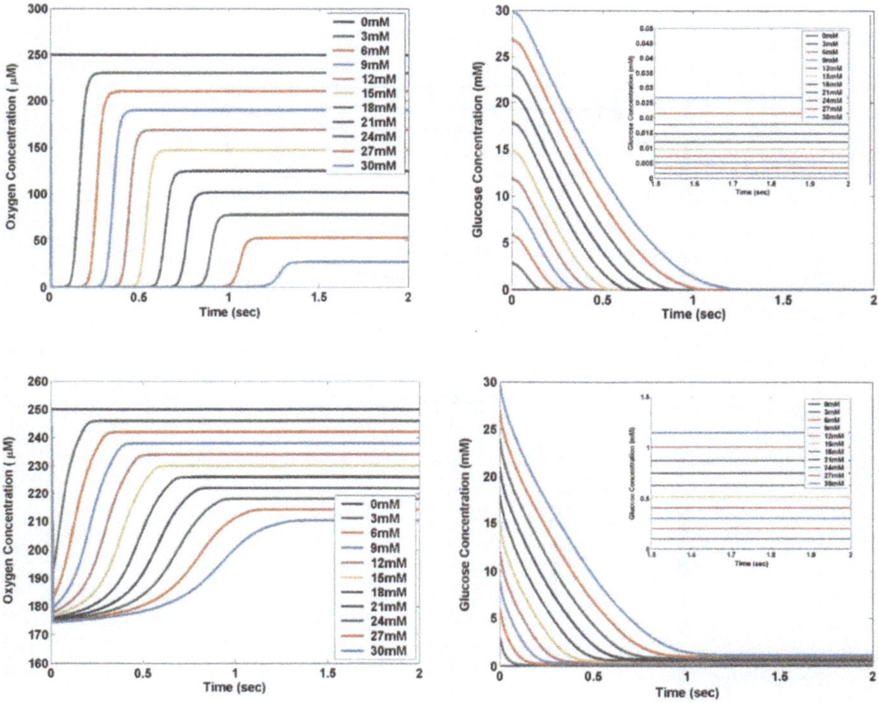

Figure 6.12. (Page 148, McShane) Oxygen (top left) and glucose (top right) temporal concentration profiles at r=r1/2; oxygen (bottom left) and glucose (bottom right) temporal concentration profiles at r=r1. The insets for glucose are exploded views of glucose levels 1.5-2 seconds following step addition.

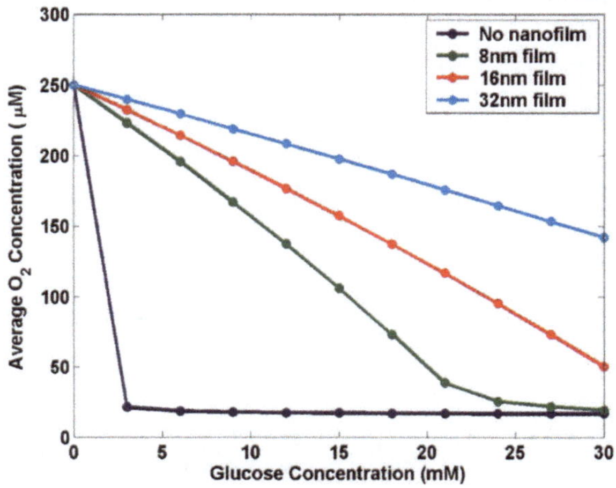

Figure 6.14. (Page 150, McShane) Steady-state sensitivity to glucose for microspheres with different nanofilm thickness.

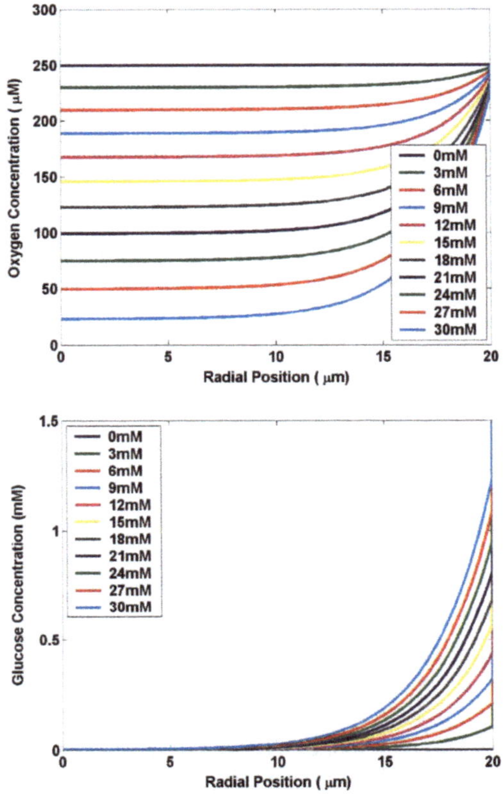

Figure 6.13. (Page 149, McShane) Steady-state spatial profiles for oxygen and glucose within spheres due to step inputs of glucose from 0-30 mM.

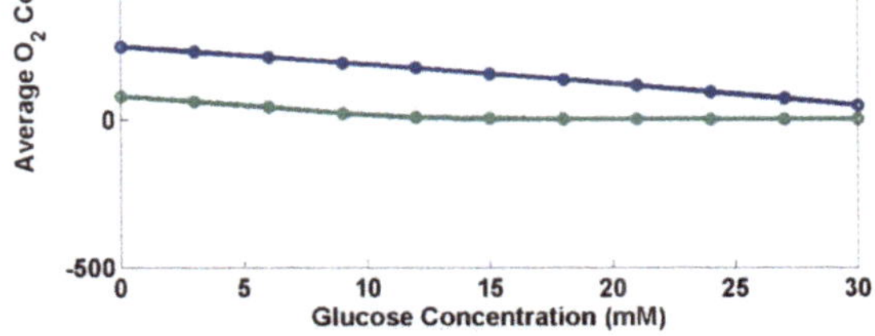

Figure 6.15. (Page 152, McShane) Average oxygen versus glucose for different bulk oxygen concentrations. The lines correspond to three different bulk oxygen levels. Top (red): 1400 μM; middle (blue): 250 μM, bottom (green): 80 μM.

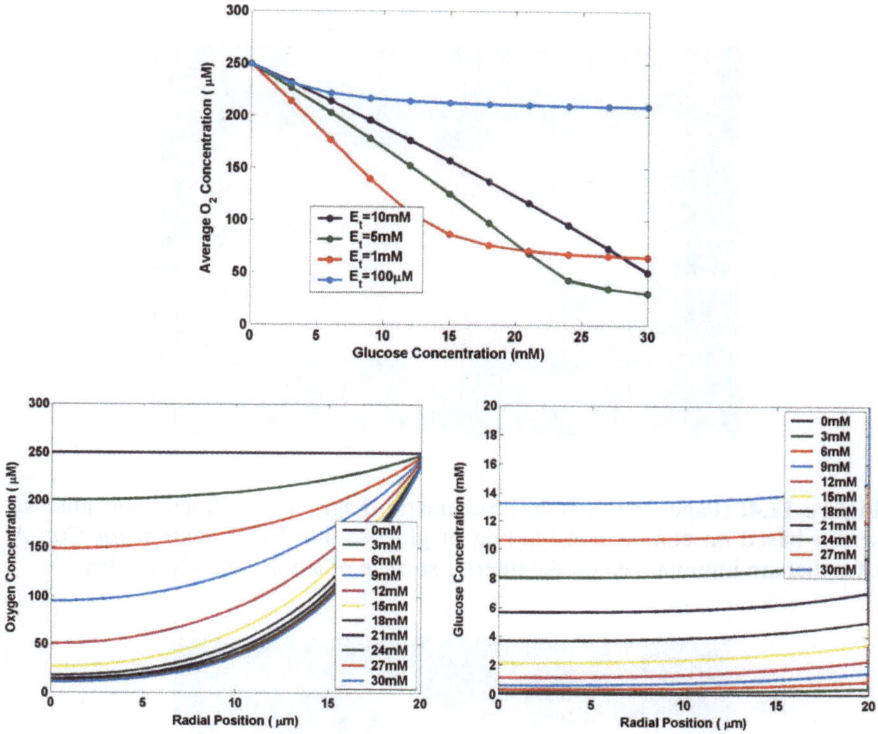

Figure 6.16. (Page 153, McShane) (Top) Average oxygen versus glucose for different enzyme concentrations and (Bottom-Left) internal oxygen and (Bottom-Left) glucose distributions for E_t=1mM.

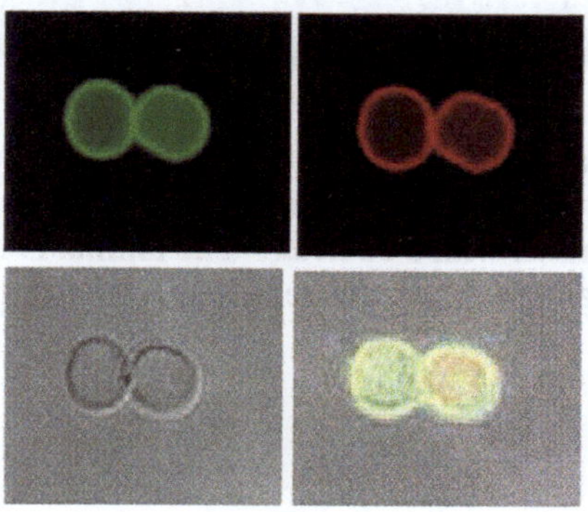

Figure 6.21. (Page 158, McShane) Confocal fluorescence (top, left: FITC, right: TRITC), phase (bottom left) and overlay (bottom right) images of polyelectrolyte capsules loaded with TRITC-apo-GOx and FITC-dextran.

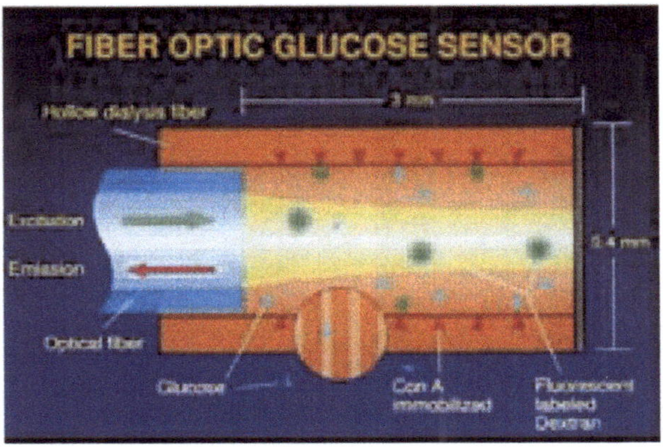

Figure 11.4. (Page 289, Schultz) Schematic diagram of a fiber optic glucose sensor based on competitive binding of glucose and FITC-dextran for Con-A sites that are immobilized on the interior surface of a hollow dialysis fiber.

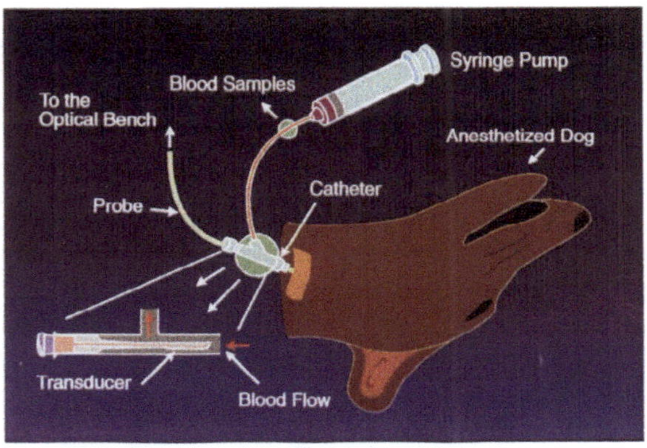

Figure 11.6. (Page 291, Schultz) Experimental setup for measuring blood glucose levels in an animal. The sensor is placed in a "t" and venous blood continually withdrawn with a catheter. The animal is injected with a bolus of a glucose solution, and then samples removed periodically for glucose analysis.

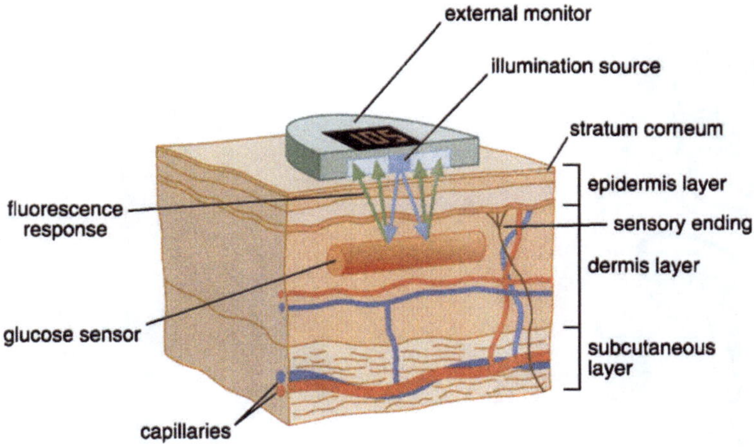

Figure 11.9. (Page 292, Schultz) Concept for a subcutaneous glucose sensor based on an optical readout principle.

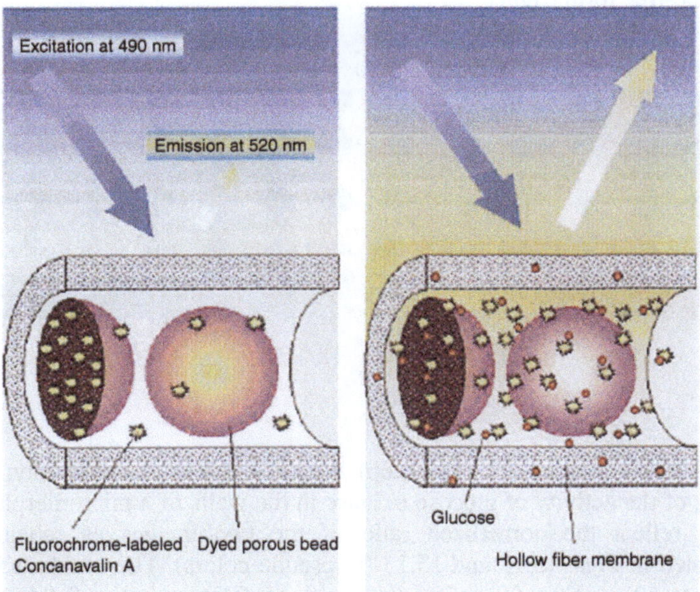

Figure 11.22. (Page 303, Schultz) Microsphere glucose biosensor. In this configuration, the analog-analyte is a porous Sephadex bead (a polymer containing pendant glucose residues). The beads are colored with dyes that prevent the excitation light from penetrating into them and another dye that blocks emission from fluorochrome-labeled Con-A in the interior of the bead. (Ballerstadt and Schultz, 2000).

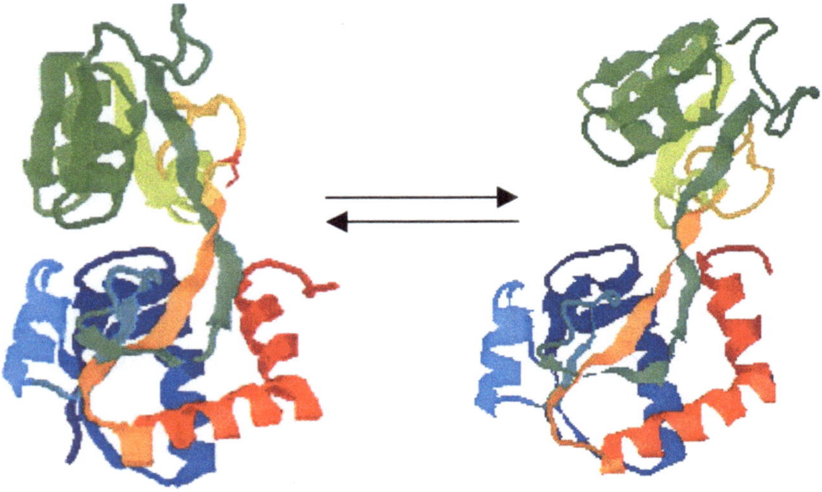

Figure 13.1. (Page 324, Tolosa and Rao) The structure of the glutamine binding protein showing the "closed" glutamine-bound form (left) and the "open" glutamine-free form (right).

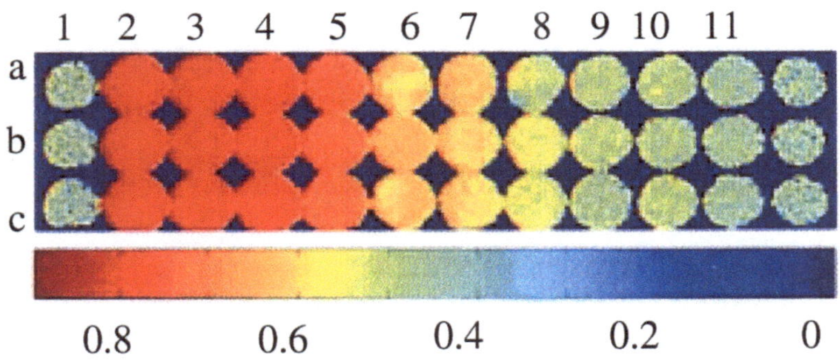

Figure 15.17. (Page 372, Duerkop) Rapid lifetime determination (RLD) imaging of the activity of glucose oxidase in the wells of a microtiterplate. The pictures reflect the normalized ratio of the two images as schematically represented in Figs. 15.12 and 15.13 (in pseudo colors). The cocktails in wells (from 1 to 12) had the following compositions: GOx activities 0 (blank), 135, 54.1, 27.1, 13.5, 5.4, 2.7, 1.35, 0.54, 0.27, 0.14 to 0.05 mUnit/mL, respectively; plus 100 μL of the EuTc stock solution, 15 μL of a 277.2 mmol L^{-1} glucose solution; total volume made up to 200 μL with MOPS buffer.

INDEX

affinity capture agents, 190
contact lens sensors
compartmentalization in, 194
specificity, 194
fluorescence measurements, 191, 192
light diffraction elements, 192, 193, 194
affinity principles, 305, 307
allophycocyanin, 191
amperometric detection, pulsed, 34
anionic fluorescent dyes, 47-87
ANS (anilino-naphthalene sulfonate), 326
open conformation, 326
anthracenediboronic acid derivative (ANDBA), 406, 412
apo-GO_x, 156, 159
aqueous micelle carriers, 222, 223
arenecarboximdes
N-phenylboronic acid derivatives of, 1-17
automated HPLC post-column detection method, 34-38. *See also* glucose monitoring
carbohydrates analysis, 34
evaporative light scattering detection (ELSD), 34
fructose, 35
mass spectrometric detection (HPLC-MS), 34
post-column derivatization techniques, 34
pre-column derivatization methods, 34
pulsed amperometric detection (PAD), 34

binding protein, glucose, 323, 324
conformation, 325
bioreactor
heterogeneous biphase, 218
particle packed, 217, 218, 219
biosensor
chemiluminogenic, 313-317
carbohydrates, 1
glucose dehydrogenase, 315
glucose oxidase (GOD), 1, 316
luminol, 314
microdialysis (MD), 315
optical sensor, 314
phenylboronic acid, 1, 2
biosensor (*cont.*)
chemiluminogenic (*cont.*)
sol-gel method, 315
surface chemiluminescence analysis, 317
competitive optical affinity, 289, 290, 291, 292, 293, 295, 296, 297
affinity based glucose biosensor systems, 299, 302, 303
affinity biosensors, 298, 299
integral biosensor proteins, 304, 305
fluorescence, 117-127
non-consuming analyte stable
glucose oxidase from *Aspergillus niger*, 119-121
*****optical
affinity principles, 305, 307
analytical feature and application, 226, 228
aqueous micelle carriers, 222, 223
bulky bi-phase bioreactor, 215, 216, 217
dissolved oxygen concentration, 223, 224
enzyme biosensor, 201-229
enzyme stability, GO_x, 224, 225
flow injection analysis, 222
in batch operation, 219
in flow mode, 220, 221
in stop-flow mode, 220
mixture of organic solvent, 222, 223
optical enzyme-based, 201-229
optical transducers, 203-207
particle packed bioreactor, 217, 218, 219
protein based, 1
biphase bioreactor
bulky, 215, 216, 217
heterogeneous, 218
bipyridinium, 56, 57, 58
meta-substiuted viologen, 58, 59
ortho-substiuted viologen, 58, 59
para-substiuted viologen, 58
bisboronic acid sensors, 2
boronic acid, 22, 387
based fluorescent reporter compounds, 380
anthracene, 381, 383, 384, 385

boronic acid (*cont.*)
  based fluorescent reporter
  compounds (*cont.*)
    chelation-enhanced quenching (CHEQ), 387, 388, 389, 390
    contact lens for glucose sensing, 390, 391
    internal charge transfer compounds, 385, 386
    polymer based sensors, 386
    pyrene based sensors, 381, 383, 384, 385
  functionalized viologen quenchers
    binding to diol, 51
  receptors, 333
  substituted viologens
    with anionic fluorescent dyes, 47-87
BP. *See* binding proteins
bulky bi-phase bioreactor, 215, 216, 217

C0 design chemosensor, 11. *See also* carbohydrate molecular probes
C4-CPB/-CyD
  binding equilibrium, 248, 249, 250
  fructose recognition by, 244, 245, 246
  in aqueous DMSO solution, 243-244
  recognition selectivity, 248, 249, 250
  response mechanism of, 246, 247, 248
capillary electrophoresis, 317-319. *See also* chemiluminescence
  ECL, 318
  laser-induced fluorescence (LIF), 317
capture agents, affinity, 190
  contact lens sensors
    compartmentalization in, 194
    specificity, 194
  fluorescence measurements, 191, 192
  light diffraction elements, 192, 193, 194
carbohydrates analysis, 34. *See also*
  automated HPLC post-column detection method
  carbohydrate sensors
    receptor, 6
    spacer, 6
  evaporative light scattering detection (ELSD), 34
  fructose, 35
  mass spectrometric detection (HPLC-MS), 34
  molecular probes
    C0 design chemosensor, 11
    N-phenyl naphthalimide fluorophore, 11
carbohydrates analysis (*cont.*)
  post-column derivatization techniques, 34
  pre-column derivatization methods, 34
  pulsed amperometric detection (PAD), 34
CAVEAT program, 393
CCA. *See* crystalline colloidal array
CE. *See* capillary electrophoresis
chalcone derivatives, 411. *See also* ICT probes
charge-transfer character, 4
chelation-enhanced quenching (CHEQ), 6, 381
chemiluminescence (CL), 312. *See also* glucose detection
  biosensor, 313-317
    carbohydrates, 1
    glucose dehydrogenase, 315
    glucose oxidase (GOD), 316
    luminol, 314
    microdialysis (MD), 315
    optical sensor, 314
    phenylboronic acid, 1, 2
    sol-gel method, 315
    surface chemiluminescence analysis, 317
  capillary electrophoresis (CE), 317-319
    ECL, 318
    laser-induced fluorescence (LIF), 317
  flow injection analysis (FIA), 313
  fructose concentration, 312
  glucose concentration, 313
  hexose concentration, 312
  periodate oxidation method, 312
  synthetic, 1
  using luminol, 312
chemosensors, fluorescent, 1
  phenylboronic acid, 1, 2
coherence tomography, optical, 169
colloidal array, crystalline, 392
competitive optical affinity glucose biosensors, 289, 290, 291, 292, 293, 295, 296, 297
  affinity based glucose biosensor systems, 299, 302, 303
  affinity biosensors, 298, 299
  integral biosensor proteins, 304, 305
concanavalin A (con A), 92-96, 118, 156, 191, 192. *See also* diabetes
  agarose beads (Sepharose), 95

concanavalin A (*cont.*)
fluorescent tags
fluorescein isothiocyanate (FITC), 95
purification, 93, 94
stability, 94
structure, 92-93
tetramethyl-rhodamine isothiocyanate (TRITC), 98
conducting polymers, organic, 365
contact lens
doped, 420
hydrogel, 91
ICT probes, 409
chalcone derivatives, 411
polyene derivative, 409
stilbene derivatives, 409
interferents, 422, 423, 424
internal properties, 412
PET probes, 412
probe leaching, 422, 423, 424
sensing probes, 412, 413, 414
based on quinolinium, 414, 415
sensors, 425
compartmentalization in, 194
specificity, 194
shelf life, 422, 423, 424
signaling probes in, 420, 421, 422
controlled porous glass (CPG), 210
crystalline colloidal array, 392
CSTBA, 405, 406

DDPBBA, 404, 405, 409
decay parameters, fluorescence intensity
for meta isomer, 15
for ortho isomer, 15
delay ratioing, phase, 368
derivatization techniques. *See also* carbohydrates analysis
post-column, 34
pre-column, 34
dextran, 92-96. *See also* diabetes
diabetes
blood glucose (sensors) level monitoring, 90
fluorescence biosensors, 117-127
interstitial fluid (ISF), 90
concanavalin A, 92-96
dextran, 92-96
finger-pricking method, 399
fluorescent based assay, 96-100
fluorescein isothiocyanate (FITC), 98, 99
diabetes (*cont.*)
fluorescent based assay (*cont.*)
fluorescence resonance energy transfer (FRET), 96, 97, 98-100
tetramethyl-rhodamine isothiocyanate (TRITC), 98, 99
glucose monitoring, 21
and fructose, 38-42
automated HPLC postcolumn detection method, 21, 34-38
boronic acid-based receptors, 34
chromophores, 21
fluorophores, 21-28
fructose, 21
hollow fiber based sensors, 100, 101
human blood plasma, 38-42
maltodextrins, 33
monosaccharide analysis, 28
sugar-induced signal transduction, 28
hyperglycemia, 117
glycated hemoglobin (HbA1c), 118
interstitial fluid glucose sensing
implantable concanavlin A based sensors, 88-111
minimally invasive modalities, 91-92
non-consuming analyte stable biosensor
glucose oxidase from *Aspergillus niger*, 119-121
non-invasive modalities, 90, 91
con A based glucose sensors, 91, 92
fluorescence resonance energy transfer (FRET), 91
fluorescent intensity detection, 92
glucose oxidase, 91
hydrogel contact lens, 91
intensity monitoring, 91
lifetime monitoring, 91
non-invasive monitoring, 164-196
specificity, 166-172
ophthalmic glucose monitoring, 399
sensing modalities, 100-110
fluorescent light through skin, 108, 109, 110
hollow fiber based sensors, 100, 101, 103, 104, 105
poly(ethylene glycol) based sensors, 106, 107
stable glucokinase, 123, 124, 125, 126
BSGK, 124, 125
from thermophilic organism *Bacillus stearothermophilus*, 124
polarization sensing, 125

diabetes (*cont.*)
  stable glucokinase (*cont.*)
    resonance energy transfer (RET), 125, 126
  thermostable glucose dehydrogenase
    apo-form of, 122
    fluorescence lifetimes, 123
    from *Thermoplasma acidophilum*, 121-123
    polarization sensing, 123
  THOR of human skin, 178
  type 1 diabetes, 89
  type 2 diabetes, 89
diffraction elements, light, 192, 193, 194
  polymerized crystalline colloidal arrays (PCCAs), 192
doped contact lenses, 420
DSTBA, 404, 405, 406, 409
dual luminophore referencing (DLR), 353
dye
  HPTS, 51
dynamic quenching, 53

electron transfer fluorescence quenching, photoinduced, 13
  ortho-substituted monoboronic acid sensor, 13
  pH effects
    fluorescence intensity decay parameters, 15
    on fluorescent decays, 14
electron transfer process, 6
electrophoresis, capillary, 317-319. *See also* chemiluminescence
  ECL, 318
  laser-induced fluorescence (LIF), 317
enzyme biosensor, 201-229
evaporative light scattering detection (ELSD), 34. *See also* carbohydrates analysis

finger-pricking method, 399
flow injection analysis (FIA), 222, 313
fluorescein isothiocyanate (FITC), 98, 99, 100, 101, 192
fluorescence
  biosensors, 117-127
  energy transfer (FRET), 320
  glucose sensing contact lenses, 320
  laser-induced (LIF), 317
fluorescence correlation spectroscopy (FCS), 352
fluorescence intensity decay parameters
  for meta isomer, 15
  for ortho isomer, 15
fluorescence polarization assay, 353
fluorescence quenching, 53
  chelation-enhanced quenching, 6
fluorescence resonance energy transfer (FRET), 91, 97, 118, 191, 320, 353, 384
  for glucose detection, 98-100
fluorescent based assay, 96-100
  fluorescein isothiocyanate (FITC), 98, 99
  fluorescence resonance energy transfer (FRET), 96, 97
    for glucose detection, 98-100
  tetramethyl-rhodamine isothiocyanate (TRITC), 98, 99
fluorescent chemosensors, 1
  phenylboronic acid, 1, 2
fluorescent dyes, anionic, 47-87
fluorescent intensity detection, 92
fluorescent molecular probes
  1,8-naphthalene dicarboximides, 2
fluorescent monitoring
  concanavalin A (Con A), 91
fluorescent sensor, 6
  chelation-enhanced quenching (CHEQ), 6
  chelation-enhanced fluorescence (CHEF), 6
  chelator group, 2
  fluorophore, 2
  glucose sensors, 51
    stable assay response, 132
  smart tattoo glucose sensor systems, 159
fluorescent tags
  fluorescein isothiocyanate (FITC), 95
fluorescent TICT sensors, 333-350. *See also* saccharides
  2-boronobenzyl-aniline (1), 346
  2-boronobenzyl-pyren-1-yl-amine (7), 349
  3-boronobenzyl-aniline (2), 347
  4-boronobenzyl-aniline (3), 347
  4-chloro-phenyl-2-boronobenzyl amine (4), 348
  4-methoxy-phenyl-2-boronobenzyl amine (5), 348
  2-boronobenzyl-naphthalen-1-yl-amine (6), 349
fluorogenic indicators, 357

fluorophores, 21-28. *See also* glucose
monitoring
formation, 21-45
hydrophobic probes, 2
resorcinarenes, 22, 23, 25
xanthenes, 25
FPIA. *See* fluorescence polarization assay
fructose
determination, 38-42
in carbohydrates analysis, 35
recognition by, 244, 245, 246
functionalized viologen, 50-56

glucose
fluorescent detection of, 352
non-invasive (NI) monitoring, 195
specificity, 166-172
optically-based affinity
biosensors, 283-307
glucose binding protein (GBP), 323, 324
as glucose sensor, 323-331
conformation, 325
glucose biosensor
competitive optical affinity, 289, 290, 291, 292, 293, 295, 296, 297
affinity based glucose biosensor systems, 299, 302, 303
affinity biosensors, 298, 299
integral biosensor proteins, 304, 305
optical
analytical feature and application, 226, 228
aqueous micelle carriers, 222, 223
based on affinity principles, 305, 307
bulky bi-phase bioreactor, 215, 216, 217
dissolved oxygen concentration, 223, 224
enzyme biosensor, 201-229
enzyme stability, GO_x, 224, 225
flow injection analysis, 222
in batch operation, 219
in flow mode, 220, 221
in stop-flow mode, 220
mixture of organic solvent, 222, 223
optical enzyme-based, 201-229
optical transducers, 203-207
particle packed bioreactor, 217, 218, 219
glucose detection, 21-45
3-dimensional scaffold, 392, 393
CAVEAT program, 393
glucose detection (*cont.*)
and fructose, 38-42
automated HPLC postcolumn detection
method, 21, 34-38, 35, 36
carbohydrates analysis, 34
xanthenes, 37
boronic acid-based receptors, 34
chemiluminescence applications, 311-322
chromophores, 21
clinical condition, 425
compliance, 425
contact lenses, 425
continuous and non-invasive, 424, 425
clear/colored contact lenses, 425
detection methods, 425
doped lenses, 425
sensor spots, 425
diagnosis from tears, 425
drug testing, 425
dual luminophore referencing
(DLR), 353
fluorescence correlation
spectroscopy, 352
fluorescence decay time, 353
fluorescence polarization assay
(FPIA), 353
fluorescence resonance energy
transfer (FRET), 353
fluorescent sensor membranes, 358, 359
fluorophores, 21-28
resorcinarenes, 22, 23, 25
xanthenes, 25
fructose, 21
gravimetric method
Fehling's reagent, 311
hollow fiber based sensors, 100, 101
human blood plasma, 38-42
imaging techniques, 368
lens feasibility, 408-413
contact lens holder, 408, 409
lens doping, 408, 409
maltodextrins, 33
minimal invasive glucose
measurement, 118
monosaccharide analysis, 28
near-infrared (NIR) spectroscopy, 377
non-invasive glucose measurement, 118
ophthalmic
smart contact lenses, 399-430
plasmonic, 259-280
fabrication of aggregates, 266, 271, 273, 274, 275, 276, 277, 278

glucose detection (*cont.*)
plasmonic (*cont.*)
gold nanoparticles, 260, 261, 262, 263, 264
plasticized PVC, 359, 360
screening, 425
sensing probes, 413
signal transduction, 56
fluorescent dye with viologen, 56
sugar with boronic acid receptor of viologen, 56
sugar-induced signal transduction, 28
time-resolved fluorescent sensing imaging, 369
activity of glucose oxidase, 371, 372
using oxygen sensors, 370
using sensor layers for hydrogen peroxide, 371
via enzymatically produced H_2O_2, 369
two-component monosaccharide sensing, 48
using boronic acid, 47, 378
and diols, 378, 379, 380
based fluorescence sensors for, 377-398
based fluorescent reporter compounds, 380
fluorescence intensity, 48
photoinduced electron transfer (PET), 48
saccharide binding, 48
using boronic acid probes
anthracenediboronic acid derivative (ANDBA), 406, 412
in solution, 402, 403
photoinduced intramolecular electron transfer mechanism (PET), 406
with intramolecular charge transfer (ICT) mechanism, 404
via GO_x and fluorescent molecular probes, 351, 353, 354-358
changes in pH, 355
oxygen consumption, 355, 356
via formation of hydrogen peroxide, 356, 357
via intrinsic fluorescence of glucose oxidase, 367, 368
high-throughput screening (HTS), 368
glucose detection (*cont.)*
via intrinsic fluorescence of glucose oxidase (*cont.*)
phase delay ratioing (PDR), 368
rapid lifetime determination (RLD), 368
sensing imaging, 368
via metal-ligand complexes
with boronic acid as recognition site, 364
via oxygen sensor membranes, 361, 362, 363
sol-gels, 361
via oxygen-sensitive nanoparticles, 367
via pH sensor layers using immobilized GO_x, 359
via sensor layers for hydrogen peroxide, 363, 364
via thin films of organic conducting polymers (OCPs), 365
photo-induced electron transfer (PET), 365, 366
glucose monitoring
fructose, 21
human blood plasma, 38-42
maltodextrins, 33
monosaccharide analysis, 28
sugar-induced signal transduction, 28
glucose oxidase, 119
immobilized, 214
glucose sensing probes, 413, 414
based on quinolinium, 414, 415
signaling mechanism, 416, 417
spectral properties, 415, 416
sugar response, 417, 418, 419, 420
glucose sensitivity, 157
glucose sensor, 48, 323-331. *See also* glucose binding protein; glucose detection
apo-GO_x, 156
cell culture, 326-328
low cost glucose sensing, 326-328
polarity sensitive probes
anilino-naphthalene sulfonate, 326
protein engineering, 324-325
smart tattoo, 131-160
enzymatic glucose sensing, 139
enzyme-based microcapsule sensors, 142-156
from nanoengineered capsules, 136-142
glucose-binding proteins in microcapsules, 156-159

glucose sensor (*cont.*)
smart tattoo (*cont.*)
LBL nanofilms, 134, 135
polyelectrolyte microcapsules, 135, 136
specificity, 132
validation, 330
glucose sensor implants
fluorescent
stable assay response, 132
glucose signaling probes, 420, 421, 422
glucose tolerance tests, oral, 172
Gluco-Watch, 320
GO. *See* glucose oxidase
gold nanoparticles
ligand functionalized, 261, 262, 263, 264
optical properties of, 260, 261

heterogeneous biphase bioreactor, 218
high-throughput screening (HTS), 368
hollow fiber based sensors, 100, 101, 103, 104, 105
FITC-dextran, 100
HPLC post-column detection method (automated), 34-38. *See also* glucose monitoring
carbohydrates analysis, 34
evaporative light scattering detection (ELSD), 34
fructose, 35
mass spectrometric detection (HPLC-MS), 34
post-column derivatization techniques, 34
pre-column derivatization methods, 34
pulsed amperometric detection (PAD), 34
HPTS, 51, 52, 53, 54, 72, 73, 74, 75, 76, 78
and methyl viologen (MV2+), 52, 53
fructose and galactose, 54, 55, 56
hydrogel contact lens, 80, 91
boronic acid, 91
hyperglycemia, 117
glycated hemoglobin (HbA1c), 118

ICT. *See* intramolecular charge transfer
implantable sensor, 119
integral biosensor proteins, 304, 305
intensity decay parameters, fluorescence
for meta isomer, 15
for ortho isomer, 15
intensity detection, fluorescent, 92
interferents, 422, 423, 424
internal charge transfer, 385
internal charge transfer compounds, 385, 386
interpenetrating networks, 80
interstitial fluid, 90
interstitial fluid glucose sensing
implantable concanavlin A based sensors, 88-111
intramolecular charge transfer, 403, 404
intramolecular charge transfer probes, 409
chalcone derivatives, 411
polyene derivative, 409
stilbene derivatives, 409
IPN. *See* interpenetrating networks
ISF. *See* interstitial fluid
isomers, positional, 13-16
PET fluorescence quenching, 13
meta-substituted monoboronic acid sensor, 13
ortho-substituted monoboronic acid sensor, 13
pH effects
fluorescence intensity decay parameters, 15
on fluorescent decays, 14
sugar effects
on fluorescent decays, 15

laser-induced fluorescence (LIF), 317
LE. *See* locally excited
leaching, probe, 422, 423, 424
light diffraction elements, 192, 193. *See also* affinity capture agents
polymerized crystalline colloidal arrays (PCCAs), 192
locally excited, 333
luminophore referencing, 353
LW fluorescence
charge-transfer character (ICT), 4

macrocycle solutions
fluorophore formation, 21-45
meal tolerance tests (MTT), 172, 188
micelle carriers, aqueous, 222, 223
microcapsules, 131-160
as smart tattoo glucose sensors, 131-160

microcapsules (*cont.*)
based on layer-by-layer assembly technique, 136
glucose-binding proteins, 156-159
polyelectrolyte, 134, 136
sensors, 142-156
design of sensors, 143-154
enzyme-loaded microcapsule sensors, 154-156
microdialysis (MD), 315
minimally invasive modalities, 91-92
molecular probes, fluorescent
1,8-naphthalene dicarboximides, 2
monoboronic acid fluorescent probes, 2, 11
N-phenyl naphthalimide fluorophore, 11
monoboronic acid saccharide sensors, 2
3-phenylboronic acid-1,8-naphthalenedicarboimide, 2
photophysical properties of, 5
synthesis, 4
chelation-enhanced quenching, 12
naphthalic anhydride chromophore, 2
N-phenyl-1,8-naphthalenedicarboximides, 3-13
electron transfer process (PET), 6
fluorescence quenching mechanism, 6
photoelectrochemical model, 6
positional isomers, 13-16
saccharide complexation results, 6-11
synthesis, 4-5
monosaccharide binding
boronic acid of viologen, 48
fluorescence emission, 48
one-component system, 48
two-component system, 48
monosaccharide detection
modulated quenching with viologens, 47-87

nanoengineered capsules, 136-142
RET, 141
naphthalic anhydride chromophore
1,8-naphthalimide, 2
N-phenylnaphthalimides, 2
non-consuming analyte stable biosensor
glucose oxidase from *Aspergillus niger*, 119-121
non-invasive (NI) modalities, 90, 91
con A based glucose sensors, 91, 92
fluorescence resonance energy transfer (FRET), 91
non-invasive (NI) modalities (*cont.*)
fluorescent intensity detection, 92
glucose oxidase, 91
hydrogel contact lens, 91
intensity monitoring, 91
lifetime monitoring, 91
non-invasive monitoring, 164-196. *See also* diabetes
affinity capture agents, 190
compartmentalization in contact lens sensors, 194, 195
fluorescence measurements, 191, 192
light diffraction elements, 192, 193, 194
specificity, 194, 195
calibration models, 172
compartmentalization of glucose values, 172
contact lens sensors, 194
fluorescence measurements, 191, 192
infrared spectroscopy of glucose, 169, 170
light diffraction elements, 192, 193, 194
meal tolerance tests (MTT), 172
near infrared spectroscopy of glucose, 169, 170, 171, 172
optical coherence tomography (OCT), 169
oral glucose tolerance tests (OGTT), 172
prediction parameters, 173
standard error of prediction (SEP), 173
scattering coefficient, 168
specificity, 166-172
tissue spectroscopy, 168, 169
thermo-optical response (THOR)
calibration models, 185, 186, 187
localized reflectance measurements, 175, 176, 177, 178
of human skin, 173, 174, 178, 179, 180, 181, 182, 183, 184, 185
prediction models, 187, 188, 189, 190
temperature-modulated, 174
N-phenyl naphthalimide fluorophore, 11. *See also* carbohydrate molecular probes
C0 design chemosensor, 11
emitting states
long wavelength (LW), 3
short wavelength (SW), 3

N-phenyl-1,8-naphthalenedicarboximides
positional isomers, 13-16
meta-substituted monoboronic acid sensor
N-phenylboronic acid arenecarboximides, 2-3. *See also* saccharide probes
virtual or C0 spacer, 2-3
N-phenylboronic acid derivatives
as saccharide probes, 1-17
virtual spacer design, 1-17

OCP. *See* organic conducting polymers
OCT. *See* optical coherence tomography
OGTT. *See* oral glucose tolerance tests
ophthalmic glucose monitoring
smart contact lenses, 399-430
optical affinity glucose biosensors,
competitive, 289, 290, 291, 292, 293, 295, 296, 297
affinity based glucose biosensor systems, 299, 302, 303
affinity biosensors, 298, 299
integral biosensor proteins, 304, 305
optical coherence tomography, 169
analytical feature and application, 226, 228
based on affinity principles, 305, 307
dissolved oxygen
concentration, 223, 224
enzyme stability
GO_x, 224, 225
performance of, 219
aqueous micelle carriers, 222, 223
flow injection analysis, 222
in batch operation, 219
in flow mode, 220, 221
in stop-flow mode, 220
mixture of organic solvent, 222, 223
optical sugar sensing, two-component
boronic acid-substituted viologens with anionic fluorescent dyes, 47-87
optical transducers, 203-207
absorbance measurement, 204
blend composites of immobilization enzyme, 214
electrogenerated chemiluminescence, 205-206
enzyme entrapped
in thin homogenous film, 210, 211, 212
within sol-gel matrix, 212, 213
optical transducers (*cont.*)
enzyme immobilized by nano-technique and with nanoparticles, 213
fluorescence measurement, 206, 207
immobilization of glucose oxidase, 207, 208
immobilizing enzyme on surface, 209, 210
by nano-technique and with nanoparticles, 213
optically based biosensors, 285
oral glucose tolerance tests, 172
organic conducting polymers (OCPs), 365
oxygen sensitive particles (OSP), 207

PAD. *See* pulsed amperometric detection
particle packed bioreactor, 217, 218, 219
heterogeneous biphase bioreactor, 218
PET. *See* photoinduced electron transfer
phase delay ratioing (PDR), 368
phenanthrolinium quenchers, 61
photoinduced electron transfer, 48, 381, 416
photoinduced electron transfer
fluorescence quenching, 13
ortho-substituted monoboronic acid sensor, 13
pH effects
fluorescence intensity decay parameters, 15
on fluorescent decays, 14
photoinduced electron transfer probes, 412
plasmonic glucose sensing, 259-280
fabrication of glucose-sensing aggregates, 266, 271, 273, 274, 275, 276, 277, 278
gold nanoparticles
ligand functionalized, 261, 262, 263, 264
optical properties of, 260, 261
polarization assay, fluorescence, 353
polarization sensing, 125
polyelectrolyte microcapsules, 136
polyene derivative, 409. *See also* ICT probes
polymeric viologens, 69
porous glass, controlled, 210
positional isomers, 13-16
PET fluorescence quenching, 13
meta-substituted monoboronic acid sensor, 13

positional isomers, (*cont.*)
  PET fluorescence quenching, (*cont.*)
    ortho-substituted monoboronic acid sensor, 13
  pH effects
    fluorescence intensity decay parameters, 15
    on fluorescent decays, 14
  sugar effects
    on fluorescent decays, 15
post-column derivatization techniques, 34
post-column HPLC detection method (automated), 34-38. *See also* glucose monitoring
  carbohydrates analysis, 34
    evaporative light scattering detection (ELSD), 34
    fructose, 35
    mass spectrometric detection (HPLC-MS), 34
    post-column derivatization techniques, 34
    pre-column derivatization methods, 34
    pulsed amperometric detection (PAD), 34
pre-column derivatization methods, 34
probe leaching, 422, 423, 424
pulsed amperometric detection, 34
pyranine. *See* HPTS

quenchers
  fluorescence, 53
    chelation enhanced quenching, 6
  functionalized viologen
    binding to diol, 51
  phenanthrolinium, 61

rapid lifetime determination (RLD), 368, 372
resonance energy transfer (RET), 125, 126
  fluorescence, 91, 97, 118, 353, 384
  using apo-GO_x, 157
resorcinarenes, 22, 23, 25. *See also* fluorophores
  boronic acids, 23

saccharide
  boronic acid receptors, 333-350
  fluorescent TICT sensors, 333-350
saccharide probes
  N-phenylboronic
    acid arenecarboximides as, 2-3
saccharide receptor unit
  boronic acid functionalized viologen quenchers, 50
saccharide recognition
  boronic acid fluorophore/cyclodextrin complexes in water, 237-256
  by boronic acid, 242-254
    C4-CPB/-CyD complex, 243-250
  fluorescent chemosensors, 1
  fluorescent sensors
    photoinduced electron transfer, 6
  glucose oxidase biosensor, 1
  phenylboronic acid probe spacer, 251, 252, 253
  supramolecular cyclodextrin complex sensors, 254, 255, 256
saccharide selectivity
  fluorophores, 2
saccharide sensors
  anthracene, 2
  stilbene-based fluorophore, 2
sensing modalities, 100-110
  fluorescent light through skin, 108, 109, 110
  hollow fiber based sensors, 100, 101, 103, 104, 105
  poly(ethylene glycol) based sensors, 106, 107
sensors
  biosensor
    carbohydrate sensors, 6
    chemiluminogenic, 313-317
    competitive optical affinity, 289, 290, 291, 292, 293, 295, 296, 297, 298, 299, 302, 303, 304, 305
    fluorescence, 117-127
    interstitial fluid (ISF), 90
    non-consuming analyte stable, 119-121
    optical, 201-229, 305, 307
    protein based, 1
    pyrene based, 381, 383, 384, 385
  bisboronic acid sensors, 2
  con A based, 91, 92
  contact lens
    compartmentalization in, 194
    specificity, 194
  glucose, 48, 323-331
    apo-GO_x, 156
    cell culture, 326-328
    low cost glucose sensing, 326-328
    polarity sensitive probes, 326

sensors (*cont.*)
glucose (*cont.*)
protein engineering, 324-325
smart tattoo, 131-160
specificity, 132
validation, 330
hollow fiber based, 100, 101
implantable, 119
microcapsule, 142-156
design of sensors, 143-154
enzyme-loaded microcapsule sensors, 154-156
monoboronic acid saccharide, 2
3-phenylboronic acid-1,8-naphthalenedicarboimide, 2, 4, 5
anthracene, 2
chelation-enhanced quenching, 12
fluorescent sensors, 6
naphthalic anhydride chromophore, 2
N-phenyl-1,8-naphthalenedicarboximides, 3-13
stilbene-based fluorophore, 2
shelf life, 422, 423, 424
signal transduction, sugar-induced, 28
smart contact lenses, 399-430
smart tattoo glucose sensors, 131-160. *See also* microcapsules
enzymatic glucose sensing, 139
enzyme-based microcapsule sensors, 142-156
design of sensors, 143-154
enzyme-loaded microcapsule sensors, 154-156
from nanoengineered capsules, 136-142
RET, 141
glucose-binding proteins in microcapsules, 156-159
LBL nanofilms, 134, 135
polyelectrolyte microcapsules, 135, 136
stable glucokinase, 123, 124, 125, 126
BSGK, 124, 125
from thermophilic organism *Bacillus stearothermophilus*, 124
polarization sensing, 125
resonance energy transfer (RET), 125, 126
static quenching, 53
Stern-Volmer equation, 424
stilbene derivatives, 409. *See also* ICT probes
CSTBA, 404, 405, 409
DSTBA, 404, 409
sugar sensing, two-component optical
boronic acid-substituted viologens with anionic fluorescent dyes, 47-87
synthetic chemosensors, 1

tattoo glucose sensors, smart, 131-160. *See also* microcapsules
enzymatic glucose sensing, 139
enzyme-based microcapsule sensors, 142-156
design of sensors, 143-154
enzyme-loaded microcapsule sensors, 154-156
from nanoengineered capsules, 136-142
RET, 141
glucose-binding proteins in microcapsules, 156-159
LBL nanofilms, 134, 135
polyelectrolyte microcapsules, 135, 136
tear glucose, 194
tetramethyl-rhodamine isothiocyanate (TRITC), 98, 192
tetramethyl-rhodamine isothiocyanate
thermo-optical response (THOR)
calibration models
at fixed source-detector distance, 186, 187
from human skin, 185, 186
localized reflectance measurements, 174, 175, 176, 177, 178
of human skin, 173, 174, 178, 179, 180
using nonlinear discriminant functions for data analysis, 180, 181, 182, 183, 184, 185
prediction models
meal tolerance (MTT), 188
of human skin, 187, 188, 189, 190
temperature-modulated, 174
thermostable glucose dehydrogenase
apo-form of, 122
fluorescence lifetimes, 123
from *Thermoplasma acidophilum*, 121-123
polarization sensing, 123
thin layer chromatography (TLC), 346
twisted internal charge transfer (TICT) sensors, fluorescent, 333-350. *See also* saccharides
2-boronobenzyl-aniline (1), 346
2-boronobenzyl-pyren-1-yl-amine (7), 349
3-boronobenzyl-aniline (2), 347
4-boronobenzyl-aniline (3), 347

twisted internal charge transfer (TICT) sensors, fluorescent (*cont.*)
    4-chloro-phenyl-2-boronobenzyl amine (4), 348
    4-methoxy-phenyl-2-boronobenzyl amine (5), 348
    2-boronobenzyl-naphthalen-1-yl-amine (6), 349
transducers, optical, 203-207
    absorbance measurement, 204
    blend composites of immobilization enzyme, 214
    electrogenerated chemiluminescence, 205-206
    enzyme entrapped
        in thin homogenous film, 210, 211, 212
        within sol-gel matrix, 212, 213
    enzyme immobilized by nano-technique and with nanoparticles, 213
    fluorescence measurement, 206, 207
    immobilization of glucose oxidase, 207, 208
    immobilizing enzyme on surface, 209, 210
two-component glucose sensing
    arylboronic acids, 52
    fluorescent reporter
        sulfonamide derivatives of HPTS, 71-80
    functionalized viologen
        boronic acid functionalized viologen quenchers, 50, 51
        HPTS, 51, 52, 53, 54
    polymeric viologens, 69
    sensing components immobilization
        glucose sensitive thin film hydrogel, 80-83
    using boronic acid-substituted viologens with anionic fluorescent dyes, 47-87
    viologen quencher
        bipyridinium, 56, 57, 58, 59
        phenanthrolinium salts, 60, 61, 69

viologens, 50-56
    boronic acid-substituted
        with anionic fluorescent dyes, 47-87
    cationic, 51
    electron accepting, 50
    functionalized, 50-56
    metasubstiuted, 58, 59
    modulated quenching, 47-87
    orthosubstiuted, 58, 59
    parasubstiuted, 58
    polymeric, 69
    positive charges, 50

water soluble fluorescent boronic acid, 387
    chelation-enhanced quenching (CHEQ), 387, 388, 389, 390

xanthenes, 25, 42

MIX
Papier aus verantwortungsvollen Quellen
Paper from responsible sources
FSC® C105338

If you have any concerns about our products,
you can contact us on
ProductSafety@springernature.com

In case Publisher is established outside the EU,
the EU authorized representative is:
Springer Nature Customer Service Center GmbH
Europaplatz 3, 69115 Heidelberg, Germany

Printed by Libri Plureos GmbH
in Hamburg, Germany